S0-AJA-848

Encyclopaedia of Mathematical Sciences
Volume 138

Invariant Theory and Algebraic Transformation Groups VIII

Subseries Editors:
Revaz V. Gamkrelidze Vladimir L. Popov

Dmitry A. Timashev

Homogeneous Spaces and Equivariant Embeddings

Dmitry A. Timashev
Lomonosov Moscow State University
Faculty of Mechanics and Mathematics
Department of Higher Algebra
Leninskie Gory 1
Moscow 119991
Russian Federation
timashev@mccme.ru

Founding editor of the Encyclopaedia of Mathematical Sciences: Revaz V. Gamkrelidze

ISSN 0938-0396
ISBN 978-3-642-18398-0 e-ISBN 978-3-642-18399-7
DOI 10.1007/978-3-642-18399-7
Springer Heidelberg Dordrecht London New York

Library of Congress Control Number: 2011926005

Mathematics Subject Classification (2010): 14L30, 14M17, 14M27, 20G15, 20G05, 22E46, 14M25, 14C20, 14C22, 14C15, 14C17, 14N10, 14N15, 14C05, 14F17, 43A85, 16S32, 14F10, 53D20, 52B20, 13A35, 13A18

Cover design: VTeX UAB, Lithuania

Printed on acid-free paper

Springer is part of Springer Science+Business Media (www.springer.com)

To my family

Acknowledgements

The author is grateful to V. L. Popov for the invitation to contribute to the series of surveys "Invariant Theory and Algebraic Transformation Groups" in "Encyclopædia of Mathematical Sciences". Many thanks are due to M. Brion, I. V. Losev, V. L. Popov, and the referees for their very valuable comments on the preliminary version of this book. The author also thanks I. V. Arzhantsev, S. Cupit-Foutou, A. Moreau, E. Sayag, E. B. Vinberg, and O. S. Yakimova for helpful discussions and comments on particular topics.

Contents

Introduction

Groups entered mathematics as transformation groups. From the works of Cayley and Klein it became clear that any geometric theory studies the properties of geometric objects that are invariant under the respective transformation group. This viewpoint culminated in the celebrated Erlangen program [Kl]. An important feature of each one of the classical geometries—affine, projective, Euclidean, spherical, and hyperbolic—is that the respective transformation group is transitive on the underlying space. Another feature of these examples is that the transformation groups are linear algebraic and their action is regular. In this way algebraic homogeneous spaces arise in geometry.

Another source for algebraic homogeneous spaces are varieties of geometric figures or tensors of certain type. Examples are provided by Grassmannians, flag varieties, varieties of conics, of triangles, of matrices with fixed rank, etc. These homogeneous spaces are of great importance in algebraic geometry. They were explored intensively, starting with the works of Chasles, Schubert, Zeuthen et al, which gave rise to the enumerative geometry and intersection theory.

Homogeneous spaces play an important rôle in representation theory, since representations of linear algebraic groups are often realized in spaces of sections or cohomologies of line (or vector) bundles over homogeneous spaces. The geometry of a homogeneous space can be used to study representations of the respective group, and conversely. Shining examples are the Borel–Weil–Bott theorem [Dem3] and Demazure's proof of the Weyl character formula [Dem1].

In the study of an algebraic homogeneous space G/H, it is often useful by standard reasons of algebraic geometry to pass to a G-equivariant completion or, more generally, to an *embedding*, i.e., a G-variety X containing a dense open orbit isomorphic to G/H.

An example is provided by the following classical problem of enumerative algebraic geometry: compute the number of plane conics tangent to 5 given ones. Equivalently, one has to compute the intersection number of certain 5 divisors on the space of conics $\mathrm{PSL}_3/\mathrm{PSO}_3$, which is an open orbit in $\mathbb{P}^5 = \mathbb{P}(\mathrm{S}^2\mathbb{C}^3)$. To solve our enumerative problem, we pass to a good compactification of $\mathrm{PSL}_3/\mathrm{PSO}_3$. Namely, consider the closure X in $\mathbb{P}^5 \times (\mathbb{P}^5)^*$ of the graph of a rational map sending a conic to the dual one. Points of X are called *complete conics*. It happens that our 5 divisors intersect the complement of the open orbit in X properly. Hence the sought number is just the intersection number of the 5 divisors in X, which is easier to compute, because X is compact.

Embeddings of homogeneous spaces arise naturally as orbit closures, when one studies arbitrary actions of algebraic groups. Such questions as normality of the orbit closure, the nature of singularities, adherence of orbits, the description of orbits in the closure of a given orbit, etc, are of importance.

Embeddings of homogeneous spaces of reductive algebraic groups are the subject of this survey. The reductivity assumption is natural for two reasons. First, reductive groups have a good structure and representation theory, and a deep theory of embeddings can be developed under this restriction. Secondly, most applications to algebraic geometry and representation theory deal with homogeneous spaces of reductive groups. However, homogeneous spaces of non-reductive groups and their embeddings are also considered. They arise naturally even in the study of reductive group actions as orbits of Borel and maximal unipotent subgroups and their closures. (An example: Schubert varieties.)

The main topics of our survey are:

- The description of all embeddings of a given homogeneous space.
- The study of geometric properties of embeddings: affinity, (quasi)projectivity, divisors and line bundles, intersection theory, singularities, etc.
- Application of homogeneous spaces and their embeddings to algebraic geometry, invariant theory, and representation theory.
- Determination of a "good" class of homogeneous spaces, for which the above problems have a good solution. Finding and studying natural invariants that distinguish this class.

Now we describe briefly the content of the survey.

In Chap. 1 we recall basic facts on algebraic homogeneous spaces and consider basic classes of homogeneous spaces: affine, quasiaffine, projective. We give group-theoretical conditions that distinguish these classes. Also bundles and fibrations over a homogeneous space G/H are considered. In particular, we compute $\mathrm{Pic}(G/H)$.

In Chap. 2 we introduce and explore two important numerical invariants of G/H—the complexity and the rank. The *complexity* of G/H is the codimension of a general B-orbit in G/H, where $B \subseteq G$ is a Borel subgroup. The *rank* of G/H is the rank of the lattice $\Lambda(G/H)$ of weights of rational B-eigenfunctions on G/H. These invariants are of great importance in the theory of embeddings. Homogeneous spaces of complexity ≤ 1 form a "good" class. It was noted by Howe [Ho] and Panyushev [Pan7] that a number of invariant-theoretic problems admitting a nice solution have a certain homogeneous space of complexity ≤ 1 in the background.

Complexity and rank may be defined for any action $G : X$. We prove some semicontinuity results for complexity and rank of G-subvarieties in X. General methods for computing complexity and rank of X were developed by Knop and Panyushev, see [Kn1] and [Pan7, §§1–2]. We describe them in this chapter, paying special attention to the case $X = G/H$. The formulæ for complexity and rank are given in terms of the geometry of the cotangent bundle T^*X and of the doubled action $G : X \times X^*$.

The general theory of embeddings developed by Luna and Vust [LV] is the subject of Chap. 3. The basic idea of Luna and Vust is to patch all embeddings $X \hookleftarrow G/H$ together in a huge prevariety and consider particular embeddings as Noetherian separated open subsets determined by certain conditions. It appears, at least for normal embeddings, that X is determined by the collection of closed G-subvarieties $Y \subseteq X$, and each Y is determined by the collection of B-stable prime divisors containing Y. This leads to a "combinatorial" description of embeddings, which can be made really combinatorial in the case of complexity ≤ 1. In this case, embeddings are classified by certain collections of convex polyhedral cones, as in the theory of toric varieties [Ful2] (which is in fact a particular case of the Luna–Vust theory). The geometry of embeddings is also reflected in these combinatorial data, as in the toric case. In fact the Luna–Vust theory is developed here in more generality as a theory of G-varieties in a given birational class (not necessarily containing an open orbit).

G-invariant valuations of the function field of G/H correspond to G-stable divisors on embeddings of G/H. They play a fundamental rôle in the Luna–Vust theory as a key ingredient of the combinatorial data used in the classification of embeddings. In Chap. 4 we explore the structure of the set of invariant valuations, following Knop [Kn3], [Kn5]. This set can be identified with a certain collection of convex polyhedral cones patched together along their common face. This face consists of *central* valuations—those that are zero on B-invariant functions. It is a solid rational polyhedral cone in $\Lambda(G/H) \otimes \mathbb{Q}$ and a fundamental domain of a crystallographic reflection group $W_{G/H}$, which is called the *little Weyl group of* G/H. The cone of central valuations and the little Weyl group are linked with the geometry of the cotangent bundle.

Spaces of complexity 0 form the most remarkable subclass of homogeneous spaces. Their embeddings are called *spherical varieties*. They are studied in Chap. 5. Grassmannians, flag varieties, determinantal varieties, varieties of conics, of complexes, and algebraic symmetric spaces are examples of spherical varieties. We give several characterizations of spherical varieties from the viewpoint of algebraic transformation groups, representation theory, and symplectic geometry. We consider important classes of spherical varieties: symmetric spaces, reductive algebraic monoids, horospherical varieties, toroidal and wonderful varieties. The Luna–Vust theory is much more developed in the spherical case by Luna, Brion, Knop, et al. We consider the structure of the Picard group of a spherical variety, the intersection theory with applications to enumerative geometry, the cohomology of coherent sheaves, and a powerful technique of Frobenius splitting, which leads to deep conclusions on the geometry and cohomology of spherical varieties by reduction to positive characteristic. A classification of spherical homogeneous spaces and their embeddings, started by Krämer and Luna–Vust, respectively, was recently completed in pure combinatorial terms (like the classification of semisimple Lie algebras).

The theory of embeddings of homogeneous spaces is relatively new and far from being complete. This survey does not cover all developments and deeper interactions with other areas. Links for further reading may be found in the bibliography. We also recommend the surveys [Kn2], [Bri6], [Bri13] on spherical varieties and [Pan7] on complexity and rank in invariant theory. A short survey paper [Tim6] covers some of the topics of this survey in a concise manner.

The reader is supposed to be familiar with basic concepts of commutative algebra, algebraic geometry, algebraic groups, and invariant theory. Our basic sources in these areas are [Ma], [Sha] and [Har2], [Hum] and [Sp3], [PV] and [MFK], respectively. More special topics are covered by Appendices.

Structure of the Survey. The text is divided into chapters, chapters are subdivided into sections, and sections are subdivided into subsections. A link 1.2 refers to Subsection (or Theorem, Lemma, Definition, etc) 2 of Section 1. We try to give proofs, unless they are too long or technical.

Notation and Conventions. We work over an algebraically closed base field $\Bbbk$. A part of our results are valid over an arbitrary characteristic, but we impose the assumption $\operatorname{char}\Bbbk = 0$ whenever it simplifies formulations and proofs. More precisely, the characteristic is assumed to be zero in §§7 (most part of), 8–11, 22–23, 29, 30 (most part of), and subsections 4.3, 16.6, 17.2–17.5, 18.2–18.6, 25.3–25.5, 26.9–26.10, 27.4–27.6, 28.3, A.2, E.3. On the other hand, we assume $\operatorname{char}\Bbbk > 0$ in §31, unless otherwise specified. Let p denote the characteristic exponent of $\Bbbk$ ($= \operatorname{char}\Bbbk$, or 1 if $\operatorname{char}\Bbbk = 0$).

Algebraic varieties are assumed to be Noetherian and separated (not necessarily irreducible).

Algebraic groups are denoted by capital Latin letters, and their tangent Lie algebras by the respective lowercase Gothic letters. We consider only *linear* algebraic groups (although more general algebraic groups, e.g., Abelian varieties, are also of great importance in algebraic geometry).

Topological terms refer to the Zariski topology, unless otherwise specified.

By a *general point* of an algebraic variety we mean a point in a certain dense open subset (depending on the considered situation), in contrast with the *generic point*, which is the dense schematic point of an irreducible algebraic variety.

Throughout the paper, G denotes a reductive connected linear algebraic group, unless otherwise specified. We may always assume that G is of simply connected type, i.e., a direct product of a torus and a simply connected semisimple group. When we study the geometry of a given homogeneous space O and embeddings of O, we often fix a base point $o \in O$ and denote by $H = G_o$ its isotropy group, thus identifying O with G/H (at least set-theoretically).

We use the following general notation.

General:

$\sqcup$ denotes a union of pairwise disjoint sets.

$\subseteq$ denotes inclusion of sets, while $\subset$ stands for strict inclusion (excluding equality).

$\lhd$ denotes inclusion of normal subgroups or ideals.

$\mathbf{1}$ is the identity map of a set under consideration.

$\mathbb{N},\mathbb{Z},\mathbb{Q},\mathbb{R},\mathbb{C}$ denote the sets of natural, integer, rational, real, and complex numbers, respectively. The sub(super)script "+" or "−" distinguishes the respective subset of non-negative (positive) or non-positive (negative) numbers, e.g., $\mathbb{Z}^+ = \mathbb{N}$, $\mathbb{Z}_+ = \mathbb{N} \sqcup \{0\}$.

$\mathbf{Z}_m = \mathbb{Z}/m\mathbb{Z}$ is the group (or ring) of residues mod m.

$\mathbb{F}_{p^k}$ is the Galois field of cardinality p^k; $\mathbb{F}_{p^\infty}$ denotes its algebraic closure.

Linear algebra:

$\mathbb{A}^n$ denotes the n-dimensional coordinate affine space.

$e_1,\dots,e_n$ denote the standard basis of $\Bbbk^n$.

$x_1,\dots,x_n$ are the standard coordinates on $\mathbb{A}^n$ or $\Bbbk^n$.

$A^\top$ is the transpose of a matrix A.

$\mathrm{L}_n(\Bbbk)$ is the algebra of $n\times n$ matrices over $\Bbbk$.

$\mathrm{GL}_n(\Bbbk),\mathrm{SL}_n(\Bbbk),\mathrm{O}_n(\Bbbk),\mathrm{SO}_n(\Bbbk),\mathrm{Sp}_n(\Bbbk)$ are the classical matrix groups: of non-degenerate, unimodular, orthogonal, unimodular orthogonal, and symplectic $n\times n$ matrices over $\Bbbk$, respectively.

$\mathrm{L}(V)$ is the algebra of linear operators on a vector space V.

$\mathrm{GL}(V),\mathrm{SL}(V),\mathrm{O}(V),\mathrm{SO}(V),\mathrm{Sp}(V)$ are the classical linear groups acting on V: general linear, special linear, orthogonal, special orthogonal, and symplectic group, respectively.

$\langle S\rangle$ is the linear span of a subset $S\subseteq V$.

V^* is the vector space dual to V.

$\Lambda^* = \mathrm{Hom}(\Lambda,\mathbb{Z})$ is the lattice dual to a lattice Λ.

$\langle\cdot,\cdot\rangle$ denotes the pairing between V and V^* or Λ and Λ^*.

$S^\perp \subseteq V^*$ (or Λ^*) is the annihilator of a subset $S\subseteq V$ (or Λ) with respect to this pairing.

$\mathbb{P}(V)$ denotes the projective space of all 1-subspaces in V.

$\mathbb{P}(X)\subseteq\mathbb{P}(V)$ is the projectivization of a subset $X\subseteq V$ stable under homotheties.

$[v]\in\mathbb{P}(V)$ is the point corresponding to a nonzero vector $v\in V$.

$\mathrm{Gr}_k(V)$ denotes the Grassmannian of k-dimensional subspaces in V.

$\mathrm{Fl}_{k_1,\dots,k_s}(V)$ is the variety of partial flags in V with subspace dimensions $k_1,\dots,k_s$.

Algebras and modules:

$A^\times$ is the unit group of an algebra A.

$\mathrm{Quot}A$ is the field of quotients of A.

$\Bbbk[S]\subseteq A$ is the $\Bbbk$-subalgebra generated by a subset $S\subseteq A$. In particular, the notation $\Bbbk[x_1,\dots,x_n]$ is used for the algebra of polynomials in the indeterminates $x_1,\dots,x_n$.

$\Bbbk[[t]]$ is the $\Bbbk$-algebra of formal power series in the indeterminate t.

$\Bbbk((t)) = \mathrm{Quot}\,\Bbbk[[t]]$ is the field of formal Laurent series in t.

$(S)\lhd A$ is the ideal generated by S.

$AS \subseteq M$ is the submodule of an A-module M generated by a subset $S \subseteq M$. This notation is also used for additive subsemigroups ($A = \mathbb{Z}_+$) and convex cones in a vector space over $\mathbb{K} = \mathbb{Q}$ or $\mathbb{R}$ ($A = \mathbb{K}_+$).

$\operatorname{Ann} S \subseteq A$ is the annihilator of $S \subseteq M$ in A.

$\mathrm{S}^\bullet M = \bigoplus_{n=0}^\infty \mathrm{S}^n M$ is the symmetric algebra of a module (sheaf) M.

$\bigwedge^\bullet M = \bigoplus_{n=0}^\infty \bigwedge^n M$ is the exterior algebra of M.

$M^{(n)}$ denotes the n-th member of the filtration of a filtered object (algebra, module, sheaf) M.

$\operatorname{gr} M$ denotes the graded object associated with a filtered object M.

M_n is the n-th homogeneous part of a graded object M.

Algebraic geometry:

$\overline{Y}$ denotes the closure of a subset Y in an algebraic variety X, unless otherwise specified.

$\mathscr{O}_X$ is the structure sheaf of X.

$\mathscr{I}_Y \lhd \mathscr{O}_X$ is the ideal sheaf of a closed subvariety (subscheme) $Y \subset X$.

$\mathscr{O}_{X,Y}$ is the local ring of an irreducible subvariety (or a schematic point) $Y \subseteq X$.

$\mathfrak{m}_{X,Y} \lhd \mathscr{O}_{X,Y}$ is the maximal ideal.

$\Bbbk[X]$ is the algebra of regular functions on X.

$\mathscr{I}(Y) \lhd \Bbbk[X]$ is the ideal of functions on X vanishing on a closed subvariety (subscheme) $Y \subseteq X$.

$\Bbbk(X)$ is the field of rational functions on an irreducible variety X.

$\operatorname{Cl} X$ is the divisor class group of X.

$\operatorname{Pic} X$ is the Picard group of X.

$\mathscr{O}(\delta) = \mathscr{O}_X(\delta)$ is the line bundle corresponding to a Cartier divisor δ on X or, more generally, the reflexive sheaf corresponding to a Weil divisor δ.

$\operatorname{div}_0 \sigma$, $\operatorname{div}_\infty \sigma$, $\operatorname{div} \sigma = \operatorname{div}_0 \sigma - \operatorname{div}_\infty \sigma$ denote the divisor of zeroes, of poles, and the full divisor of a rational section σ of a line bundle on X (e.g., $\sigma \in \Bbbk(X)$), respectively.

$X_\sigma = \{x \in X \mid \sigma(x) \neq 0\}$ is the non-vanishing locus of a section σ of a line bundle on X (including the case $\sigma \in \Bbbk[X]$).

φ^* denotes the pullback of functions, divisors, sheaves, etc, along a morphism $\varphi : X \to Y$.

φ_* is the pushforward along φ (whenever it exists).

$\mathrm{R}^i \varphi_* \mathscr{F}$ is the i-th higher direct image of a sheaf $\mathscr{F}$ on X.

$\mathrm{H}^i(X, \mathscr{F})$ denotes the i-th cohomology space of $\mathscr{F}$. In particular, $\mathrm{H}^0(X, \mathscr{F})$ is the space of global sections of $\mathscr{F}$.

X^{reg} is the regular (smooth) locus of X.

$T_x X$, $T_x^* X$ are the tangent, resp. cotangent, space to X at $x \in X$.

$d_x\varphi : T_xX \to T_{\varphi(x)}Y$ is the differential of $\varphi : X \to Y$ at $x \in X$.

$\Omega_X^\bullet = \bigwedge^\bullet \Omega_X^1$ is the sheaf of differential forms on X.

$\omega_X = \Omega_X^{\dim X}$ is the canonical sheaf on X.

Groups, Lie algebras, and actions:

e is the unity element of a group G.

$Z(G)$ denotes the center of G.

$G' = [G,G]$ is the commutator subgroup of G.

G^0 is the unity component of an algebraic group G.

$\mathrm{R}(G)$ denotes the radical of G.

$\mathrm{R_u}(G)$ is the unipotent radical of G.

$\mathfrak{X}(G)$ is the character group of G, i.e., the group of homomorphisms $G \to \Bbbk^\times$ written additively.

$\mathfrak{X}^*(G)$ is the set of (multiplicative) one-parameter subgroups of G, i.e., homomorphisms $\Bbbk^\times \to G$.

$\mathrm{Ad} = \mathrm{Ad}_G : G \to \mathrm{GL}(\mathfrak{g})$ is the adjoint representation of G.

$G : M$ denotes an action of a group G on a set M. As a rule, it is a regular action of an algebraic group on an algebraic variety.

G_x is the isotropy group (= stabilizer) in G of a point $x \in M$.

Gx is the G-orbit of x.

$N_G(S) = \{g \in G \mid gS = S\}$ is the normalizer of a subset $S \subseteq M$. In particular, this notation is used for the normalizer of a subgroup.

$Z_G(S) = \{g \in G \mid gx = x,\ \forall x \in S\}$ is the centralizer of S (e.g., of a subgroup).

M^G is the set of fixed elements under an action $G : M$.

$M^{(G)}$ is the set of all (nonzero) G-eigenvectors in a linear representation $G : M$.

$M_\chi = M_\chi^{(G)} \subseteq M$ is the subspace of G-eigenvectors of the weight $\chi \in \mathfrak{X}(G)$.

$\mathfrak{z}(\mathfrak{g})$ denotes the center of a Lie algebra $\mathfrak{g}$.

$\mathfrak{g}' = [\mathfrak{g},\mathfrak{g}]$ is the commutator subalgebra.

$\mathfrak{n}_\mathfrak{g}(S) = \{\xi \in \mathfrak{g} \mid \xi S \subseteq S\}$ is the Lie algebra normalizer of a subspace S in a $\mathfrak{g}$-module M. In particular, this notation is used for the normalizer of a Lie subalgebra.

$\mathfrak{z}_\mathfrak{g}(S) = \{\xi \in \mathfrak{g} \mid \xi x = 0,\ \forall x \in S\}$ is the Lie algebra centralizer of a subset $S \subseteq M$ (e.g., of a Lie subalgebra).

ξx is the velocity vector of $\xi \in \mathfrak{g}$ at $x \in M$, i.e., the image of ξ under the differential of the orbit map $G \to Gx$, $g \mapsto gx$, where G is an algebraic group and M is an algebraic G-variety. In characteristic zero, $\mathfrak{g}x := \{\xi x \mid \xi \in \mathfrak{g}\} = T_xGx$.

$\mathrm{Aut}_G M$ denotes the group of G–equivariant automorphisms of a G-set (variety, module, algebra, ...) M.

$\mathrm{Hom}_G(M,N)$ is the space of G–equivariant homomorphisms of G-modules (or sheaves) $M \to N$.

Other notation is gradually introduced in the text, see the Notation Index.

Chapter 1
Algebraic Homogeneous Spaces

In this chapter, G denotes an *arbitrary* linear algebraic group (not supposed to be either connected nor reductive), and $H \subseteq G$ a closed subgroup. We begin in §1 with the definition of an algebraic homogeneous space G/H as a geometric quotient, and prove its quasiprojectivity. We also prove some elementary facts on tangent vectors and G-equivariant automorphisms of G/H. In §2, we describe the structure of G-fibrations over G/H and compute $\operatorname{Pic}(G/H)$. Some related representation theory is discussed there: induction, multiplicities, the structure of $\Bbbk[G]$. Basic classes of homogeneous spaces are considered in §3. We prove that G/H is projective if and only if H is parabolic, and consider criteria of affinity of G/H. Quasiaffine G/H correspond to observable H, which may be defined by several equivalent conditions (see Theorem 3.12).

1 Homogeneous Spaces

1.1 Basic Definitions.

Definition 1.1. An algebraic group action $G : O$ is *transitive* if for any $x, y \in O$ there exists $g \in G$ such that $y = gx$. In this situation, O is said to be a *homogeneous space*.

A *pointed homogeneous space* is a pair (O, o), where O is a homogeneous space and $o \in O$. The natural map $\pi : G \to O$, $g \mapsto go$, is called the *orbit map*.

A basic property of algebraic group actions is that each orbit is a locally closed subvariety and hence a homogeneous space in the sense of Definition 1.1. Homogeneous spaces are always smooth and quasiprojective, by Sumihiro's Theorem C.7. The next definition provides a universal construction of algebraic homogeneous spaces.

Definition 1.2. The *(geometric) quotient* of G modulo H is the space G/H equipped with the quotient topology and a structure sheaf $\mathscr{O}_{G/H}$ which is the direct image of the sheaf $\mathscr{O}_G^H$ of H-invariant (with respect to the H-action on G by right translations) regular functions on G.

D.A. Timashev, *Homogeneous Spaces and Equivariant Embeddings*,
Encyclopaedia of Mathematical Sciences 138, DOI 10.1007/978-3-642-18399-7_1,

Theorem 1.3. (1) $(G/H, \mathscr{O}_{G/H})$ *is a quasiprojective homogeneous algebraic variety.*
(2) *For any pointed homogeneous space* (O, o) *such that* $G_o \supseteq H$*, the orbit map* $\pi : G \to O$ *factors through* $\bar{\pi} : G/H \to O$.
(3) $\bar{\pi}$ *is an isomorphism if and only if* $G_o = H$ *and* π *is separable.*

Proof. To prove (1), we use the following theorem of Chevalley [Hum, 11.2]:

> There exists a rational G-module V and a 1-dimensional subspace $L \subseteq V$ such that
>
> $$H = N_G(L) = \{g \in G \mid gL = L\},$$
> $$\mathfrak{h} = \mathfrak{n}_{\mathfrak{g}}(L) = \{\xi \in \mathfrak{g} \mid \xi L \subseteq L\}.$$

Let $x \in \mathbb{P}(V)$ correspond to L; then it follows that $H = G_x$ and $\mathfrak{h} = \operatorname{Ker} \mathrm{d}_x\pi$, where $\pi : G \to Gx$ is the orbit map. By a dimension argument, $\mathrm{d}_x\pi$ is surjective, whence π is separable. Further, Gx is homogeneous, whence smooth, and π is smooth [Har2, III.10.4], whence open [Har2, Ch. III, Ex. 9.1].

Let $U \subseteq Gx$ be an open subset. We claim that each $f \in \Bbbk[\pi^{-1}(U)]^H$ is the pullback of some $h \in \Bbbk[U]$. Indeed, consider the rational map $\varphi = (\pi, f) : G \dashrightarrow Gx \times \mathbb{A}^1$ and put $Z = \overline{\varphi(G)}$. The projection $Z \to Gx$ is separable and generically bijective, whence birational. Therefore $f \in \varphi^*\Bbbk[Z]$ descends to $h \in \Bbbk(U)$, $f = \pi^* h$. If h has the nonzero divisor of poles $D \subset U$, then f has the nonzero divisor of poles $\pi^* D \subset \pi^{-1}(U)$, a contradiction. It follows that $Gx \simeq G/H$ is a geometric quotient.

The universal property (2) is an obvious consequence of the definition. Moreover, any morphism $\varphi : G \to Y$ constant on H-orbits factors through $\bar{\varphi} : G/H \to Y$.

Finally, (3) follows from the separability of the quotient map $G \to G/H$: π is separable if and only if $\bar{\pi}$ is so, and $G_o = H$ means that $\bar{\pi}$ is bijective, whence birational and, by equivariance and homogeneity, is an isomorphism. □

Remark 1.4. In (2), if $G_o = H$ and π is not separable, then $\bar{\pi}$ is bijective purely inseparable and finite [Hum, 4.3, 4.6]. The schematic fiber $\pi^{-1}(o)$ is then a non-reduced group subscheme of G containing H as the reduced part. The homogeneous space O is uniquely determined by this subscheme [DG, III, §3], [Jan, I, 5.6].

Remark 1.5. If $H \lhd G$, then G/H is equipped with the structure of a linear algebraic group with usual properties of the quotient group. Indeed, in the notation of Chevalley's theorem, we may replace V by $\bigoplus_{\chi \in \mathfrak{X}(H)} V_\chi$ and consider the natural linear action $G : \mathrm{L}(V)$ by conjugation. The subspace $E = \prod \mathrm{L}(V_\chi)$ of operators preserving each V_χ is G-stable, and the image of G in $\mathrm{GL}(E)$ is isomorphic to G/H. See [Hum, 11.5] for details.

Definition 1.6. More generally, the *geometric quotient* of a G-variety X is defined as an algebraic variety isomorphic (as a ringed space) to the orbit space X/G equipped with the quotient topology and the structure sheaf $\mathscr{O}_{X/G}$ which is the direct image of the sheaf $\mathscr{O}_X^G$ of G-invariant regular functions on X, see [PV, 4.2].

In contrast with Theorem 1.3(1), the geometric quotient does not always exist in general: a necessary condition, e.g., is that all G-orbits must be closed and, if X is irreducible, they all must have the same dimension. However the geometric quotient always exists for a sufficiently small G-stable open subset of X [PV, Th. 4.4]. This quotient variety is a model of the field of invariant functions $\Bbbk(X)^G$ (provided X is irreducible). Such a model, considered up to birational equivalence, is called a *rational quotient* of X by G [PV, 2.4].

1.2 Tangent Spaces and Automorphisms. Recall that the *isotropy representation* for an action $G : X$ at $x \in X$ is the natural representation $G_x : T_xX$ by differentials of translations. For a quotient, the isotropy representation has a simple description:

Proposition 1.7. *$T_{eH}G/H \simeq \mathfrak{g}/\mathfrak{h}$ as H-modules.*

The isomorphism is given by the differential of the (separable) quotient map $G \to G/H$. The right-hand representation of H is the quotient of the adjoint representation of H in $\mathfrak{g}$.

Now we describe the group $\operatorname{Aut}_G(G/H)$ of G-equivariant automorphisms of G/H.

Proposition 1.8. *$\operatorname{Aut}_G(G/H) \simeq N_G(H)/H$ is an algebraic group acting on G/H regularly and freely. The action $N_G(H)/H : G/H$ is induced by the action $N_G(H) : G$ by right translations: $(nH)(gH) = gn^{-1}H$, $\forall g \in G$, $n \in N_G(H)$.*

Proof. The regularity of the action $N_G(H)/H : G/H$ is a consequence of the universal property of quotients. Clearly, this action is free. Conversely, if $\varphi \in \operatorname{Aut}_G(G/H)$, then $\varphi(eH) = nH$, and $n \in N_G(H)$, because the φ-action preserves stabilizers. Finally, $\varphi(gH) = g\varphi(eH) = gnH$, $\forall g \in G$. □

2 Fibrations, Bundles, and Representations

2.1 Homogeneous Bundles. The concept of associated bundle is fundamental in topology. We consider its counterpart in algebraic geometry in a particular case.

Let Z be an H-variety. Then H acts on $G \times Z$ by $h(g,z) = (gh^{-1}, hz)$.

Definition 2.1. The quotient set $G *_H Z = (G \times Z)/H$, equipped with the quotient topology and a structure sheaf which is the direct image of the sheaf of H-invariant regular functions, becomes a ringed space. It is called the *homogeneous fiber bundle* over G/H associated with Z.

The G-action on $G \times Z$ by left translations of the first factor commutes with the H-action and induces a G-action on $G *_H Z$. We denote by $g * z$ the image of (g,z) in $G *_H Z$ and identify $e * z$ with z. The embedding $Z \hookrightarrow G *_H Z$, $z \mapsto e * z$, solves the universal problem for H-equivariant morphisms of Z into G-spaces.

The homogeneous bundle $G *_H Z$ is G-equivariantly fibered over G/H with fibers gZ, $g \in G$. The fiber map is $g * z \mapsto gH$. This explains the terminology.

Theorem 2.2 ([B-B2], [PV, 4.8]). *If Z is covered by H-stable quasiprojective open subsets, then $G *_H Z$ is an algebraic G-variety, and the fiber map $G *_H Z \to G/H$ is locally trivial in étale topology.*

The proof is based on the fact that the fibration $G \to G/H$ is locally trivial in étale topology [Se1]. We shall always suppose that the assumption of the theorem is satisfied when we consider homogeneous bundles. The assumption is satisfied, e.g., if Z is quasiprojective, or normal and H is connected (by Sumihiro's Theorem C.7). If H is reductive and Z is affine, then $G *_H Z \simeq (G \times Z)/\!/H$ (see Appendix D) is affine.

The universal property of homogeneous bundles implies that any G-variety mapped onto G/H is a homogeneous bundle over G/H. More precisely, a G-equivariant map $\varphi : X \to G/H$ induces a bijective G-map $G *_H Z \to X$, where $Z = \varphi^{-1}(eH)$. If φ is separable, then $X \simeq G *_H Z$. In particular, any G-subvariety $Y \subseteq G *_H Z$ is G-isomorphic to $G *_H (Y \cap Z)$.

Since homogeneous bundles are locally trivial in étale topology, a number of local properties such as smoothness, normality, rationality of singularities, etc, are transferred from Z to $G *_H Z$ and back again. The next lemma indicates when a homogeneous bundle is trivial.

Lemma 2.3. *$G *_H Z \simeq G/H \times Z$ as G-varieties if the H-action on Z extends to a G-action.*

Proof. The isomorphism is given by $g * z \mapsto (gH, gz)$. □

If the fiber is an H-module, then the homogeneous bundle is locally trivial in Zariski topology. By the above, any G-vector bundle over G/H is G-isomorphic to $G *_H M$ for some finite-dimensional rational H-module M. The respective sheaf of sections $\mathscr{L}(M)$ is described in the following way.

Proposition 2.4. *For any open subset $U \subseteq G/H$, we have $\mathrm{H}^0(U, \mathscr{L}(M)) \simeq \mathrm{Mor}_H(\pi^{-1}(U), M)$, where $\pi : G \to G/H$ is the quotient map.*

Proof. It is easy to see that the pullback of $G *_H M \to G/H$ under π is a trivial vector bundle $G \times M \to G$. Hence for each $\sigma \in \mathrm{H}^0(U, \mathscr{L}(M))$ we have $\pi^*\sigma \in \mathrm{Mor}(\pi^{-1}(U), M)$, and clearly $\pi^*\sigma$ is H-equivariant. Conversely, any H-morphism $\pi^{-1}(U) \to M$ induces a section $U \to G *_H M$ by the universal property of the quotient. □

If $H : M$ is an infinite-dimensional rational module, we may *define* a quasicoherent sheaf $\mathscr{L}(M) = \mathscr{L}_{G/H}(M)$ on G/H by the formula of Proposition 2.4 [Jan, I.5.8–5.9]. The functor $\mathscr{L}_{G/H}(\cdot)$ establishes an equivalence between the category of rational H-modules and that of G-sheaves on G/H.

Any G-line bundle over G/H is G-isomorphic to $G *_H \Bbbk_\chi$, where $\Bbbk_\chi = \Bbbk$ with the H-action via a character $\chi \in \mathfrak{X}(H)$. This yields an isomorphism $\mathfrak{X}(H) \xrightarrow{\sim} \mathrm{Pic}_G(G/H)$, $\chi \mapsto \mathscr{L}(\chi) = \mathscr{L}(\Bbbk_\chi)$. The kernel of the forgetful homomorphism $\mathrm{Pic}_G(G/H) \to \mathrm{Pic}(G/H)$ consists of characters that correspond to different G-linearizations of the trivial line bundle $G/H \times \Bbbk$ over G/H. If G is connected, then these characters are exactly the restrictions to H of characters of G (see Lemma 2.3 and (C.1)).

Consider a universal cover $\widetilde{G} \to G$ (see Appendix C). By $\widetilde{H}$ denote the inverse image of H in $\widetilde{G}$; then $G/H \simeq \widetilde{G}/\widetilde{H}$. Since any line bundle over G/H is $\widetilde{G}$-linearizable (Corollary C.5), we obtain the following theorem of Popov.

Theorem 2.5 ([Po2], [KKV, §3]). $\operatorname{Pic}_G(G/H) \simeq \mathfrak{X}(H)$. *If G is connected, then* $\operatorname{Pic} G/H \simeq \mathfrak{X}(\widetilde{H})/\operatorname{Res}^{\widetilde{G}}_{\widetilde{H}} \mathfrak{X}(\widetilde{G})$.

Here $\operatorname{Res}^{\widetilde{G}}_{\widetilde{H}}$ denotes the restriction from $\widetilde{G}$ to $\widetilde{H}$.

Example 2.6. Let G be a connected reductive group, and let $B \subseteq G$ be a Borel subgroup. Then $\operatorname{Pic} G/B$ is isomorphic to the weight lattice of the root system of G.

Let X be a G-variety, and $Z \subseteq X$ an H-stable closed subvariety. By the universal property, we have a G-equivariant map $\mu : G *_H Z \to X$, $\mu(g * z) = gz$.

Proposition 2.7. *If H is parabolic, then μ is proper and GZ is closed in X.*

Proof. The map μ factors as $\mu : G *_H Z \overset{\iota}{\hookrightarrow} G *_H X \simeq G/H \times X$ (Lemma 2.3) $\overset{\pi}{\to} X$, where ι is a closed embedding and π is a projection along a complete variety by Theorem 3.3. □

Example 2.8. Let $\mathfrak{N} \subseteq \mathfrak{g}$ be the set of nilpotent elements and let $U = \mathrm{R_u}(B)$, a maximal unipotent subgroup of G. Then the map $G *_B \mathfrak{u} \to \mathfrak{N}$ is proper and birational, see, e.g., [PV, 5.6] or [McG, 7.1]. This is a well-known Springer's resolution of singularities of $\mathfrak{N}$.

2.2 Induction and Restriction. Now we discuss some representation theory related to homogeneous spaces and to vector bundles over them.

We always deal with rational modules over algebraic groups (see Appendix C) and often drop the word "rational". As usual in representation theories, we may define functors of induction and restriction on categories of rational modules. Let H act on G by right translations, and let M be an H-module.

Definition 2.9. A G-module $\operatorname{Ind}^G_H M = \operatorname{Mor}_H(G,M) \simeq (\Bbbk[G] \otimes M)^H$ is said to be *induced* from $H : M$ to G. It is a rational G-$\Bbbk[G/H]$-module. By definition, we have $\operatorname{Ind}^G_H M = \mathrm{H}^0(G/H, \mathscr{L}(M))$.

A G-module N considered as an H-module is denoted by $\operatorname{Res}^G_H N$.

Example 2.10. $\operatorname{Ind}^G_H \Bbbk = \Bbbk[G/H]$, where $\Bbbk$ is the trivial H-module. More generally, $\operatorname{Ind}^G_H \Bbbk_\chi = \Bbbk[G]^{(H)}_{-\chi}$, $\forall \chi \in \mathfrak{X}(H)$.

Clearly, Ind^G_H is a left exact functor from the category of rational H-modules to that of rational G-modules. The functor Res^G_H is exact. We collect basic properties of induction in the following

Theorem 2.11. (1) *If M is a G-module, then* $\operatorname{Ind}^G_H M \simeq \Bbbk[G/H] \otimes M$.
(2) (Frobenius reciprocity) *For rational modules $G : N$, $H : M$, we have*

$$\operatorname{Hom}_G(N, \operatorname{Ind}^G_H M) \simeq \operatorname{Hom}_H(\operatorname{Res}^G_H N, M).$$

(3) *For any H-module M,* $(\mathrm{Ind}_H^G M)^G \simeq M^H$.
(4) *If M,N are rational algebras, then* (1) *and* (3) *are isomorphisms of algebras, and* (2) *holds for equivariant algebra homomorphisms.*

Proof. (1) The isomorphism $\iota : \mathrm{Mor}_H(G,M) \xrightarrow{\sim} \mathrm{Mor}(G/H,M)$ is given by $\iota(m)(gH) = g\cdot m(g)$, $\forall m \in \mathrm{Mor}_H(G,M)$. The inverse mapping is $\mu \mapsto m$, $m(g) = g^{-1}\mu(gH)$, $\forall \mu \in \mathrm{Mor}(G/H,M)$.
(2) The isomorphism is given by the map $\Phi \mapsto \varphi$, $\forall \Phi : N \to \mathrm{Mor}_H(G,M)$, where $\varphi : N \to M$ is defined by $\varphi(n) = \Phi(n)(e)$, $\forall n \in N$. The inverse map $\varphi \mapsto \Phi$ is given by $\Phi(n)(g) = \varphi(g^{-1}n)$.
(3) Any G-invariant H-equivariant morphism $G \to M$ is constant, and its image lies in M^H. Alternatively, one may apply the Frobenius reciprocity to $N = \Bbbk$.
(4) It is easy. □

Remark 2.12. The union of (1) and (3) yields the following assertion: if M is a G-module, then $(\Bbbk[G/H] \otimes M)^G \simeq M^H$. This is often called the *transfer principle*, because it allows transfer of information from $\Bbbk[G/H]$ to M^H. For example, if G is reductive, $\Bbbk[G/H]$ is finitely generated, and $M = A$ is a finitely generated G-algebra, then A^H is finitely generated. Other applications are discussed in Appendix D. A good treatment of induced modules and the transfer principle can be found in [Gr2].

2.3 Multiplicities. We are interested in the G-module structure of $\Bbbk[G/H]$ and of global sections of line bundles over G/H.

For any two rational G-modules V,M ($\dim V < \infty$), put

$$m_V(M) = \dim \mathrm{Hom}_G(V,M),$$

the *multiplicity* of V in M. If V is simple and M completely reducible (e.g., G is an algebraic torus or a reductive group in characteristic zero), then $m_V(M)$ is the number of occurrences of V in a decomposition of M into simple summands.

For any G-variety X and a G-line bundle $\mathscr{L} \to X$, we abbreviate:

$$m_V(X) = m_V(\Bbbk[X]), \qquad m_V(\mathscr{L}) = m_V(\mathrm{H}^0(X,\mathscr{L})).$$

Here is a particular case of Frobenius reciprocity:

Corollary 2.13. $m_V(G/H) = \dim(V^*)^H$, $m_V(\mathscr{L}(\chi)) = \dim(V^*)^{(H)}_{-\chi}$.

Proof. We have $\mathrm{H}^0(G/H,\mathscr{L}(\chi)) = \mathrm{Ind}_H^G \Bbbk_\chi$, whence

$$\mathrm{Hom}_G\big(V,\mathrm{H}^0(G/H,\mathscr{L}(\chi))\big) = \mathrm{Hom}_H(V,\Bbbk_\chi) = (V^*)^{(H)}_{-\chi}.$$

The first equality follows by taking $\chi = 0$. □

2.4 Regular Representation. A related problem is to describe the module structure of $\Bbbk[G]$ (=the so-called *regular representation*). Namely, G itself is acted on by $G \times G$ via $(g_1,g_2)g = g_1 g g_2^{-1}$. Hence $\Bbbk[G]$ is a $(G \times G)$-algebra.

Every finite-dimensional G-module V generates a $(G \times G)$-stable subspace $\mathrm{M}(V) \subset \Bbbk[G]$ spanned by matrix entries $f_{\omega,v}(g) = \langle \omega, gv \rangle$ ($v \in V$, $\omega \in V^*$) of the representation $G \to \mathrm{GL}(V)$. Clearly $\mathrm{M}(V)$ is the image of a $(G \times G)$-module homomorphism $V^* \otimes V \to \Bbbk[G]$, $\omega \otimes v \mapsto f_{\omega,v}$, and $\mathrm{M}(V) \simeq V^* \otimes V$ is a simple $(G \times G)$-module whenever V is simple.

Matrix entries behave well with respect to algebraic operations:

$$\mathrm{M}(V) + \mathrm{M}(V') = \mathrm{M}(V \oplus V'), \qquad \mathrm{M}(V) \cdot \mathrm{M}(V') = \mathrm{M}(V \otimes V'). \tag{2.1}$$

The inversion of G sends $\mathrm{M}(V)$ to $\mathrm{M}(V^*)$.

Proposition 2.14. $\Bbbk[G] = \bigcup \mathrm{M}(V)$, *where* V *runs through all finite-dimensional* G-*modules.*

Proof. Take any finite-dimensional G-submodule $V \subset \Bbbk[G]$ with respect to the G-action by right translations. We claim that $V \subseteq \mathrm{M}(V)$. Indeed, let $\omega \in V^*$ be defined by $\langle \omega, v \rangle = v(e)$, $\forall v \in V$; then $v(g) = f_{\omega,v}(g), \forall v \in V, g \in G$. □

Theorem 2.15. *Suppose that* $\operatorname{char} \Bbbk = 0$ *and* G *is reductive. Then there is a* $(G \times G)$-*module isomorphism*

$$\Bbbk[G] = \bigoplus \mathrm{M}(V) \simeq \bigoplus V^* \otimes V,$$

where V *runs through all simple* G-*modules.*

Proof. All the $\mathrm{M}(V) \simeq V^* \otimes V$ are pairwise non-isomorphic simple $(G \times G)$-modules. By Proposition 2.14 and (2.1) they span the whole of $\Bbbk[G]$. □

Remark 2.16. Corollary 2.13 can be derived from Theorem 2.15 by taking H-(semi)invariants from the right.

2.5 Hecke Algebras. The dual object to the coordinate algebra of G provides a version of the group algebra for algebraic groups.

Definition 2.17. The *(algebraic) group algebra* of G is $\mathscr{A}(G) = \Bbbk[G]^*$ equipped with the multiplication law coming from the comultiplication in $\Bbbk[G]$.

For finite G we obtain the usual group algebra. Generally, $\mathscr{A}(G)$ can be described by finite-dimensional approximations. The group algebra $\mathscr{A}(V)$ of a finite-dimensional G-module V is defined as the linear span of the image of G in $\mathrm{L}(V)$. Note that $\mathscr{A}(V)$ is the $(G \times G)$-module dual to $\mathrm{M}(V)$. We have $\mathscr{A}(V) = \mathrm{L}(V)$ whenever V is simple. Given a subquotient module V' of V, there are a canonical inclusion $\mathrm{M}(V') \subseteq \mathrm{M}(V)$ and a canonical epimorphism $\mathscr{A}(V) \twoheadrightarrow \mathscr{A}(V')$. Therefore the algebras $\mathscr{A}(V)$ form an inverse system over all V ordered by the relation of being a subquotient. It readily follows from Proposition 2.14 that $\mathscr{A}(G) \simeq \varprojlim \mathscr{A}(V)$. One deduces that $\mathscr{A}(G)$ is a universal ambient algebra containing both G and $\mathrm{U}\mathfrak{g}$, the (restricted) enveloping algebra of $\mathfrak{g}$ [DG, II, §§6,7], [Jan, I, §7].

Definition 2.18. The algebra $\mathscr{A}(G/H)$ of all G-equivariant linear endomorphisms of $\Bbbk[G/H]$ is called the *Hecke algebra* of G/H, or of (G,H).

Remark 2.19. If $\operatorname{char} \Bbbk = 0$ and G is reductive, then $\mathscr{A}(V) = \prod \mathrm{L}(V_i)$ over all simple G-modules V_i occurring in V with positive multiplicity. Furthermore, $\mathscr{A}(G) = \prod \mathrm{L}(V_i)$ and $\mathscr{A}(G/H) = \prod \mathrm{L}(V_i^H)$ over all simple V_i by Theorem 2.15 and Schur's lemma.

Proposition 2.20 (E. B. Vinberg). *If* $\operatorname{char} \Bbbk = 0$ *and* H *is reductive, then* $\mathscr{A}(G/H) \simeq \mathscr{A}(G)^{H \times H}$. *In particular, the above notation is compatible for* $H = \{e\}$.

Proof. First consider the case $H = \{e\}$. The algebra $\mathscr{A}(V)$ acts on $\mathscr{A}(V)^* = \mathrm{M}(V)$ by right translations: $af(x) = f(xa)$, $\forall a, x \in \mathscr{A}(V)$, $f \in \mathrm{M}(V)$. These actions commute with the G-action by left translations and merge together into a G-equivariant linear $\mathscr{A}(G)$-action on $\Bbbk[G]$.

Conversely, every G-equivariant linear map $\varphi : \Bbbk[G] \to \Bbbk[G]$ preserves all the spaces $\mathrm{M}(V)$. Indeed, it follows from the proof of Proposition 2.14 by applying the inversion that $W \subseteq \mathrm{M}(W^*)$ for any G-submodule $W \subset \Bbbk[G]$. On the other hand, for $W = \mathrm{M}(V)$ one has $\mathrm{M}(W) \subseteq \mathrm{M}(V^* \oplus \cdots \oplus V^*) = \mathrm{M}(V^*)$ whence, by applying the inversion, one easily deduces that $W = \mathrm{M}(W^*)$ and $\varphi W \subseteq \mathrm{M}((\varphi W)^*) \subseteq \mathrm{M}(W^*) = W$.

The restriction of φ to $\mathrm{M}(V)$ is the right translation by some $a_V \in \mathscr{A}(V)$. These a_V give rise to $a \in \mathscr{A}(G)$ representing φ on $\Bbbk[G]$. Hence the group algebra coincides with the Hecke algebra of G.

In the general case, every linear G-endomorphism φ of $\Bbbk[G]^H$ extends to a unique $a \in \mathscr{A}(G)^{H \times H}$, which annihilates the right-H-invariant complement of $\Bbbk[G]^H$ in $\Bbbk[G]$. □

2.6 Weyl Modules. Now we discuss applications of homogeneous line bundles to the representation theory of reductive groups. We start with fixing some notation [Hum, Ch. IX–XI], [OV, Ch. 4].

Let G be a connected reductive group, $B \subseteq G$ a Borel subgroup, $U = \mathrm{R_u}(B)$ a maximal unipotent subgroup, and $T \subseteq B$ a maximal torus. Recall the root decomposition

$$\mathfrak{g} = \mathfrak{t} \oplus \bigoplus_{\alpha \in \Delta} \mathfrak{g}^{\alpha},$$

where $\Delta = \Delta_G \subset \mathfrak{X}(T)$ denotes the root system of G with respect to T and $\mathfrak{g}^{\alpha}$ denote the root subspaces. Let $\alpha^\vee \in \mathfrak{X}^*(T)$ denote the respective coroots and put $\Delta^\vee = \{\alpha^\vee \mid \alpha \in \Delta\}$. The set of positive roots Δ_+ consists of α such that $\mathfrak{g}^{\alpha}$ span $\mathfrak{u}$. Let $\Pi = \Pi_G \subseteq \Delta_+$ denote the base set of simple roots and $\Pi^\vee = \{\alpha^\vee \mid \alpha \in \Pi\}$. The set of negative roots $\Delta_- = \Delta \setminus \Delta_+ = -\Delta_+$ consists of α such that $\mathfrak{g}^{\alpha}$ span $\mathfrak{u}^-$, the Lie algebra of $U^- = \mathrm{R_u}(B^-)$, where B^- is the opposite Borel subgroup of G intersecting B in T. The choice of positive roots defines the dominant Weyl chamber

$$\mathbf{C} = \mathbf{C}(\Delta_+) = \{\lambda \in \mathfrak{X}(T) \otimes \mathbb{Q} \mid \langle \lambda, \alpha^\vee \rangle \geq 0, \ \forall \alpha \in \Delta_+\}.$$

Let $\mathfrak{X}_+ = \mathfrak{X}(T) \cap \mathbf{C}$ denote the semigroup of dominant weights.

Let $W = N_G(T)/T$ be the Weyl group of G with respect to T. It contains root reflections r_α ($\alpha \in \Delta$) acting on $\mathfrak{X}(T)$ as $r_\alpha(\lambda) = \lambda - \langle \lambda, \alpha^\vee \rangle \alpha$ and is generated by reflections corresponding to simple roots. Let w_G denote the longest element of W, i.e., the unique element mapping positive roots to negative ones. Sometimes it is convenient to denote elements of W and their representatives in $N_G(T)$ by the same letters.

Parabolic subgroups of G containing a given Borel subgroup B are usually called *standard*. They are parameterized by subsets of simple roots $I \subseteq \Pi$. The Lie algebra of the respective parabolic $P = P_I$ is

$$\mathfrak{p} = \mathfrak{t} \oplus \bigoplus_{\alpha \in \Delta_+ \cup \Delta_I} \mathfrak{g}^\alpha,$$

where $\Delta_I \subseteq \Delta$ is the root subsystem spanned by $I \subseteq \Pi$. There is a unique Levi decomposition $P = P_u \rightthreetimes L$, where $P_u = \mathrm{R_u}(P)$ and $L = L_I \subseteq P_I$ is a unique Levi subgroup containing a given maximal torus $T \subseteq B$, sometimes called a *standard Levi subgroup* of G by abuse of language. The root system of L_I with respect to T is Δ_I. The opposite parabolic $P^- = P_I^- \supseteq B^-$ associated with I intersects P_I in L_I and has the standard Levi decomposition $P^- = P_u^- \rightthreetimes L$, where $P_u^- = \mathrm{R_u}(P^-)$.

Every G-module V contains a B-eigenvector v, by the Lie–Kolchin theorem. If V is generated by v as a G-module, then v is called a *highest vector* of V. A highest vector is unique up to proportionality and its weight $\lambda \in \mathfrak{X}(B) = \mathfrak{X}(T)$ is the *highest weight* of V, i.e., all other T-weights of V are obtained from λ by subtracting positive roots, whence $\lambda \in \mathfrak{X}_+$ [Hum, 31.2]. Likewise, we use the term "*lowest vector* (*weight*)" of V for (the weight of) a B^--eigenvector generating V as a G-module. All other T-weights of V are obtained from the lowest one by adding positive roots. In the above notation, the lowest weight of V is $w_G\lambda$ and a lowest vector is w_Gv.

In particular, any simple G-module V contains a unique, up to proportionality, B-eigenvector (highest vector) of weight $\lambda \in \mathfrak{X}_+$ (highest weight). Conversely, for every $\lambda \in \mathfrak{X}_+$ there is a unique, up to isomorphism, simple G-module V of highest weight λ [Hum, 31.3–31.4]. Thus isomorphism classes of simple G-modules are indexed by dominant weights. The highest weight of V^* is $\lambda^* = -w_G\lambda$.

By Corollary 2.13,

$$m_V(\mathscr{L}_{G/B}(\mu)) = \dim(V^*)^{(B)}_{-\mu} = \begin{cases} 1, & \mu = -\lambda^*, \\ 0, & \text{otherwise.} \end{cases}$$

It follows that $V^*(\lambda) = \mathrm{Ind}_B^G \Bbbk_{-\lambda}$ contains a unique simple G-module (of highest weight λ^*) whenever $\lambda \in \mathfrak{X}_+$, otherwise $V^*(\lambda) = 0$. The dual G-module $V(\lambda) = V_G(\lambda) = (\mathrm{Ind}_B^G \Bbbk_{-\lambda})^*$ is called a *Weyl module* [Jan, II.2].

Put $m_\lambda(M) = m_{V(\lambda)}(M)$ for brevity.

Proposition 2.21. $m_\lambda(M) = \dim M_\lambda^{(B)}$.

Proof. As G/B is a projective variety (Theorem 3.3), $V(\lambda) = \mathrm{H}^0(G/B, \mathscr{L}(-\lambda))^*$ is finite-dimensional. If $\dim M < \infty$, then

$$\mathrm{Hom}_G(V(\lambda), M) \simeq \mathrm{Hom}_G(M^*, V^*(\lambda)) \simeq \mathrm{Hom}_B(M^*, \Bbbk_{-\lambda}) \simeq M_\lambda^{(B)}.$$

However, any rational G-module M is a union of finite-dimensional submodules. □

Thus $V(\lambda)$ can be characterized as the universal covering G-module of highest weight λ: the generating highest vector $v_\lambda \in V(\lambda)$ is given by evaluation of $\mathrm{H}^0(G/B, \mathscr{L}(-\lambda))$ at eB.

By Corollary 2.13, we have

$$m_\lambda(G/H) = \dim V^*(\lambda)^H, \qquad m_\lambda(\mathscr{L}_{G/H}(\chi)) = \dim V^*(\lambda)^{(H)}_{-\chi}. \tag{2.2}$$

In characteristic zero, complete reducibility yields:

Borel–Weil theorem. *If* $\operatorname{char} \Bbbk = 0$, *then* $V(\lambda)$ *is a simple* G-*module of highest weight* λ *and* $V^*(\lambda) \simeq V(\lambda^*)$.

Furthermore, Theorem 2.15 yields

$$\Bbbk[G] \simeq \bigoplus_{\lambda \in \mathfrak{X}_+} V(\lambda^*) \otimes V(\lambda). \tag{2.3}$$

In arbitrary characteristic, Formula (2.3) is no longer true, but $\Bbbk[G]$ possesses a "good" $(G \times G)$-module filtration with factors $V^*(\lambda) \otimes V^*(\lambda^*)$ [Don], [Jan, II.4.20].

Notice that all the dual Weyl modules are combined in a multigraded algebra

$$\Bbbk[G/U] = \bigoplus_{\lambda \in \mathfrak{X}_+} \Bbbk[G]^{(B)}_\lambda = \bigoplus_{\lambda \in \mathfrak{X}_+} V^*(\lambda) \tag{2.4}$$

called the *covariant algebra* of G. The covariant algebra is an example of a multiplicity-free G-algebra, in the sense of the following

Definition 2.22. A G-module M is said to be *multiplicity-free* if $m_\lambda(M) \leq 1$, $\forall \lambda \in \mathfrak{X}_+$.

The multiplication in the covariant algebra has a nice property:

Lemma 2.23. $V^*(\lambda) \cdot V^*(\mu) = V^*(\lambda + \mu)$.

Proof. The inclusion "$\subseteq$" in the lemma is obvious since the $V^*(\lambda)$ are the homogeneous components of $\Bbbk[G/U]$ with respect to an algebra grading. In characteristic zero, the reverse inclusion stems from the fact that the $V^*(\lambda)$ are simple G-modules and $\Bbbk[G/U]$ is an integral domain. In positive characteristic it was proved by Ramanan and Ramanathan [Jan, II.14.20]. □

3 Classes of Homogeneous Spaces

3.1 Reductions. We answer the following question: when is a homogeneous space O projective or (quasi)affine? First, we reduce the question to a property of the pair (G,H), where $H = G_o$ is the stabilizer of a point $o \in O$.

Lemma 3.1. *O is projective, resp. (quasi)affine, if and only if G/H has this property.*

Proof. We may assume that $\operatorname{char}\Bbbk = p > 0$. The natural map $\varphi : G/H \to O$ is finite bijective purely inseparable (Remark 1.4). We deduce the assertions on projectivity and affinity by [Har2, III, Ex. 4.2].

For quasiaffinity, we argue as follows. Without loss of generality, G may be assumed connected. First note that $\Bbbk(G/H)^{p^s} \subseteq \varphi^*\Bbbk(O)$ for some $s \geq 0$. Furthermore, $\Bbbk[G/H]^{p^s} \subseteq \varphi^*\Bbbk[O]$ (If $f \in \Bbbk(O) \setminus \Bbbk[O]$, then $\varphi^* f$ has poles on G/H.) Assume that O is an open subset of an affine variety Y. Let B be the integral closure of $\varphi^*\Bbbk[Y]$ in $\Bbbk(G/H)$. Then B is finitely generated, and $X = \operatorname{Spec} B$ contains G/H as an open subset. Conversely, if G/H is open in an affine variety X, then $A = \Bbbk[X] \cap \varphi^*\Bbbk(O)$ is finite over $\Bbbk[X]^{p^s}$. Hence A is finitely generated, and $X = \operatorname{Spec} A$ contains O as an open subset. □

In the sequel, we may assume that $O = G/H$.

Lemma 3.2. *If $G \supseteq H \supseteq K$ and G/H, H/K are projective, resp. (quasi)affine, then G/K is projective, resp. (quasi)affine.*

Proof. The natural map $\varphi : G/K \to G/H$ transforms after a faithfully flat base change $G \to G/H$ to the projection $\pi : G/K \times_{G/H} G \simeq H/K \times G \to G$. If H/K is projective (resp. affine), then π is proper (resp. affine), whence φ is proper (resp. affine). If in addition G/H is complete (resp. affine), then G/K is also complete (resp. affine). Another proof for projective and affine cases relies on Theorems 3.3 and 3.9 below. In the quasiaffine case, the lemma follows from Theorem 3.12(4). □

3.2 Projective Homogeneous Spaces.

Theorem 3.3. *G/H is projective if and only if H is parabolic.*

Proof. If G/H is projective, then a Borel subgroup $B \subseteq G$ has a fixed point $gH \in G/H$, by the Borel Fixed Point Theorem [Hum, 21.2]. Hence $H \supseteq g^{-1}Bg$ is parabolic.

To prove the converse, consider a faithful representation $G : V$. The induced action of G on the variety of complete flags in V has a closed orbit. Its stabilizer B is solvable, and we may assume that $B \subseteq H$. By Lemma 3.1, G/B is complete, and hence G/H is complete. □

3.3 Affine Homogeneous Spaces. A group-theoretical characterization of affine homogeneous spaces is not known at the moment. We give several sufficient conditions of affinity and a criterion for reductive G. The following easy lemma is well known.

Lemma 3.4 ([Bor2, 4.10]). *The orbits of a unipotent group G on an affine variety X are closed, whence affine.*

Proof. For any $x \in X$, consider closed affine subvarieties $Y = \overline{Gx} \subseteq X$ and $Z = Y \setminus Gx$. Since $\mathscr{I}(Z) \lhd \Bbbk[Y]$ is a G-submodule, the Lie–Kolchin theorem implies $\exists f \in \mathscr{I}(Z)^G$, $f \neq 0$. However f is a nonzero constant on Gx, whence on Y. Thus $Z = \emptyset$. □

Theorem 3.5. *G/H is affine if G is solvable.*

Proof. We may assume that G, H are connected. First suppose that G is unipotent. Take a representation $G : V$ such that $\exists v \in V : G[v] \simeq G/H$. Then H normalizes $\langle v \rangle$. But $\mathfrak{X}(H) = 0$, whence $G_v = H$ and $Gv \simeq G/H$. We conclude by Lemma 3.4.

In the general case, $G = T \leftthreetimes U$ and $H = S \leftthreetimes V$, where U, V are unipotent radicals and T, S are maximal tori of G, H. We have $U \supseteq V$ and may assume that $T \supseteq S$. It is easy to see that $G/H \simeq T *_S U/V = (T \times U/V)/\!/S$ is affine. □

The following notion is often useful in the theory of homogeneous spaces.

Definition 3.6. We say that H is *regularly embedded* in G if $\mathrm{R_u}(H) \subseteq \mathrm{R_u}(G)$.

For example, any subgroup of a solvable group is regularly embedded. The next theorem generalizes Theorem 3.5.

Theorem 3.7. *G/H is affine if H is regularly embedded in G.*

Proof. As $\mathrm{R_u}(G)$ is normal in G, the quotient $G/\mathrm{R_u}(G)$ is affine. By Theorem 3.5, $\mathrm{R_u}(G)/\mathrm{R_u}(H)$ is affine. Thence by Lemma 3.2, $G/\mathrm{R_u}(H)$ is affine. By the Main Theorem of GIT (see Appendix D), $G/H = (G/\mathrm{R_u}(H))/\!/(H/\mathrm{R_u}(H))$ is affine, because $H/\mathrm{R_u}(H)$ is reductive. □

Weisfeiler proved [Wei] that any subgroup H of a connected group G is regularly embedded in some parabolic subgroup $P \subseteq G$. (See also [Hum, 30.3].) Thus G/H is a fibration with the projective base G/P and affine fiber P/H.

The following theorem is often called *Matsushima's criterion*. It was proved for $\Bbbk = \mathbb{C}$ by Matsushima [Mat] and Onishchik [Oni1], and in the general case by Richardson [Ri1].

Theorem 3.8. *G/H is affine if H is reductive. If G is reductive, the converse is also true.*

Proof. If H is reductive, then by the Main Theorem of GIT, $G/H \simeq G/\!/H$ is affine. For a simple proof of the converse see [Lu2] ($\operatorname{char}\Bbbk = 0$) or [Ri1]. □

The lack of a group-theoretical criterion of affinity is partially compensated by a cohomological criterion.

Theorem 3.9. *G/H is affine if and only if Ind_H^G is exact.*

Proof. Recall that $\mathrm{Ind}_H^G(M) = \mathrm{H}^0(G/H, \mathscr{L}(M))$, the sheaf $\mathscr{L}(M)$ is quasicoherent, and the functor $\mathscr{L}(\cdot)$ is exact. If G/H is affine, then by Serre's criterion [Har2, III.3.7] Ind_H^G is exact. For a proof of the converse, see [Gr2, §6]. □

3.4 Quasiaffine Homogeneous Spaces. The class of quasiaffine homogeneous spaces is of interest in invariant theory and representation theory. If G/H is quasiaffine, then the subgroup H is called *observable*. Observable subgroups are exactly the stabilizers of vectors in rational G-modules, since any quasiaffine G-variety can be equivariantly embedded in a G-module [PV, 1.2].

Example 3.10. By Chevalley's theorem, H is observable if $\mathfrak{X}(H) = 0$. In particular a unipotent subgroup is observable.

Example 3.11 ([BHM]). If $\mathrm{R}(H)$ is nilpotent, then H is observable.

It is easy to see that an intersection of observable subgroups is again observable. Therefore, for any $H \subseteq G$, there exists a smallest observable subgroup $\widehat{H} \subseteq G$ containing H. It is called the *observable hull* of H. Clearly, for any rational G-module M we have $M^H = M^{\widehat{H}}$. This property illustrates the importance of observable subgroups in invariant theory, see [PV, 3.7].

We give several characterizations of observable subgroups in the next theorem, essentially due to Białynicki-Birula, Hochschild, and Mostow [BHM].

Theorem 3.12. *The following conditions are equivalent:*

(1) *G/H is quasiaffine.*
(2) *G^0/H^0 is quasiaffine.*
(3) $\operatorname{Quot}\Bbbk[G/H] = \Bbbk(G/H)$.
(4) *Any finite-dimensional H-module is embedded as an H-submodule in a finite-dimensional G-module.*
(5) $\forall \chi \in \mathfrak{X}(H): \ \Bbbk[G]_\chi \neq 0 \implies \Bbbk[G]_{-\chi} \neq 0$. *(In other words, the semigroup of weights of H-eigenfunctions on G is actually a group.)*

Proof. (1) $\Longrightarrow$ (3) is obvious.
(3) $\Longrightarrow$ (1) We have $\Bbbk[G/H] = \Bbbk[G/\widehat{H}] \implies \Bbbk(G/H) = \Bbbk(G/\widehat{H})$, whence $H = \widehat{H}$.
(1) $\Longleftrightarrow$ (2) We may assume that G is connected. The map $G/H^0 \to G/H$ is a Galois covering with the Galois group $\Gamma = H/H^0$. If G/H is open in an affine variety X, then G/H^0 is open in $Y = \operatorname{Spec} A$, where A is the integral closure of $\Bbbk[X]$ in $\Bbbk(G/H^0)$. Conversely, if G/H^0 is open in affine Y, then G/H is open in $X = Y/\Gamma$.
(1) $\Longrightarrow$ (5) For a nonzero $g \in \Bbbk[G]_\chi$, consider its zero set $Z \subset G$. The quotient morphism $\pi : G \to G/H$ maps Z onto a proper closed subset of G/H. Hence $\exists f \in \Bbbk[G/H] : f|_{\pi(Z)} = 0$. By Nullstellensatz, $\pi^* f^n = gh$ for some $n \in \mathbb{N}$, $h \in \Bbbk[G]_{-\chi}$.
(5) $\Longrightarrow$ (4) First note that a 1-dimensional H-module $W = \Bbbk_\chi$ can be embedded in a G-module V if and only if $\Bbbk[G]_\chi \neq 0$. (Any function $f_{\omega,v}$, where $v \in W$, $\omega \in V^*$, belongs to $\Bbbk[G]_\chi$.)

Now, for any finite-dimensional H-module W, consider the embedding $W \hookrightarrow \operatorname{Mor}(H,W)$ taking each $w \in W$ to the orbit morphism $g \mapsto gw$, $g \in H$. It is H-equivariant with respect to the H-action on $\operatorname{Mor}(H,W)$ by right translations of an argument. The restriction of morphisms yields a surjection $\operatorname{Mor}(G,W) \to \operatorname{Mor}(H,W)$, and we may choose a finite-dimensional H-submodule $N \subset \operatorname{Mor}(G,W)$ mapped onto W. Embed N into a finite-dimensional G-submodule $M \subset \operatorname{Mor}(G,W)$ and put

$U = \mathrm{Ker}(N \to W)$, $m = \dim U$. Then $\bigwedge^m U \hookrightarrow \bigwedge^m M$ and $W \otimes \bigwedge^m U \hookrightarrow \bigwedge^{m+1} M$. By (5) and the above remark, $\bigwedge^m U^*$ is embedded in a G-module. We conclude by noting that $W \simeq (W \otimes \bigwedge^m U) \otimes \bigwedge^m U^*$.
(4) $\Longrightarrow$ (1) By Chevalley's theorem, $H = G_{[v]}$ for some vector v in a G-module V. As an H-module, $\langle v \rangle \simeq \Bbbk_\chi$ for some $\chi \in \mathfrak{X}(H)$. Then $\Bbbk_{-\chi}$ can be embedded in a G-module, i.e., there exists a G-module W and $w \in W$ such that H acts on w via $-\chi$. It follows that $G_{v \otimes w} = H$. □

Surprisingly, quasiaffine homogeneous spaces admit a group-theoretical characterization. Recall that a *quasiparabolic* subgroup of a connected group is the stabilizer of a highest vector in an irreducible representation.

Theorem 3.13 ([Sukh]). *$H \subseteq G$ is observable if and only if H^0 is regularly embedded in a quasiparabolic subgroup of G^0.*

Chapter 2
Complexity and Rank

We retain the general conventions of our survey. In particular, G denotes a reductive connected linear algebraic group. We begin with local structure theorems, which claim that a G-variety may be covered by affine open subsets stable under parabolic subgroups of G, and describe the structure of these subsets. In §5, we define two numerical invariants of a G-variety related to the action of a Borel subgroup of G: the complexity and the rank. We reduce their computation to a general orbit on X (i.e., a homogeneous space) and prove some basic results including the semicontinuity of complexity and rank with respect to G-subvarieties. We also introduce the notion of the weight lattice and consider the connection of complexity with the growth of multiplicities in coordinate algebras and spaces of sections of line bundles. The relation of complexity and modality of an action is considered in §6. In §7, we introduce the class of horospherical varieties defined by the property that all isotropy groups contain a maximal unipotent subgroup of G. The computation of complexity and rank is fairly simple for them. On the other hand, any G-variety can be contracted to a horospherical one of the same complexity and rank.

General formulæ for complexity and rank are obtained in §8 as a by-product of the study of the cotangent action $G : T^*X$ and the doubled action $G : X \times X^*$. These formulæ involve generic stabilizers of these actions. The particular case of a homogeneous space $X = G/H$ is considered in §9. In §10, we classify homogeneous spaces of complexity and rank ≤ 1. An application to the problem of decomposing tensor products of representations is considered in §11. Decomposition formulæ are obtained from the description of the G-module structure of coordinate algebras on double cones of small complexity.

4 Local Structure Theorems

4.1 Locally Linearizable Actions. Those algebraic group actions can be effectively studied which are more or less reduced to linear or projective representations

D.A. Timashev, *Homogeneous Spaces and Equivariant Embeddings*,
Encyclopaedia of Mathematical Sciences 138, DOI 10.1007/978-3-642-18399-7_2,

and their restrictions to stable subvarieties in representation spaces. Therefore it is natural to restrict our attention to the following class of actions.

Definition 4.1. A regular algebraic group action $G : X$ (or a G-variety X) is *locally linearizable* if X can be covered by G-stable quasiprojective open subsets X_i such that $G : X_i$ is the restriction of the projective representation of G in an ambient projective space.

Example 4.2 ([Po3]). Consider a rational projective curve X obtained from $\mathbb{P}^1$ by identifying $0, \infty \in \mathbb{P}^1$ in an ordinary double point (a Cartesian leaf). A $\Bbbk^\times$-action on $\mathbb{P}^1$ with the fixed points $0, \infty$ goes down to X. This action is not locally linearizable. (Otherwise, there is a $\Bbbk^\times$-stable hyperplane section of X in an ambient $\mathbb{P}^n$ that does not contain the double point. But there are no other $\Bbbk^\times$-fixed points on X.)

The reason why the action of Example 4.2 fails to be locally linearizable is non-normality of X.

Example 4.3. If X is a G-stable subvariety of a normal G-variety (e.g., X is itself normal) and G is connected, then $G : X$ is locally linearizable by Sumihiro's Theorem C.7.

The normalization or the equivariant Chow lemma [Sum, Th. 2] reduce the study of arbitrary algebraic group actions to locally linearizable ones. In the sequel, only locally linearizable actions are considered unless otherwise specified.

4.2 Local Structure of an Action. Now let G be a connected reductive group. Fix a Borel subgroup $B \subseteq G$ and a maximal torus $T \subseteq B$.

In order to describe the local structure of locally linearizable G-actions, we begin with a helpful technical construction in characteristic zero due to Brion–Luna–Vust and Grosshans.

Let V be a finite-dimensional G-module with a B^--eigenvector v, and let $\omega \in V^*$ be the dual B-eigenvector such that $\langle v, \omega \rangle \neq 0$. Let $P = G_{[\omega]} = P_{\mathrm{u}} \rtimes L$ be the projective stabilizer of ω, where $L \supseteq T$ is a Levi subgroup and $P_{\mathrm{u}} = \mathrm{R_u}(P)$. Then $P^- = G_{[v]} = P_{\mathrm{u}}^- \rtimes L$ is the opposite parabolic to P, where $P_{\mathrm{u}}^- = \mathrm{R_u}(P^-)$.

Put $\mathring{V} = V \setminus \langle \omega \rangle^\perp$, $F = \langle \mathfrak{p}_{\mathrm{u}}^- \omega \rangle^\perp$, and $\mathring{F} = F \cap \mathring{V}$.

Lemma 4.4 ([BLV], [Gr1]). *In characteristic zero, the action $P : \mathring{V}$ gives rise to an isomorphism*

$$P_{\mathrm{u}} \times \mathring{F} \simeq P *_L \mathring{F} \xrightarrow{\sim} \mathring{V}.$$

Proof. Consider level hyperplanes $V_c = \{x \in V \mid \langle x, \omega \rangle = c\}$, $F_c = F \cap V_c$. We have

$$\mathring{V} = \bigsqcup_{c \neq 0} V_c = \Bbbk^\times v + V_0,$$

and similarly for $\mathring{F}$. Note that $F_0 = \langle \mathfrak{g}\omega \rangle^\perp$ is P-stable. Affine hyperplanes $V_c \subset V$ and $V_c/F_0 \subset V/F_0$ are P-stable. It suffices to show that the induced action $P_{\mathrm{u}} : V_c/F_0$ is transitive and free whenever $c \neq 0$. But $V_0 = \mathfrak{p}_{\mathrm{u}} v \oplus F_0$, whence $cv + F_0$ has the dense P_{u}-orbit in V_c/F_0 with the trivial stabilizer. It remains to note that all orbits of a unipotent group on an affine variety are closed (Lemma 3.4). □

Corollary 4.5. $\mathbb{P}(\mathring{V}) = \mathbb{P}(V)_\omega \simeq P *_L \mathbb{P}(\mathring{F}) \simeq P_u \times (F_0 \otimes \langle v \rangle^*)$.

Theorem 4.6 ([Kn3, 1.2], [BLV]). *Let X be a G-variety, let $\mathscr{L}$ be a G-line bundle on X whose power is generated by global sections, and let $\omega \in \mathrm{H}^0(X, \mathscr{L})^{(B)}$ be an eigensection. Let $P = G_{[\omega]}$ be the projective stabilizer of ω, with a Levi decomposition $P = P_u \rightthreetimes L$, $P_u = \mathrm{R_u}(P)$, $L \supseteq T$. Then the open subset $\mathring{X} = X_\omega$ is P-stable and satisfies the following properties:*

(1) *The action $P_u : \mathring{X}$ is proper (i.e., the map $P_u \times \mathring{X} \to \mathring{X} \times \mathring{X}$, $(g,x) \mapsto (gx,x)$ is proper) and has a geometric quotient.*
(2) *There exists a T-stable closed subvariety $Z \subseteq \mathring{X}$ such that the natural map $P_u \times Z \to \mathring{X}$, $(g,z) \mapsto gz$, and the quotient map $Z \to \mathring{X}/P_u$ are finite and surjective.*
(2)′ *In characteristic zero, Z may be chosen to be L-stable and such that the action $P : \mathring{X}$ gives rise to an isomorphism*

$$P_u \times Z \simeq P *_L Z \xrightarrow{\sim} \mathring{X}.$$

Proof. We will assume that $\operatorname{char} \Bbbk = 0$. Replacing $\mathscr{L}$ by a power, we may choose a finite-dimensional G-submodule $M \subseteq \mathrm{H}^0(X, \mathscr{L})$ containing ω such that $\mathscr{L}$ is generated by sections from M. Put $V = M^*$ and consider the natural G-equivariant morphism $\varphi : X \to \mathbb{P}(V)$ given by the sections in M. The pullback along φ reduces all assertions to the case $X = \mathbb{P}(V)$, where they stem from Corollary 4.5. The same argument applies in positive characteristic, but Corollary 4.5 has to be replaced by a weaker assertion, see [Kn3, 1.2]. □

Theorem 4.7 ([Kn1, 2.3], [Kn3, §2], [BLV]). *Let X be a locally linearizable irreducible G-variety and let $Y \subseteq X$ be a G-stable irreducible subvariety. Then there exists a unique parabolic subgroup $P = P(Y) \supseteq B$ with a Levi decomposition $P = P_u \rightthreetimes L$, $P_u = \mathrm{R_u}(P)$, $L \supseteq T$, and a T-stable locally closed affine subvariety $Z \subseteq X$ such that:*

(1) *$\mathring{X} = PZ$ is an affine open subset of X satisfying the properties (1)–(2)′ from Theorem 4.6.*
(2) *$\mathring{Y} = Y \cap \mathring{X} \neq \emptyset$, and the kernel L_0 of the natural action $L = P/P_u : \mathring{Y}/P_u$ contains L'.*
(3) *The quotient torus $A = L/L_0$ acts on $\mathring{Y}/P_u$ freely, so that $\mathring{Y}/P_u \simeq A \times C$, with A acting on C trivially. In characteristic zero, $Y \cap Z \simeq A \times C$.*

In particular, let $P = P(X)$ be the smallest stabilizer of a B-stable divisor in X. Then there exists a T-stable (L-stable if $\operatorname{char} \Bbbk = 0$) locally closed affine subset $Z \subseteq X$ such that $\mathring{X} = PZ$ is an open affine subset of X, the natural maps $P_u \times Z \to \mathring{X}$, $Z \to \mathring{X}/P_u$ are finite and surjective (are isomorphisms if $\operatorname{char} \Bbbk = 0$), and $\mathring{X}/P_u \simeq A \times C$, where $A = L/L_0$ is a quotient torus of L acting on C trivially.

Remark 4.8. In positive characteristic, L_0 may be a non-reduced group subscheme of L, cf. Remark 1.4.

Proof. Replacing X by an open G-subvariety, we may assume that X is quasiprojective, Y is closed in X, and there is a very ample G-line bundle $\mathscr{L}$ on X. Then X is G-equivariantly embedded in $\mathbb{P}(V)$, where $V^* \subseteq \mathrm{H}^0(X,\mathscr{L})$ is a certain finite-dimensional G-submodule.

Let $\overline{X},\overline{Y}$ be the closures of X,Y in $\mathbb{P}(V)$. We can find an eigensection $\omega \in \mathrm{H}^0(\overline{X},\mathscr{O}(d))^{(B)}$ that vanishes on $\overline{X}\setminus X$ and on any given closed B-subvariety $D \subset Y$, but not on Y. (In characteristic zero, take a nonzero B-eigenform of some degree d in the ideal of $(\overline{X}\setminus X)\cup D$ in the homogeneous coordinate ring of $(\overline{X}\setminus X)\cup\overline{Y}$, and extend it to $\overline{X}$ by complete reducibility of G-modules. In positive characteristic, use Corollary D.2 instead of complete reducibility.) Now Theorem 4.6 yields (1) for $\mathring{X} = \overline{X}_\omega$.

If we choose for D a (maybe reducible) B-stable divisor in Y whose stabilizer P is the smallest possible one, then any $(B\cap L)$-stable divisor of $\mathring{Y}/P_{\mathrm{u}}$ is L-stable. It follows that each $(B\cap L)$-semiinvariant function in $\Bbbk[\mathring{Y}]^{P_{\mathrm{u}}}$ is L-semiinvariant, whence L' acts on $\mathring{Y}/P_{\mathrm{u}}$ trivially. Taking D sufficiently large, we may replace $\mathring{Y}$ by an open subset with the P_{u}-quotient L-isomorphic to $A \times C$.

To complete the proof, note that P is uniquely determined by the conditions of the theorem. Namely, $P = P(Y)$ equals the smallest stabilizer of a B-stable divisor in Y. □

Example 4.9. Let $X = \mathbb{P}(\mathrm{S}^2\Bbbk^{n*})$ be the space of quadrics in $\mathbb{P}^{n-1}$, where $\operatorname{char}\Bbbk \neq 2$. Then $G = \mathrm{GL}_n(\Bbbk)$ acts on X by linear variable changes with the orbits $O_1,\dots,O_n$, where O_r is the set of quadrics of rank r, and $\overline{O_1} \subset \dots \subset \overline{O_n} = X$. Choose the standard Borel subgroup $B \subseteq G$ of upper-triangular matrices and the standard maximal torus $T \subseteq B$ of diagonal matrices.

(1) Put $Y = O_1$, the unique closed G-orbit in X, which consists of double hyperplanes. In the notation of Lemma 4.4 and Theorem 4.6, we have $V = \mathrm{S}^2\Bbbk^{n*} \ni v = x_1^2$, $V^* = \mathrm{S}^2\Bbbk^n \ni \omega = e_1^2$. Then P is a standard parabolic subgroup of matrices of the form

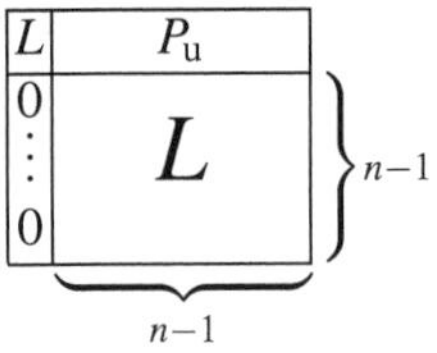

$$\underbrace{\left.\begin{array}{|c|c|}\hline L & P_{\mathrm{u}} \\ \hline \begin{matrix}0\\ \vdots\\ 0\end{matrix} & L \\ \hline \end{array}\right\}}_{n-1} n-1$$

(We indicate the Levi decomposition of P in the figure.) $\mathring{V}$ is the set of quadratic forms $q = cx_1^2 + \cdots$, $c \neq 0$, $\mathfrak{p}_{\mathrm{u}}v = \{a_{12}x_1x_2 + \cdots + a_{1n}x_1x_n \mid a_{ij} \in \Bbbk\}$, and F is the space of forms $q = cx_1^2 + q'(x_2,\dots,x_n)$, where $c \in \Bbbk$ and q' is a quadratic form in $x_2,\dots,x_n$.

Now $\mathring{X}$ is the set of quadrics given by an equation $x_1^2 + \cdots = 0$, Z consists of quadrics with an equation $x_1^2 + q'(x_2,\dots,x_n) = 0$, and $Y \cap Z = \{[x_1^2]\}$. Lemma 4.4 or Theorem 4.6 says that every quadratic form with nonzero coefficient at x_1^2 can be moved by P_{u}, i.e., by a linear change of x_1, to a form containing no products x_1x_j, $j > 1$. This is the first step in the Lagrange method of transforming a quadric to a normal form.

(2) More generally, put $Y = O_r$. It is easy to see that $P(Y)$ is the group of matrices of the form

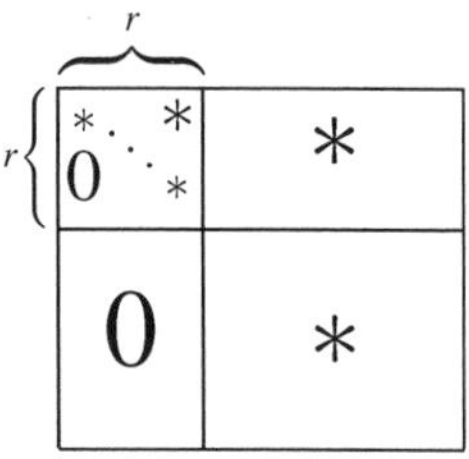

Clearly, $P(Y) = G_{[\omega]}$, where ω is the product of the first r upper-left corner minors of the matrix of a quadratic form. Then $\mathring{X}$ is the set of quadrics, where ω does not vanish, i.e., having non-degenerate intersection with all subspaces $\{x_k = \cdots = x_n = 0\}$, $k \leq r+1$. Further, Z consists of quadrics with an equation $c_1x_1^2 + \cdots + c_rx_r^2 + q'(x_{r+1}, \dots, x_n) = 0$, $c_i \neq 0$, and $Y \cap Z = \{[c_1x_1^2 + \cdots + c_rx_r^2] \mid c_i \neq 0\}$. The Levi subgroup $L = (\Bbbk^\times)^r \times \mathrm{GL}_{n-r}(\Bbbk)$ acts on $Y \cap Z$ via the first factor, and $Y \cap Z = (\Bbbk^\times)^r \times \{[x_1^2 + \cdots + x_r^2]\}$. Theorems 4.6–4.7 say that each quadric with nonzero first r upper-left corner minors transforms by a unitriangular linear variable change to the form $c_1x_1^2 + \cdots + c_rx_r^2 + q'(x_{r+1}, \dots, x_n)$—this is nothing else but the Gram–Schmidt orthogonalization method.

4.3 Local Structure Theorem of Knop. A refined version of the local structure theorem was proved by F. Knop in characteristic zero.

Let X be an irreducible normal G-variety. We call any formal $\Bbbk$-linear combination of prime Cartier divisors on X a *$\Bbbk$-divisor*. Let $\delta = a_1D_1 + \cdots + a_sD_s$ be a B-stable $\Bbbk$-divisor, and let $P = P[D]$ be the stabilizer of its support $D = D_1 \cup \cdots \cup D_s$. Replacing G by a finite cover, we may assume that the line bundles $\mathscr{O}(D_i)$ are G-linearized (Corollary C.5). Let $\eta_i \in \mathrm{H}^0(X, \mathscr{O}(D_i))^{(B)}$ be the sections of B-weights λ_i such that $\operatorname{div} \eta_i = D_i$, and set $\lambda_\delta = \sum a_i\lambda_i$. We say that δ is *regular* if $\langle \lambda_\delta, \alpha^\vee \rangle \neq 0$ for any root α such that $\mathfrak{g}^\alpha \subseteq \mathfrak{p}_u$. (For example, any effective B-stable Cartier divisor is regular.)

Define a morphism $\psi_\delta : X \setminus D \to \mathfrak{g}^*$ by the formula

$$\langle \psi_\delta(x), \xi \rangle = \sum a_i \frac{\xi \eta_i}{\eta_i}(x), \qquad \forall \xi \in \mathfrak{g}.$$

Theorem 4.10 ([Kn5, 2.3]). *The map ψ_δ is a P-equivariant fibration over the P-orbit of λ_δ considered as a linear function on a maximal torus $\mathfrak{t} \subseteq \mathfrak{b}$ and extended to $\mathfrak{g}$ by putting $\langle \lambda_\delta, \mathfrak{g}^\alpha \rangle = 0$, $\forall \alpha$. The stabilizer $L = P_{\lambda_\delta}$ is the Levi subgroup of P containing T. In particular, $X \setminus D \simeq P *_L Z$, where $Z = \psi_\delta^{-1}(\lambda_\delta)$.*

Other versions of the local structure theorem can be found in [BL] and in Subsections 13.3, 15.3, 20.3, and 29.1.

5 Complexity and Rank of G-varieties

5.1 Basic Definitions. As before, G is a connected reductive group with a fixed Borel subgroup B, a maximal unipotent subgroup $U = \mathrm{R_u}(B)$, and a maximal torus $T \subseteq B$. Let X be an irreducible G-variety.

Definition 5.1. The *complexity* $c_G(X)$ of the action $G : X$ is the codimension of a general B-orbit in X. By the lower semicontinuity of the function $x \mapsto \dim Bx$, $c_G(X) = \min_{x\in X} \operatorname{codim} Bx$. By the Rosenlicht theorem [PV, 2.3], $c_G(X) = \operatorname{tr.deg} \Bbbk(X)^B$.

The *weight lattice* $\Lambda(X)$ (resp. the *weight semigroup* $\Lambda_+(X)$) is the set of weights of all rational (regular) B-eigenfunctions on X. It is a sublattice in the character lattice $\mathfrak{X}(B) = \mathfrak{X}(T)$ (a submonoid in the monoid $\mathfrak{X}_+$ of dominant weights, respectively).

The integer $r_G(X) = \operatorname{rk} \Lambda(X)$ is called the *rank* of $G : X$.

We usually drop the subscript G in the notation of complexity and rank.

Remark 5.2. The reductivity of G is not used in the definition. In fact, we use the notions of complexity and rank for a non-reductive group action, namely the B-action on a B-stable subvariety in X, in 6.2. However, the properties of complexity and rank, discussed below, rely on the reductivity assumption essentially.

Complexity, rank, and the weight lattice are birational invariants of an action. Replacing X by a G-birationally equivalent variety, we may always assume that X is locally linearizable, normal, quasiprojective, or smooth, when required. These invariants are very important in studying the geometry of the action $G : X$ and the related representation and compactification theory. Here we examine the most basic properties of complexity and rank.

Example 5.3. Let X be a projective homogeneous G-space. By the Bruhat decomposition, U has a dense orbit in X (a *big cell*). Hence $c(X) = r(X) = 0$.

Example 5.4. Assume that $G = T$ and let T_0 be the kernel of the action $T : X$. Then X contains an open T-stable subset $\mathring{X} = T/T_0 \times Z$ [PV, 2.6, 7.2]. Hence $c(X) = \dim Z = \dim \Bbbk(X)^T$, $\Lambda(X) = \mathfrak{X}(T/T_0)$, $r(X) = \dim T/T_0$.

Example 5.5. Let G act on $X = G$ by left or right translations. Then $c(X) = \dim G/B = \dim U$ is the number of positive roots of G. By (2.4), $\Lambda_+(X) = \mathfrak{X}_+$.

It is easy to prove the following

Proposition 5.6. $c(X) + r(X) = \min_{x\in X} \operatorname{codim} Ux = \operatorname{tr.deg} \Bbbk(X)^U$.

(Just apply Example 5.4 to the T-action on the rational quotient of X by U.)

5.2 Complexity and Rank of Subvarieties. Complexity and weight lattice (semigroup) are monotonous by inclusion. More precisely, we have

Theorem 5.7 ([Kn7, 2.3]). *For any closed irreducible G-subvariety $Y \subseteq X$, $c(Y) \le c(X)$, $r(Y) \le r(X)$, and the equalities hold if and only if $Y = X$. Furthermore, $\Lambda(Y) \subseteq \frac{1}{q}\Lambda(X)$ and, if X is affine, then $\Lambda_+(Y) \subseteq \frac{1}{q}\Lambda_+(X)$, where q is a sufficiently big power of p.*

The proof relies on a helpful lemma of Knop:

Lemma 5.8. *Suppose that $G : X$ is locally linearizable. Let $Y \subseteq X$ be an irreducible G-subvariety. Then*

$$\forall f \in \Bbbk(Y)^{(B)} \ \exists \widetilde{f} \in \mathscr{O}_{X,Y}^{(B)} \ \exists q = p^n : \ f^q = \widetilde{f}|_Y.$$

Proof. We may assume that X is a closed G-subvariety in a projective space and Y is closed in X. Let $\widehat{X}, \widehat{Y}$ be the cones over X, Y. These cones are $\widehat{G} = G \times \Bbbk^\times$-stable ($\Bbbk^\times$ acts by homotheties), and f is pulled back to $\Bbbk(\widehat{Y})^{(\widehat{B})}$, where $\widehat{B} = B \times \Bbbk^\times$. By Lemma D.7, $f = F_1/F_2$, where $F_i \in \Bbbk[\widehat{Y}]^{(\widehat{B})}$ are homogeneous B-semiinvariant polynomials. By Corollary D.2, $F_i^q = \widetilde{F}_i|_{\widehat{Y}}$ for some $\widetilde{F}_i \in \Bbbk[\widehat{X}]^{(\widehat{B})}$, $q = p^n$. Now $\widetilde{f} = \widetilde{F}_1/\widetilde{F}_2$ is pulled down to a rational B-eigenfunction on X such that $\widetilde{f}|_Y = f^q$. □

Proof of the theorem. Applying normalization, we may assume that $G : X$ is locally linearizable. Lemma 5.8 implies that $q\Lambda(Y) \subseteq \Lambda(X)$ and that $\Bbbk(Y)^B$ is a purely inseparable extension of the residue field of $\mathscr{O}_{X,Y}^B$, whence the inequalities and the inclusion of weight lattices. The inclusion of weight semigroups stems from Corollary D.2.

Now suppose that $c(Y) = c(X)$ and $r(Y) = r(X)$. As in Lemma 5.8, we may assume that X, Y are closed G-subvarieties in a projective space and consider the cones $\widehat{X}, \widehat{Y}$ over X, Y. We have $c_{\widehat{G}}(\widehat{X}) = c_G(X)$ and $r_{\widehat{G}}(\widehat{X}) = r_G(X) + 1$, in view of an exact sequence

$$0 \longrightarrow \Lambda(X) \longrightarrow \widehat{\Lambda}(\widehat{X}) \longrightarrow \mathfrak{X}(\Bbbk^\times) = \mathbb{Z} \longrightarrow 0,$$

where $\widehat{\Lambda}$ is the weight lattice relative to $\widehat{B}$. Similar equalities hold for $\widehat{Y}$.

By assumption and Proposition 5.6, $\operatorname{tr.deg} \Bbbk(\widehat{Y})^U = \operatorname{tr.deg} \Bbbk(\widehat{X})^U$. But a rational U-invariant function on an affine variety is the quotient of two U-invariant polynomials (Lemma D.7), whence $\Bbbk[\widehat{Y}]^U$ and $\Bbbk[\widehat{X}]^U$ have the same transcendence degree. By Lemma D.1, $\Bbbk[\widehat{Y}]^U$ is a purely inseparable integral extension of $\Bbbk[\widehat{X}]^U|_{\widehat{Y}}$, whence $\Bbbk[\widehat{X}]^U$ restricts to $\widehat{Y}$ injectively. Therefore the ideal of $\widehat{Y}$ contains no nonzero U-invariants, and hence is zero. It follows that $\widehat{Y} = \widehat{X}$, whence $Y = X$. □

On the other side, there is a general procedure of "enlarging" a variety which preserves complexity, rank, and the weight lattice.

Definition 5.9. Let G, G_0 be connected reductive groups. We say that a G-variety X is obtained from a G_0-variety X_0 by *parabolic induction* if $X = G *_Q X_0$, where $Q \subseteq G$ is a parabolic subgroup acting on X_0 via an epimorphism $Q \twoheadrightarrow G_0$.

Proposition 5.10. $c_G(X) = c_{G_0}(X_0)$, $r_G(X) = r_{G_0}(X_0)$, $\Lambda(X) = \Lambda(X_0)$.

The proof is easy.

The weight lattice is actually an attribute of a general G- (and even B-) orbit.

Proposition 5.11 ([Kn7, 2.6]). $\Lambda(X) = \Lambda(Gx)$ *for all x in an open subset of X.*

Proof. Replacing X by its rational quotient by U, we reduce the problem to the case $G = B = T$ and apply Example 5.4. □

Corollary 5.12. *The function $x \mapsto r(Gx)$ is lower semicontinuous on X.*

Using Lemma 5.8, Arzhantsev proved the following

Proposition 5.13 ([Arzh, Pr. 1]). *The function $x \mapsto c(Gx)$ is lower semicontinuous on X.*

5.3 Weight Semigroup. In the affine case, the weight semigroup is a more subtle invariant of an action than the weight lattice.

Proposition 5.14. *For quasiaffine X, $\Lambda(X) = \mathbb{Z}\Lambda_+(X)$.*

Proof. We have $\Bbbk(X) = \operatorname{Quot}\Bbbk[X]$. By Lemma D.7, for any $\lambda \in \Lambda(X)$ and any $f \in \Bbbk(X)^{(B)}_{\lambda}$, there exist $f_i \in \Bbbk[X]^{(B)}_{\lambda_i}$ $(i = 1,2)$ such that $f = f_1/f_2$, whence $\lambda = \lambda_1 - \lambda_2$, $\lambda_i \in \Lambda_+(X)$. □

Proposition 5.15. *For affine X, the semigroup $\Lambda_+(X)$ is finitely generated.*

Proof. The semigroup $\Lambda_+(X)$ is the semigroup of weights for the T-weight decomposition of $\Bbbk[X]^U$, the latter algebra being finitely generated by Theorem D.5(1). □

5.4 Complexity and Growth of Multiplicities. The complexity controls the growth of multiplicities in the spaces of global sections of G-line bundles on X.

Theorem 5.16. (1) *If X is affine and $\Bbbk[X]^G = \Bbbk$ (e.g., X contains an open G-orbit), then $c(X)$ is the minimal integer c such that $m_{n\lambda}(X) = O(n^c)$ for every dominant weight λ.*
(2) *If X is projective, then $c(X)$ is the minimal integer c such that $m_{n\lambda}(\mathscr{L}^{\otimes n}) = O(n^c)$ for any G-line bundle $\mathscr{L}$ on X and any dominant weight λ.*
(3) *The assertion* (2), *resp.* (1), *extends to an arbitrary, resp. quasiaffine, homogeneous space X.*

Proof. First we prove that $c = c(X)$ yields the above estimate for multiplicities.

(1) By Proposition 2.21, $m_\lambda(X) = \dim \Bbbk[X]^U_\lambda$. Replacing X by $X/\!/U$, we may assume that $G = B = T$. Put

$$A = \bigoplus_{n \geq 0} \Bbbk[X]_{n\lambda}.$$

Then $A \simeq (\Bbbk[X] \otimes \Bbbk[t])^T$, where the indeterminate t has the T-weight $-\lambda$. The function field $K = \operatorname{Quot} A$ is purely transcendental of degree 1 over $K^B \subseteq \Bbbk(X)^B$, whence $\operatorname{tr.deg} K \leq c+1$.

By the above, A is a finitely generated positively graded algebra of Krull dimension $d \leq c+1$. We conclude by a standard result of dimension theory that $\dim A_n = O(n^{d-1})$.

(2) We have $m_{n\lambda}(\mathscr{L}^{\otimes n}) = \dim \mathrm{H}^0(X, \mathscr{L}^{\otimes n})^U_{n\lambda}$. There exists a very ample G-line bundle $\mathscr{M}$ such that the line bundle $\mathscr{L} \otimes \mathscr{M}$ is also very ample, and for nonzero $\sigma \in \mathrm{H}^0(X, \mathscr{M})^U_{\mu}$, we have an inclusion

$$\mathrm{H}^0(X, \mathscr{L}^{\otimes n})^U_{n\lambda} \hookrightarrow \mathrm{H}^0(X, (\mathscr{L} \otimes \mathscr{M})^{\otimes n})^U_{n(\lambda+\mu)}$$

provided by $(\cdot \otimes \sigma^n)$. Thus it suffices to consider very ample $\mathscr{L}$. Consider the respective projective embedding of X, and let $\widehat{X}$ be the affine cone over X. Then $\widehat{G} = G \times \Bbbk^{\times}$ acts on $\widehat{X}$ (here $\Bbbk^{\times}$ acts by homotheties), and $m_{n\lambda}(\mathscr{L}^{\otimes n}) = m_{(n\lambda,n)}(\widehat{X})$ (at least for $n \gg 0$). The assertion (1) applied to $\widehat{X}$ yields the desired estimate.
(3) For homogeneous X we use a different argument. Without loss of generality we may assume that $X = G/H$. (Passing to the covering quotient space preserves the complexity and may only increase multiplicities.) For any line bundle $\mathscr{L} = \mathscr{L}(\chi)$ on G/H we have $m_{\lambda}(\mathscr{L}) = \dim V^*(\lambda)^{(H)}_{-\chi}$ by (2.2). The idea is to embed $V^*(\lambda)^{(H)}_{-\chi}$ into the dual of a certain B-submodule of $V(\lambda)$ and to estimate the dimension of the latter.

Let $o = eH$ be the base point of G/H. We may assume that $\operatorname{codim} Bo = c$. If $c > 0$, then there exists a minimal parabolic $P_1 \supset B$ which does not stabilize $\overline{Bo}$, whence $\operatorname{codim} P_1 o = c - 1$. Proceeding in the same way, we construct a sequence of minimal parabolics $P_1, \dots, P_c \supset B$ such that $\overline{P_c \cdots P_1 o} = G/H$. Hence $P_c \cdots P_1 H$ is dense in G and $\dim P_1 \cdots P_c = \dim B + c$.

Let W be the Weyl group of G and let $s_i \in W$ be the simple reflections corresponding to P_i. It follows that $w = s_1 \cdots s_c$ is a reduced decomposition and $P_1 \cdots P_c = D_w := \overline{BwB}$ [Hum, §29], cf. Example 6.6. Therefore $P_1 \cdots P_c / B = S_w := D_w / B \subseteq G/B$ is a Schubert subvariety of dimension c.

Denote by $v_{\lambda} \in V(\lambda)$ a highest vector. The B-submodule $V_w(\lambda) \subseteq V(\lambda)$ generated by wv_{λ} is called a *Demazure module*. One has $V_w(\lambda) = \langle P_1 \cdots P_c v_{\lambda} \rangle = \mathrm{H}^0(S_w, \mathscr{L}_{G/B}(-\lambda))^*$ [Jan, 14.19].

A crucial observation is that the pairing between $V^*(\lambda)$ and $V(\lambda)$ yields an embedding $V^*(\lambda)^{(H)}_{-\chi} \hookrightarrow V_w(\lambda)^*$. Indeed, if $\omega \in V^*(\lambda)^{(H)}_{-\chi}$ vanishes on $V_w(\lambda)$, then we have: $\langle \omega, P_1 \cdots P_c v_{\lambda} \rangle = 0 \implies \langle G\omega, v_{\lambda} \rangle = \langle P_c \cdots P_1 \omega, v_{\lambda} \rangle = 0 \implies \langle \omega, V(\lambda) \rangle = \langle \omega, G v_{\lambda} \rangle = 0 \implies \omega = 0$.

Thus we have $m_{\lambda}(\mathscr{L}) \leq \dim V_w(\lambda)$. Hence $m_{n\lambda}(\mathscr{L}^{\otimes n}) \leq \dim V_w(n\lambda)$. As S_w is a projective variety of dimension c, the dimension of $V_w(n\lambda) = \mathrm{H}^0(S_w, \mathscr{L}_{G/B}(-\lambda)^{\otimes n})^*$ grows as $O(n^c)$.

Now we prove that the estimate for multiplicities cannot be improved. Let $f_1, \dots, f_c$ be a transcendence base of $\Bbbk(X)^B$. Take a very ample line bundle $\mathscr{L}$ on X. Replacing $\mathscr{L}$ by a power and using Lemma D.7 one finds B-eigensections $\sigma_0, \dots, \sigma_c \in \mathrm{H}^0(X, \mathscr{L})^{(B)}_{\lambda}$ (for some $\lambda \in \mathfrak{X}_+$) such that $f_i = \sigma_i / \sigma_0$, $i = 1, \dots, c$. The σ_i are algebraically independent in the algebra

$$A = \bigoplus_{n \geq 0} \mathrm{H}^0(X, \mathscr{L}^{\otimes n})^U_{n\lambda},$$

whence

$$m_{n\lambda}(\mathscr{L}^{\otimes n}) = \dim A_n \geq \binom{n+c}{c} \sim n^c.$$

If X is quasiaffine, then one may take $\mathscr{L} = \mathscr{O}_X$. □

Example 5.17. In the settings of Example 5.5, $c(G)$ is the number of positive roots and $m_{n\lambda}(G) = \dim V^*(n\lambda)$ is a polynomial in n of degree $\leq c(G)$, with equality for general λ, by the Weyl dimension formula [Bou2, Ch. VIII, §9, n°2].

A stronger result can be proved for homogeneous spaces. Fix an arbitrary norm $|\cdot|$ on $\mathfrak{X}(B) \otimes \mathbb{R}$.

Theorem 5.18 ([Tim5]). *The complexity $c(O)$ of a homogeneous space O is the minimal integer c such that $m_\lambda(\mathscr{L}) = O(|\lambda|^c)$ over all $\lambda \in \mathfrak{X}_+$ and all G-line bundles $\mathscr{L}$ over O. This estimate is uniform over all homogeneous G-spaces O with $c(O) = c$. For quasiaffine O it suffices to consider only $m_\lambda(O)$ (i.e., trivial $\mathscr{L}$).*

The proof is based on the same ideas as in Theorem 5.16(3).

Remark 5.19. For G-varieties of complexity ≤ 1, more precise results on multiplicities are obtained, see Theorem 25.1, Proposition 16.2, and Corollary 16.3.

6 Complexity and Modality

6.1 Modality of an Action. The notion of modality was introduced in the works of Arnold on the theory of singularities. The modality of an action is the maximal number of parameters in a continuous family of orbits. More precisely:

Definition 6.1. Let $H : X$ be an algebraic group action. If X is irreducible, then the integer

$$d_H(X) = \min_{x \in X} \operatorname{codim}_X Hx = \operatorname{tr.deg} \Bbbk(X)^H$$

is called the *generic modality* of the action. The *modality* of $H : X$ is the number $\operatorname{mod}_H X = \max_{Y \subseteq X} d_H(Y)$, where Y runs through H-stable irreducible subvarieties of X.

Note that $c(X) = d_B(X)$.

It may happen that the modality is greater than the generic modality of an action. For example, the natural action $\mathrm{GL}_n(\Bbbk) : \mathrm{L}_n(\Bbbk)$ by left multiplication has an open orbit, whereas its modality equals $[n^2/4]$. Indeed, $\mathrm{L}_n(\Bbbk)$ is covered by finitely many locally closed $\mathrm{GL}_n(\Bbbk)$-stable subsets $Y_{i_1,\dots,i_k}$, where $Y_{i_1,\dots,i_k}$ is the set of matrices of rank k with linearly independent columns $i_1, \dots, i_k$. An orbit in $Y_{i_1,\dots,i_k}$ depends on $k(n-k)$ parameters, which are the coefficients of linear expressions of the remaining $n-k$ columns by the columns $i_1, \dots, i_k$. The maximal number of parameters is obtained for $k = [n/2]$.

Replacing $\mathrm{GL}_n(\Bbbk)$ by the group B of non-degenerate upper-triangular matrices and $\mathrm{L}_n(\Bbbk)$ by the space $\overline{B}$ of all upper-triangular matrices shows that the same thing may happen for a solvable group action. The action $B : \overline{B}$ has an open orbit, but infinitely many orbits in its complement.

6.2 Complexity and B-modality. Remarkably, for an irreducible G-variety X and the restricted action $B : X$, the modality equals the generic modality (=the complexity) of the action. This result was obtained by Vinberg [Vin1] with the aid of Popov's technique of contracting to a horospherical variety (cf. 7.3). We present a proof due to Knop [Kn7], who developed some earlier ideas of Matsuki. A basic tool is an action of a certain monoid on the set of B-stable subvarieties.

Let W be the Weyl group of G. By the Bruhat decomposition, the only irreducible closed $(B \times B)$-stable subvarieties in G are the closures of the Bruhat double cosets $D_w = \overline{BwB}$, $w \in W$. (Here $B \times B$ acts on G by left/right multiplication.)

Definition 6.2 ([Kn7, §2], [RS1]). The *Richardson–Springer monoid* (*RS-monoid*) of G is the set of all D_w, $w \in W$, with the multiplication as of subsets in G. Equivalently, the RS-monoid is the set W with a new multiplication $*$ defined by $D_{v*w} = D_v D_w$. We denote the set W equipped with this product by W^*.

Clearly, W^* is an associative monoid with the unity e. It is easy to describe W^* by generators and relations. Namely, W is defined by generators $s_1, \dots, s_l$ (simple reflections) and relations $s_i^2 = e$ and

$$\underbrace{s_i s_j s_i \cdots}_{n_{ij} \text{ terms}} = \underbrace{s_j s_i s_j \cdots}_{n_{ij} \text{ terms}} \qquad \text{(braid relations)},$$

where (n_{ij}) is the Coxeter matrix of W. The monoid W^* has the same generators and relations $s_i^2 = s_i$ and braid relations. If $w = s_{i_1} \cdots s_{i_n}$ is a reduced decomposition of $w \in W$, then $w = s_{i_1} * \cdots * s_{i_n}$ in W^*. All these assertions follow from standard facts on multiplication of Bruhat double cosets in G [Hum, §29].

Let $\mathfrak{B}(X)$ be the set of all closed irreducible B-stable subvarieties in X. The RS-monoid acts on $\mathfrak{B}(X)$ in a natural way: for any $Z \in \mathfrak{B}(X)$, $w \in W$, the set $w * Z := D_w Z$ is B-stable, irreducible, and closed as the image of $D_w *_B Z$ under the proper morphism $G *_B X \simeq G/B \times X \to X$.

Proposition 6.3. *$c(w * Z) \geq c(Z)$, $r(w * Z) \geq r(Z)$ for any $Z \in \mathfrak{B}(X)$.*

Here we use the notions of complexity and rank for B-varieties, see Remark 5.2.

Proof. It suffices to consider the case of a simple reflection $w = s_i$. In this case, $D_w = P_i$ is the respective minimal parabolic subgroup of G. If Z is P_i-stable, there is nothing to prove. Otherwise, the map $P_i *_B Z \to P_i Z$ is generically finite, and we may replace $s_i * Z$ by $P_i *_B Z$ and, further, by an open subset $Bs_iB *_B Z = B *_{B_i} s_i Z$, where $B_i = B \cap s_i B s_i^{-1}$. Therefore the complexity (rank) of $s_i * Z$ equals the complexity (resp. rank) of $s_i Z$ with respect to the B_i-action or of Z with respect to the action of $s_i^{-1} B s_i \cap B$. The assertion follows. □

Theorem 6.4 ([Kn7, 2.3, 2.4]). *For any B-stable irreducible subvariety $Y \subseteq X$, we have $c(Y) \leq c(X)$, $r(Y) \leq r(X)$. In particular,* $\operatorname{mod}_H(X) = d_H(X)$, *where $H = B$ or U.*

Proof. Follows from Proposition 6.3, Theorem 5.7, and Proposition 5.6. □

G-varieties of complexity zero (i.e., having a dense B-orbit) are called *spherical*. They are discussed in Chap. 5.

Corollary 6.5 ([Vin1], [Bri1]). *Every spherical variety contains finitely many B-orbits.*

6.3 Adherence of B-orbits. In the spherical case, elements of $\mathfrak{B}(X)$ are just B-orbit closures. The set of all B-orbits on a spherical variety, identified with $\mathfrak{B}(X)$, is an interesting combinatorial object. It is finite and partially ordered by the adherence relation $\preceq$ (= inclusion of orbit closures). This partial order is compatible with the action of the RS-monoid and with the dimension function in the following sense:

(1) $O \preceq s_i * O$;
(2) $O_1 \preceq O_2 \implies s_i * O_1 \preceq s_i * O_2$;
(3) $O_1 \prec O_2 \implies \dim O_1 < \dim O_2$;
(4) $O \prec s_i * O \implies \dim(s_i * O) = \dim O + 1$;
(5) (*One step property*) $(s_i * O)_{\preceq} = W_i * O_{\preceq}$, where $W_i = \{e, s_i\}$ is a minimal standard Coxeter subgroup in W, and $O_{\preceq} = \{O' \in \mathfrak{B}(X) \mid O' \preceq O\}$ is the set of orbits in the closure of O.

This compatibility imposes strong restrictions on the adherence of B-orbits on a spherical homogeneous space $X = G/H$. By (5), it suffices to know the closures of the minimal orbits, i.e., of $O \in \mathfrak{B}(G/H)$ such that $O \neq w * O'$, $\forall O' \neq O$, $w \in W^*$. If all minimal orbits have the same dimension then they are closed.

Example 6.6. For $H = B$, the B-orbits are the Schubert cells $B(wo) \subset G/B$, $w \in W$, $o = eB$, and their closures are the Schubert subvarieties $S_w = D_w/B$ in G/B. By standard facts on the multiplication of Bruhat double cosets, $Bo = \{o\}$ is the unique minimal B-orbit. Whence

$$\begin{aligned} S_w = \overline{s_{i_1} * \cdots * s_{i_n} * Bo} \qquad & (w = s_{i_1} \cdots s_{i_n} \text{ is a reduced decomposition}) \\ = W_{i_1} * \cdots * W_{i_n} * Bo = P_{i_1} \cdots P_{i_n} o = (B \sqcup B s_{i_1} B) \cdots (B \sqcup B s_{i_n} B) o = \\ = \bigsqcup B s_{j_1} \cdots s_{j_k} Bo = \bigsqcup_{v = s_{j_1} \cdots s_{j_k}} B(vo) \end{aligned}$$

over all subsequences $(j_1, \dots, j_k)$ of $(i_1, \dots, i_n)$. This is a well-known description of the Bruhat order on W.

Example 6.7. If G/H is a symmetric space, i.e., H is a fixed point set of an involution, up to connected components, then G/H is spherical (Theorem 26.14) and all minimal B-orbits have the same dimension [RS1]. A complete description of B-orbits, of the W^*-action, and of the adherence relation is obtained in [RS1] (cf. Proposition 26.20).

Example 6.8. For $H = TU'$, the space G/H is spherical, but the minimal B-orbits have different dimensions. However, the adherence of B-orbits is completely determined by the W^*-action with the aid of properties (1)–(5). The set $\mathfrak{B}(G/H)$, the W^*-action, and the adherence relation are described in [Tim1].

Conjecture ([Tim1]). *For any spherical homogeneous space O, there is a unique partial order on $\mathfrak{B}(O)$ satisfying* (1)–(5).

6.4 Complexity and G-modality. By Theorem 6.4, the complexity of a G-variety equals the maximal number of parameters determining a continuous family of B-orbits in X. Generally, continuous families of G-orbits depend on a lesser number of parameters. However, a result of Akhiezer shows that the complexity of a G-action is the maximal modality in the class of all actions birationally G-isomorphic to the given one.

Theorem 6.9 ([Akh3]). *There exists a G-variety X' birationally G-isomorphic to X such that* $\operatorname{mod}_G X' = c(X)$.

Proof. Let $f_1, \dots, f_c$ be a transcendence base of $\Bbbk(X)^B/\Bbbk$. We may replace X by a birationally G-isomorphic normal projective variety. Consider an ample G-line bundle $\mathscr{L}$ on X. Replacing $\mathscr{L}$ by a power, we may find a section $\sigma_0 \in \mathrm{H}^0(X, \mathscr{L})^{(B)}$ such that $\operatorname{div}\sigma_0 \geq \operatorname{div}_\infty f_i$ for $\forall i$. Put $\sigma_i = f_i \sigma_0 \in \mathrm{H}^0(X, \mathscr{L})^{(B)}$.

Take a G-module M generated by a highest vector m_0 and such that there is a homomorphism $\psi_i : M \to \mathrm{H}^0(X, \mathscr{L})$, $\psi_i(m_0) = \sigma_i$. (E.g., $M = V(\lambda_0)$, where λ_0 is the common B-weight of all σ_i.) Let $m_0, \dots, m_n$ be a basis in M consisting of T-eigenvectors with the weights $\lambda_0, \dots, \lambda_n$. Let $E = \langle e_0, \dots, e_c \rangle$ be a trivial G-module of dimension $c+1$. A homomorphism $\psi : E \otimes M \to \mathrm{H}^0(X, \mathscr{L})$, $e_i \otimes m \mapsto \psi_i(m)$, gives rise to a rational G-equivariant map $\varphi : X \dashrightarrow \mathbb{P}((E \otimes M)^*)$. In projective coordinates,

$$\psi(x) = [\cdots : \psi_i(m_j)(x) : \cdots].$$

Take a one-parameter subgroup $\gamma \in \mathfrak{X}^*(T)$ such that $\langle \alpha, \gamma \rangle > 0$ for each positive root α. If there exists $\sigma_i(x) \neq 0$, then

$$\begin{aligned} \psi(\gamma(t)x) &= [\cdots : t^{-\langle \lambda_j, \gamma \rangle} \psi_i(m_j)(x) : \cdots] \\ &= [\cdots : t^{\langle \lambda_0 - \lambda_j, \gamma \rangle} \psi_i(m_j)(x) : \cdots] \longrightarrow [\sigma_0(x) : \cdots : \sigma_c(x) : 0 : \cdots : 0] \end{aligned}$$

as $t \to 0$, because $\lambda_0 - \lambda_j$ is a positive linear combination of positive roots for every $j > 0$. Thus

$$\lim_{t \to 0} \gamma(t)\psi(x) = \big([\sigma_0(x) : \cdots : \sigma_c(x)], [m_0^*]\big) \in \mathbb{P}(E^*) \times \mathbb{P}(M^*) \hookrightarrow \mathbb{P}((E \otimes M)^*)$$

(the Segre embedding), where $m_0^*, \dots, m_n^*$ is the dual basis of M^*.

Let $X' \subseteq X \times \mathbb{P}((E \otimes M)^*)$ be the closure of the graph of φ. By the above, $Y = X' \cap (X \times \mathbb{P}(E^*) \times \mathbb{P}(M^*))$ contains points of the form

$$y_0 = \lim_{t \to 0} \gamma(t)(x, \psi(x)) = \big(\lim_{t \to 0} \gamma(t)x, [\sigma_0(x) : \cdots : \sigma_c(x)], [m_0^*]\big).$$

The G-invariant projection $Y \to \mathbb{P}(E^*)$ maps y_0 to $[\sigma_0(x) : \cdots : \sigma_c(x)]$, and hence is dominant, because $f_i = \sigma_i/\sigma_0$ are algebraically independent in $\Bbbk(X)$. Thence the generic modality of any component of Y dominating $\mathbb{P}(E^*)$ is greater than or equal to c. □

Corollary 6.10 ([Akh2]). *A homogeneous space O is spherical if and only if any G-variety X with an open orbit isomorphic to O has finitely many G-orbits.*

7 Horospherical Varieties

7.1 Horospherical Subgroups and Varieties. There is a nice class of G-varieties, which is easily accessible for study from the viewpoint of the local structure, complexity, and rank.

Definition 7.1. A subgroup $S \subseteq G$ is *horospherical* if S contains a maximal unipotent subgroup of G. A G-variety X is called *horospherical* if the stabilizer of any point in X is horospherical. In other words, $X = GX^U$.

Remark 7.2. In the definition, it suffices to require that the stabilizer of a general point is horospherical. Indeed, this implies that GX^U is dense in X. On the other hand, X^U is B-stable, whence GX^U is closed by Proposition 2.7.

Example 7.3. Consider a Lobachevsky space L^n in the hyperbolic realization, i.e., L^n is the upper pole of a hyperboloid $\{x \in \mathbb{R}^{n+1} \mid (x,x) = 1\}$ in an $(n+1)$-dimensional pseudo-Euclidean space $\mathbb{R}^{n+1}$ of signature $(1,n)$ [AVS, 1.5]. Recall that a *horosphere* in L^n is a hypersurface perpendicular to a pencil of parallel lines. These lines intersect the absolute $\partial L^n \subseteq \mathbb{RP}^n$ in a point $[v]$ called the *center* of the horosphere, where $v \in \mathbb{R}^{n+1}$ is an isotropic vector such that the horosphere consists of $x \in L^n$ with $(x,v) = 1$ [AVS, 2.2]. The group $(\operatorname{Isom} L^n)^0 = \mathrm{O}^0_{1,n}$ acts transitively on the set of horospheres in L^n. Fix a horosphere $H^{n-1} \subset L^n$ and let $[e_1] \in \partial L^n$ be its center. The stabilizer P of $[e_1]$ is a parabolic subgroup of $\mathrm{O}^0_{1,n}$. Take a line $\ell \subset L^n$ orthogonal to H^{n-1}. It intersects ∂L^n in two points $[e_1], [e_{n+1}]$, so that $(e_1, e_{n+1}) = 1$. The group P contains a one-parameter subgroup A acting in $\langle e_1, e_{n+1} \rangle$ by hyperbolic rotations and trivially on $\langle e_1, e_{n+1} \rangle^\perp = \langle e_2, \dots, e_n \rangle$ (we choose an orthonormal basis of the complement). Then A acts on ℓ by translations and the complementary subgroup $S = P'$ is the stabilizer of H^{n-1}. In the matrix form (in the basis $e_1, \dots, e_{n+1}$),

$$S = \left\{ \left. \begin{array}{|c|c|c|} \hline 1 & u^\top C & u^\top u \\ \hline 0 & & \\ \vdots & C & u \\ 0 & & \\ \hline 0 & 0 \cdots 0 & 1 \\ \hline \end{array} \;\right|\; C \in \mathrm{SO}_{n-1},\ u \in \mathbb{R}^{n-1} \right\}.$$

Recall that H^{n-1} carries a Euclidean geometry [AVS, Th. 2.3], and $S = (\operatorname{Isom} H^{n-1})^0$, where $\mathrm{R_u}(S)$ acts by translations and a Levi subgroup of S by rotations with an origin fixed. Clearly, $S(\mathbb{C})$ is a horospherical subgroup of $\mathrm{SO}_{1,n}(\mathbb{C})$, which explains the terminology.

For any parabolic $P \subseteq G$, let $P = P_{\mathrm{u}} \rtimes L$ be a Levi decomposition, and L_0 be any intermediate subgroup between L and L'. Then a subgroup $S = P_{\mathrm{u}} \rtimes L_0$ is horospherical. Conversely:

Lemma 7.4 ([Kn1, 2.1]). *Let $S \subseteq G$ be a horospherical subgroup. Then $P = N_G(S)$ is parabolic, and for a Levi decomposition $P = P_{\mathrm{u}} \rtimes L$ we have $S = P_{\mathrm{u}} \rtimes L_0$, where $L' \subseteq L_0 \subseteq L$.*

Proof. Embed S regularly in a parabolic $P \subseteq G$. Since S is horospherical, $\mathrm{R_u}(S) = P_{\mathrm{u}}$, and S/P_{u} contains a maximal unipotent subgroup of $P/P_{\mathrm{u}} \simeq L$, whence $S/P_{\mathrm{u}} \simeq L_0 \supseteq L'$. Now it is clear that $S = P_{\mathrm{u}} \rtimes L_0$ and $P = N_G(S)$, because P normalizes S and $N_G(S)$ normalizes $\mathrm{R_u}(S)$. □

Definition 7.5. A *standard reductive subgroup* is a subgroup $L_0 \subseteq G$ such that $L' \subseteq L_0 \subseteq L$, where $L \supseteq T$ is the standard Levi subgroup in a standard parabolic $P \supseteq B$.

By Lemma 7.4 conjugacy classes of horospherical subgroups are in bijection with standard reductive subgroups in G.

In the sequel, assume that $\operatorname{char} \Bbbk = 0$ for simplicity.

Horospherical varieties can be characterized in terms of the properties of multiplication in the algebra of regular functions. For any G-module M and any $\lambda \in \mathfrak{X}_+$, let $M_{(\lambda)}$ denote the isotypic component of type λ in M.

Proposition 7.6. *A quasiaffine G-variety X is horospherical if and only if $\Bbbk[X]_{(\lambda)} \cdot \Bbbk[X]_{(\mu)} \subseteq \Bbbk[X]_{(\lambda+\mu)}$, $\forall \lambda, \mu \in \Lambda_+(X)$.*

The implication "if" stems from [Po5, Pr. 8(3)] while "only if" is deduced from the inclusion $\Bbbk[X] \subseteq \Bbbk[G/U] \otimes \Bbbk[X^U]$ given by the natural surjective map $G/U \times X^U \to X$ and from Lemma 2.23. See also Theorem 21.10.

The local structure of a horospherical action is simple.

Proposition 7.7 ([Kn1, 2.2]). *An irreducible horospherical G-variety X contains an open G-stable subset $\mathring{X} \simeq G/S \times C$, where $S \subseteq G$ is horospherical and $G : C$ is trivial.*

Proof. By a theorem of Richardson [PV, Th. 7.1], Levi subgroups of stabilizers of general points on X are conjugate, and unipotent radicals of stabilizers form a continuous family of subgroups in G. Now it is clear from Lemma 7.4 that a horospherical subgroup cannot be deformed outside its conjugacy class, whence stabilizers of general points are all conjugate to a certain $S \subseteq G$. Replacing X by an open G-stable subset yields $X \simeq G *_P X^S$, where $P = N_G(S)$ [PV, 2.8]. But P acts on X^S via a torus P/S, and hence X^S is locally P-isomorphic to $P/S \times C$, where P acts on C trivially [PV, 2.6]. □

7.2 Horospherical Type. To any irreducible G-variety X, one can relate a certain horospherical subgroup of G. Recall that, by Theorem 4.7, X contains an open affine subset $\mathring{X} \simeq P_u \times A \times C$, where P_u is the unipotent radical of a parabolic subgroup $P = P(X)$, which is the normalizer of a general B-orbit in X, and $A = L/L_0$ is a quotient torus of a Levi subgroup $L \subseteq P$. Then $S(X) = P_u \rightthreetimes L_0$ is the normalizer of a general U-orbit in X.

Definition 7.8. The *horospherical type* of X is, up to conjugation, the opposite horospherical subgroup $S = S(X)^- = P_u^- \rightthreetimes L_0$, where P_u^- is the unipotent radical of the opposite parabolic subgroup P^- intersecting P in L.

Example 7.9. The horospherical type of a horospherical homogeneous space G/S is S, because G contains an open "big cell" $P_u \times L \times P_u^-$, where $P^- = N_G(S) = P_u^- \rightthreetimes L$. For general horospherical varieties, the horospherical type is (the conjugation class of) the stabilizer of general position (Proposition 7.7).

Complexity, rank and weight lattice can be read off the horospherical type. Namely, it follows from Theorem 4.7 that $c(X) = \dim X - \dim G + \dim S$, $\Lambda(X) = \mathfrak{X}(A)$, $r(X) = \dim A$, where $A = P^-/S$.

7.3 Horospherical Contraction. Every G-action degenerates to a horospherical one of the same type. A construction of such degeneration, called the *horospherical contraction*, was suggested by Popov [Po5]. We review the horospherical contraction in characteristic zero, referring to [Gr2, §15] for arbitrary characteristic.

First consider an affine G-variety X. Choose a one-parameter subgroup $\gamma \in \mathfrak{X}^*(T)$ such that $\langle\gamma,\lambda\rangle \geq 0$ for any $\lambda \in \mathfrak{X}_+$ and $\langle\gamma,\alpha\rangle > 0$ for any $\alpha \in \Delta_+$. Then $\Bbbk[X]^{(n)} = \bigoplus_{\langle\gamma,\lambda\rangle\leq n} \Bbbk[X]_{(\lambda)}$ is a G-stable filtration of $\Bbbk[X]$. The algebra $\operatorname{gr}\Bbbk[X]$ is finitely generated and has no nilpotents. It is easy to see using Proposition 7.6 that $X_0 = \operatorname{Spec}\operatorname{gr}\Bbbk[X]$ is a horospherical variety of the same type as X. Moreover, $\Bbbk[X_0]^U \simeq \Bbbk[X]^U$ and $\Bbbk[X_0] \simeq \Bbbk[X]$ as G-modules. (Note that $S(X)$ may be described as the common stabilizer of all $f \in \Bbbk[X]^{(B)}$.)

Furthermore, X_0 may be described as the zero-fiber of a flat family over $\mathbb{A}^1$ with a general fiber X. Namely, put $R = \bigoplus_{n=0}^{\infty} \Bbbk[X]^{(n)} t^n \subseteq \Bbbk[X][t]$ and $E = \operatorname{Spec} R$. The natural morphism $\delta : E \to \mathbb{A}^1$ is flat and $(G \times \Bbbk^\times)$-equivariant, where G acts on $\mathbb{A}^1$ trivially and $\Bbbk^\times$ acts by homotheties. Now $\delta^{-1}(t) \simeq X$, $\forall t \neq 0$, and $\delta^{-1}(0) \simeq X_0$.

If X is an arbitrary G-variety, then we may replace it by a birationally G-isomorphic projective variety, build an affine cone $\widehat{X}$ over X, and perform the above construction for $\widehat{X}$. Passing again to a projectivization and taking a sufficiently small open G-stable subset, we obtain

Proposition 7.10 ([Kn1, 2.7]). *There exists a smooth $(G \times \Bbbk^\times)$-variety E and a smooth $(G \times \Bbbk^\times)$-morphism $\delta : E \to \mathbb{A}^1$ such that $X_t := \delta^{-1}(t)$ is G-isomorphic to an open smooth G-stable subset of X for any $t \neq 0$, and X_0 is a smooth horospherical variety of the same type as X.*

8 Geometry of Cotangent Bundles

To a smooth G-variety X, we relate a Hamiltonian G-action on the cotangent bundle T^*X equipped with a natural symplectic structure. Remarkably, the invariants of $G : X$ introduced in §5 are closely related to the symplectic geometry of T^*X and to the respective moment map. In particular, one obtains effective formulæ for complexity and rank involving symplectic invariants of $G : T^*X$. This theory was developed by F. Knop in [Kn1]. To the end of this chapter, we assume that $\operatorname{char}\Bbbk = 0$.

8.1 Symplectic Structure. Let X be a smooth irreducible variety. A standard symplectic structure on T^*X [AG, Ch. 2, 2.1] is given by a 2-form $\omega = \mathrm{d}\mathbf{q} \wedge \mathrm{d}\mathbf{p} = \sum \mathrm{d}q_i \wedge \mathrm{d}p_i$, where $\mathbf{q} = (q_1, \dots, q_n)$ is a tuple of local coordinates on X, and $\mathbf{p} = (p_1, \dots, p_n)$ is an impulse, i.e., tuple of dual coordinates in a cotangent space. In a coordinate-free form, $\omega = -\mathrm{d}\ell$, where a 1-form ℓ on T^*X is given by $\langle \ell(\alpha), v\rangle = \langle \alpha, \mathrm{d}\pi(v)\rangle$, $\forall v \in T_\alpha T^*X$, and $\pi = \pi_X : T^*X \to X$ is the canonical projection.

This symplectic structure defines the Poisson bracket of functions on T^*X. Another way to define this Poisson structure is to consider the sheaf $\mathscr{D}_X$ of differential operators on X. There is an increasing filtration $\mathscr{D}_X = \bigcup \mathscr{D}_X^{(m)}$ by the order of a differential operator and the isomorphism $\operatorname{gr}\mathscr{D}_X \simeq \mathrm{S}^\bullet \mathscr{T}_X = \pi_* \mathscr{O}_{T^*X}$ given by the symbol map, where $\mathscr{T}_X$ is the sheaf of vector fields on X. Since $\operatorname{gr}\mathscr{D}_X$ is commutative, the commutator in $\mathscr{D}_X$ induces the Poisson bracket on $\mathscr{O}_{T^*X}$ by the rule

$$\{\partial_1 \bmod \mathscr{D}_X^{(m-1)}, \partial_2 \bmod \mathscr{D}_X^{(n-1)}\} = [\partial_1, \partial_2] \bmod \mathscr{D}_X^{(m+n-2)}, \ \forall \partial_1 \in \mathscr{D}_X^{(m)},\ \partial_2 \in \mathscr{D}_X^{(n)}.$$

8.2 Moment Map. If X is a G-variety, then the symplectic structure on T^*X is G-invariant and, for every $\xi \in \mathfrak{g}$, the velocity field of ξ on T^*X has a Hamiltonian $H_\xi = \xi_*$, the respective velocity field on X considered as a linear function on T^*X [AG, Ch. 3, 3.1], [Vin3, II.2.1]. Furthermore, the action $G : T^*X$ is Hamiltonian, i.e., the map $\xi \mapsto H_\xi$ is a homomorphism of $\mathfrak{g}$ to the Poisson algebra of functions on T^*X [AG, Ch. 3, 3.1], [Vin3, II.2.1]. The dual morphism $\Phi = \Phi_X : T^*X \to \mathfrak{g}^*$,

$$\langle \Phi(\alpha), \xi\rangle = H_\xi(\alpha) = \langle \alpha, \xi x\rangle, \qquad \forall \alpha \in T_x^*X,\ \xi \in \mathfrak{g},$$

is called the *moment map*. By $M_X \subseteq \mathfrak{g}^*$ we denote the closure of its image. Also set $L_X = M_X /\!/ G$.

The moment map is G-equivariant [AG, Ch. 3, 3.1], [Vin3, II.2.3]. Clearly $\langle \mathrm{d}_\alpha \Phi(v), \xi\rangle = \omega(\xi\alpha, v)$, $\forall v \in T_\alpha T^*X$, $\xi \in \mathfrak{g}$. It follows that $\operatorname{Ker}\mathrm{d}_\alpha\Phi = (\mathfrak{g}\alpha)^\angle$, $\operatorname{Im}\mathrm{d}_\alpha\Phi = (\mathfrak{g}_\alpha)^\perp$, where $^\angle$ and $^\perp$ denote the skew-orthocomplement and the annihilator in $\mathfrak{g}^*$, respectively.

Example 8.1. If $X = G/H$, then $T^*X = G *_H \mathfrak{h}^\perp$ and $\Phi(g * \alpha) = g\alpha$, $\forall g \in G$, $\alpha \in \mathfrak{h}^\perp$. Thus $M_{G/H} = \overline{G\mathfrak{h}^\perp}$. In the general case, for all $x \in X$, the moment map restricted to $T^*X|_{Gx}$ factors as $\Phi : T^*X|_{Gx} \twoheadrightarrow T^*Gx \to M_{Gx} \subseteq M_X$. We shall see below that, for general x, $M_{Gx} = M_X$. All maps $T^*Gx \to M_X$ patch together in the localized moment map, see 8.3.

The cohomomorphism Φ^* exists in two versions—a commutative and a non-commutative one. Let $\mathrm{U}\mathfrak{g}$ denote the universal enveloping algebra of $\mathfrak{g}$, and let $\mathscr{D}(X)$ be the algebra of differential operators on X. The action $G : X$ induces a homomorphism $\Phi^* = \Phi_X^* : \mathrm{U}\mathfrak{g} \to \mathscr{D}(X)$ mapping each $\xi \in \mathfrak{g}$ to a 1-order differential operator ξ_* on X. The map Φ^* is a homomorphism of filtered algebras, and the associated graded map

$$\mathrm{gr}\,\Phi^* : \mathrm{gr}\,\mathrm{U}\mathfrak{g} \simeq \mathrm{S}^\bullet\mathfrak{g} = \Bbbk[\mathfrak{g}^*] \longrightarrow \mathrm{gr}\,\mathscr{D}(X) \subseteq \Bbbk[T^*X], \qquad \xi \mapsto H_\xi,$$

is the commutative version of the cohomomorphism. The isomorphism $\mathrm{gr}\,\mathrm{U}\mathfrak{g} \simeq \mathrm{S}^\bullet\mathfrak{g}$ is provided by the Poincaré–Birkhoff–Witt theorem, and the embedding $\mathrm{gr}\,\mathscr{D}(X) \subseteq \Bbbk[T^*X]$ is the symbol map.

There is a natural symplectic structure on each coadjoint orbit $G\beta \subset \mathfrak{g}^*$ given by the *Kirillov form* $\omega(\xi\beta, \eta\beta) = \langle [\xi,\eta], \beta \rangle$, $\forall \xi, \eta \in \mathfrak{g}$ [AG, Ch. 2, 2.4]. The Poisson brackets on G-orbits merge together into a Poisson bracket on $\Bbbk[\mathfrak{g}^*] \simeq \mathrm{S}^\bullet\mathfrak{g}$, which can be defined on generators by $\{\xi,\eta\} = [\xi,\eta]$, $\forall \xi,\eta \in \mathfrak{g}$ [Vin3, II.1.5]. The map $\mathrm{gr}\,\Phi^*$ is a homomorphism of Poisson algebras, i.e., it respects Poisson brackets.

8.3 Localization. On the sheaf level, we have the homomorphisms $\Phi^* : \mathscr{O}_X \otimes \mathfrak{g} \to \mathscr{T}_X$, $\mathscr{O}_X \otimes \mathrm{U}\mathfrak{g} \to \mathscr{D}_X$. Let $\mathscr{G}_X = \Phi^*(\mathscr{O}_X \otimes \mathfrak{g})$ denote the *action sheaf* (generated by velocity fields), let $\mathscr{F}_X$ be the $\mathscr{O}_X$-subalgebra in $\mathrm{S}^\bullet\mathscr{T}_X$ generated by $\mathscr{G}_X$, and let $\mathscr{U}_X = \Phi^*(\mathscr{O}_X \otimes \mathrm{U}\mathfrak{g})$. Clearly,

$$T^{\mathfrak{g}}X := \mathrm{Spec}_{\mathscr{O}_X} \mathscr{F}_X = \overline{\mathrm{Im}(\pi \times \Phi)} \subseteq X \times \mathfrak{g}^*.$$

The moment map factors as

$$\Phi : T^*X \longrightarrow T^{\mathfrak{g}}X \xrightarrow{\overline{\Phi}} \mathfrak{g}^*.$$

The (non-empty) fibers of $\pi \times \Phi$ are affine translates of the conormal spaces to G-orbits. General fibers of $T^{\mathfrak{g}}X \to X$ are the cotangent spaces $\mathfrak{g}_x^\perp = T_x^*Gx$ to general orbits. The morphism $\overline{\Phi}$ is called the *localized moment map* [Kn6, §2].

Example 8.2. If X is a smooth completion of a homogeneous space $O = G/H$, then $T^{\mathfrak{g}}X \supset T^*O$ and $\overline{\Phi}$ is a proper map extending $\Phi : T^*O \to \mathfrak{g}^*$. Thus one compactifies the moment map of a homogeneous cotangent bundle.

Definition 8.3. A smooth G-variety X is called *pseudo-free* if $\mathscr{G}_X$ is locally free or $T^{\mathfrak{g}}X$ is a vector bundle over X. In other words, the rational map $X \dashrightarrow \mathrm{Gr}_k(\mathfrak{g}) \simeq \mathrm{Gr}_{n-k}(\mathfrak{g}^*)$, $x \mapsto [\mathfrak{g}_x] \mapsto [\mathfrak{g}_x^\perp]$ ($n = \dim\mathfrak{g}$, $k = \dim G_x$ for $x \in X$ in general position) extends to X, i.e., generic isotropy subalgebras degenerate at the boundary to specific limits.

Example 8.4. For trivial reasons, X is pseudo-free if the action $G : X$ is generically free. Also, X is pseudo-free if all G-orbits in X have the same dimension: in this case $T^{\mathfrak{g}}X = \bigsqcup_{Gx \subseteq X} T^*Gx$ [Kn6, 2.3].

It is instructive to note that every G-variety X has a pseudo-free resolution of singularities $\check{X} \to X$: just consider the closure of the graph of $X \dashrightarrow \mathrm{Gr}_k(\mathfrak{g})$ and take for $\check{X}$ an equivariant desingularization of this closure.

It is easy to see that $\operatorname{gr}\Phi^*$ behaves well on the sheaf level on pseudo-free varieties.

Proposition 8.5 ([Kn6, 2.6]). *If X is pseudo-free, then the filtrations on $\mathscr{U}_X$ induced from* $\mathrm{U}\mathfrak{g}$ *and $\mathscr{D}_X$ coincide, and* $\operatorname{gr}\Phi^* : \mathscr{O}_X \otimes \mathrm{S}^\bullet\mathfrak{g} \to \mathrm{S}^\bullet \mathscr{T}_X$ *is surjective onto* $\operatorname{gr}\mathscr{U}_X = \mathscr{F}_X \simeq \mathrm{S}^\bullet\mathscr{G}_X$.

8.4 Logarithmic Version. Sometimes it is useful to replace the usual cotangent bundle by its logarithmic version [Dan, §15], [Oda, 3.1]. Let $D \subset X$ be a divisor with normal crossings (which means that the components of D are smooth and intersect transversally). The sheaf $\Omega^1_X(\log D)$ of differential 1-forms with at most logarithmic poles along D is locally generated by $\mathrm{d}f/f$ with f invertible outside D. It contains Ω^1_X and is locally free. The respective vector bundle $T^*X(\log D)$ is said to be the *logarithmic cotangent bundle*. The dual vector bundle $TX(-\log D)$, called the *logarithmic tangent bundle*, corresponds to the subsheaf $\mathscr{T}_X(-\log D) \subset \mathscr{T}_X$ of vector fields preserving the ideal sheaf $\mathscr{I}_D \lhd \mathscr{O}_X$.

If D is G-stable, then the velocity fields of G on X are tangent to D, i.e., $\mathscr{G}_X \subseteq \mathscr{T}_X(-\log D)$. By duality, we obtain the logarithmic moment map $\Phi : T^*X(\log D) \to T^{\mathfrak{g}}X \to \mathfrak{g}^*$ extending the usual one on $T^*(X \setminus D)$.

8.5 Image of the Moment Map. We are going to describe the structure of M_X. We do it first for horospherical varieties. Then we contract any G-variety to a horospherical one and show that this contraction does not change M_X.

Remark 8.6. We assume that X is smooth in order to use the notions of symplectic geometry. However, the moment map may be defined for any G-variety X. As the definition is local and M_X is a G-birational invariant of X, we may always pass to a smooth open subset of X, and conversely to a (maybe singular) G-embedding of a smooth G-variety.

Let X be a horospherical variety of type S. It is clear from Proposition 7.7 that $M_X = M_{G/S}$. The moment map factors as

$$\Phi_{G/S} : G *_S \mathfrak{s}^\perp \xrightarrow{\pi_A} G *_{P^-} \mathfrak{s}^\perp \xrightarrow{\varphi} \mathfrak{g}^*,$$

where $P^- = N_G(S)$, and π_A is the quotient map modulo $A = P^-/S$. By Proposition 2.7, the map φ is proper, whence $M_{G/S} = G\mathfrak{s}^\perp$. Let us identify $\mathfrak{g}^*$ with $\mathfrak{g}$ via a non-degenerate G-invariant inner product on $\mathfrak{g}$ given by $(\xi,\eta) = \operatorname{tr}\xi\eta$, where we assume that $G \subseteq \mathrm{GL}_n(\Bbbk)$, $\mathfrak{g} \subseteq \mathfrak{gl}_n(\Bbbk)$. Then $\mathfrak{s}^\perp$ is identified with $\mathfrak{a} \oplus \mathfrak{p}_{\mathrm{u}}^-$ and is retracted onto $\mathfrak{a}$ by a certain one-parameter subgroup of $Z(L)$. It follows that $\mathfrak{a} \simeq \mathfrak{a}^*$ intersects all closed G-orbits in $M_{G/S}$, and $L_{G/S} = \pi_G(\mathfrak{a}^*)$, where $\pi_G : \mathfrak{g}^* \to \mathfrak{g}^*/\!/G$ is the quotient map.

Finally, general fibers of φ are finite. Indeed, it suffices to find at least one finite fiber. But $\varphi^{-1}(G\mathfrak{p}_{\mathrm{u}}^-) = G *_{P^-} \mathfrak{p}_{\mathrm{u}}^-$ maps onto $G\mathfrak{p}_{\mathrm{u}}^-$ with finite general fibers by a theorem of Richardson [McG, 5.1].

We sum up in

Theorem 8.7. *For any horospherical G-variety X of type S, the natural map $G *_{P^-} \mathfrak{s}^\perp \to M_X = G\mathfrak{s}^\perp$ is generically finite, proper and surjective, and $L_X = \pi_G(\mathfrak{a}^*)$.*

We have already seen that horospherical varieties, their cotangent bundles and moment maps are easily accessible for study. A deep result of Knop says that the closure of the image of the moment map depends only on the horospherical type.

Theorem 8.8 ([Kn1, 5.4]). *Assume that X is a G-variety of horospherical type S. Then $M_X = M_{G/S}$.*

In the physical language, the idea of the proof is to apply quantum technique to classical theory. We study the homomorphism $\Phi_X^* : \mathrm{U}\mathfrak{g} \to \mathscr{D}(X)$ and show that its kernel $\mathscr{J}_X \lhd \mathrm{U}\mathfrak{g}$ depends only on the horospherical type. Then we deduce that $J_X = \operatorname{Ker} \operatorname{gr} \Phi_X^* = \mathscr{I}(M_X) \lhd \Bbbk[\mathfrak{g}^*]$ depends only on the type of X, which is the desired assertion. We retain the notation of Proposition 7.10.

Lemma 8.9. $\mathscr{J}_X = \mathscr{J}_{X_0} = \mathscr{J}_{G/S}$.

In the affine case, lemma stems from a G-module isomorphism $\Bbbk[X] \simeq \Bbbk[X_0]$. Indeed, Φ_X^* depends only on the G-module structure of $\Bbbk[X]$. The general case is deduced from the affine one by the technique of affine cones and some additional arguments [Kn1, 5.1].

Put $\mathscr{M}_X = \operatorname{Im} \Phi_X^* \subseteq \mathscr{D}(X)$. By the previous lemma, $\mathscr{M}_X \simeq \mathscr{M}_{X_0} \simeq \mathscr{M}_{G/S}$.

Lemma 8.10. $\operatorname{gr} \mathscr{M}_{G/S}$ *is a finite* $\Bbbk[\mathfrak{g}^*]$*-module.*

Proof. By Proposition 1.8, $A = P^-/S$ acts on G/S by G-automorphisms. Therefore $\mathscr{M}_{G/S} \subseteq \mathscr{D}(G/S)^A$ and $\operatorname{gr} \mathscr{M}_{G/S} \subseteq \Bbbk[T^*G/S]^A = \Bbbk[G *_{P^-} \mathfrak{s}^\perp]$, the latter being a finite $\Bbbk[\mathfrak{g}^*]$-module by Theorem 8.7. □

The restriction maps $\mathscr{M}_E \xrightarrow{\sim} \mathscr{M}_{X_t}$ agree with Φ^* and do not raise the order of a differential operator. We identify $\mathscr{M}_E$ and $\mathscr{M}_{X_t}$ via this isomorphism and denote by $\operatorname{ord}_E \partial$ ($\operatorname{ord}_{X_t} \partial$) the order of a differential operator ∂ on E (resp. on X_t).

Theorem 8.8 follows from

Lemma 8.11. *On* $\mathscr{M}_X \simeq \mathscr{M}_E \simeq \mathscr{M}_{X_0}$, $\operatorname{ord}_X \partial = \operatorname{ord}_E \partial = \operatorname{ord}_{X_0} \partial$ *for any* ∂.

Proof. The first equality is clear, because an open subset of E is G-isomorphic to $X \times \Bbbk^\times$. It follows from Lemma 8.10 that the orders of a given differential operator on E and on X_0 do not differ very much. Indeed, choose $\partial_1, \dots, \partial_s \in \mathscr{M}_{X_0} = \mathscr{M}_{G/S}$ representing generators of $\operatorname{gr} \mathscr{M}_{X_0} = \operatorname{gr} \mathscr{M}_{G/S}$ over $\Bbbk[\mathfrak{g}^*]$. Put $d_i = \operatorname{ord}_{X_0} \partial_i$ and $d = \max_i \operatorname{ord}_E \partial_i$. If $\operatorname{ord}_{X_0} \partial = n$, then $\partial = \sum u_i \partial_i$ for some $u_i \in \mathrm{U}^{(n-d_i)}\mathfrak{g}$, and hence $\operatorname{ord}_E \partial \le n + d$.

However, if $\operatorname{ord}_{X_0} \partial < \operatorname{ord}_E \partial$, then $\operatorname{ord}_{X_0} \partial^{d+1} < \operatorname{ord}_E \partial^{d+1} - d$, a contradiction. □

Remark 8.12. For an alternative proof of Theorem 8.8 in the quasiaffine case, which does not use “quantization”, see Remark 23.4.

The explicit description of M_X in terms of the horospherical type allows invariant-theoretic properties of the action $G : M_X$ to be examined.

In the above notation, put $M = Z_G(\mathfrak{a}) \supseteq L$. Every G-orbit in M_X is of the form $G\xi$, $\xi \in \mathfrak{s}^{\perp} = \mathfrak{a} \oplus \mathfrak{p}_{\mathrm{u}}^{-}$. Consider the Jordan decomposition $\xi = \xi_{\mathrm{s}} + \xi_{\mathrm{n}}$, where ξ_{s} is semisimple and ξ_{n} is nilpotent. The (unique) closed orbit in $\overline{G\xi}$ is $G\xi_{\mathrm{s}}$. Moving ξ by P_{u}^{-}, we may assume that $\xi_{\mathrm{s}} \in \mathfrak{a}$. If ξ is a general point, then $G_{\xi_{\mathrm{s}}} = M$, $\mathfrak{z}_{\mathfrak{g}}(\xi_{\mathrm{s}}) = \mathfrak{m}$, thence $\xi_{\mathrm{n}} \in \mathfrak{m} \cap \mathfrak{p}_{\mathrm{u}}^{-}$.

The concept of a general point can be specified as follows: consider the principal open stratum $\mathfrak{a}^{\mathrm{pr}} \subseteq \mathfrak{a}$ obtained by removing all proper intersections with kernels of roots and with W-translates of $\mathfrak{a}$. Then $G_{\xi} = M$, $\forall \xi \in \mathfrak{a}^{\mathrm{pr}}$, and G-orbits intersect $\mathfrak{a}^{\mathrm{pr}}$ in orbits of a finite group $W(\mathfrak{a}) = N_G(\mathfrak{a})/M$ acting freely on $\mathfrak{a}^{\mathrm{pr}}$. Furthermore,

$$M_X^{\mathrm{pr}} := \pi_G^{-1}\pi_G(\mathfrak{a}^{\mathrm{pr}}) \simeq G *_{N_G(\mathfrak{a})} (\mathfrak{a}^{\mathrm{pr}} + \mathfrak{M}), \qquad \text{where } \mathfrak{M} = N_G(\mathfrak{a})(\mathfrak{m} \cap \mathfrak{p}_{\mathrm{u}}^{-})$$

is a nilpotent cone in $\mathfrak{m}$.

Definition 8.13. The Hamiltonian action $G : T^*X$ is said to be *symplectically stable* if the action $G : M_X$ is stable, i.e., general G-orbits in M_X are closed. By the above discussion, symplectic stability is equivalent to $M = L$.

This class of actions is wide enough.

Proposition 8.14. *If X is quasiaffine, then T^*X is symplectically stable.*

Proof. The horospherical contraction X_0 and a typical orbit G/S therein are quasi-affine, too. If $M \supset L$, then there is a root α with respect to the maximal torus $T \subseteq L$ such that $\alpha|_{\mathfrak{a}} = 0$ and $\mathfrak{g}^{\alpha} \subseteq \mathfrak{p}_{\mathrm{u}}$. Then $[\mathfrak{g}^{\alpha}, \mathfrak{g}^{-\alpha}] \subseteq \mathfrak{l}_0$. Let $\mathfrak{s}_{\alpha} = \mathfrak{g}^{\alpha} \oplus [\mathfrak{g}^{\alpha}, \mathfrak{g}^{-\alpha}] \oplus \mathfrak{g}^{-\alpha}$ be the corresponding $\mathfrak{sl}_2$-subalgebra of $\mathfrak{g}$. Then $\mathfrak{s}_{\alpha} \cap \mathfrak{s} = [\mathfrak{g}^{\alpha}, \mathfrak{g}^{-\alpha}] \oplus \mathfrak{g}^{-\alpha}$ is a Borel subalgebra in $\mathfrak{s}_{\alpha}$, and an orbit in G/S of the respective subgroup $S_{\alpha} \subseteq G$ is isomorphic to $\mathbb{P}^1$, a contradiction with quasiaffinity. □

Remark 8.15. A symplectically stable action is stable, i.e., general orbits of $G : T^*X$ are closed. Indeed, Φ is smooth along $\Phi^{-1}(\xi)$ for general $\xi \in M_X$, whence $(\mathfrak{g}\alpha)^{\angle} = \operatorname{Ker} d_{\alpha}\Phi = T_{\alpha}\Phi^{-1}(\xi)$ have one and the same dimension $\dim T^*X - \dim M_X$ for all $\alpha \in \Phi^{-1}(\xi)$. It follows that all orbits over $G\xi$ are closed in $\Phi^{-1}(G\xi)$.

8.6 Corank and Defect. The Hamiltonian G-action on T^*X provides two important invariants:

Definition 8.16. The *defect* $\operatorname{def} T^*X$ is the defect of the symplectic form restricted to a general G-orbit.

The *corank* $\operatorname{cork} T^*X$ is the rank of the symplectic form on the skew-orthogonal complement to the tangent space of a general G-orbit.

In other words,

$$\operatorname{def} T^*X = \dim(\mathfrak{g}\alpha)^{\angle} \cap \mathfrak{g}\alpha,$$
$$\operatorname{cork} T^*X = \dim(\mathfrak{g}\alpha)^{\angle}/(\mathfrak{g}\alpha)^{\angle} \cap \mathfrak{g}\alpha$$

for general $\alpha \in T^*X$.

The cohomomorphism $\operatorname{gr}\Phi^*$ maps $\Bbbk[\mathfrak{g}^*]^G$ onto a Poisson-commutative subalgebra $\mathscr{A}_X \subseteq \Bbbk[T^*X]^G$ isomorphic to $\Bbbk[L_X]$. Skew gradients of functions in $\mathscr{A}_X$ commute, are G-stable, and both skew-orthogonal and tangent to G-orbits. Indeed, for any $f \in \mathscr{A}_X$, $\alpha \in T^*X$, $\mathrm{d}f$ is zero on $\mathfrak{g}\alpha$ (since f is G-invariant) and on $(\mathfrak{g}\alpha)^\angle = \operatorname{Ker} \mathrm{d}_\alpha \Phi$ (because f is pulled back under Φ).

Those skew gradients generate a flow of G-automorphisms preserving G-orbits on T^*X, which is called *G-invariant collective motion.* The restriction of this flow to $G\alpha$ is a connected Abelian subgroup (in fact, a torus for general α, see §23) $A_\alpha \subseteq \operatorname{Aut}_G(G\alpha)$ with the Lie algebra $\mathfrak{a}_\alpha \subseteq \mathfrak{n}_\mathfrak{g}(\mathfrak{g}_\alpha)/\mathfrak{g}_\alpha$.

For general α, $\Phi^{-1}\Phi(G\alpha)$ are level varieties for $\mathscr{A}_X$, because G-invariant regular functions separate general G-orbits in $M_X \subseteq \mathfrak{g}^*$. It follows that

$$\operatorname{Ker} \mathrm{d}_\alpha \mathscr{A}_X = T_\alpha \Phi^{-1}\Phi(G\alpha) = (\mathfrak{g}\alpha) + (\mathfrak{g}\alpha)^\angle, \qquad \text{and}$$
$$\mathfrak{a}_\alpha = (\mathfrak{g}\alpha)^\angle \cap (\mathfrak{g}\alpha) = T_\alpha(G\alpha \cap \Phi^{-1}\Phi(\alpha)) \simeq \mathfrak{g}_{\Phi(\alpha)}/\mathfrak{g}_\alpha.$$

In particular, $\mathfrak{g}_{\Phi(\alpha)} \supset \mathfrak{g}_\alpha \supset \mathfrak{g}'_{\Phi(\alpha)}$.

The defect of $G : T^*X$ is the dimension of the invariant collective motion: $\operatorname{def} T^*X = \dim \mathfrak{a}_\alpha$ for general $\alpha \in T^*X$.

8.7 Cotangent Bundle and Geometry of an Action. The next theorem links the geometry of X and the symplectic geometry of T^*X.

Theorem 8.17 ([Kn1, 7.1]). *Put* $n = \dim X$, $c = c(X)$, $r = r(X)$. *Then* $\dim M_X = 2n - 2c - r$, $\operatorname{def} T^*X = d_G(M_X) = r$, $d_G(T^*X) = 2c + r$, $\operatorname{cork} T^*X = 2c$.

Proof. In the notation of Definition 7.8, we have a decomposition $\mathfrak{g} = \mathfrak{p}_\mathrm{u} \oplus \mathfrak{l}_0 \oplus \mathfrak{a} \oplus \mathfrak{p}_\mathrm{u}^-$, where $\mathfrak{s} = \mathfrak{l}_0 \oplus \mathfrak{p}_\mathrm{u}^-$ is (the Lie algebra of) the horospherical type of X, and $\mathfrak{s}^\perp$ is identified with $\mathfrak{a} \oplus \mathfrak{p}_\mathrm{u}^-$ via $\mathfrak{g} \simeq \mathfrak{g}^*$. By Theorem 4.7, $\dim \mathfrak{p}_\mathrm{u} = \dim \mathfrak{p}_\mathrm{u}^- = n - c - r$ whence, by Theorems 8.7–8.8, $\dim M_X = \dim G/P^- + \dim \mathfrak{s}^\perp = \dim \mathfrak{p}_\mathrm{u} + \dim \mathfrak{p}_\mathrm{u}^- + \dim \mathfrak{a} = 2n - 2c - r$, and $d_G(M_X) = \dim L_X = \dim \mathfrak{a} = r$. For general $\alpha \in T^*X$, we have $d_G(T^*X) = \operatorname{codim} G\alpha = \dim (\mathfrak{g}\alpha)^\angle = \dim \Phi^{-1}\Phi(\alpha) = 2n - \dim M_X = 2c + r$, and $\operatorname{def} T^*X = \dim G\alpha \cap \Phi^{-1}\Phi(\alpha) = \dim G_{\Phi(\alpha)}/G_\alpha = \dim G\alpha - \dim G\Phi(\alpha) = \dim M_X - \dim G\Phi(\alpha) = d_G(M_X) = r$. Finally, $\operatorname{cork} T^*X = d_G(T^*X) - \operatorname{def} T^*X = 2c$. □

Another application of the horospherical contraction and of the moment map is the existence of the stabilizer of general position for the G-action in a cotangent bundle.

Theorem 8.18 ([Kn1, §8]). *Stabilizers in G of general points in T^*X are conjugate to a stabilizer of the open orbit of $M \cap S$ in $\mathfrak{m} \cap \mathfrak{p}_\mathrm{u}^-$, in the above notation. In the symplectically stable (e.g., quasiaffine) case, generic stabilizers of $G : T^*X$ are conjugate to L_0.*

Proof. We prove the first assertion for horospherical X. The general case is derived form the horospherical one with the aid of the horospherical contraction using Theorem 8.8 and the invariant collective motion, see [Kn1, 8.1], or Remark 23.8 for the symplectically stable case.

We may assume that $X = G/S$. A generic stabilizer of $G : T^*G/S$ equals a generic stabilizer of $S : \mathfrak{s}^\perp = \mathfrak{a} \oplus \mathfrak{p}_u^-$. Take a general point $\xi \in \mathfrak{s}^\perp$ and let $\xi = \xi_s + \xi_n$ be the Jordan decomposition. Moving ξ by P_u^-, we may assume that ξ_s is a general point in $\mathfrak{a}$. Then $S_{\xi_s} = M \cap S$, $\xi_n \in \mathfrak{m} \cap \mathfrak{p}_u^-$, and $S_\xi = (M \cap S)_{\xi_n}$ is the stabilizer of a general point in $\mathfrak{m} \cap \mathfrak{p}_u^-$. But $M \cap S$ has the same orbits in $\mathfrak{m} \cap \mathfrak{p}_u^-$ as a parabolic subgroup $M \cap P^- \subseteq M$, because these two groups differ by a central torus in M. By [McG, Th. 5.1], $M \cap P^-$ has an open orbit in the Lie algebra of its unipotent radical $\mathfrak{m} \cap \mathfrak{p}_u^-$, which proves the first assertion of the theorem.

If X is symplectically stable, then $M = L$, whence the second assertion. □

The last two theorems reduce the computation of complexity and rank to studying generic orbits and stabilizers of a *reductive* group. Namely, it suffices to know generic G-modalities of T^*X and M_X. We have a formula

$$2c(X) + r(X) = d_G(T^*X) = 2\dim X - \dim G + \dim G_*, \tag{8.1}$$

where G_* is the stabilizer of general position for $G : T^*X$. For quasiaffine X,

$$r(X) = \operatorname{rk} G - \operatorname{rk} G_*. \tag{8.2}$$

Furthermore, $\Lambda(X)$ is the group of characters of T vanishing on $T \cap G_*$, where G_* is a certain standard reductive subgroup in the conjugacy class of generic stabilizers. For homogeneous spaces, everything is reduced even to *representations* of reductive groups, see §9.

Remark 8.19. While for computing complexity and rank of a quasiaffine G-variety it suffices to know the isomorphism class of G_*, in order to compute the weight lattice one has to determine G_* as a standard reductive subgroup of G. Caution is required here, because there may exist different conjugate standard reductive subgroups in G, cf. Example 9.3.

8.8 Doubled Actions. Now we explain another approach to computing complexity and rank based on the theory of doubled actions [Pan1], [Pan7, Ch. 1]. This approach is parallel to Knop's one and coincides with the latter in the case of G-modules.

Let θ be a Weyl involution of G relative to T, i.e., an involution of G acting on T as an inversion. Then $\theta(P) = P^-$ for any parabolic $P \supseteq B$.

Example 8.20. If $G = \mathrm{GL}_n(\Bbbk)$, or $\mathrm{SL}_n(\Bbbk)$, and T is the standard diagonal torus, then we may put $\theta(g) = (g^\top)^{-1}$.

Definition 8.21. The *dual* G-variety X^* is a copy of X equipped with a twisted G-action: $gx^* = (\theta(g)x)^*$, where $x \mapsto x^*$ is a fixed isomorphism $X \xrightarrow{\sim} X^*$.

The diagonal action $G : X \times X^*$ is called the *doubled action* with respect to $G : X$.

Remark 8.22. If V is a G-module, then V^* is the dual G-module with a fixed linear G^θ-isomorphism $V \xrightarrow{\sim} V^*$. Similarly, $\mathbb{P}(V)^* \simeq \mathbb{P}(V^*)$. If $X \subseteq \mathbb{P}(V)$ is a quasiprojective G-variety, then $X^* \subseteq \mathbb{P}(V^*)$.

Remark 8.23. For a G-module V, the doubled G-variety $V \oplus V^*$ is nothing else but the cotangent bundle T^*V.

The following theorems, due to Panyushev, are parallel to Theorems 8.18 and 8.17.

Theorem 8.24. *Stabilizers in G of general points in $X \times X^*$ are conjugate to L_0.*

Theorem 8.25. *Let G_* be the stabilizer of general position for $G : X \times X^*$. Then*

$$2c(X) + r(X) = d_G(X \times X^*) = 2\dim X - \dim G + \dim G_*, \tag{8.3}$$

$$r(X) = \operatorname{rk} G - \operatorname{rk} G_*, \tag{8.4}$$

and $\Lambda(X)$ is the group of characters of T which vanish on $T \cap G_*$, under appropriate choice of G_* as a standard reductive subgroup of G.

Proofs. Consider an open embedding $P_u \times Z \hookrightarrow X$ from Theorem 4.7. Then $P_u^- \times Z^* \hookrightarrow X^*$, where Z^* is the dual L-variety to Z. One deduces [Pan7, 1.2.2] that $\overline{G(Z \times Z^*)} = X \times X^*$ and the stabilizers in G of general points in $Z \times Z^*$ are contained in L and hence equal to L_0. This proves Theorem 8.24. Theorem 8.25 follows from Theorem 8.24 and the equalities $\dim G/L_0 = 2\dim P_u + \dim A = 2\dim X - 2c(X) - r(X)$. □

Example 8.26. If $X = G/P$ is a projective homogeneous space, then $X^* = G/P^-$ and the stabilizer of general position for $G : X \times X^*$ equals $P \cap P^- = L$.

We have a nice invariant-theoretic property of doubled actions on affine varieties.

Theorem 8.27 ([Pan6, 1.6], [Pan7, 1.3.13]). *If X is affine, then general G-orbits on $X \times X^*$ are closed.*

For a G-module V, a stabilizer of general position for the doubled G-module (or the cotangent bundle) $V \oplus V^*$ can be found by an effective recursive algorithm relying on the Brion–Luna–Vust construction (see 4.2).

Retain the notation of Lemma 4.4. The orbit $G(v + \omega)$ is closed in $V \oplus V^*$ (which is easy to prove, e.g., using Luna's criterion [PV, 6.11]) and a subspace $N = F_0 \oplus F_0^* \oplus \langle v + \omega \rangle$ is a slice module at $v + \omega$, i.e., complementary to $\mathfrak{g}(v + \omega) = \mathfrak{p}_u v \oplus \mathfrak{p}_u^- \omega \oplus \langle v - \omega \rangle$ and stable under $G_{v+\omega} \subset L$. Since LN is dense in $F \oplus F^*$, the étale slice theory [Lu2], [PV, §6] implies that $G(F \oplus F^*)$ is dense in $V \oplus V^*$ and a stabilizer of general position for $G : V \oplus V^*$ equals that for $L : F \oplus F^*$.

Replacing $G : V$ by $L : F$, we apply the Brion–Luna–Vust construction again, and so on. We obtain a descending sequence of Levi subgroups $L_i \subseteq G$ and L_i-submodules $F_i \subseteq V$. As the semisimple rank of L_i decreases, the sequence terminates and, on the final s-th step, L_s' acts on F_s trivially. Then G_* is just the kernel of $L_s : F_s$.

Example 8.28. Let $G = \mathrm{Spin}_{10}(\Bbbk)$ and $V = V(\omega_4)$ be one of its half-spinor representations. Here and below ω_i denote the fundamental weights. In the notation of [OV], the positive roots of G are $\varepsilon_i - \varepsilon_j$, $\varepsilon_i + \varepsilon_j$, the simple roots are $\alpha_i = \varepsilon_i - \varepsilon_{i+1}$, $\alpha_5 = \varepsilon_4 + \varepsilon_5$, where $1 \le i < j \le 5$ and $\pm\varepsilon_1, \dots, \pm\varepsilon_5$ are the T-weights of the tautological representation $V(\omega_1)$ of $\mathrm{SO}_{10}(\Bbbk)$. The weights of V are

$(\pm\varepsilon_1 \pm \cdots \pm \varepsilon_5)/2$, where the number of minuses is odd, $\omega_4 = (\varepsilon_1 + \cdots + \varepsilon_4 - \varepsilon_5)/2$, $\omega_4^* = \omega_5 = (\varepsilon_1 + \cdots + \varepsilon_5)/2$, and $\omega_i = \varepsilon_1 + \cdots + \varepsilon_i$ for $i = 1,2,3$.

Take a highest vector $v_1^* \in V^*$ and let $P_1 \subset G$ be the stabilizer of $[v_1^*]$. Its Levi subgroup L_1 has $\alpha_1, \dots, \alpha_4$ as simple roots. The weights of $(\mathfrak{p}_1^-)_{\mathrm{u}} v_1^*$ are of the form $(\pm\varepsilon_1 \pm \cdots \pm \varepsilon_5)/2$ with 2 minuses, and the weights of $F_1 = ((\mathfrak{p}_1^-)_{\mathrm{u}} v_1^*)^\perp$ have 1 or 5 minuses. Clearly, L_1 is of type $\mathbf{A}_4$, and F_1 is the direct sum of a 1-dimensional L_1-submodule of the weight $(-\varepsilon_1 - \cdots - \varepsilon_5)/2$ and of a 5-dimensional L_1-submodule with highest weight $(\varepsilon_1 + \cdots + \varepsilon_4 - \varepsilon_5)/2$.

Take a highest vector v_2^* in the 5-dimensional summand of F_1^* with highest weight $(\varepsilon_1 - \varepsilon_2 - \cdots - \varepsilon_5)/2$, and let $P_2 \subset L_1$ be the stabilizer of $[v_2^*]$. The second Levi subgroup $L_2 \subset P_2$ has $\alpha_2, \alpha_3, \alpha_4$ as simple roots, the weights of $(\mathfrak{p}_2^-)_{\mathrm{u}} v_2^*$ are $(-\varepsilon_1 \pm \varepsilon_2 \pm \cdots \pm \varepsilon_5)/2$ with exactly 1 plus, and $F_2 = ((\mathfrak{p}_2^-)_{\mathrm{u}} v_2^*)^\perp$ has the weights $(-\varepsilon_1 + \varepsilon_2 + \cdots + \varepsilon_5)/2$, $(-\varepsilon_1 - \varepsilon_2 - \cdots - \varepsilon_5)/2$.

It is easy to see that L_2' is exactly the kernel of $L_2 : F_2$. Thus our algorithm terminates, and we obtain $G_* = L_2' \simeq \mathrm{SL}_4$, $\Lambda(V) = \mathfrak{X}(T/T \cap L_2') = \langle \omega_1, \omega_5 \rangle$, $r(V) = 2$, and $c(V) = 0$. Moreover, $\Bbbk[V]_1 = V^* = V(\omega_5)$ and $\Bbbk[V]_2 = \mathrm{S}^2 V(\omega_5) = V(2\omega_5) \oplus V(\omega_1)$, whence $\Lambda_+(V) = \langle \omega_1, \omega_5 \rangle$.

9 Complexity and Rank of Homogeneous Spaces

9.1 General Formulæ. We apply the methods developed in §8 to computing complexity and rank of homogeneous spaces.

The cotangent bundle of G/H is identified with $G *_H \mathfrak{h}^\perp$, where $\mathfrak{h}^\perp \simeq (\mathfrak{g}/\mathfrak{h})^*$ is the annihilator of $\mathfrak{h}$ in $\mathfrak{g}^*$. The representation $H : \mathfrak{h}^\perp$ is the *coisotropy representation* at $eH \in G/H$. If we identify $\mathfrak{g}$ with $\mathfrak{g}^*$ via a non-degenerate G-invariant inner product on $\mathfrak{g}$ as in 8.5, then $\mathfrak{h}^\perp$ is just the orthogonal complement of $\mathfrak{h}$.

If H is reductive, then $\mathfrak{g} = \mathfrak{h} \oplus \mathfrak{h}^\perp$ as H-modules. In particular, the isotropy and coisotropy representations are isomorphic.

The following theorem is a reformulation of Theorems 8.17–8.18 for homogeneous spaces.

Theorem 9.1. *Generic stabilizers of the coisotropy representation are all conjugate to a certain subgroup $H_* \subseteq H$. For the complexity and rank of G/H, we have the equations:*

$$\begin{aligned} 2c(G/H) + r(G/H) &= d_H(\mathfrak{h}^\perp) = \dim \mathfrak{h}^\perp - \dim H + \dim H_* && (9.1)\\ &= \dim G - 2\dim H + \dim H_*, \\ r(G/H) &= \dim G_\alpha - \dim H_\alpha, && (9.2)\\ 2c(G/H) &= \dim G\alpha - 2\dim H\alpha, && (9.3) \end{aligned}$$

where $\alpha \in \mathfrak{h}^\perp$ is a general point considered as an element of $\mathfrak{g}^*$. If H is observable (e.g., reductive), then $H_* = L_0$ is the Levi subgroup of the horospherical type of G/H,

$$r(G/H) = \operatorname{rk} G - \operatorname{rk} H_*, \tag{9.4}$$

and $\Lambda(G/H)$ is the group of characters of T vanishing on $T \cap H_*$.

Proof. Generic modalities and stabilizers of the actions $G : T^*G/H$ and $H : \mathfrak{h}^\perp$ coincide. This implies all assertions except (9.2), (9.3). We have $\Phi(e * \alpha) = \alpha$ and $G_{e*\alpha} = H_\alpha$, whence the r.h.s. of (9.2) is the dimension of the invariant collective motion, which yields (9.2). Subtracting (9.2) from (9.1) yields (9.3). □

Formulæ (9.1), (9.4) are most helpful, especially for reductive H, because stabilizers of general position for reductive group representations are known, e.g., from Elashvili's tables [Ela1], [Ela2].

Example 9.2. Let $G = \mathrm{Sp}_{2n}(\Bbbk)$, $n \geq 2$, and $H = \mathrm{Sp}_{2n-2}(\Bbbk)$ be the stabilizer of a general pair of vectors in a symplectic space, say $e_1, e_{2n} \in \Bbbk^{2n}$, where a (standard) symplectic form on $\Bbbk^{2n}$ is $\omega = \sum_{i=1}^n x_i \wedge x_{2n+1-i}$. Then the adjoint representations are $\mathfrak{g} \simeq \mathrm{S}^2\Bbbk^{2n}$, $\mathfrak{h} \simeq \mathrm{S}^2\Bbbk^{2n-2}$, and $\mathfrak{g} \simeq \mathrm{S}^2\Bbbk^2 \oplus (\Bbbk^2 \otimes \Bbbk^{2n-2}) \oplus \mathrm{S}^2\Bbbk^{2n-2}$ as an H-module, where $\Bbbk^2 = \langle e_1, e_{2n} \rangle$ and $\Bbbk^{2n-2} = \langle e_2, \dots, e_{2n-1} \rangle$. Hence the coisotropy representation of H equals $\Bbbk^{2n-2} \oplus \Bbbk^{2n-2} \oplus \Bbbk^3$, where $\Bbbk^{2n-2}$ is the tautological and $\Bbbk^3$ a trivial representation of H.

Clearly, $H_* = \mathrm{Sp}_{2n-4}(\Bbbk)$ is the stabilizer of $e_2, e_{2n-1} \in \Bbbk^{2n-2}$. It is a standard reductive subgroup of G with respect to the choice of the standard Borel subgroup of upper-triangular matrices $B \subset G$ and the standard diagonal torus $T = \{t = \operatorname{diag}(t_1, \dots, t_n, t_n^{-1}, \dots, t_1^{-1})\} \subset B$. There are no other standard reductive subgroups in G conjugate to H_*. It follows that $r(G/H) = 2$ and $2c(G/H) + r(G/H) = 2(2n-2) + 3 - (n-1)(2n-1) + (n-2)(2n-3) = 4$, whence $c(G/H) = 1$. Furthermore, $\Lambda(G/H) = \mathfrak{X}(T/T \cap H_*) = \langle \omega_1, \omega_2 \rangle$, where $\omega_i(t) = t_1 \cdots t_i$ denote the fundamental weights of G with respect to the standard Borel subgroup .

Example 9.3. Let $G = \mathrm{SL}_n(\Bbbk)$, $H = \mathrm{GL}_{n-1}(\Bbbk)$. We may assume that H normalizes the subspaces $\langle e_1 \rangle \simeq (\Bbbk_{\det})^*$, $\langle e_2, \dots, e_n \rangle \simeq \Bbbk^{n-1}$ of $\Bbbk^n$. It is easy to see that $\mathfrak{h}^\perp \simeq (\Bbbk_{\det} \otimes \Bbbk^{n-1}) \oplus (\Bbbk_{\det})^* \otimes (\Bbbk^{n-1})^*$ as an H-module. Here $H_* = \{\operatorname{diag}(s, C, s) \mid C \in \mathrm{GL}_{n-2}(\Bbbk),\ s^2 = 1/\det C\}$ is the common stabilizer of $x_1 \otimes e_n \in \Bbbk_{\det} \otimes \Bbbk^{n-1}$ and $e_1 \otimes x_n \in (\Bbbk_{\det})^* \otimes (\Bbbk^{n-1})^*$. Therefore $r(G/H) = 1$, and $2c(G/H) + r(G/H) = 2(n-1) - (n-1)^2 + (n-2)^2 = 1$, whence $c(G/H) = 0$.

However, there are three different standard reductive subgroups in G conjugate to H_*, which are obtained by permuting the diagonal blocks. To choose the right one, note that G/H is embedded in $\mathfrak{g}$ as the orbit of $\xi = \operatorname{diag}(n-1, -1, \dots, -1)$, whence $\mathfrak{g}^* \hookrightarrow \Bbbk[G/H]$. It follows that $\Lambda(G/H) \ni \omega_1 + \omega_{n-1}$, the highest root of $\mathfrak{g}$. Here $\omega_i(t) = t_1 \cdots t_i$ denote the fundamental weights of the standard diagonal torus $T = \{t = \operatorname{diag}(t_1, \dots, t_n) \mid t_1 \cdots t_n = 1\} \subset G$ with respect to the standard Borel subgroup of upper-triangular matrices $B \subset G$. Since $\omega_1 + \omega_{n-1}$ must vanish on $T \cap H_*$, the above choice of H_* among conjugate standard reductive subgroups is the only possible one, and $\Lambda(G/H) = \mathfrak{X}(T/T \cap H_*) = \langle \omega_1 + \omega_{n-1} \rangle$.

9.2 Reduction to Representations. Theorem 9.1 reduces the computation of complexity and rank of affine homogeneous spaces to finding stabilizers of general position for reductive group representations. The next theorem does the same thing for arbitrary homogeneous spaces.

Consider a regular embedding $H \subseteq Q$ in a parabolic subgroup $Q \subseteq G$. Let $K \subseteq M$ be Levi subgroups and let $H_{\mathrm{u}} \subseteq Q_{\mathrm{u}}$ be the unipotent radicals of H and Q. Clearly, K acts on $Q_{\mathrm{u}}/H_{\mathrm{u}}$ by conjugations, and this action is isomorphic to a linear action $K : \mathfrak{q}_{\mathrm{u}}/\mathfrak{h}_{\mathrm{u}}$ [Mon].

We may assume that $M \supseteq T$ and $B \subseteq Q^-$. Then $B(M) = B \cap M$ is a Borel subgroup in M. We may assume that $eK \in M/K$ is a general point, i.e., $M_* = K \cap \theta(K)$ is a stabilizer of general position for $M : M/K \times (M/K)^*$, and $B(M_*) = B(M) \cap K$ is a stabilizer of general position for $B(M) : M/K$. By Theorem 8.24, M_* and $B(M_*)$ are normalized by T, and $B(M_*)^0$ is a Borel subgroup in M_*^0. M_* may be non-connected, but it is a direct product of M_*^0 and a finite subgroup of T. The notions of complexity, rank and weight lattice generalize to M_*-actions immediately.

Theorem 9.4 ([Pan4, 1.2], [Pan7, 2.5.20]). *With the above choice of H, Q, and K among conjugates,*

$$c_G(G/H) = c_M(M/K) + c_{M_*}(Q_{\mathrm{u}}/H_{\mathrm{u}}), \tag{9.5}$$
$$r_G(G/H) = r_M(M/K) + r_{M_*}(Q_{\mathrm{u}}/H_{\mathrm{u}}), \tag{9.6}$$

and there is a canonical exact sequence of weight lattices

$$0 \longrightarrow \Lambda_M(M/K) \longrightarrow \Lambda_G(G/H) \longrightarrow \Lambda_{M_*}(Q_{\mathrm{u}}/H_{\mathrm{u}}) \longrightarrow 0. \tag{9.7}$$

Proof. As $B \subseteq Q^-$, the B- and even U-orbit of eQ is open in G/Q. Hence codimensions of general orbits and weight lattices for the actions $B : G/H \simeq G *_Q Q/H$ and $B \cap Q = B(M) : Q/H$ are equal. Further, $Q/H \simeq M *_K Q_{\mathrm{u}}/H_{\mathrm{u}}$. It follows with our choice of K that the codimension of a general $B(M)$-orbit in M/K is the sum of the codimension of a general $B(M)$-orbit in M/K and of a general $B(M_*)$-orbit in $Q_{\mathrm{u}}/H_{\mathrm{u}}$, whence (9.5).

Furthermore, stabilizers of general position of the actions $B : G/H$, $B(M) : Q/H$, $B(M_*) : Q_{\mathrm{u}}/H_{\mathrm{u}}$ are all equal to $B(L_0) = B \cap L_0$, where L_0 is the standard Levi subgroup of the horospherical type of G/H. The equalities $\Lambda(G/H) = \Lambda(B/B(L_0))$, $\Lambda(M/K) = \Lambda(B(M)/B(M_*))$, $\Lambda(Q_{\mathrm{u}}/H_{\mathrm{u}}) = \Lambda(B(M_*)/B(L_0))$ imply (9.7) and (9.6). □

Thus the computation of complexity and rank of G/H is performed in two steps:

(1) Compute the group $M_* \subseteq K$, which is by Theorem 8.18 a stabilizer of general position for the coisotropy representation $K : \mathfrak{k}^{\perp}$ (the orthocomplement in $\mathfrak{m}$). This can be done using, e.g., Elashvili's tables.
(2) Compute the stabilizer of general position for $M_* : \mathfrak{q}_{\mathrm{u}}/\mathfrak{h}_{\mathrm{u}} \oplus (\mathfrak{q}_{\mathrm{u}}/\mathfrak{h}_{\mathrm{u}})^*$ using, e.g., an algorithm at the end of 8.8.

Complexity and rank are read off these stabilizers.

Example 9.5. Let $G = \mathrm{Sp}_{2n}(\Bbbk)$, $n \geq 3$, and H be the stabilizer of a general triple of vectors in a symplectic space, say $e_1, e_{2n-1}, e_{2n} \in \Bbbk^{2n}$, in the notation of Example 9.2. We choose $K = \mathrm{Sp}_{2n-4}(\Bbbk)$, the stabilizer of $e_1, e_2, e_{2n-1}, e_{2n} \in \Bbbk^{2n}$, for a Levi subgroup of H. The unipotent radical of $\mathfrak{h}$ is

$$\mathfrak{h}_{\mathrm{u}} = \left\{ \left(\begin{array}{cc|c|c} 0 & 0 & 0 & 0 \\ 0 & 0 & & \\ \hline 0 & & & \\ \vdots & v & 0 & 0 \\ 0 & & & \\ \hline 0 & w & -v^{\top}\Omega & 0\ 0 \\ 0 & 0 & 0 \cdots 0 & 0\ 0 \end{array} \right) \middle|\ v \in \Bbbk^{2n-4},\ w \in \Bbbk \right\},$$

where Ω is the matrix of the symplectic form on $\Bbbk^{2n-4} = \langle e_3, \ldots, e_{2n-2} \rangle$. For Q we take the stabilizer of a flag $\langle e_{2n} \rangle \subset \langle e_{2n-1}, e_{2n} \rangle \subset \Bbbk^{2n}$. Then

$$M = \left\{ \left(\begin{array}{cc|c|cc} t_1 & 0 & 0 & 0 & \\ 0 & t_2 & & & \\ \hline 0 & & C & 0 & \\ \hline 0 & & 0 & t_2^{-1} & 0 \\ & & & 0 & t_1^{-1} \end{array} \right) \middle|\ t_1, t_2 \in \Bbbk^{\times},\ C \in \mathrm{Sp}_{2n-4}(\Bbbk) \right\}$$

and

$$\mathfrak{q}_{\mathrm{u}} = \left\{ \left(\begin{array}{cc|c|cc} 0 & 0 & 0 & 0 & \\ x & 0 & & & \\ \hline u & v & 0 & 0 & \\ \hline y & w & -v^{\top}\Omega & 0 & 0 \\ z & y & -u^{\top}\Omega & -x & 0 \end{array} \right) \middle|\ x, y, z, w \in \Bbbk,\ u, v \in \Bbbk^{2n-4} \right\}.$$

Clearly, $M/K \simeq (\Bbbk^{\times})^2$, and $M_* = K = \mathrm{Sp}_{2n-4}(\Bbbk)$ acts on $\mathfrak{q}_{\mathrm{u}}$ by left multiplication of $u, v \in \Bbbk^{2n-4}$ by $C \in \mathrm{Sp}_{2n-4}(\Bbbk)$. It follows that $\mathfrak{q}_{\mathrm{u}}/\mathfrak{h}_{\mathrm{u}} = \Bbbk^{2n-4} \oplus \Bbbk^3$, a sum of the tautological and a trivial $\mathrm{Sp}_{2n-4}(\Bbbk)$-module.

We deduce that $c(M/K) = 0$, $r(M/K) = 2$, and $\Lambda(M/K) = \langle \omega_1, \omega_2 \rangle$. A generic stabilizer of $M_* : \mathfrak{q}_{\mathrm{u}}/\mathfrak{h}_{\mathrm{u}} \oplus (\mathfrak{q}_{\mathrm{u}}/\mathfrak{h}_{\mathrm{u}})^*$ equals $\mathrm{Sp}_{2n-6}(\Bbbk)$ (=the stabilizer of e_1, e_2, e_3, e_{2n-2}, e_{2n-1}, e_{2n}), whence $\Lambda(\mathfrak{q}_{\mathrm{u}}/\mathfrak{h}_{\mathrm{u}}) = \langle \overline{\omega}_3 \rangle$, where $\overline{\omega}_3$ is the first fundamental weight of $\mathrm{Sp}_{2n-4}(\Bbbk)$ or, equivalently, the restriction to the diagonal torus in $\mathrm{Sp}_{2n-4}(\Bbbk)$ of the third fundamental weight ω_3 of $\mathrm{Sp}_{2n}(\Bbbk)$. It follows that $r(\mathfrak{q}_{\mathrm{u}}/\mathfrak{h}_{\mathrm{u}}) = 1$, $2c(\mathfrak{q}_{\mathrm{u}}/\mathfrak{h}_{\mathrm{u}}) + r(\mathfrak{q}_{\mathrm{u}}/\mathfrak{h}_{\mathrm{u}}) = 2(2n-4+3) - (n-2)(2n-3) + (n-3)(2n-5) = 7$, hence $c(\mathfrak{q}_{\mathrm{u}}/\mathfrak{h}_{\mathrm{u}}) = 3$. We conclude that $c(G/H) = r(G/H) = 3$, and $\Lambda(G/H) = \langle \omega_1, \omega_2, \omega_3 \rangle$.

10 Spaces of Small Rank and Complexity

The term “complexity” is justified by the fact that homogeneous spaces of small complexity are more accessible for study. In particular, a good compactification theory can be developed for homogeneous spaces of complexity ≤ 1, see Chap. 3. On the other hand, rank and complexity are not completely independent invariants of a homogeneous space. In this section, we discuss the interactions between rank and complexity, paying special attention to homogeneous spaces of small rank and complexity.

10.1 Spaces of Rank ≤ 1. We begin with a simple (and valid in any characteristic) proposition.

Proposition 10.1. *$r(G/H) = 0$ if and only if H is parabolic, i.e., G/H is projective.*

Proof. The “only if” implication follows from the Bruhat decomposition, cf. Example 5.3. Conversely, if $r(G/H) = 0$, then H contains a maximal torus of G. Replacing H by a conjugate, we may assume that $H \supseteq T$ and $B \cap H$ has the minimal possible dimension. We claim that $H \supseteq B^-$. Otherwise, there is a simple root α such that $\mathfrak{h} \not\supseteq \mathfrak{g}^{-\alpha}$. Let $P_\alpha = N_\alpha \rightthreetimes L_\alpha$ be the respective minimal parabolic subgroup with the Levi decomposition such that $L_\alpha \supseteq T$. Then $B = N_\alpha \rightthreetimes B_\alpha$, where $B_\alpha = B \cap L_\alpha$ is a Borel subgroup in L_α, and $(H \cap L_\alpha)^0 = B_\alpha$ or T. In both cases, we may replace B_α by a conjugate Borel subgroup $\widetilde{B}_\alpha$ in L_α so that $\dim H \cap \widetilde{B}_\alpha < \dim H \cap B_\alpha$. Then for $\widetilde{B} = \widetilde{B}_\alpha N_\alpha$ we have $\dim H \cap \widetilde{B} < \dim H \cap B$, a contradiction. □

In particular, homogeneous spaces of rank zero have complexity zero. This can be generalized to the following general inequality between complexity and rank.

Theorem 10.2 ([Pan1, 2.7], [Pan7, 2.2.10]). *$2c(G/H) \leq \operatorname{Cox} G \cdot r(G/H)$, where $\operatorname{Cox} G$ is the maximum of the Coxeter numbers of simple components of G.*

Observe that if G is a simple group and $H = \{e\}$, then the inequality becomes an equality, since $c(G) = \dim U$, $r(G) = \operatorname{rk} G$, and $\operatorname{Cox} G = 2\dim U/\operatorname{rk} G$. This inequality is rather rough, and various examples create an impression that the majority of homogeneous spaces have either small complexity or large rank. In particular, Panyushev proved that $r(G/H) = 1$ implies $c(G/H) \leq 1$.

Proposition 10.3 ([Pan5]). *If $r(G/H) = 1$, then either $c(G/H) = 0$, or G/H is obtained from a homogeneous $\mathrm{SL}_2(\Bbbk)$-space with finite stabilizer by parabolic induction, whence $c(G/H) = 1$.*

Spherical homogeneous spaces of rank 1 where classified by Akhiezer [Akh1] and Brion [Bri5], see Proposition 30.17 and Table 30.1. The above proposition says that, besides the spherical case, there is only one essentially new example of rank 1, namely $\mathrm{SL}_2(\Bbbk)$ acting on itself by left multiplications. (Factorizing by a finite group preserves complexity and rank.) The proof (and the classification) is based on Theorem 9.4. Homogeneous spaces of rank 1 are also characterized in terms of equivariant completions, see Proposition 30.17 and Remark 30.18.

10.2 Spaces of Complexity ≤ 1. Homogeneous spaces of small complexity are much more numerous. Here we discuss classification results in the case where H is reductive, i.e., G/H is affine. See §30 for a conceptual approach to classifying arbitrary spherical homogeneous spaces. For simple G, affine homogeneous spaces of complexity 0 were classified by Krämer [Krä] and of complexity 1 by Panyushev [Pan2], [Pan7, Ch. 3]. A complete classification of spherical affine homogeneous spaces was obtained by Mikityuk [Mik] and Brion [Bri2], with a final stroke put by Yakimova [Yak1]. We exhibit their results in Tables 10.1–10.3, see also Table 26.3. Classification of affine homogeneous spaces of complexity 1 was completed by Arzhantsev and Chuvashova [AC]. Since the computation of complexity and rank of a given homogeneous space represents no difficulties by Theorems 9.1–9.4, the main problem of classification is to "cut off infinity".

Clearly, complexity and rank of G/H do not change if we replace G by a finite cover and/or H by a subgroup of finite index. Thus complexity and rank depend only on the local isomorphism class of G/H, i.e., on the pair $(\mathfrak{g}, \mathfrak{h})$. Therefore we may assume that H is connected.

If G is not semisimple, then it decomposes into an almost direct product $G = G' \cdot Z$, where G' is its (semisimple) commutator subgroup and Z is the connected center of G. It is easy to see that the complexities of G/H, G/HZ, and $G'/(HZ \cap G')$ are equal. Therefore it suffices to solve the classification problem for semisimple G.

An initial arithmetical restriction on a subgroup $H \subseteq G$ such that $c(G/H) \leq c$ is that

$$\dim H \geq \dim U - c. \tag{10.1}$$

A more subtle restriction is based on the notion of d-decomposition [Pan2]. A triple of reductive groups (L, L_1, L_2) is called a *d-decomposition* if $d_{L_1 \times L_2}(L) = d$, where $L_1 \times L_2$ acts on L by left and right multiplications. Clearly, (L, L_1, L_2) remains a d-decomposition if one permutes L_1, L_2 or replaces them by conjugates. Also, $d_{L_1}(L/L_2) = d_{L_2}(L/L_1) = d$. By [Lu1], general orbits of each one of the actions $L_1 \times L_2 : L$, $L_1 : L/L_2$, $L_2 : L/L_1$ are closed. In particular, 0-decompositions are indeed decompositions: $L = L_1 \cdot L_2$. They were classified by Onishchik [Oni2]. Some special kinds of 1-decompositions of classical groups occurring in the classification of homogeneous spaces of complexity 1 (see below) were described by Panyushev [Pan2].

Let $H \subseteq F$ be reductive subgroups of G, and let F_* be the stabilizer of general position for the coisotropy representation $F : \mathfrak{f}^{\perp}$. We may assume that $o = eF$ is a general point for the B-action on G/F, so that $\dim Bo$ is maximal and B_o^0 is a Borel subgroup in F_*^0. We have immediately:

Proposition 10.4.

$$c_G(G/H) = c_G(G/F) + c_{F_*}(F/H) \geq c_G(G/F) + d_{F_*}(F/H).$$

In particular, if $c(G/H) \leq c$, then (F, F_*, H) is a d-decomposition for some $d \leq c$.

The latter assertion is the keystone in a method of classifying affine spherical homogeneous spaces of simple groups suggested by Mikityuk and extended by Panyushev to the case of complexity one. Let us explain its core.

Let G be a simple algebraic group. Maximal connected reductive subgroups $F \subset G$ are known due to Dynkin [GOV, Ch. 6, §3]. We choose among them those with $c(G/F) \leq 1$ and search for reductive $H \subset F$ such that still $c(G/H) \leq 1$.

If G is exceptional, then either $c(G/F) = 0$ or $c(G/F) \geq 2$. For spherical F, sorting out those $H \subset F$ which satisfy (10.1) gives only 4 new subgroups with $c(G/H) \leq 1$ (Nos. 11 of Table 10.1 and 15–17 of Table 10.2).

If G is classical, then inequality (10.1) gives a finite list of subgroups. Again $c(G/F) \neq 1$ with only one exception $G = \mathrm{Sp}_4(\Bbbk)$, $F = \mathrm{SL}_2(\Bbbk)$ embedded in $\mathrm{Sp}_4(\Bbbk)$ by a 4-dimensional irreducible representation (No. 8 of Table 10.2). Here (10.1) becomes an equality, and F cannot be reduced. Sorting out $H \subset F$ with $c(G/H) \leq 1$ is based on (10.1) and on the fact that (F, F_*, H) is a decomposition or 1-decomposition. Here we find 22 new subgroups (Nos. 1–9 of Table 10.1 and 1–7, 9–14 of Table 10.2).

If G/H is a symmetric space, i.e., $H = (G^\theta)^0$, where θ is an involutive automorphism of G, then $c(G/H) = 0$ (Theorem 26.14). Symmetric spaces are considered in §26 and classified in Table 26.3.

Up to a local isomorphism, all non-symmetric affine homogeneous spaces of simple groups with complexity 0 are listed in Table 10.1 and those of complexity 1 in Table 10.2. In the tables, we use the following notation.

We assume that G is a simply connected group covering the one in the column "G" and $H \subset G$ is connected and maps onto the one in the column "H".

Fundamental weights of simple groups are numbered as in [OV]. By ω_i we denote fundamental weights of G, by $\omega_i', \omega_i'', \dots$ those of simple components of H, and by ε_i basic characters of the central torus of H. We drop an index for a group of rank 1.

The column "$H \hookrightarrow G$" describes the embedding in terms of the restriction to H of the minimal representation of G (the tautological representation for classical groups). We use the multiplicative notation for representations: irreducible representations are indicated by their highest weights expressed in basic weights multiplicatively (i.e., products instead of sums, powers instead of multiples, 1 for the trivial one-dimensional representation, etc), and "+" stands for the sum of representations.

The rank of G/H is indicated in the column "$r(G/H)$", and the column "$\Lambda_+(G/H)$" contains a minimal system of generators of the weight semigroup (except for No. 2 of Table 10.1, where all elements of the weight semigroup are given).

Now we describe spherical affine homogeneous spaces of semisimple groups.

We say that G/H is *decomposable* if, up to a local isomorphism, $G = G_1 \times G_2$, $H = H_1 \times H_2$, and $H_i \subseteq G_i$, $i = 1, 2$. Clearly, $G/H = G_1/H_1 \times G_2/H_2$ is spherical if and only if G_i/H_i are spherical. Thus it suffices to classify indecomposable spherical spaces.

Let $H \subseteq G$ be a reductive subgroup. We say that G/H is *strictly indecomposable* if G/H' is still indecomposable. All strictly indecomposable spherical affine homogeneous spaces of semisimple (non-simple) groups are listed in Table 10.3. The

Table 10.1 Spherical affine homogeneous spaces of simple groups

No.	G	H	$H \hookrightarrow G$	$r(G/H)$	$\Lambda_+(G/H)$
1	SL_n	$\mathrm{SL}_m \times \mathrm{SL}_{n-m}$ $(m < n/2)$	$\omega_1' + \omega_1''$	$m+1$	$\omega_1 + \omega_{n-1}, \dots, \omega_{m-1} + \omega_{n-m+1}, \omega_m, \omega_{n-m}$
2	SL_{2n+1}	$\mathrm{Sp}_{2n} \times \Bbbk^\times$	$\omega_1'\varepsilon + \varepsilon^{-2n}$	$2n-1$	$\sum k_i\omega_i$ $\sum\limits_{i\ \mathrm{odd}} (2n+1-i)k_i = \sum\limits_{i\ \mathrm{even}} ik_i$
3	SL_{2n+1}	Sp_{2n}	$\omega_1' + 1$	$2n$	$\omega_1, \dots, \omega_{2n}$
4	Sp_{2n}	$\mathrm{Sp}_{2n-2} \times \Bbbk^\times$	$\omega_1' + \varepsilon + \varepsilon^{-1}$	2	$2\omega_1, \omega_2$
5	SO_{2n+1}	GL_n	$\omega_1'\varepsilon + \omega_{n-1}'\varepsilon^{-1} + 1$	n	$\omega_1, \dots, \omega_{n-1}, 2\omega_n$
6	SO_{4n+2}	SL_{2n+1}	$\omega_1' + \omega_{2n}'$	$n+1$	$\omega_2, \omega_4, \dots, \omega_{2n}, \omega_{2n+1}$
7	SO_{10}	$\mathrm{Spin}_7 \times \mathrm{SO}_2$	$\omega_3' + \varepsilon + \varepsilon^{-1}$	4	$2\omega_1, \omega_2, \omega_4 + \omega_5$ $\omega_1 + 2\omega_4, \omega_1 + 2\omega_5$
8	SO_9	Spin_7	$\omega_3' + 1$	2	ω_1, ω_4
9	SO_8	$\mathbf{G}_2$	$\omega_1' + 1$	3	$\omega_1, \omega_3, \omega_4$
10	SO_7	$\mathbf{G}_2$	ω_1'	1	ω_3
11	$\mathbf{E}_6$	$\mathbf{D}_5$	$\omega_1' + \omega_5' + 1$	3	$\omega_1, \omega_5, \omega_6$
12	$\mathbf{G}_2$	$\mathbf{A}_2$	$\omega_1' + \omega_2' + 1$	1	ω_1

column "$H \hookrightarrow G$" describes the embedding in the following way. White vertices of a diagram denote simple factors of G and black vertices denote factors of H. (Some factors may vanish for small n.) If a factor H_j of H projects non-trivially to a factor G_i of G, then the respective vertices are joined by an edge. The product of those H_j which project to G_i is a spherical subgroup in G_i, and its embedding in G_i is described in Table 10.1. It follows from Tables 10.1, 10.3 that $\dim Z(H) \le 1$ for all strictly indecomposable spherical homogeneous spaces G/H.

Now assume that G/H is indecomposable, but not strictly. Then, up to a local isomorphism, $G = G_1 \times \cdots \times G_s$ and $H' = H_1' \times \cdots \times H_s'$, where H_i are the projections of H to G_i, and G_i/H_i are strictly indecomposable. Furthermore, $H_i = H_i'Z_i$, where Z_i is a one-dimensional central torus, and $H = H'Z$, where $Z \subset Z_1 \times \cdots \times Z_s$ is a subtorus. Since G/H is indecomposable, Z cannot be decomposed as $Z' \times Z''$, where Z', Z'' are the projections of Z to the products of two disjoint sets of factors Z_i.

If G_i/H_i is spherical and $B_i \subset G_i$ is a Borel subgroup such that $\dim B_i \cap H_i$ is minimal, then $\mathfrak{g}_i = \mathfrak{b}_i + \mathfrak{h}_i'$ if G_i/H_i' is spherical, or $\mathfrak{g}_i = (\mathfrak{b}_i + \mathfrak{h}_i') \oplus \mathfrak{z}_i$ otherwise. It follows that G/H is spherical if and only if all G_i/H_i are spherical and Z projects onto the product of those Z_i for which G_i/H_i' is not spherical. This completes the classification.

11 Double Cones

The theory of complexity and rank can be applied to a fundamental problem of representation theory: decompose a tensor product of two simple G-modules into irreducibles. The idea is to realize this tensor product as a G-submodule in the coordinate algebra of a certain affine G-variety—a double cone—and to compute the

Table 10.2 Affine homogeneous spaces of simple groups with complexity 1

No.	G	H	$H \hookrightarrow G$	$r(G/H)$	$\Lambda_+(G/H)$
1	SL_{2n}	$\mathrm{SL}_n \times \mathrm{SL}_n$	$\omega_1' + \omega_1''$	n	$\omega_1 + \omega_{2n-1}, \dots, \omega_{n-1} + \omega_{n+1}, \omega_n$
2	SL_n $(n \geq 3)$	$\mathrm{SL}_{n-2} \times (\Bbbk^\times)^2$	$\omega_1' \varepsilon_1 \varepsilon_2 + \varepsilon_1^{2-n} + \varepsilon_2^{2-n}$	3 $(n > 3)$	$\omega_1 + \omega_{n-1}, \omega_2 + \omega_{n-2}$ $2\omega_1 + \omega_{n-2}, \omega_2 + 2\omega_{n-1}$
				2 $(n = 3)$	$\omega_1 + \omega_2, 3\omega_1, 3\omega_2$
3	SL_n $(n \geq 5)$	$\mathrm{SL}_{n-2} \times \Bbbk^\times$	$\omega_1' \varepsilon + \varepsilon^{d_1} + \varepsilon^{d_2}$ $d_1 \neq d_2$ $d_1 + d_2 = 2 - n$	4	
4	SL_6	$\mathrm{Sp}_4 \times \mathrm{SL}_2 \times \Bbbk^\times$	$\omega_1' \varepsilon + \omega'' \varepsilon^{-2}$	5	
5	Sp_{2n}	Sp_{2n-2}	$\omega_1' + 2$	2	ω_1, ω_2
6	Sp_{2n} $(n \geq 3)$	$\mathrm{Sp}_{2n-4} \times \mathrm{SL}_2 \times \mathrm{SL}_2$	$\omega_1' + \omega'' + \omega'''$	3 $(n > 3)$	$\omega_1 + \omega_3, \omega_2, \omega_4$
				2 $(n = 3)$	$\omega_1 + \omega_3, \omega_2$
7	Sp_{2n}	SL_n	$\omega_1' + \omega_{n-1}'$	n	$2\omega_1, \dots, 2\omega_{n-1}, \omega_n$
8	Sp_4	SL_2	ω'^3	2	$4\omega_1, 3\omega_2$ $4\omega_1 + 2\omega_2, 6\omega_1 + 3\omega_2$
9	SO_n $(n \geq 4)$	SO_{n-2}	$\omega_1' + 2$	2	ω_1, ω_2
10	SO_{2n+1}	SL_n	$\omega_1' + \omega_{n-1}' + 1$	n	$\omega_1, \dots, \omega_n$
11	SO_{4n} $(n \geq 2)$	SL_{2n}	$\omega_1' + \omega_{2n-1}'$	n	$\omega_2, \omega_4, \dots, \omega_{2n}$
12	SO_{11}	$\mathrm{Spin}_7 \times \mathrm{SO}_3$	$\omega_3' + \omega''^2$	5	$2\omega_1, 2\omega_2, \omega_3, \omega_4$ $\omega_1 + 2\omega_5, \omega_2 + 2\omega_5$ $\omega_1 + \omega_2 + 2\omega_5$
13	SO_{10}	Spin_7	$\omega_3' + 2$	4	$\omega_1, \omega_2, \omega_4, \omega_5$
14	SO_9	$\mathbf{G}_2 \times \mathrm{SO}_2$	$\omega_1' + \varepsilon + \varepsilon^{-1}$	4	
15	$\mathbf{E}_6$	$\mathbf{B}_4 \times \Bbbk^\times$	$(\omega_1' + 1)\varepsilon^2 + \omega_4' \varepsilon^{-1} + \varepsilon^{-4}$	5	
16	$\mathbf{E}_7$	$\mathbf{E}_6$	$\omega_1' + \omega_5' + 2$	3	$\omega_1, \omega_2, \omega_6$
17	$\mathbf{F}_4$	$\mathbf{D}_4$	$\omega_1' + \omega_3' + \omega_4' + 2$	2	ω_1, ω_2

algebra of U-invariants on a double cone, which yields the G-module structure of the whole coordinate algebra. In cases of small complexity, the algebra of U-invariants can be effectively computed.

11.1 HV-cones and Double Cones. We may and will assume that G is a simply connected semisimple group. Recall that we work in characteristic zero.

Definition 11.1. Let $\lambda \in \mathfrak{X}_+$ be a dominant weight and let $v_{\lambda^*} \in V(\lambda^*)$ be a highest vector. A cone $C(\lambda) = \overline{Gv_{\lambda^*}} \subseteq V(\lambda^*)$ is called the *cone of highest vectors* (*HV-cone*). Clearly, $C(\lambda) = \overline{Gv_{-\lambda}}$, where $v_{-\lambda} \in V(\lambda^*)$ is a lowest vector.

The projectivization of $C(\lambda)$ is a projective homogeneous space $G/P(\lambda^*) \simeq G/P(\lambda)^-$, where $P(\lambda)$ denotes the stabilizer of $[v_\lambda] \in \mathbb{P}(V(\lambda))$.

The following assertions on HV-cones are well known [VP, §2], cf. §28.

Proposition 11.2. (1) *$C(\lambda) = Gv_{-\lambda} \cup \{0\}$ is a normal conical variety in $V(\lambda^*)$.*

Table 10.3 Spherical affine homogeneous spaces of semisimple groups

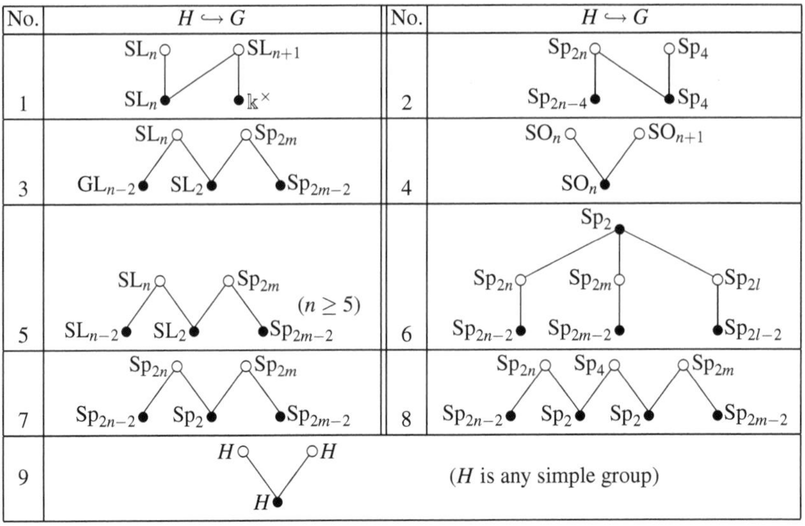

No.	$H \hookrightarrow G$	No.	$H \hookrightarrow G$
1	SL_n, SL_{n+1}; SL_n, $\Bbbk^\times$	2	Sp_{2n}, Sp_4; Sp_{2n-4}, Sp_4
3	SL_n, Sp_{2m}; GL_{n-2}, SL_2, Sp_{2m-2}	4	SO_n, SO_{n+1}; SO_n
5	SL_n, Sp_{2m}; SL_{n-2}, SL_2, Sp_{2m-2} $(n \geq 5)$	6	Sp_{2n}, Sp_{2m}, Sp_{2l}; Sp_2, Sp_{2n-2}, Sp_{2m-2}, Sp_{2l-2}
7	Sp_{2n}, Sp_{2m}; Sp_{2n-2}, Sp_2, Sp_{2m-2}	8	Sp_{2n}, Sp_4, Sp_{2m}; Sp_{2n-2}, Sp_2, Sp_2, Sp_{2m-2}
9	H, H; H		(H is any simple group)

(2) $\Bbbk[C(\lambda)]_n \simeq V(n\lambda)$ *as a G-module.*
(3) $C(\lambda)$ *is factorial if and only if* λ *is a fundamental weight.*

Proof. As $G[v_{-\lambda}]$ is closed, $Gv_{-\lambda}$ is the punctured cone over $G[v_{-\lambda}]$, whence the equality in (1). Lemma 2.23 implies (2), since $\Bbbk[C(\lambda)] \subseteq \Bbbk[G/U]$ is generated by $V^*(\lambda^*) \simeq V(\lambda)$. It follows that $\Bbbk[C(\lambda)] = \bigoplus_{n\geq 0} \mathrm{H}^0(G/P(\lambda)^-, \mathscr{L}(\lambda)^{\otimes n})$ is integrally closed. We have

$$\mathrm{Cl}\, C(\lambda) = \mathrm{Pic}(G/G_{v_{-\lambda}}) = \mathfrak{X}(G_{v_{-\lambda}}) = \mathfrak{X}(P(\lambda)^-)/\langle\lambda\rangle,$$

whence (3). □

Remark 11.3. Proposition 11.2 extends to arbitrary characteristic if one replaces $V(n\lambda)$ by $V^*(n\lambda^*)$ in (2).

Definition 11.4. A variety $Z(\lambda,\mu) = C(\lambda) \times C(\mu)$ is said to be a *double cone.*

The group $\widehat{G} = G \times (\Bbbk^\times)^2$ acts on $Z(\lambda,\mu)$ in a natural way, where the factors $\Bbbk^\times$ act by homotheties. Thus $\Bbbk[Z(\lambda,\mu)]$ is bigraded and

$$\Bbbk[Z(\lambda,\mu)]_{n,m} = V(n\lambda) \otimes V(m\mu).$$

The algebra $\Bbbk[Z(\lambda,\mu)]^U$ is finitely generated (Theorem D.5(1)) and $(\mathfrak{X}_+ \times \mathbb{Z}_+^2)$-graded, and it is clear from the above that the knowledge of its (polyhomogeneous) generators and syzygies provides immediately a series of decomposition rules for $V(n\lambda) \otimes V(m\mu)$. Namely, the highest vectors of irreducible summands in $V(n\lambda) \otimes V(m\mu)$ are (linearly independent) products of generators of total bidegree (n,m).

11.2 Complexity and Rank. The smaller is the $\widehat{G}$-complexity of $Z(\lambda,\mu)$, the simpler is the structure of $\Bbbk[Z(\lambda,\mu)]^U$. Say, if $Z(\lambda,\mu)$ is $\widehat{G}$-spherical, then $Z(\lambda,\mu)/\!/U$ is a toric $\widehat{T}$-variety, where $\widehat{T}=T\times(\Bbbk^\times)^2$. Hence $\Bbbk[Z(\lambda,\mu)]^U$ is the semigroup algebra of the weight semigroup of $Z(\lambda,\mu)$ (cf. Example 15.8). If in addition $Z(\lambda,\mu)$ is factorial, then $Z(\lambda,\mu)/\!/U$ is a factorial toric variety, and hence $\Bbbk[Z(\lambda,\mu)]^U$ is freely generated by $\widehat{T}$-eigenfunctions of linearly independent weights. This yields a very simple decomposition rule, see 11.4.

Therefore it is important to have a transparent method for computing complexity and rank of double cones. By the theory of doubled actions (8.8), the problem reduces to computing the stabilizer of general position for the doubling $Z\times Z^*$ of a double cone $Z=Z(\lambda,\mu)$. This was done by Panyushev in [Pan3], see also [Pan7, Ch. 6]. Here are his results.

Let $L(\lambda)$ be the Levi subgroup of $P(\lambda)$ containing T. The character λ extends to $L(\lambda)$. Put $G(\lambda)=\operatorname{Ker}\lambda\subset L(\lambda)$. Denote by $G(\lambda,\mu)$ the stabilizer of general position for $G(\lambda):G/G(\mu)$ and by $L(\lambda,\mu)$ the stabilizer of general position for $L(\lambda):G/L(\mu)$. Recall from [Lu1] that general orbits of both these actions are closed, and hence the codimension of a general orbit equals the dimension of a categorical quotient:

$$\dim G(\lambda)\backslash\!\backslash G/\!/G(\mu)=\dim G+\dim G(\lambda,\mu)-\dim G(\lambda)-\dim G(\mu),\qquad(11.1)$$

$$\dim L(\lambda)\backslash\!\backslash G/\!/L(\mu)=\dim G+\dim L(\lambda,\mu)-\dim L(\lambda)-\dim L(\mu),\qquad(11.2)$$

where $L_1\backslash\!\backslash L/\!/L_2$ denotes the categorical quotient of the action $L_1\times L_2:L$ by left and right multiplication. Also put

$$\mathbb{P}(Z)=\mathbb{P}(C(\lambda))\times\mathbb{P}(C(\mu))\simeq G/P(\lambda)^-\times G/P(\mu)^-.$$

Theorem 11.5. (1) *The stabilizers of general position for the doubled actions* $G:Z\times Z^*$, $G:\mathbb{P}(Z)\times\mathbb{P}(Z^*)$, $\widehat{G}:Z\times Z^*$ *are equal to* $G(\lambda,\mu)$, $L(\lambda,\mu)$ *and* $\widehat{G}(\lambda,\mu)=\{(g,\lambda(g),\mu(g))\mid g\in L(\lambda,\mu)\}$, *respectively.*
(2) *Put* $V(\lambda,\mu)=\mathfrak{p}(\lambda)_{\mathrm{u}}\cap\mathfrak{p}(\mu)_{\mathrm{u}}$. *Then* $G(\lambda,\mu)$ *and* $L(\lambda,\mu)$ *are equal to the stabilizers of general position for the doubled actions* $G(\lambda)\cap G(\mu):V(\lambda,\mu)\oplus V(\lambda,\mu)^*$, $L(\lambda)\cap L(\mu):V(\lambda,\mu)\oplus V(\lambda,\mu)^*$, *respectively.*

The proof of (1) uses the following

Lemma 11.6. *The stabilizers of general position for the doubled actions* $G:C(\lambda)\times C(\lambda^*)$, $G:\mathbb{P}(C(\lambda))\times\mathbb{P}(C(\lambda^*))$, $G\times\Bbbk^\times:C(\lambda)\times C(\lambda^*)$ *(where* $\Bbbk^\times$ *acts on* $C(\lambda)$ *by homotheties) are equal to* $G(\lambda)$, $L(\lambda)$ *and* $\widehat{G}(\lambda)=\{(g,\lambda(g))\mid g\in L(\lambda)\}$, *respectively.*

Proof. Observe that both $z=(v_{-\lambda},v_\lambda)\in C(\lambda)\times C(\lambda^*)$ and $[z]\in\mathbb{P}(V(\lambda^*)\oplus V(\lambda))$ have stabilizer $G(\lambda)$ in G, whence $\operatorname{codim}Gz=1$, and $\overline{G\langle z\rangle}=C(\lambda)\times C(\lambda^*)$. It follows that $G(\lambda)$ is the stabilizer of general position in G. Other assertions are proved similarly (cf. Example 8.26). □

Proof of Theorem 11.5. (1) We have $Z\times Z^*=(C(\lambda)\times C(\lambda^*))\times(C(\mu)\times C(\mu^*))$, and similarly for $\mathbb{P}(Z)\times\mathbb{P}(Z^*)$. The stabilizer of general position for any diagonal

action $L : X_1 \times X_2$ can be computed in two steps: first find the stabilizers of general position L_i for the actions $L : X_i$, $i = 1,2$, and then find the stabilizer of general position for $L_1 : L/L_2$. It remains to apply Lemma 11.6 for $L = G$ or $\widehat{G}$ and $X_1 = C(\lambda) \times C(\lambda^*)$ or $\mathbb{P}(C(\lambda)) \times \mathbb{P}(C(\lambda^*))$, $X_2 = C(\mu) \times C(\mu^*)$ or $\mathbb{P}(C(\mu)) \times \mathbb{P}(C(\mu^*))$.
(2) One can prove (2) using Luna's slice theorem [Lu2], [PV, Th. 6.1], if one observes that the $L(\lambda)$-orbit of $eL(\mu) \in G/L(\mu)$ and the $G(\lambda)$-orbit of $eG(\mu) \in G/G(\mu)$ are closed, and computes the slice module. However, the proof also stems from the theory of doubled actions. It suffices to prove that the actions $B : Z$ (or $B : \mathbb{P}(Z)$) and $B \cap G(\lambda) \cap G(\mu) : V(\lambda,\mu)$ (resp. $B \cap L(\lambda) \cap L(\mu) : V(\lambda,\mu)$) have the same stabilizers of general position. (These are Borel subgroups in the generic stabilizers of double actions.) For computing these stabilizers, we apply the algorithm at the end of 8.8.

We have $Z \subseteq V = V(\lambda^*) \oplus V(\mu^*)$. Take a B-eigenvector $\omega = (v_\lambda, 0) \in V^*$ and put $\mathring{Z} = Z_\omega$. By Lemma 4.4, $\mathring{Z} \simeq P(\lambda)_{\mathrm{u}} \times Z'$, where $Z' \simeq \Bbbk^\times v_{-\lambda} \times C(\mu)$ as an $L(\lambda)$-variety. Now take $\omega' = (0, v_\mu)$ and put $\mathring{Z}' = Z'_{\omega'}$. Then $\mathring{Z}' \simeq [L(\lambda) \cap P(\mu)_{\mathrm{u}}] \times Z''$, where $Z'' = \Bbbk^\times v_{-\lambda} \times \Bbbk^\times v_{-\mu} \times V(\lambda,\mu)$ as an $L(\lambda) \cap L(\mu)$-variety. This proves our claim on stabilizers of general position. □

We shall denote by c, r, Λ (resp. $\widehat{c}, \widehat{r}, \widehat{\Lambda}$) the complexity, rank and the weight lattice of a G- (resp. $\widehat{G}$-) action. Since maximal unipotent subgroups of G and $\widehat{G}$ coincide, it follows from Proposition 5.6 that

$$c(Z) + r(Z) = \widehat{c}(Z) + \widehat{r}(Z). \tag{11.3}$$

It is also clear that $\widehat{c}(Z) \leq c(Z) \leq \widehat{c}(Z) + 2$. Since an open subset $Gv_{-\lambda} \times Gv_{-\mu} \subset Z$ is a G-equivariant principal $(\Bbbk^\times)^2$-bundle over $\mathbb{P}(Z)$, and $\widehat{\Lambda}(Z) \subseteq \mathfrak{X}(\widehat{G}) = \mathfrak{X}(G) \oplus \mathbb{Z}^2$ projects onto $\mathbb{Z}^2$ with the kernel $\Lambda(\mathbb{P}(Z))$, we have

$$\widehat{c}(Z) = c(\mathbb{P}(Z)), \tag{11.4}$$
$$\widehat{r}(Z) = r(\mathbb{P}(Z)) + 2. \tag{11.5}$$

Theorem 11.5, together with Theorem 8.25, yields

Theorem 11.7. *The following formulæ are valid:*

$$2c(Z) + r(Z) = 2 + \dim G(\lambda)\backslash\backslash G/\!/G(\mu), \qquad r(Z) = \operatorname{rk} G - \operatorname{rk} G(\lambda,\mu),$$
$$2\widehat{c}(Z) + \widehat{r}(Z) = 2 + \dim L(\lambda)\backslash\backslash G/\!/L(\mu), \qquad \widehat{r}(Z) = \operatorname{rk} \widehat{G} - \operatorname{rk} \widehat{G}(\lambda,\mu).$$

For the proof, just note that $\dim C(\lambda) = (\dim G - \dim G(\lambda) + 1)/2 = (\dim \widehat{G} - \dim L(\lambda))/2$ and recall (11.1)–(11.2).

Corollary 11.8. *The numbers $c, r, \widehat{c}, \widehat{r}$ do not change if one transposes λ and μ or replaces λ (or μ) by the dual weight λ^* (resp. μ^*).*

Indeed, the doubled G-variety $Z \times Z^*$ and the generic stabilizers $G(\lambda,\mu)$, $L(\lambda,\mu)$ do not change.

Corollary 11.9. *For $\mu = \lambda$ or λ^*,*

$$c(Z) = c(G/G(\lambda)) + 1, \qquad r(Z) = r(G/G(\lambda)),$$
$$\widehat{c}(Z) = c(G/L(\lambda)), \qquad \widehat{r}(Z) = r(G/L(\lambda)) + 2.$$

Proof. We may assume that $\mu = \lambda^*$. It follows from (the proof of) Lemma 11.5 that a general orbit of $G : Z$ has codimension 1 and is isomorphic to $G/G(\lambda)$, and $G : \mathbb{P}(Z)$ has an open orbit isomorphic to $G/L(\lambda)$. Now apply (11.4)–(11.5). □

11.3 Factorial Double Cones of Complexity ≤ 1. Now we restrict our attention to factorial double cones. By Proposition 11.2(3), $Z(\lambda,\mu)$ is factorial if and only if λ,μ are fundamental weights. We shall write $C(i)$, $Z(i,j)$, $P(i)$, ... instead of $C(\omega_i)$, $Z(\omega_i,\omega_j)$, $P(\omega_i)$, ... For all simple groups G and all pairs of fundamental weights ω_i,ω_j, complexities and ranks of $Z(i,j)$ with respect to the G- and $\widehat{G}$-actions were computed in [Pan3]. All pairs of fundamental weights (ω_i,ω_j) such that $\widehat{c}(Z(i,j)) = 0,1$ are listed, up to the transposition of i,j and an automorphism of the Dynkin diagram, in Tables 11.1–11.2. Fundamental weights are numbered as in [OV].

Suppose that $\Bbbk[Z(i,j)]^U$ is minimally generated by bihomogeneous eigenfunctions $f_1,\dots,f_r$ of weights $\lambda_1,\dots,\lambda_r$ and bidegrees $(n_1,m_1),\dots,(n_r,m_r)$. We may assume that f_1,f_2 have the weights ω_i,ω_j and bidegrees $(1,0)$, $(0,1)$. The weights of other generators and their bidegrees are indicated in the columns "Weights" and "Degrees", respectively. Here we assume that $\omega_k = 0$ whenever $k \neq 1,\dots,\mathrm{rk}\,G$.

We already noted in 11.2 that, if $\widehat{c}(Z(i,j)) = 0$, then $f_1,\dots,f_r$ are algebraically independent and $(\lambda_1,n_1,m_1),\dots,(\lambda_r,n_r,m_r)$ are linearly independent. If $\widehat{c}(Z(i,j)) = 1$, then $Z(i,j)/\!/U$ is a hypersurface [Pan3, 6.5] and the (unique) syzygy between $f_1,\dots,f_r$ is of the form $P+Q+R=0$, where P,Q,R are all monomials in $f_1,\dots,f_r$ of the same weight λ_0 and bidegree (n_0,m_0) indicated in the column "Syzygy" of Table 11.2.

It follows from the classification that, if $i = j$, then $\widehat{c}(Z(i,j)) = 0$ and $m_k = n_k = 1$ for $k = 3,\dots,r$. Hence $\widehat{T}$-eigenspaces in $\Bbbk[Z(i,i)]^U$ are one-dimensional, and the involution transposing the factors of $Z(i,i) = C(i)\times C(i)$ multiplies each $\widehat{T}$-eigenfunction f of bidegree (n,n) by $p(f) = \pm 1$. We call $p(f)$ the *parity* of f. The parities of generators are given in the column "Parity" of Table 11.1. If $f = f_1^{k_1}\cdots f_r^{k_r}$ ($k_1 = k_2$), then $p(f) = p(k_3,\dots,k_r) := p(f_3)^{k_3}\cdots p(f_r)^{k_r}$.

Table 11.1 Spherical double cones (factorial case)

G	Pair	Weights	Degrees	Parity
$\mathbf{A}_l$	(ω_i,ω_j) $i \leq j$	$\omega_{i-k}+\omega_{j+k}$ $k = 1,\dots,\min(i,l+1-j)$	$(1,1)$	$(-1)^k$ for $i = j$
$\mathbf{B}_l$	(ω_1,ω_1)	$0,\omega_2$	$(1,1)$	$1,-1$
	(ω_1,ω_j) $2 \leq j \leq l-2$	$\omega_{j-1},\omega_{j+1}$ ω_j	$(1,1)$ $(2,1)$	
	(ω_1,ω_{l-1})	$\omega_{l-2},2\omega_l$ ω_{l-1}	$(1,1)$ $(2,1)$	
	(ω_1,ω_l)	ω_l ω_{l-1}	$(1,1)$ $(1,2)$	

Table 11.1 (continued)

G	Pair	Weights	Degrees	Parity
	(ω_l, ω_l)	ω_k $k = 0, \ldots, l-1$	$(1,1)$	$(-1)^{k(k+1)/2}$
$\mathbf{C}_l$	(ω_1, ω_1)	$0, \omega_2$	$(1,1)$	-1
	(ω_1, ω_j) $2 \leq j \leq l-1$	$\omega_{j-1}, \omega_{j+1}$ ω_j	$(1,1)$ $(2,1)$	
	(ω_1, ω_l)	ω_{l-1} ω_l	$(1,1)$ $(2,1)$	
	(ω_l, ω_l)	$2\omega_k$ $k = 0, \ldots, l-1$	$(1,1)$	$(-1)^{l-k}$
$\mathbf{D}_l$	(ω_1, ω_1)	$0, \omega_2$	$(1,1)$	$1, -1$
	(ω_1, ω_j) $2 \leq j \leq l-3$	$\omega_{j-1}, \omega_{j+1}$ ω_j	$(1,1)$ $(2,1)$	
	(ω_1, ω_{l-2})	$\omega_{l-3}, \omega_{l-1} + \omega_l$ ω_{l-2}	$(1,1)$ $(2,1)$	
	(ω_1, ω_{l-1})	ω_l	$(1,1)$	
	(ω_l, ω_l)	ω_{l-2k} $1 \leq k \leq l/2$	$(1,1)$	$(-1)^k$
	(ω_{l-1}, ω_l)	ω_{l-2k-1} $1 \leq k \leq (l-1)/2$	$(1,1)$	
	(ω_2, ω_{l-1})	$\omega_{l-1}, \omega_1 + \omega_l$ ω_{l-2}	$(1,1)$ $(1,2)$	
$\mathbf{D}_l$ $l \geq 6$	(ω_3, ω_{l-1})	$\omega_1 + \omega_{l-1}, \omega_2 + \omega_l, \omega_l$ $\omega_{l-3}, \omega_1 + \omega_{l-2}$ $\omega_2 + \omega_{l-2}$	$(1,1)$ $(1,2)$ $(2,2)$	
$\mathbf{D}_5$	(ω_3, ω_4)	$\omega_1 + \omega_4, \omega_2 + \omega_5, \omega_5$ $\omega_2, \omega_1 + \omega_3$	$(1,1)$ $(1,2)$	
$\mathbf{E}_6$	(ω_1, ω_1)	ω_2, ω_5	$(1,1)$	$-1, 1$
	(ω_1, ω_2)	$\omega_1 + \omega_5, \omega_3, \omega_6$ $\omega_2 + \omega_5, \omega_4$	$(1,1)$ $(2,1)$	
	(ω_1, ω_4)	$\omega_2, \omega_5, \omega_5 + \omega_6$ ω_3, ω_6	$(1,1)$ $(2,1)$	
	(ω_1, ω_5)	$0, \omega_6$	$(1,1)$	
	(ω_1, ω_6)	ω_1, ω_4 ω_2	$(1,1)$ $(2,1)$	
$\mathbf{E}_7$	(ω_1, ω_1)	$0, \omega_2, \omega_6$	$(1,1)$	$-1, -1, 1$
	(ω_1, ω_6)	ω_1, ω_7 ω_2	$(1,1)$ $(2,1)$	
	(ω_1, ω_7)	$\omega_2, \omega_5, \omega_6$ ω_3, ω_7 ω_4	$(1,1)$ $(2,1)$ $(2,2)$	

11.4 Applications to Representation Theory. For spherical $Z(i,j)$, the algebra $\Bbbk[Z(i,j)]^U$ was computed by Littelmann [Lit]. He observed that a simple structure of $\Bbbk[Z(i,j)]^U$ leads to the following decomposition rules:

Tensor products.

$$V(n\omega_i) \otimes V(m\omega_j) = \bigoplus_{k_1(n_1,m_1)+\cdots+k_r(n_r,m_r)=(n,m)} V(k_1\lambda_1 + \cdots + k_r\lambda_r).$$

Table 11.2 Double cones of complexity one (factorial case)

G	Pair	Weights	Degrees	Syzygy
$\mathbf{B}_l$ $l \geq 4$	(ω_2, ω_l)	$\omega_1+\omega_l, \omega_l$ $\omega_{l-2}, \omega_{l-1}, \omega_1+\omega_{l-1}$ $\omega_1+\omega_{l-1}$	$(1,1)$ $(1,2)$ $(2,2)$	$\omega_1+\omega_{l-1}+\omega_l$ $(2,3)$
$\mathbf{B}_3$	(ω_2, ω_3)	$\omega_1+\omega_3, \omega_3$ $\omega_1, \omega_2, \omega_1+\omega_2$	$(1,1)$ $(1,2)$	$\omega_1+\omega_2+\omega_3$ $(2,3)$
$\mathbf{C}_l$ $l \geq 4$	(ω_2, ω_l)	$\omega_{l-2}, \omega_1+\omega_{l-1}$ $\omega_1+\omega_{l-1}, 2\omega_1+\omega_l, \omega_l$ $2\omega_{l-1}$	$(1,1)$ $(2,1)$ $(2,2)$	$2\omega_1+2\omega_{l-1}+\omega_l$ $(4,3)$
$\mathbf{C}_3$	(ω_2, ω_3)	$\omega_1, \omega_1+\omega_2$ $2\omega_1+\omega_3, \omega_3$ $2\omega_2$	$(1,1)$ $(2,1)$ $(2,2)$	$2\omega_1+2\omega_2+\omega_3$ $(4,3)$
$\mathbf{D}_6$	(ω_4, ω_5)	$\omega_2+\omega_5, \omega_5, \omega_1+\omega_6, \omega_3+\omega_6$ $\omega_4, \omega_2+\omega_4, \omega_2, \omega_1+\omega_3$	$(1,1)$ $(1,2)$	$\omega_2+\omega_4+\omega_5$ $(2,3)$
$\mathbf{E}_7$	(ω_1, ω_2)	$\omega_1, \omega_1+\omega_6, \omega_3, \omega_7$ $\omega_2+\omega_6, \omega_2, \omega_5, \omega_6$	$(1,1)$ $(2,1)$	$\omega_1+\omega_2+\omega_6$ $(3,2)$

Symmetric and exterior squares.

$$\mathrm{S}^2 V(n\omega_i) = \bigoplus_{\substack{k_1+k_3+\cdots+k_r=n \\ p(k_3,\ldots,k_r)=1}} V(2k_1\omega_i + k_3\lambda_3 + \cdots + k_r\lambda_r),$$

$$\bigwedge^2 V(n\omega_i) = \bigoplus_{\substack{k_1+k_3+\cdots+k_r=n \\ p(k_3,\ldots,k_r)=-1}} V(2k_1\omega_i + k_3\lambda_3 + \cdots + k_r\lambda_r).$$

Restriction.

$$\mathrm{Res}^G_{L(i)} V_G(m\omega_j) = \bigoplus_{k_2m_2+\cdots+k_rm_r=m} V_{L(i)}(k_2(\lambda_2 - n_2\omega_i) + \cdots + k_r(\lambda_r - n_r\omega_i)).$$

Proofs. The first two rules stem immediately from the above discussion. Indeed, highest vectors in $\Bbbk[Z(i,j)]^U_{n,m} = V(n\omega_i) \otimes V(m\omega_j)$ are proportional to monomials $f = f_1^{k_1} \cdots f_r^{k_r}$ with $k_1(n_1,m_1) + \cdots + k_r(n_r,m_r) = (n,m)$. The transposition of the factors of $Z(i,i)$ transposes the factors of $\Bbbk[Z(i,i)]^U_{n,n} = V(n\omega_i)^{\otimes 2}$, and f is (skew)symmetric if and only if $p(f) = 1$ (resp. -1).

To prove the restriction rule, observe that $Z(i,i)_{f_1} = C(i)_{f_1} \times C(j) = P(i) *_{L(i)} (\Bbbk^\times v_{-\omega_i} \times C(j)) = P(i)_{\mathrm{u}} \times \Bbbk^\times v_{-\omega_i} \times C(j)$. Hence $\Bbbk[Z(i,j)]^U_{f_1} \simeq \Bbbk[\Bbbk^\times v_{-\omega_i} \times C(j)]^{U \cap L(i)} \simeq \Bbbk[f_1, f_1^{-1}] \otimes \Bbbk[C(j)]^{U \cap L(i)}$, and $f_2, \ldots, f_r$ restrict to a free system of generators $\bar{f}_l(y) = f_l(v_{-\omega_i}, y)$ of $\Bbbk[C(j)]^{U \cap L(i)}$. It remains to remark that $\Bbbk[C(j)]_m \simeq V_G(m\omega_j)$, $U \cap L(i)$ is a maximal unipotent subgroup of $L(i)$, and $\bar{f}_l$ have T-eigenweights $\lambda_l - n_l\omega_i$:

$$t\bar{f}_l(y) = f_l(v_{-\omega_i}, t^{-1}y) = \omega_i(t)^{-n_l} f_l(t^{-1}v_{-\omega_i}, t^{-1}y) = \lambda_l(t)\omega_i(t)^{-n_l}\bar{f}_l(y).$$

$\square$

For the cases $\widehat{c}(Z(i,j))=1$, the algebra $\Bbbk[Z(i,j)]^U$ was computed in [Pan7, 6.5]. A decomposition rule for tensor products is of the form:

$$V(n\omega_i)\otimes V(m\omega_j) = \bigoplus_{k_1(n_1,m_1)+\cdots+k_r(n_r,m_r)=(n,m)} V(k_1\lambda_1+\cdots+k_r\lambda_r)$$
$$- \bigoplus_{l_1(n_1,m_1)+\cdots+l_r(n_r,m_r)=(n-n_0,m-m_0)} V(\lambda_0+l_1\lambda_1+\cdots+l_r\lambda_r).$$

(Here "−" is an operation in the Grothendieck group of G-modules.)

Example 11.10. Suppose that $G=\mathrm{SL}_d(\Bbbk)$. Consider a double cone $Z(1,1)$. We have $L(1)=\mathrm{GL}_{d-1}(\Bbbk)$, $G(1)=\mathrm{SL}_{d-1}(\Bbbk)$, $V(1,1)=(\Bbbk^{d-1}\otimes\Bbbk_{\det})^*$, and $L(1,1)$ consists of matrices of the form

$$\left(\begin{array}{c|c} \begin{matrix} t & 0 \\ 0 & t \end{matrix} & 0 \\ \hline 0 & * \end{array}\right) \qquad (t\in\Bbbk^\times).$$

Its subgroup $G(1,1)$ is defined by $t=1$. Hence $r=2$, $\widehat{r}=3$, $c=1$, $\widehat{c}=0$, and

$$\begin{aligned}
\Lambda(Z(1,1)) &= \langle\varepsilon_1,\varepsilon_2\rangle = \langle\omega_1,\omega_2\rangle,\\
\Lambda(\mathbb{P}(Z(1,1))) &= \langle\varepsilon_2-\varepsilon_1\rangle = \langle\omega_2-2\omega_1\rangle,\\
\widehat{\Lambda}(Z(1,1)) &= \Lambda(\mathbb{P}(Z(1,1)))+\langle(\omega_1,1,0),(\omega_1,0,1)\rangle\\
&= \langle(\omega_2,1,1),(\omega_1,1,0),(\omega_1,0,1)\rangle.
\end{aligned}$$

(Here ε_i is the T-weight of $e_i\in\Bbbk^d$, $\omega_i=\varepsilon_1+\cdots+\varepsilon_i$.)

Since $V(\omega_1)^{\otimes 2}=(\Bbbk^d)^{\otimes 2}=\mathrm{S}^2\Bbbk^d\oplus\bigwedge^2\Bbbk^d=V(2\omega_1)\oplus V(\omega_2)$, the algebra $\Bbbk[Z(1,1)]^U$ contains eigenfunctions of the weights $(\omega_1,1,0),(\omega_1,0,1),(\omega_2,1,1)$, and a function of the weight $(\omega_2,1,1)$ has parity -1. Clearly, these three functions are algebraically independent (because their weights are linearly independent) and compose a part of a minimal generating system of $\Bbbk[Z(1,1)]^U$. Since $\dim Z(1,1)/\!\!/U=3$, they generate $\Bbbk[Z(1,1)]^U$.

As a corollary, we obtain decomposition formulæ:

$$\begin{aligned}
V(n\omega_1)\otimes V(m\omega_1) &= \bigoplus_{0\le k\le\min(n,m)} V((n+m-2k)\omega_1+k\omega_2),\\
\mathrm{S}^2V(n\omega_1) &= \bigoplus_{0\le k\le n/2} V((2n-4k)\omega_1+2k\omega_2),\\
\textstyle\bigwedge^2V(n\omega_1) &= \bigoplus_{0\le k<n/2} V((2n-4k-2)\omega_1+(2k+1)\omega_2).
\end{aligned}$$

For $d=2$, these are Clebsch–Gordan formulæ.

11.5 Spherical Double Cones. All $\widehat{G}$-spherical double cones $Z = Z(\lambda, \mu)$ were classified by Stembridge [Ste]. By (11.4), $\widehat{c}(Z)$ depends only on the parabolics $P(\lambda), P(\mu)$, i.e., on the supports of λ, μ with respect to fundamental weights. (The *support* of λ is the set of fundamental weights occurring in the decomposition of λ with nonzero coefficients). When the support of λ is reduced, $P(\lambda)$ increases, whence $\widehat{c}(Z)$ may only decrease. Therefore it suffices to find all pairs of maximal possible supports such that $\widehat{c}(Z) = 0$ for all simple groups.

The case of one-element supports, i.e., where λ, μ are multiples of fundamental weights, is already covered by Table 11.1. All remaining pairs of maximal supports, up to the transposition and an automorphism of the Dynkin diagram, are listed in Table 11.3. Fundamental weights of simple groups are numbered as in [OV].

Note that $Z(\lambda, \mu)$ is spherical if and only if $V(n\lambda) \otimes V(m\mu)$ is multiplicity-free for all n, m (see §25).

Table 11.3 Spherical double cones (non-factorial case)

G	$\mathbf{A}_l$				$\mathbf{D}_l$					$\mathbf{E}_6$
λ	ω_1	ω_2	ω_i	ω_i	ω_1	ω_l	ω_l	ω_l	ω_l	ω_1
μ	$\omega_1, \dots, \omega_l$	ω_i, ω_j	ω_1, ω_j	ω_j, ω_{j+1}	ω_i, ω_l	ω_1, ω_{l-1}	ω_1, ω_l	ω_{l-1}, ω_l	ω_1, ω_2	ω_1, ω_5

Chapter 3
General Theory of Embeddings

Equivariant embeddings of homogeneous spaces are one of the main topics of this survey. The general theory of them was developed by D. Luna and Th. Vust in a fundamental paper [LV]. However it was noticed in [Tim2] that the whole theory admits a natural exposition in a more general framework, which is discussed in this chapter. The generically transitive case differs from the general one by the existence of a smallest G-variety of a given birational type, namely, a homogeneous space.

In §12 we discuss the general approach of Luna and Vust based on patching all G-varieties of a given birational class together in one huge prevariety and studying particular G-varieties as open subsets in it. An important notion of a B-chart arising in such a local study is considered in §13. A B-chart is a B-stable affine open subset of a G-variety, and any normal G-variety is covered by (finitely many) G-translates of B-charts. B-charts and their "admissible" collections corresponding to coverings of G-varieties are described in terms of *colored data* composed of B-stable divisors and G-invariant valuations of a given function field. This leads to a "combinatorial" description of normal G-varieties in terms of colored data, obtained in §14. In the cases of complexity ≤ 1, considered in §§15–16, this description is indeed combinatorial, namely, in terms of polyhedral cones, their faces, fans, and other objects of combinatorial convex geometry.

Divisors on G-varieties are studied in §17. We give criteria for a divisor to be Cartier, finitely generated and ample, and we describe global sections in terms of colored data. Aspects of the intersection theory on a G-variety are discussed in §18, including the rôle of B-stable cycles and a formula for the degree of an ample divisor.

12 The Luna–Vust Theory

12.1 Equivariant Classification of G-varieties. The fundamental problem of classifying algebraic varieties has an equivariant analogue: to describe up to a G-isomorphism all irreducible varieties equipped with an action of a connected algebraic group G. A birational classification of G-varieties (with a given field of G-invariant

D.A. Timashev, *Homogeneous Spaces and Equivariant Embeddings*,
Encyclopaedia of Mathematical Sciences 138, DOI 10.1007/978-3-642-18399-7_3,

functions) may be obtained in terms of Galois cohomology [PV, §2]. The second, "biregular", part of the problem may be formulated as follows: to describe all G-actions in a given birational class. More precisely, let K be a fixed function field (i.e., a finitely generated extension of $\Bbbk$), and let G act on K birationally. In other words, K is the function field on some irreducible G-variety X. We say that K is a *G-field* and X is a *G-model* of K. The problem is to classify all G-models of K in terms involving certain invariants of K itself (such as valuations etc).

Remark 12.1. If $K^G = \Bbbk$ or, equivalently, the G-action on each G-model of K is generically transitive, then there is a minimal G-model O, which is embedded as a dense orbit in any other G-model of K. The homogeneous space O determines and is determined by K completely. So the problem may be thought of as classifying all G-equivariant embeddings of O in terms of invariants of O itself.

A general approach to this problem was introduced by Luna and Vust [LV]. They considered only embeddings of homogeneous spaces. We will follow [Tim2] in our more general point of view.

12.2 Universal Model. All models of K may be glued together into one huge scheme $\mathbb{X} = \mathbb{X}(K)$. By definition, points of $\mathbb{X}$ are local rings that are localizations of finitely generated $\Bbbk$-algebras with quotient field K. Any model X of K (i.e., a variety with $\Bbbk(X) = K$) may be considered as a subset of $\mathbb{X}$, and such subsets define a base of the *Zariski topology* on $\mathbb{X}$. The structure sheaves $\mathscr{O}_X$ are patched together in a structure sheaf $\mathscr{O}_{\mathbb{X}}$. A local ring $\mathscr{O}_{\mathbb{X},Y}$ of $Y \in \mathbb{X}$ in the sense of this sheaf is exactly the ring defining Y as a point of $\mathbb{X}$.

The scheme $\mathbb{X}$ is irreducible, but neither Noetherian nor separated. It can be considered as a prevariety if we consider only closed points $x \in \mathbb{X}$ (i.e., such that the residue field $\Bbbk(x) = \mathscr{O}_x/\mathfrak{m}_x$ of the respective local ring $\mathscr{O}_x = \mathscr{O}_{\mathbb{X},x}$ equals k). Non-closed schematic points are identified with closed irreducible subvarieties $Y \subseteq \mathbb{X}$.

We distinguish in $\mathbb{X}$ open subsets $\mathbb{X}^{\mathrm{reg}}, \mathbb{X}^{\mathrm{norm}}, \dots$ of smooth, normal, ... points.

From this point of view, a model of K is nothing but a Noetherian separated open subset $X \subseteq \mathbb{X}$.

The birational G-action on K permutes local subrings of K, which yields an action $G : \mathbb{X}$. However this is not an action in the category of schemes or prevarieties. But the action map $\alpha : G \times \mathbb{X} \to \mathbb{X}$ is rational and induces an embedding of function fields $\alpha^* : K = \Bbbk(\mathbb{X}) \hookrightarrow \Bbbk(G \times \mathbb{X}) = \Bbbk(G) \cdot K$. (Here $\Bbbk(G) \cdot K = \mathrm{Quot}(\Bbbk(G) \otimes_k K)$ is a free composite of fields.) It is obvious that G acts on a G-stable open subset $X \subseteq \mathbb{X}$ regularly if and only if $\alpha^*(\mathscr{O}_{X,x}) \subseteq \mathscr{O}_{G \times X, e \times x}$, $\forall x \in X$.

Denote by $\mathbb{X}_G$ the set of those $x \in \mathbb{X}$ whose local rings $\mathscr{O}_{\mathbb{X},x}$ are mapped by α^* to $\mathscr{O}_{G \times \mathbb{X}, e \times x}$.

Proposition 12.2. *$\mathbb{X}_G$ is the largest open subset of $\mathbb{X}$ on which G acts regularly.*

Proof. We have only to prove that $\mathbb{X}_G$ is open. In other words, if $\mathscr{O}_{\mathbb{X},x} \hookrightarrow \mathscr{O}_{G \times \mathbb{X}, e \times x}$ for $x = x_0$, then the same thing holds in a neighborhood of x_0. Let $X = \operatorname{Spec} A$ be an affine neighborhood of x_0, where $A = \Bbbk[f_1, \dots, f_s]$ is a finitely generated algebra with quotient field K. Then $\mathscr{O}_{\mathbb{X},x}$ is a localization of A at the maximal ideal of x_0. By assumption, $\alpha^*(f_i)$ are defined in a neighborhood E of $e \times x_0$ (one and the same for $i = 1, \dots, s$), and hence α restricts to a regular map $E \to X$. The set of all $x \in X$ such that $e \times x \in E$ is a neighborhood of x_0. In this neighborhood, we have $\alpha(e \times x) = x$, because this holds generically on $\mathbb{X}$. This yields the assertion. □

Observe that $\mathfrak{g}$ acts on K by derivations (along velocity fields on $\mathbb{X}_G$ or on any other G-model).

Proposition 12.3 ([LV, 1.4]). *In characteristic zero, $x \in \mathbb{X}_G$ if and only if $\mathscr{O}_x$ is $\mathfrak{g}$-stable.*

In particular, if $A \subset K$ is a $\mathfrak{g}$-stable finitely generated subalgebra, then any localization of A is $\mathfrak{g}$-stable and consequently $X = \operatorname{Spec} A \subseteq \mathbb{X}_G$.

Example 12.4. Let $G = \Bbbk$ act on $\mathbb{A}^1$ by translations. This yields a birational action $G : K = \Bbbk(t)$, so that $\alpha^*(t) = u + t$ (u is a coordinate on G). A cuspidal curve $X \subset \mathbb{A}^2$ (the Neile parabola) defined by the equation $y^2 = x^3$ becomes a model of K if we put $t = y/x$. The local ring of the singular point $x_0 = (0,0) \in X$ consists of rational functions in $x = t^2$, $y = t^3$ whose denominators have nonzero constant term. But $\alpha^*(t^k) = u^k + ku^{k-1}t + \cdots$ is not defined at $0 \times x_0 \in G \times \mathbb{X}$ (at least when $\operatorname{char} \Bbbk$ does not divide k), because t is not defined at x_0. Therefore $x_0 \notin \mathbb{X}_G$. All other points of X are in $\mathbb{X}_G$, because they are identified via the normalization map $t \mapsto (t^2, t^3)$ with the respective points of $\mathbb{A}^1$, where G acts regularly.

The standard basic vector $\xi \in \mathfrak{g} = \Bbbk$ acts on K as $\mathrm{d}/\mathrm{d}t$, and $\xi(t^k) = kt^{k-1} \in \mathscr{O}_{x_0}$ if $k > 2$. But $\xi x = 2t \notin \mathscr{O}_{x_0}$ if $\operatorname{char} \Bbbk \neq 2$, in accordance with Proposition 12.3. However in characteristic 2, the algebra $\Bbbk[X] = \Bbbk[x, y]$ is $\mathfrak{g}$-stable and Proposition 12.3 is not applicable.

Example 12.5. Another example of this kind is the birational action of $G = \Bbbk^n$ by translations on a blow-up X of $\mathbb{A}^n$ at 0. All points in the complement to the exceptional divisor are in $\mathbb{X}_G$, since they come from $\mathbb{A}^n$, where G acts regularly. In a neighborhood of a point x_0 on the exceptional divisor, X is defined by local equations $x_i = x_1 y_i$ ($1 < i \leq n$) in $\mathbb{A}^n \times \mathbb{A}^{n-1}$. We have $\alpha^*(x_i) = x_i + u_i$, where u_i are coordinates on G, and $\alpha^*(y_i) = (x_i + u_i)/(x_1 + u_1) = (x_1 y_i + u_i)/(x_1 + u_1)$ are not defined at $0 \times x_0$. Hence $x_0 \notin \mathbb{X}_G$. On the other hand, the standard basic vectors $\xi_1, \dots, \xi_n$ of $\mathfrak{g} = \Bbbk^n$ act on $K = \Bbbk(x_1, \dots, x_n)$ as $\partial/\partial x_1, \dots, \partial/\partial x_n$, and

$$\xi_i y_j = \begin{cases} 1/x_1, & i = j > 1, \\ -y_j/x_1, & i = 1 < j, \\ 0, & \text{otherwise}, \end{cases}$$

so that not all $\xi_i y_j$ are in $\mathscr{O}_{x_0}$.

These two examples are typical in a sense that one obtains "bad" birational actions if one blows up or contracts G-nonstable subvarieties in a variety with a "good" (regular) action.

By Proposition 12.2, a G-model of K is nothing but a G-stable Noetherian separated open subset of $\mathbb{X}_G$. The next theorem gives a method of constructing G-models as "G-spans", which we use in §13.

Theorem 12.6 ([LV, 1.5]). *Assume that $\mathring{X}$ is an open subset in $\mathbb{X}_G$. Then $X = G\mathring{X}$ is Noetherian (separated) if and only if $\mathring{X}$ is Noetherian (separated).*

Proof. If $\mathring{X}$ is Noetherian, then $G \times \mathring{X}$ is Noetherian, and $X = \alpha(G \times \mathring{X})$ is Noetherian, too. If X is not separated, then the diagonal $\operatorname{diag} X$ is not closed in $X \times X$, and the non-empty G-stable subset $\overline{\operatorname{diag} X} \setminus \operatorname{diag} X$ contains an orbit O. Clearly, O intersects the two open subsets $\mathring{X} \times X$ and $X \times \mathring{X}$ of $X \times X$. Since O is irreducible, $O \cap (\mathring{X} \times X) \cap (X \times \mathring{X}) = O \cap (\mathring{X} \times \mathring{X}) \subseteq \overline{\operatorname{diag} \mathring{X}} \setminus \operatorname{diag} \mathring{X}$ is non-empty, whence $\mathring{X}$ is not separated.

The converse implications are obvious. □

Example 12.7. Let $G = \Bbbk^\times$ act on $\mathbb{A}^1$ by homotheties. Here $K = \Bbbk(t)$ and a generator ξ of $\mathfrak{g} = \Bbbk$ acts on K as $t(\mathrm{d}/\mathrm{d}t)$. Put $x = t/(1+t)^2$, $y = t/(1+t)^3$. Then $t \mapsto (x,y)$ is a birational map of $\mathbb{A}^1$ to the Cartesian leaf $\mathring{X} \subset \mathbb{A}^2$ defined by the equation $x^3 = xy - y^2$. This map provides a biregular isomorphism of $\mathbb{A}^1 \setminus \{0,-1\}$ onto $\mathring{X} \setminus \{x_0\}$, where $x_0 = (0,0)$ is the singular point of $\mathring{X}$. Therefore $\mathring{X} \setminus \{x_0\} \subseteq \mathbb{X}_G$. One can verify by direct computation that $\alpha^*(x), \alpha^*(y) \in \mathscr{O}_{1 \times x_0}$, whence $x_0 \in \mathbb{X}_G$. In characteristic zero, the situation is simpler, because $\xi x = 2y - x$, $\xi y = y - 3x^2$ imply that the algebra $A = \Bbbk[x,y]$ is $\mathfrak{g}$-stable, and hence $\mathring{X} = \operatorname{Spec} A \subset \mathbb{X}_G$. Put $X = G\mathring{X}$. Then G acts on X with 2 orbits $X \setminus \{x_0\}$ and $\{x_0\}$ (an ordinary double point), cf. Example 4.2.

12.3 Germs of Subvarieties. In the study of the local geometry of a variety X in a neighborhood of an (irreducible) subvariety Y, we may replace X by any open subset intersecting Y, thus arriving at the notion of a *germ* of a variety in (a neighborhood of) a subvariety. If X is a model of K, then a germ of X in Y is essentially the local ring $\mathscr{O}_{X,Y}$ or the respective schematic point of $\mathbb{X}$.

Definition 12.8. A *G-germ* (of K) is a G-fixed schematic point of $\mathbb{X}_G$ (or a G-stable irreducible subvariety of $\mathbb{X}_G$). The set of all G-germs is denoted by ${}_G\mathbb{X}$; a similar notation ${}_GX$ is used for an arbitrary open subset $X \subseteq \mathbb{X}_G$. A G-model X such that a given G-germ is contained in ${}_GX$ (i.e., intersects X in a G-stable subvariety Y) is called a *geometric realization* or a *model* of the G-germ.

Every G-germ admits a geometric realization: just take an affine neighborhood $\mathring{X} \subseteq \mathbb{X}_G$ of the germ and put $X = G\mathring{X}$. If $X \subseteq \mathbb{X}_G$ is G-stable, then X and ${}_GX$ determine each other. The Zariski topology is induced on ${}_G\mathbb{X}$, with ${}_GX$ the open subsets. It is straightforward [LV, 6.1] that X is Noetherian if and only if ${}_GX$ is Noetherian.

Remark 12.9. In characteristic zero, a germ of X in Y is a G-germ if and only if its local ring $\mathscr{O}_{X,Y}$ is G- and $\mathfrak{g}$-stable (cf. Proposition 12.3).

Germs of normal G-models in G-stable prime divisors play an important rôle in the Luna–Vust theory. They are identified with the respective G-invariant valuations of K, called *G-valuations*, see Definition 19.1 and Proposition 19.8. The set $\mathscr{V} = \mathscr{V}(K)$ of all G-valuations of $K/\Bbbk$ is studied in Chap. 4, see also §§13–16 below.

Definition 12.10. The *support* $\mathscr{S}_Y$ of a G-germ Y is the set of $v \in \mathscr{V}$ such that the valuation ring $\mathscr{O}_v$ dominates $\mathscr{O}_{\mathbb{X},Y}$ (i.e., v has center Y in any geometric realization).

The support of a G-germ is non-empty: e.g., if $X \supseteq Y$ is an arbitrary geometric realization of the G-germ, and v is the valuation corresponding to a component of the exceptional divisor in the normalized blow-up of X along Y, then $v \in \mathscr{S}_Y$.

12.4 Morphisms, Separation, and Properness. Here is a version of the valuative criterion of separation.

Theorem 12.11. *A G-stable open subset $X \subseteq \mathbb{X}_G$ is separated if and only if the supports of all its G-germs are disjoint.*

Proof. The closure $\overline{\operatorname{diag} X}$ of the diagonal $\operatorname{diag} X \subseteq X \times X$ is a G-model of K. The projections of $X \times X$ to the factors induce birational regular G-maps $\pi_i : \overline{\operatorname{diag} X} \to X$ $(i = 1,2)$. If X is not separated and $Y \subseteq \overline{\operatorname{diag} X} \setminus \operatorname{diag} X$ is a G-orbit, then the orbits $Y_i = \pi_i(Y)$ are distinct for $i = 1,2$. But $\mathscr{S}_{Y_1} \cap \mathscr{S}_{Y_2} \supseteq \mathscr{S}_Y \neq \emptyset$, a contradiction.

The converse implication follows from the valuative criterion of separation (Appendix B). □

Assume that $K' \subseteq K$ is a subfield containing $\Bbbk$. We have a natural dominant rational map $\varphi : \mathbb{X} \dashrightarrow \mathbb{X}' = \mathbb{X}(K')$. If $X \subseteq \mathbb{X}$, $X' \subseteq \mathbb{X}'$ are models of K, K', then $\varphi : X \to X'$ is regular if and only if for any $x \in X$ there exists an $x' \in X'$ such that $\mathscr{O}_x$ dominates $\mathscr{O}_{x'}$. This x' is necessarily unique (because X' is separated), and $x' = \varphi(x)$.

Now assume that K' is a G-subfield of K. Suppose that X and X' are G-models of K and K'.

Proposition 12.12. *The natural rational map $\varphi : X \to X'$ is regular if and only if for any G-germ $Y \in {}_GX$ there exists a (necessarily unique) G-germ $Y' \in {}_GX'$ such that $\mathscr{O}_{X,Y}$ dominates $\mathscr{O}_{X',Y'}$.*

Proof. If $\mathscr{O}_{X,Y}$ dominates $\mathscr{O}_{X',Y'}$, then there exist finitely generated subalgebras $A \supseteq A'$ such that $\mathscr{O}_{X,Y}$ and $\mathscr{O}_{X',Y'}$ are their respective localizations. Localizing A' and A sufficiently, we may assume that $\mathring{X} = \operatorname{Spec} A$ and $\mathring{X}' = \operatorname{Spec} A'$ are open subsets of X and X' intersecting Y and Y', respectively. The regular map $\mathring{X} \to \mathring{X}'$ extends to the regular map $G\mathring{X} \to G\mathring{X}'$. Since $Y \subseteq X$ is arbitrary, these maps paste together in the regular map $X \to X'$.

The converse implication is obvious: just put $Y' = \overline{\varphi(Y)}$. □

The restriction of a G-valuation of K to K' is a G-valuation, and any G-valuation of K' can be extended to a G-valuation of K (Corollary 19.6). Thus the restriction map $\varphi_* : \mathscr{V}(K) \to \mathscr{V}(K')$ is well defined and surjective. If $\varphi : X \to X'$ is a regular map, then $\varphi_*(\mathscr{S}_Y) \subseteq \mathscr{S}_{Y'}$ for any G-germ $Y \subseteq X$ and $Y' = \overline{\varphi(Y)}$.

Here is a version of the valuative criterion of properness.

Theorem 12.13. *A morphism $\varphi : X \to X'$ is proper if and only if*

$$\bigcup_{Y \subseteq X} \mathscr{S}_Y = \varphi_*^{-1}\left(\bigcup_{Y' \subseteq X'} \mathscr{S}_{Y'}\right).$$

Proof. The implication "only if" follows from the valuative criterion of properness (Appendix B). Normalization reduces the proof of the implication "if" to the case of locally linearizable actions.

To prove the implication in this case, we extend φ to a proper G-morphism $\overline{\varphi}$: $\overline{X} \to X'$ such that $X \subseteq \overline{X}$ is an open subset as follows. There exist G-equivariant completions $\overline{X}, \overline{X'}$ of X, X' [Sum, Th. 3]. Replace $\overline{X}$ by the closure of the graph of the rational map $\overline{X} \dashrightarrow \overline{X'}$. This yields a proper morphism $\overline{\varphi} : \overline{X} \to \overline{X'}$ extending φ. Then replace $\overline{X}$ by $\overline{\varphi}^{-1}(X')$.

Now if φ is not proper, then there exists a G-orbit $Y_0 \subseteq \overline{X} \setminus X$. Since $\overline{X}$ is separated, $\mathscr{S}_{Y_0} \cap \mathscr{S}_Y = \emptyset$, $\forall Y \subseteq X$, but $\varphi_*(\mathscr{S}_{Y_0}) \subseteq \mathscr{S}_{Y'}$, where $Y' = \overline{\varphi}(Y_0) \subseteq X'$, a contradiction. □

Corollary 12.14. *X is complete if and only if $\bigcup_{Y \subseteq X} \mathscr{S}_Y = \mathscr{V}$ (i.e., each G-valuation has center on X).*

13 B-charts

13.1 B-charts and Colored Equipment. From now on, G is assumed to be reductive of simply connected type and all G-models to be normal, i.e., to lie in $\mathbb{X}_G^{\mathrm{norm}}$. We have seen in §12 that a G-model X is given by a Noetherian set ${}_GX$ of G-germs whose supports are disjoint. The Noether property means that X is covered by G-spans of finitely many "simple", e.g., affine, open subsets $\mathring{X} \subseteq X$. An important class of such "local charts" is introduced in

Definition 13.1. A *B-chart* of X is a B-stable affine open subset of X. Generally, a *B-chart* is a B-stable affine open subset $\mathring{X} \subset \mathbb{X}_G^{\mathrm{norm}}$.

It follows from the local structure theorem (Theorem 4.7) that any G-germ $Y \in {}_GX$ admits a B-chart $\mathring{X} \subseteq X$ intersecting Y. Therefore X is covered by finitely many B-charts and their translates. Thus it is important to obtain a compact description for B-charts. We describe their coordinate algebras in terms of their B-stable divisors.

Definition 13.2. Denote by $\mathscr{D} = \mathscr{D}(K)$ the set of prime divisors on X that are not G-stable. The valuation corresponding to a divisor $D \in \mathscr{D}$ is denoted by v_D. Prime divisors that are B-stable but not G-stable, i.e., elements of $\mathscr{D}^B$, are called *colors.*

Let $K_B \subseteq K$ be the subalgebra of rational functions with B-stable divisor of poles on X.

Remark 13.3. The sets $\mathscr{D}$, $\mathscr{D}^B$ and K_B do not depend on the choice of a G-model X. Indeed, a non-G-stable prime divisor on $\mathbb{X}_G^{\mathrm{norm}}$ intersects any G-model $X \subseteq \mathbb{X}_G^{\mathrm{norm}}$, and K_B consists of rational functions defined on $\mathbb{X}_G^{\mathrm{norm}}$ everywhere outside a B-stable divisor.

Since $K_B \supseteq \Bbbk[\mathring{X}]$ for any B-chart $\mathring{X}$, it follows that $\operatorname{Quot} K_B = K$.

The pair $(\mathscr{V}, \mathscr{D}^B)$ is said to be the *colored equipment* (of K). It is in terms of colored equipment that B-charts, G-germs and G-models are described, as we shall see below.

Remark 13.4. In the case $K = \Bbbk(O)$ and $O = G/H$, $\mathscr{D}^B$ may be computed as follows. Each $D \in \mathscr{D}$ determines a G-line bundle $\mathscr{L}(\chi) = \mathscr{L}(\Bbbk_\chi)$ over O and a section $\eta \in \mathrm{H}^0(O, \mathscr{L}(\chi)) = \Bbbk[G]^{(H)}_{-\chi}$ defined uniquely up to multiplication by an invertible function on O, i.e., by a scalar multiple of a character of G [KKV, 1.2]. The section η may be regarded as an equation for the preimage of D under the orbit map $G \to O$. Since D is prime, η is indecomposable in the multiplicative semigroup $\Bbbk[G]^{(H)}/\Bbbk^\times\mathfrak{X}(G)$.

Each $f \in \Bbbk(O)$ decomposes as $f = \eta^d \eta_1^{d_1} \cdots \eta_s^{d_s} \mu$, where $\mu \in \mathfrak{X}(G), d, d_1, \dots, d_s \in \mathbb{Z}$ and $\eta, \eta_1, \dots, \eta_s \in \Bbbk[G]^{(H)}$ are pairwise coprime indecomposables. Then $v_D(f) = d$.

Finally, D is a color if and only if η is a $(B \times H)$-eigenfunction. Therefore $\mathscr{D}^B$ is in bijection with the set of generators of $\Bbbk[G]^{(B\times H)}/\Bbbk^\times\mathfrak{X}(G)$.

The "dual" object is the multiplicative group $K^{(B)}$ of rational B-eigenfunctions. There is an exact sequence

$$1 \longrightarrow (K^B)^\times \longrightarrow K^{(B)} \longrightarrow \Lambda \longrightarrow 0, \tag{13.1}$$

where $\Lambda = \Lambda(K)$ is the weight lattice (of any G-model) of K.

In the sequel, we frequently use Knop's approximation Lemma 19.12, which is crucial for reducing various questions to B-eigenfunctions. In particular, it implies that G-valuations are determined uniquely by their restriction to $K^{(B)}$ (Corollary 19.13).

13.2 Colored Data. Let $\mathring{X}$ be a B-chart. Then $\mathscr{A} = \Bbbk[\mathring{X}]$ is an integrally closed finitely generated algebra; in particular, it is a Krull ring. (See [Ma, §12] for the definition and properties of Krull rings.) Therefore

$$\mathscr{A} = \bigcap \mathscr{O}_{\mathring{X},D} \text{ (over all prime divisors } D \subset \mathring{X}) = \bigcap_{w\in\mathscr{W}} \mathscr{O}_w \cap \bigcap_{D\in\widetilde{\mathscr{R}}} \mathscr{O}_{v_D},$$

where $\mathscr{O}_v$ is the valuation ring of v, $\mathscr{W} \subseteq \mathscr{V}$, $\mathscr{R} \subseteq \mathscr{D}^B$, and $\widetilde{\mathscr{R}} = \mathscr{R} \sqcup (\mathscr{D} \setminus \mathscr{D}^B)$. Here the G-valuations $w \in \mathscr{W}$ are determined up to a rational multiple, and we shall ignore this indeterminacy, thus passing to a "projectivization" of $\mathscr{V}$. In particular, we may assume that the group of values of every $w \in \mathscr{W}$ is exactly $\mathbb{Z} \subset \mathbb{Q}$.

The pair $(\mathscr{W}, \mathscr{R})$ is said to be the *colored data* of $\mathring{X}$. A B-chart is uniquely determined by its colored data. Taking another B-chart changes $\mathscr{W}$ and $\mathscr{R}$ by finitely many elements. Hence all possible $\mathscr{W} \sqcup \mathscr{R}$ lie in a certain distinguished class **CD** of equivalent subsets of $\mathscr{V} \sqcup \mathscr{D}^B$ with respect to the equivalence relation "differ by finitely many elements".

Conversely, if $\mathscr{W} \subseteq \mathscr{V}$, $\mathscr{R} \subseteq \mathscr{D}^B$, and $\mathscr{W} \sqcup \mathscr{R} \in$ **CD**, then

$$\mathscr{A} = \mathscr{A}(\mathscr{W}, \mathscr{R}) = \bigcap_{w\in\mathscr{W}} \mathscr{O}_w \cap \bigcap_{D\in\widetilde{\mathscr{R}}} \mathscr{O}_{v_D} \tag{13.2}$$

is a Krull ring. Indeed, for each $f \in K$ almost all (i.e., all but finitely many) valuations from $\mathscr{W} \sqcup \mathscr{R}$ vanish on f, since it is true for colored data of B-charts, and hence for any subset in the class **CD**.

Example 13.5. $K_B = \mathscr{A}(\emptyset, \emptyset)$.

Remark 13.6. Here and below, we identify prime divisors and respective valuations. To emphasize that valuations (divisors) and functions are thought of as "dual" to each other, we write $\langle \mathscr{V}_0, f \rangle \geq 0$ for $\mathscr{V}_0 \subseteq \mathscr{V} \sqcup \mathscr{D}$ if $v(f) \geq 0$, $\forall v \in \mathscr{V}_0$ ($v = v_D$ for $D \in \mathscr{D}$), and so on.

Proposition 13.7. (1) *All valuations from $\widetilde{\mathscr{R}}$ are essential for $\mathscr{A}$ (i.e., cannot be removed from the r.h.s. of* (13.2)*).*
(2) *A valuation $w \in \mathscr{W}$ is essential for $\mathscr{A}$ if and only if*

$$\exists f \in K^{(B)} : \ \langle \mathscr{W} \sqcup \mathscr{R} \setminus \{w\}, f \rangle \geq 0, \ w(f) < 0. \tag{W}$$

Proof. (1) Let X be a smooth G-model of K. Consider the G-line bundle $\mathscr{L} = \mathscr{O}_X(D)$, where $D \in \widetilde{\mathscr{R}}$, and let $\eta \in \mathrm{H}^0(X, \mathscr{L})$ be a section with $\operatorname{div} \eta = D$. Put $f = g\eta/\eta$, where $g \in G$, $gD \neq D$. Then $v_D(f) = -1$, $\langle \widetilde{\mathscr{R}} \setminus \{D\}, f \rangle \geq 0$, and $\langle \mathscr{W}, f \rangle = 0$ by Corollary 19.7. Thus v_D is essential for $\mathscr{A}$.
(2) Assume that $w \in \mathscr{W}$ is essential for $\mathscr{A}$; then $\exists f \in K : \ \langle \mathscr{W} \sqcup \widetilde{\mathscr{R}} \setminus \{w\}, f \rangle \geq 0$, $w(f) < 0$. Applying Lemma 19.12, we replace f by a B-eigenfunction and obtain (W). The converse implication is obvious. □

Theorem 13.8. (1) $\operatorname{Quot} \mathscr{A} = K$ *if and only if*

$$\forall \mathscr{V}_0 \subseteq \mathscr{W} \sqcup \mathscr{R}, \ \mathscr{V}_0 \textit{ finite}, \ \exists f \in K^{(B)} : \ \langle \mathscr{W} \sqcup \mathscr{R}, f \rangle \geq 0, \ \langle \mathscr{V}_0, f \rangle > 0. \tag{C}$$

(2) *$\mathscr{A}$ is finitely generated if and only if*

$$\mathscr{A}^U = \Bbbk\big[f \in K^{(B)} \bigm| \langle \mathscr{W} \sqcup \mathscr{R}, f \rangle \geq 0\big] \textit{ is finitely generated.} \tag{F}$$

(3) *Under the equivalent conditions of* (1)–(2), $\mathring{X} = \operatorname{Spec} \mathscr{A}$ *is a B-chart.*

Proof. (1) Assume that $\operatorname{Quot} \mathscr{A} = K$. We may assume that $\mathscr{V}_0 = \{v\}$; then there exists $f \in \mathscr{A} \subseteq K_B$ such that $v(f) > 0$. Applying Lemma 19.12, we replace f by an element of $\mathscr{A}^{(B)}$ and obtain (C). Conversely, assume that (C) is true and $h \in K_B$. Then we take $\mathscr{V}_0 = \{v \in \mathscr{W} \sqcup \mathscr{R} \mid v(h) < 0\}$ and, multiplying h by f^n for $n \gg 0$ (killing the poles), we fall into $\mathscr{A}$. Hence $K_B \subseteq \operatorname{Quot} \mathscr{A}$, and this yields $K = \operatorname{Quot} \mathscr{A}$.
(2) Let X be a smooth G-model of K. Take an effective divisor on X with support $\{D \subset X \mid v_D \in \mathscr{V} \setminus \mathscr{W}\} \sqcup (\mathscr{D}^B \setminus \mathscr{R})$ and consider the corresponding section $\eta \in \mathrm{H}^0(X, \mathscr{L})^{(B)}$ of the G-line bundle $\mathscr{L}$. Consider an algebra $R = \bigoplus_{n \geq 0} R_n$, where

$$R_n = \{\sigma \in \mathrm{H}^0(X, \mathscr{L}^{\otimes n}) \mid \sigma/\eta^n \in \mathscr{A}\}.$$

Then $\mathscr{A} = \bigcup \eta^{-n} R_n \subseteq \operatorname{Quot} R$. Since every G-valuation of K can be extended to a G-valuation of $\operatorname{Quot} R$ (Corollary 19.6), we see that

$$R_n = \{\sigma \in \mathrm{H}^0(X, \mathscr{L}^{\otimes n}) \mid \forall w \in \mathscr{W} : w(\sigma) \geq nw(\eta)\}$$

is G-stable.

Though $\mathscr{A}$ is not a G-algebra, it is very close to being a G-algebra, so that Theorem D.5(1) extends to this case. Replacing K with $K \cap \operatorname{Quot} R$, we may assume that $\operatorname{Quot} \mathscr{A} = K$.

If $\mathscr{A}$ is finitely generated, then it is easy to construct a finitely generated graded G-subalgebra $S \subseteq R$ such that $\eta \in S_1$ and $\mathscr{A} = \bigcup \eta^{-n} S_n$. By Theorem D.5(1), S^U and also $\mathscr{A}^U = \bigcup \eta^{-n} S_n^U$ are finitely generated.

Conversely, if $\mathscr{A}^U$ is finitely generated, then we construct a finitely generated integrally closed graded G-subalgebra $S \subseteq R$ such that $\eta \in S_1$, $\operatorname{Quot} S = \operatorname{Quot} R$, and $\mathscr{A}^U = \bigcup \eta^{-n} S_n^U$. Then $\mathscr{A}' := \bigcup \eta^{-n} S_n = \mathscr{A}(\mathscr{W}', \mathscr{R})$ (for some $\mathscr{W} \subseteq \mathscr{W}' \subseteq \mathscr{V}$) is finitely generated, and $(\mathscr{A}')^U = \mathscr{A}^U$. However for any $f \in \mathscr{A}$ and $w' \in \mathscr{W}'$ we have $w'(f) \geq 0$ (since otherwise f could be replaced with an element of $\mathscr{A}^{(B)} \setminus (\mathscr{A}')^{(B)}$ by Lemma 19.12, a contradiction). We conclude that $\mathscr{A} = \mathscr{A}'$.

(3) In characteristic zero, just note that $\mathscr{A}$ is $\mathfrak{g}$-stable, because all $\mathscr{O}_v$ ($v \in \mathscr{W} \sqcup \widetilde{\mathscr{R}}$) are. In general, since $\mathscr{A}$ is finitely generated, it follows that $\mathscr{A} = \Bbbk[\eta^{-n} M]$ for a finite-dimensional G-submodule $M \ni \eta^n$ in some R_n. Let X' be the closure of the image of the natural rational map $X \dashrightarrow \mathbb{P}(M^*)$. Then X' is a G-model and $\mathring{X} = \operatorname{Spec} \mathscr{A} \subseteq X'$. □

Corollary 13.9. *A pair* $(\mathscr{W}, \mathscr{R})$ *from* **CD** *is the colored data of a B-chart if and only if conditions* (C),(F),(W) *are satisfied.*

Note that elements of $\mathscr{D}^B \setminus \mathscr{R}$ are exactly the irreducible components of $X \setminus \mathring{X}$, where $X = G\mathring{X}$. A B-chart is G-stable if and only if $\mathscr{R} = \mathscr{D}^B$.

Corollary 13.10. *Affine normal G-models are in bijection with colored data* $(\mathscr{W}, \mathscr{D}^B)$ *satisfying* (C),(F),(W).

Remark 13.11. In this section, we never use an *a priori* assumption that G-invariant valuations from $\mathscr{W}$ are geometric, cf. Remark 19.3.

13.3 Local Structure. The local structure of B-charts is well understood.

The subgroup $P = N_G(\mathring{X})$ is a parabolic containing B. We have $P = P[\mathscr{D}^B \setminus \mathscr{R}] = \bigcap_{D \in \mathscr{D}^B \setminus \mathscr{R}} P[D]$, where $P[D]$ is the stabilizer of D. In the case $K = \Bbbk(G/H)$, if $\eta \in \Bbbk[G]^{(B \times H)}_{\lambda, \chi}$ is an equation of D, then $P[D] = P(\lambda)$ is the parabolic associated with λ.

Let $P = P_{\mathrm{u}} \leftthreetimes L$ be the Levi decomposition ($L \supseteq T$).

In §17, we prove that the divisor $X \setminus \mathring{X}$ is ample on $X = G\mathring{X}$ (Corollary 17.20). Now Theorem 4.6 implies the following

Proposition 13.12 ([Tim3]).

(1) *The action* $P_{\mathrm{u}} : \mathring{X}$ *is proper and has a geometric quotient* $\mathring{X}/P_{\mathrm{u}} = \operatorname{Spec} \Bbbk[\mathring{X}]^{P_{\mathrm{u}}}$.

(2) *There exists a T-stable closed affine subvariety* $Z \subseteq \mathring{X}$ *such that* $\mathring{X} = PZ$ *and the natural maps* $P_{\mathrm{u}} \times Z \to \mathring{X}$, $Z \to \mathring{X}/P_{\mathrm{u}}$ *are finite and surjective.*

(2)′ *In characteristic zero, Z is L-stable, and the P-action on* $\mathring{X}$ *induces an isomorphism*

$$P_{\mathrm{u}} \times Z \simeq P *_L Z \xrightarrow{\sim} \mathring{X}.$$

14 Classification of G-models

14.1 G-germs. We begin with a description of G-germs in terms of colored data.

Consider a G-germ $Y \in {}_G\mathbb{X}^{\mathrm{norm}}$. Let $\mathscr{V}_Y \subseteq \mathscr{V}$, $\mathscr{D}_Y \subseteq \mathscr{D}$ be the subsets corresponding to all prime divisors on $\mathbb{X}_G^{\mathrm{norm}}$ containing Y. The pair $(\mathscr{V}_Y, \mathscr{D}_Y^B)$ is said to be the *colored data* of the G-germ.

If the G-germ intersects a B-chart $\mathring{X}$, then $\mathring{Y} = Y \cap \mathring{X}$ is the center of any $v \in \mathscr{S}_Y$, i.e., $v|_{k[\mathring{X}]} \geq 0$, and the ideal $\mathscr{I}(\mathring{Y}) \lhd \Bbbk[\mathring{X}]$ is given by $v > 0$. Conversely, if a G-valuation $v \in \mathscr{V}$ is non-negative on $\Bbbk[\mathring{X}]$, then it determines a G-germ intersecting $\mathring{X}$. If $(\mathscr{W}, \mathscr{R})$ is the colored data of $\mathring{X}$, then $\mathscr{V}_Y \subseteq \mathscr{W}$, $\mathscr{D}_Y^B \subseteq \mathscr{R}$.

Proposition 14.1 ([Kn3, 3.8]).

(1) *A G-germ is uniquely determined by its colored data.*
(2) *A G-valuation v is in $\mathscr{S}_Y$ if and only if*

$$\forall f \in K^{(B)} : \langle \mathscr{V}_Y \sqcup \mathscr{D}_Y^B, f \rangle \geq 0 \implies v(f) \geq 0, \quad \text{and if} > \text{occurs in the l.h.s., then } v(f) > 0. \tag{S}$$

Proof. Choose a geometric realization $X \supseteq Y$ and a B-chart $\mathring{X} \subseteq X$ intersecting Y. Let $\mathfrak{m}_{X,Y}$ denote the maximal ideal in $\mathscr{O}_{X,Y}$.

(2) Observe that for $f \in K^{(B)}$ we have $f \in \mathscr{O}_{X,Y}^{(B)} \iff \langle \mathscr{V}_Y \sqcup \mathscr{D}_Y^B, f \rangle \geq 0$ and $f \in \mathfrak{m}_{X,Y}^{(B)}$ if and only if one of these inequalities is strict.

If $\mathscr{O}_v$ dominates $\mathscr{O}_{X,Y}$, then (S) is satisfied. Conversely, if there exists $f \in \mathscr{O}_{X,Y}$ such that $v(f) < 0$, then, applying Lemma 19.12, we replace f by a B-eigenfunction and see that (S) fails. Therefore $\mathscr{O}_v \supseteq \mathscr{O}_{X,Y} \supseteq \Bbbk[\mathring{X}]$ and v has center $Y' \supseteq Y$ on X. If $Y' \neq Y$, then for all $v' \in \mathscr{S}_Y$ there is $f \in \mathfrak{m}_{X,Y}$ such that $v'(f) > 0$, $v(f) = 0$. Replacing f by a B-eigenfunction again, we obtain a contradiction with (S). Thus $\mathscr{O}_v$ dominates $\mathscr{O}_{X,Y}$.
(1) Since $\Bbbk[\mathring{X}] \subseteq \mathscr{A} = \mathscr{A}(\mathscr{V}_Y, \mathscr{D}_Y^B) \subseteq \mathscr{O}_{X,Y}$, the local ring $\mathscr{O}_{X,Y}$ is the localization of $\mathscr{A}$ in the ideal $I_Y = \mathscr{A} \cap \mathfrak{m}_{X,Y}$. Take any $v \in \mathscr{S}_Y$; then I_Y is defined in $\mathscr{A}$ by $v > 0$. But $\mathscr{S}_Y$ is determined by $(\mathscr{V}_Y, \mathscr{D}_Y^B)$. □

Now we describe G-germs in a given B-chart $\mathring{X} = \operatorname{Spec} \mathscr{A}$, $\mathscr{A} = \mathscr{A}(\mathscr{W}, \mathscr{R})$.

Theorem 14.2. (1) *$v \in \mathscr{V}$ has a center on $\mathring{X}$ if and only if*

$$\forall f \in K^{(B)} : \langle \mathscr{W} \sqcup \mathscr{R}, f \rangle \geq 0 \implies v(f) \geq 0. \tag{V}$$

(2) *Assume that $v \in \mathscr{S}_Y$. A G-valuation $w \in \mathscr{W}$ belongs to $\mathscr{V}_Y$ if and only if*

$$\forall f \in K^{(B)} : \langle \mathscr{W} \sqcup \mathscr{R}, f \rangle \geq 0,\ v(f) = 0 \implies w(f) = 0. \tag{V'}$$

Similarly, $D \in \mathscr{R}$ belongs to $\mathscr{D}_Y^B$ if and only if

$$\forall f \in K^{(B)} : \langle \mathscr{W} \sqcup \mathscr{R}, f \rangle \geq 0, v(f) = 0 \implies v_D(f) = 0. \tag{D'}$$

Proof. (1) v has a center if and only if $v|_{\mathscr{A}} \geq 0$. This clearly implies (V). On the other hand, if $f \in \mathscr{A}$, $v(f) < 0$, then, applying Lemma 19.12, we replace f by an element of $\mathscr{A}^{(B)}$ and see that (V) is false.
(2) Assume that $w \in \mathscr{W}$ (or $D \in \mathscr{R}$) belongs to $\mathscr{V}_Y$ (or $\mathscr{D}_Y^B$); then every function $f \in \mathscr{A}$ not vanishing on $\mathring{Y}$ (i.e., $v(f) = 0$) does not vanish on the respective B-stable divisor of $\mathring{X}$ as well (i.e., $w(f) = 0$ or $v_D(f) = 0$). For $f \in \mathscr{A}^{(B)}$, we obtain (V$'$) (or (D$'$)).

Conversely, assume that $w \notin \mathscr{V}_Y$ (or $D \notin \mathscr{D}_Y^B$); then there exists $f \in \mathscr{A}$ vanishing on the respective B-stable divisor of $\mathring{X}$ (i.e., $w(f) > 0$ or $v_D(f) > 0$) but not on $\mathring{Y}$ (i.e., $v(f) = 0$). Applying Lemma 19.12, we replace f by an element of $\mathscr{A}^{(B)}$ and see that (V$'$) (or (D$'$)) is false. □

14.2 G-models. Summing up, we can construct every normal G-model in the following way:

(1) Take a finite collection of colored data $(\mathscr{W}_i, \mathscr{R}_i)$ in **CD** satisfying (C),(F). Decrease $\mathscr{W}_i$ if necessary so as to satisfy (W). These colored data determine finitely many B-charts $\mathring{X}_i$.
(2) Compute from $(\mathscr{W}_i, \mathscr{R}_i)$ via conditions (V),(V$'$),(D$'$) the collection of colored data $(\mathscr{V}_Y, \mathscr{D}_Y^B)$ of G-germs Y intersecting $\mathring{X}_i$.
(3) Compute the supports $\mathscr{S}_Y$ from $(\mathscr{V}_Y, \mathscr{D}_Y^B)$ using (S).

The G-models $X_i = G\mathring{X}_i$ may be glued together in a G-model X if and only if the supports $\mathscr{S}_Y$ obtained at Step (3) are disjoint (Theorem 12.11). The collection ${}_GX$ of G-germs is given by Step (2) as the collection of their colored data, which is called the *colored data* of X.

Remark 14.3. We notice in addition that the collection of covering B-charts $\mathring{X}_i$ is of course not uniquely determined. Furthermore, one G-germ may have a lot of different B-charts. For example, in the notation of Theorem 14.2, we may consider a principal open subset $\mathring{X}_f = \{x \mid f(x) \neq 0\}$ in $\mathring{X}$, where $f \in \mathscr{A}^{(B)}$, $v(f) = 0$ (to avoid cutting $\mathring{Y}$ off), i.e., pass from $\mathscr{A}$ to a localization $\mathscr{A}_f$. This corresponds to removing from $(\mathscr{W} \setminus \mathscr{V}_Y) \sqcup (\mathscr{R} \setminus \mathscr{D}_Y^B)$ a finite set $\mathscr{W}_0 \sqcup \mathscr{R}_0$ of those valuations that are positive on f. By (V$'$) and (D$'$), this set may contain any finite number of elements from $(\mathscr{W} \setminus \mathscr{V}_Y) \sqcup (\mathscr{R} \setminus \mathscr{D}_Y^B)$. In particular, if $(\mathscr{W} \setminus \mathscr{V}_Y) \sqcup (\mathscr{R} \setminus \mathscr{D}_Y^B)$ is finite, then there exists a *minimal* B-chart $\mathring{X}_Y$ with $\mathscr{W} = \mathscr{V}_Y$, $\mathscr{R} = \mathscr{D}_Y^B$.

Parabolic induction does not change $K^{(B)}$ and $\mathscr{V}$, while $\mathscr{D}^B$ is extended by finitely many colors, whose valuations vanish on K^B, see Proposition 20.13. The G-germs of an induced variety are induced from those of the original variety, and it is easy to prove the following result:

Proposition 14.4. *Parabolic induction does not change the colored data of a G-model.*

15 Case of Complexity 0

A practical use of the theory developed in the preceding sections depends on whether the colored equipment of a G-field is accessible for computation and operation or not. It was noted already in [LV] that there is no hope to obtain a transparent classification of G-models from the general description in §14 (maybe excepting particular examples) if the complexity is > 1. On the other hand, if the complexity is ≤ 1, then an explicit solution to the classification problem is obtained. An appropriate language to operate with the colored equipment is that of convex polyhedral geometry.

15.1 Combinatorial Description of Spherical Varieties. We shall write $c(K)$, $r(K)$ for the complexity, resp. rank, of (any G-model of) K. If $c(K)=0$, then any G-model contains an open B-orbit, hence an open G-orbit O. Homogeneous spaces of complexity zero (=spherical spaces) and their embeddings are studied in details in Chap. 5. Here we classify the normal embeddings of a given spherical homogeneous space in the framework of the Luna–Vust theory. This classification was first obtained by Luna and Vust [LV, 8.10]. For a modern self-contained exposition, see [Kn2], [Bri13, §3].

Let O be a spherical homogeneous space with the base point $o \in O$, $H = G_o$, and $K = \Bbbk(O)$. Since $K^B = \Bbbk$, it follows from Corollary 19.13 and the exact sequence (13.1) that G-valuations are identified by restriction to $K^{(B)}$ with $\mathbb{Q}$-linear functionals on the lattice $\Lambda = \Lambda(O)$. The set $\mathscr{V}$ is a convex solid polyhedral cone in $\mathscr{E} = \mathrm{Hom}(\Lambda, \mathbb{Q})$ (Theorem 21.1), which is cosimplicial in characteristic zero (Theorem 22.13). The set $\mathscr{D}^B$ consists of irreducible components of the complement to the dense B-orbit in O, and hence is finite. The restriction to $K^{(B)}$ yields a map $\varkappa : \mathscr{D}^B \to \mathscr{E}$, which is in general not injective.

Remark 15.1. If $O = G/H$, $\mathscr{D}^B = \{D_1, \dots, D_s\}$, and $\eta_1, \dots, \eta_s \in \Bbbk[G]^{(B\times H)}$ are the respective indecomposable elements of biweights $(\lambda_1, \chi_1), \dots, (\lambda_s, \chi_s)$, then (λ_i, χ_i) are linearly independent. (Otherwise, there is a linear dependence $\sum d_i(\lambda_i, \chi_i) = 0$, and $f = \eta_1^{d_1} \cdots \eta_s^{d_s}$ is a non-constant B-invariant rational function on O.) If $f = \eta_1^{d_1} \cdots \eta_s^{d_s} \in K_\lambda^{(B)}$, then $\sum d_i\lambda_i = \lambda$, $\sum d_i\chi_i = 0$, and $\langle \varkappa(D_i), \lambda \rangle = v_{D_i}(f) = d_i$.

Definition 15.2. The space $\mathscr{E}$ equipped with the cone $\mathscr{V} \subseteq \mathscr{E}$ and with the map $\varkappa : \mathscr{D}^B \to \mathscr{E}$ is the *colored space* (of O).

Now we consider the structure of colored data and reorganize them in a more convenient way. The proofs are straightforward, as soon as we interpret B-eigenfunctions as linear functionals on $\mathscr{E}$.

The class **CD** consists of finite sets.

Condition (C) means that $\mathscr{W} \cup \varkappa(\mathscr{R})$ generates a strictly convex cone $\mathscr{C} = \mathscr{C}(\mathscr{W}, \mathscr{R})$ in $\mathscr{E}$ and $\varkappa(\mathscr{R}) \not\ni 0$.

Condition (W) means that the elements of $\mathscr{W}$ are exactly the generators of those edges of $\mathscr{C}$ that do not intersect $\varkappa(\mathscr{R})$.

Condition (F) holds automatically: $\mathscr{A}^U$ is the semigroup algebra of $\Lambda \cap \mathscr{C}^\vee$, where $\mathscr{C}^\vee = \{\lambda \in \mathscr{E}^* \mid \langle \mathscr{C}, \lambda \rangle \geq 0\}$ is the dual cone to $\mathscr{C}$. Since $\mathscr{C}^\vee$ is finitely generated, the semigroup $\Lambda \cap \mathscr{C}^\vee$ is finitely generated by Gordan's lemma [Ful2, 1.2, Pr. 1].

Condition (V) means that $v \in \mathscr{C}$.

Conditions (V′) and (D′) say that $\mathscr{V}_Y$ and $\mathscr{D}_Y^B$ consist of those elements of $\mathscr{W} \sqcup \mathscr{R}$ which lie in the face $\mathscr{C}_Y = \mathscr{C}(\mathscr{V}_Y, \mathscr{D}_Y^B) \subseteq \mathscr{C}$ containing v in its (relative) interior.

Condition (S) means that $v \in \mathscr{V} \cap \operatorname{int}\mathscr{C}_Y$, where int denotes the relative interior.

Observe that every G-germ Y has a minimal B-chart $\mathring{X}_Y$ with $\mathscr{C} = \mathscr{C}_Y$, $\mathscr{R} = \mathscr{D}_Y^B$ (Remark 14.3). It suffices to consider only such charts.

Definition 15.3. A *colored cone* in $\mathscr{E}$ is a pair $(\mathscr{C}, \mathscr{R})$, where $\mathscr{R} \subseteq \mathscr{D}^B$, $\varkappa(\mathscr{R}) \not\ni 0$, and $\mathscr{C}$ is a strictly convex cone generated by $\varkappa(\mathscr{R})$ and finitely many vectors from $\mathscr{V}$.

A colored cone $(\mathscr{C}, \mathscr{R})$ is *supported* if $\operatorname{int}\mathscr{C} \cap \mathscr{V} \neq \emptyset$.

A *face* of $(\mathscr{C}, \mathscr{R})$ is a colored cone $(\mathscr{C}', \mathscr{R}')$, where $\mathscr{C}'$ is a face of $\mathscr{C}$ and $\mathscr{R}' = \mathscr{R} \cap \varkappa^{-1}(\mathscr{C}')$.

A *colored fan* is a finite set of supported colored cones which is closed under passing to a supported face and such that different cones intersect in faces inside $\mathscr{V}$.

Theorem 15.4. (1) *B-charts are in bijection with colored cones in $\mathscr{E}$.*
(2) *G-germs are in bijection with supported colored cones.*
(3) *Normal G-models are in bijection with colored fans.*
(4) *Every G-model X contains finitely many G-orbits. If $Y_1, Y_2 \subseteq X$ are two G-orbits, then $Y_1 \preceq Y_2$ if and only if $(\mathscr{C}_{Y_2}, \mathscr{D}_{Y_2}^B)$ is a face of $(\mathscr{C}_{Y_1}, \mathscr{D}_{Y_1}^B)$.*

Corollary 15.5. *Affine normal G-models are in bijection with colored cones of the form $(\mathscr{C}, \mathscr{D}^B)$.*

Corollary 15.6. *O is (quasi)affine if and only if $\varkappa(\mathscr{D}^B)$ can be separated from $\mathscr{V}$ by a hyperplane (resp. does not contain 0 and spans a strictly convex cone).*

Corollary 15.7. *A G-model is complete if and only if its colored fan covers the whole of $\mathscr{V}$.*

Example 15.8 (Toric varieties). Suppose that $G = B = T$ is a torus. We may assume that $O = T$, $H = \{e\}$. Here $\mathscr{V} = \mathscr{E}$ (Example 20.2) and there are no colors. Hence embeddings of T are in bijection with fans in $\mathscr{E}$, where a *fan* is a finite set of strictly convex polyhedral cones which is closed under passing to a face and such that different cones intersect in faces. Every embedding X of T contains finitely many T-orbits, which correspond to cones in the fan. For any orbit $Y \subseteq X$, the union X_Y of all orbits containing Y in their closures is the minimal T-chart of Y determined by $\mathscr{C}_Y$. We have $\Bbbk[X_Y] = \Bbbk[\mathfrak{X}(T) \cap \mathscr{C}_Y^\vee] \subseteq \Bbbk[T]$. X is complete if and only if its fan is the subdivision of the whole $\mathscr{E}$.

Equivariant embeddings of a torus are called *toric varieties*. Due to their nice combinatorial description, toric varieties are a good testing site for various concepts and problems of algebraic geometry. Their theory is well developed, see [Dan], [Oda], [Ful2].

Other examples of spherical varieties are considered in Chap. 5.

15.2 Functoriality. Now we discuss the functoriality of colored data.

Let $\varphi : O \to \overline{O}$ be a G-morphism of homogeneous spaces. Denote $\overline{K} = \Bbbk(\overline{O})$ and by $(\overline{\mathscr{E}}, \overline{\mathscr{V}}, \overline{\mathscr{D}}^B, \overline{\varkappa})$ the colored space of $\overline{O}$. The map φ induces an embedding $\varphi^* : \overline{K} \hookrightarrow K$ and a linear map $\varphi_* : \mathscr{E} \twoheadrightarrow \overline{\mathscr{E}}$. We have $\varphi_*(\mathscr{V}) = \overline{\mathscr{V}}$. If $\mathscr{D}^B_\varphi$ is the set of colors in O mapping dominantly to $\overline{O}$, then there is a canonical surjection $\varphi_* : \mathscr{D}^B \setminus \mathscr{D}^B_\varphi \twoheadrightarrow \overline{\mathscr{D}}^B$ such that $\overline{\varkappa}\varphi_* = \varphi_*\varkappa$.

Definition 15.9. A colored cone $(\mathscr{C}, \mathscr{R})$ in $\mathscr{E}$ *dominates* a colored cone $(\overline{\mathscr{C}}, \overline{\mathscr{R}})$ in $\overline{\mathscr{E}}$ if $\varphi_*(\operatorname{int}\mathscr{C}) \subseteq \operatorname{int}\overline{\mathscr{C}}$ and $\varphi_*(\mathscr{R} \setminus \mathscr{D}^B_\varphi) \subseteq \overline{\mathscr{R}}$. A colored fan $\mathscr{F}$ in $\mathscr{E}$ *dominates* a colored fan $\overline{\mathscr{F}}$ in $\overline{\mathscr{E}}$ if each cone from $\mathscr{F}$ dominates a cone from $\overline{\mathscr{F}}$.

The *support* of $\mathscr{F}$ is $\operatorname{Supp}\mathscr{F} = \bigcup_{(\mathscr{C},\mathscr{R})\in\mathscr{F}} \mathscr{C} \cap \mathscr{V}$. (Observe that $\{\mathscr{C} \cap \mathscr{V} \mid (\mathscr{C}, \mathscr{R}) \in \mathscr{F}\}$ is a polyhedral subdivision of $\operatorname{Supp}\mathscr{F}$.)

The next theorem is deduced from the results of 12.4.

Theorem 15.10 ([Kn2, 5.1–5.2]). *Let $X, \overline{X}$ be the embeddings of $O, \overline{O}$ determined by fans $\mathscr{F}, \overline{\mathscr{F}}$. Then φ extends to a morphism $X \to \overline{X}$ if and only if $\mathscr{F}$ dominates $\overline{\mathscr{F}}$. Furthermore, $\varphi : X \to \overline{X}$ is proper if and only if* $\operatorname{Supp}\mathscr{F} = \varphi_*^{-1}(\operatorname{Supp}\overline{\mathscr{F}}) \cap \mathscr{V}$.

Proof. If $\mathscr{O}_{X,Y}$ dominates $\mathscr{O}_{\overline{X},\overline{Y}}$, then clearly $\varphi_*(\mathscr{D}^B_Y \setminus \mathscr{D}^B_\varphi) \subseteq \overline{\mathscr{D}}^B_{\overline{Y}}$ and $\varphi_*(\mathscr{S}_Y) \subseteq \mathscr{S}_{\overline{Y}}$ or, equivalently, $(\mathscr{C}_Y, \mathscr{D}^B_Y)$ dominates $(\mathscr{C}_{\overline{Y}}, \overline{\mathscr{D}}^B_{\overline{Y}})$. Conversely, if $(\mathscr{C}_Y, \mathscr{D}^B_Y)$ dominates $(\mathscr{C}_{\overline{Y}}, \overline{\mathscr{D}}^B_{\overline{Y}})$ for some $Y \subseteq X$, $\overline{Y} \subseteq \overline{X}$, then $\mathscr{A} = \mathscr{A}(\mathscr{V}_Y, \mathscr{D}^B_Y) \supseteq \overline{\mathscr{A}} = \mathscr{A}(\overline{\mathscr{V}}_{\overline{Y}}, \overline{\mathscr{D}}^B_{\overline{Y}})$ (this is verified using Lemma 19.12) and $I_{\overline{Y}} = I_Y \cap \overline{\mathscr{A}}$, where $I_Y = \mathscr{A} \cap \mathfrak{m}_{X,Y}$ is defined in $\mathscr{A}$ by $v > 0$, $\forall v \in \mathscr{S}_Y$. Hence $\mathscr{O}_{X,Y}$ dominates $\mathscr{O}_{\overline{X},\overline{Y}}$. Thus the first assertion follows by Proposition 12.12.

A criterion of properness is a reformulation of Theorem 12.13. □

Overgroups of H can be classified in terms of the colored space.

Definition 15.11. A *colored subspace* of $\mathscr{E}$ is a pair $(\mathscr{E}_0, \mathscr{R}_0)$, where $\mathscr{R}_0 \subseteq \mathscr{D}^B$ and $\mathscr{E}_0 \subseteq \mathscr{E}$ is a subspace generated as a cone by $\varkappa(\mathscr{R}_0)$ and some vectors from $\mathscr{V}$.

For example, $(\mathscr{E}_\varphi, \mathscr{D}^B_\varphi)$ is a colored subspace, where $\mathscr{E}_\varphi = \operatorname{Ker}\varphi_*$ [Bri13, 3.4].

Theorem 15.12 ([Kn2, 5.4]). *The correspondence $\overline{H} \mapsto (\mathscr{E}_\varphi, \mathscr{D}^B_\varphi)$ is an order-preserving bijection between overgroups of H with $\overline{H}/H$ connected and colored subspaces of $\mathscr{E}$.*

Example 15.13. If $H = B$, then $\mathscr{E} = \mathscr{V} = 0$ and $\mathscr{D}^B$ is the set of Schubert divisors on G/B, which are in bijection with the simple roots. Hence an overgroup of B is determined by a subset of simple roots—a well-known classification of parabolics, cf. 2.6.

More generally, parabolic overgroups $P \supseteq H$ are in bijection with subsets $\mathscr{R}_0 \subseteq \mathscr{D}^B$ such that $\varkappa(\mathscr{R}_0) \cup \mathscr{V}$ generates $\mathscr{E}$ as a cone. Indeed, P is parabolic $\iff r(G/P) = 0 \iff \mathscr{E}(G/P) = 0 \iff \mathscr{E}_\varphi = \mathscr{E}$.

One may consider *generalized* colored fans, dropping the assumption that colored cones are strictly convex and their colors do not map to 0. (These are exactly the preimages of usual colored fans in quotients by colored subspaces.) Then there is a bijection between dominant separable G-maps $O \to X$ to normal G-varieties and generalized colored fans [Kn2, 5.5].

15.3 Orbits and Local Geometry. Now we derive some properties of G-orbits (due to Brion) and the local geometry of a spherical embedding.

Proposition 15.14. *Suppose that X is an embedding of O and $Y \subseteq X$ an irreducible G-subvariety. Then $c(Y) = 0$, $r(Y) = r(X) - \dim \mathscr{C}_Y = \operatorname{codim} \mathscr{C}_Y$, and $\Lambda(Y) = \mathscr{C}_Y^{\perp} \cap \Lambda(X)$ up to p-torsion.*

Proof. By Theorem 5.7, Y is spherical. By Lemma 5.8, for any $f \in \Bbbk(Y)^{(B)}$ there is $\widetilde{f} \in \Bbbk(X)^{(B)}$ such that $\widetilde{f}|_Y = f^q$, where q is a sufficiently big power of p. It remains to note that $\widetilde{f}$ is defined and nonzero on Y if and only if $\langle \mathscr{V}_Y \cap \mathscr{D}_Y^B, \widetilde{f} \rangle = 0$, i.e., the B-eigenweight of $\widetilde{f}$ lies in $\mathscr{C}_Y^{\perp}$. □

In this section we deal with normal spherical varieties. The following result concerns orbits in non-normal spherical varieties.

Proposition 15.15. *Let X be a locally linearizable spherical G-variety. Then the normalization map $\widetilde{X} \to X$ is bijective on the sets of G-orbits.*

Proof. Standard arguments reduce the claim to the case where X is affine. The G-orbits $Y \subseteq X$ are in bijection with the G-stable prime ideals $I = \mathscr{I}(Y) \lhd \Bbbk[X]$.

Note that I is uniquely determined by $I^U \lhd \Bbbk[X]^U$. Indeed, if $I \not\supseteq J \lhd \Bbbk[X]$ is another G-stable prime ideal, then we choose nonzero $f \in (J/I \cap J)^U \lhd (\Bbbk[X]/I)^U$ and lift a certain power f^q $(q = p^n)$ to $J^U \setminus I^U$ by Lemma D.1, thus proving that $I^U \not\supseteq J^U$.

Hence G-orbits in X are in bijection with certain T-stable prime ideals in $\Bbbk[X]^U$ or with certain T-orbits in a toric variety $X/\!/U$. By Lemma D.6, $\widetilde{X}/\!/U$ is the normalization of $X/\!/U$. This reduces the claim to the case $G = T$.

In this case $\Bbbk[X] = \Bbbk[\Lambda_+(X)]$ is a semigroup algebra. It is easy to prove that every T-stable prime ideal in $\Bbbk[X]$ is spanned by the weights in $\Lambda_+(X) \cap (\mathscr{C} \setminus \mathscr{C}')$, where $\mathscr{C}$ is the cone spanned by $\Lambda_+(X)$ and $\mathscr{C}'$ is a face of $\mathscr{C}$. But $\Lambda_+(\widetilde{X}) = \mathbb{Z}\Lambda_+(X) \cap \mathscr{C}$ spans the same cone $\mathscr{C}$, whence the claim. □

Remark 15.16. The proposition is false for non locally linearizable actions, as Example 4.2 shows.

The local structure of B-charts is given by Proposition 13.12. For a minimal B-chart $\mathring{X}_Y$, the description can be refined.

Let $P = P[\mathscr{D}^B \setminus \mathscr{D}_Y^B]$ and $P = P_{\mathrm{u}} \rtimes L$ be its Levi decomposition ($L \supseteq T$). Theorem 4.7 yields

Theorem 15.17. *There is a T-stable closed subvariety $Z \subseteq \mathring{X}_Y$ such that:*

(1) *The natural maps $P_{\mathrm{u}} \times Z \to \mathring{X}_Y$ and $Z \to \mathring{X}/P_{\mathrm{u}}$ are finite and surjective.*

(2) *Put* $\mathring{Y} = Y \cap \mathring{X}_Y$. *Then* $\mathring{Y}/P_u \simeq L/L_0$, *where* $L_0 \supseteq L'$.
(3) *In characteristic zero,* Z *is* L*-stable,* $\mathring{X}_Y \simeq P *_L Z \simeq P_u \times Z$, $Y \cap Z \simeq L/L_0$, *and there exists an* L_0*-stable subvariety* $Z_0 \subseteq Z$ *transversal to* $Y \cap Z$ *at a fixed point* z *and such that* $Z \simeq L *_{L_0} Z_0$. *The varieties* Z *and* Z_0 *are affine and spherical, and* $r(Z) = r(O)$, $r(Z_0) = \dim \mathscr{C}_Y$.

The isomorphism $Z \simeq L *_{L_0} Z_0$ stems, e.g., from Luna's slice theorem [Lu2], [PV, Th. 6.1], or is proved directly: since L/L_0 is a torus, $\Bbbk[Y \cap Z]$ can be lifted to $\Bbbk[Z]$ as an L-stable subalgebra, whence an equivariant retraction $Z \to Y \cap Z \simeq L/L_0$ with a fiber Z_0.

Corollary 15.18. $\dim Y = \operatorname{codim} \mathscr{C}_Y + \dim P_u$

Remark 15.19. In characteristic zero, there is a bijection $f \leftrightarrow f|_Z$ between B-eigenfunctions on X and $(B \cap L)$-eigenfunctions on Z, which preserve the order along a divisor. Hence $\mathscr{C}_Y = \mathscr{C}_{Y \cap Z}$ and $\mathscr{D}_Y^B \supseteq \mathscr{D}_{Y \cap Z}^B$. However some colors on X may become L-stable divisors on Z ("a discoloration").

Theorem 15.20. *In characteristic zero, all irreducible* G*-subvarieties* $Y \subseteq X$ *are normal and have rational singularities (in particular, they are Cohen–Macaulay).*

Proof. By the local structure theorem, we may assume that X is affine. Then $X /\!/ U$ is an affine toric variety and $Y /\!/ U$ a T-stable subvariety. It is well known [Ful2, 3.1, 3.5] that $Y /\!/ U$ is a normal toric variety and has rational singularities. By Theorem D.5(3), the same is true for Y. □

A spherical embedding defined by a fan whose colored cones have no colors is called *toroidal*. In particular, toric varieties are toroidal. Conversely, the local structure theorem readily implies that toroidal varieties are "locally toric" (Theorem 29.1). This is the reason for most nice geometric properties which distinguish toroidal varieties among arbitrary spherical varieties. Toroidal varieties are discussed in §29.

16 Case of Complexity 1

16.1 Generically Transitive and One-parametric Cases. Here we obtain the classification of normal G-models in the case $c(K) = 1$. This case splits in two subcases:

(1) *Generically transitive case*: $d_G(K) = 0$. Here any G-model contains a dense G-orbit O of complexity 1.
(2) *One-parametric case*: $d_G(K) = 1$. Here general G-orbits in any G-model are spherical and form a one-parameter family. (In fact, all G-orbits are spherical by Proposition 5.13.)

We are interested mainly in the generically transitive case. However the one-parametric case might be of interest, e.g., in studying deformations of spherical homogeneous spaces and their embeddings. There are differences between these two cases (e.g., in the description of colors), but the description of G-models is uniform [Tim2].

16.2 Hyperspace. First we describe the colored equipment.

Since $c(K) = 1$, there is a (unique) non-singular projective curve C such that $K^B = \Bbbk(C)$. General B-orbits on a G-model of K are parameterized by an open subset of C. In the generically transitive case, $K \subseteq \Bbbk(G)$ is unirational, because G is a rational variety, which is proved by considering the "big cell" in G. Whence $C = \mathbb{P}^1$ by the Lüroth theorem.

Definition 16.1. For any $x \in C$ consider the half-space $\mathscr{E}_{x,+} = \mathbb{Q}_+ \times \mathscr{E}$. The *hyperspace* (of K) is the union $\breve{\mathscr{E}}$ of all $\mathscr{E}_{x,+}$ glued together along their common boundary hyperplane $\mathscr{E}$, called the *center* of $\breve{\mathscr{E}}$. More formally,

$$\breve{\mathscr{E}} = \bigsqcup_{x \in C} \{x\} \times \mathscr{E}_{x,+} \Big/ \sim$$

where $(x,h,\ell) \sim (x',h',\ell')$ if and only if $x = x'$, $h = h'$, $\ell = \ell'$ or $h = h' = 0$, $\ell = \ell'$.

Since Λ is a free Abelian group, the exact sequence (13.1) splits. Fix a splitting $\mathbf{f} : \Lambda \to K^{(B)}$, $\lambda \mapsto \mathbf{f}_\lambda$.

If v is a geometric valuation of K, then $v|_{K^{(B)}}$ is determined by a triple (x,h,ℓ), where $x \in C$, $h \in \mathbb{Q}_+$ satisfy $v|_{K^B} = hv_x$ and $\ell = v|_{\mathbf{f}(\Lambda)} \in \mathscr{E} = \operatorname{Hom}(\Lambda, \mathbb{Q})$. Therefore $v|_{K^{(B)}} \in \breve{\mathscr{E}}$. Thus $\mathscr{V}$ is embedded in $\breve{\mathscr{E}}$, and we have a map $\varkappa : \mathscr{D}^B \to \breve{\mathscr{E}}$ (restriction to $K^{(B)}$). We say that $(\breve{\mathscr{E}}, \mathscr{V}, \mathscr{D}^B, \varkappa)$ is the *colored hyperspace*. The valuation v and the respective divisor are called *central* if $v|_{K^{(B)}} \in \mathscr{E}$.

By Theorems 20.3, 21.1, and Corollary 22.14, $\mathscr{V}_x = \mathscr{V} \cap \mathscr{E}_{x,+}$ is a convex solid polyhedral cone in $\mathscr{E}_{x,+}$, simplicial in characteristic 0, and $\mathscr{Z} = \mathscr{V} \cap \mathscr{E}$ is a convex solid cone in $\mathscr{E}$.

By Corollary 20.5, the set $\mathscr{D}^B_x = \mathscr{D}^B \cap \varkappa^{-1}(\mathscr{E}_{x,+})$ is finite for each $x \in C$. In particular, the set of central colors is finite.

For an arbitrary G-model X, consider the rational B-quotient map $\pi : X \dashrightarrow C$ separating general B-orbits. Thus general B-orbits determine a one-parameter family of B-stable prime divisors on X parameterized by an open subset $\mathring{C} \subseteq C$. Decreasing $\mathring{C}$ if necessary, we may assume that these divisors $D_x = \pi^*(x)$ are pullbacks of points $x \in \mathring{C}$ and do not occur in $\operatorname{div} \mathbf{f}_\lambda$ ($\lambda \in \Lambda$). Then $\varkappa(D_x) = \varepsilon_x := (1,0) \in \mathscr{E}_{x,+}$. Clearly, $\{D_x \mid x \in \mathring{C}\} \in \mathbf{CD}$.

In the generically transitive case, $\pi : O \dashrightarrow \mathbb{P}^1$ is determined by a one-dimensional linear system of colors. In other words, there is a G-line bundle $\mathscr{L}$ on O and a two-dimensional subspace M of B-eigensections of $\mathscr{L}$ which defines this linear system. Elements of M are homogeneous coordinates on $\mathbb{P}^1 = \mathbb{P}(M^*)$. If $O = G/H$, then $\mathscr{L} = \mathscr{L}(\chi_0)$ and $M = \Bbbk[G]^{(B \times H)}_{(\lambda_0, -\chi_0)}$ for some $\lambda_0 \in \mathfrak{X}_+$, $\chi_0 \in \mathfrak{X}(H)$. Except for finitely many lines, M consists of indecomposable elements corresponding to generic colors.

Indecomposable elements of M and the respective colors are called *regular*. A regular color $D_x = \pi^*(x)$ is represented in $\breve{\mathscr{E}}$ by a vector $(1,\ell) \in \mathscr{E}_{x,+}$, and $\ell = 0$ for all but finitely many x.

In addition, there is a finite set of one-dimensional subspaces $\Bbbk[G]^{(B\times H)}_{(\lambda_i,-\chi_i)}$, $i = 1,\dots,s$, consisting of indecomposable elements that correspond to other colors. If $\eta_i \in \Bbbk[G]^{(B\times H)}_{(\lambda_i,-\chi_i)}$ divides some $\eta \in \Bbbk[G]^{(B\times H)}_{(\lambda_0,-\chi_0)}$, then $D_i = \operatorname{div}\eta_i$ is represented in $\breve{\mathscr{E}}$ by $(h_i,\ell_i) \in \mathscr{E}_{x,+}$, where $\operatorname{div}\eta = \pi^*(x)$ and h_i is the multiplicity of η_i in η (or of D_i in $\pi^*(x)$). Such η_i and D_i are called *subregular*. Other η_i are called *central* (since D_i are central).

The above description of colors allows the computation of multiplicities in the spaces of global sections of G-line bundles on $O = G/H$.

Proposition 16.2 ([Tim5, §3]). *For any $\chi \in \mathfrak{X}(H)$ and $\lambda \in \mathfrak{X}_+$, let k_0 be the maximal integer such that $(\lambda,-\chi) = \sum_{i=0}^{s} k_i(\lambda_i,-\chi_i) + (\mu,-\mu)$, where $k_i \geq 0$ and $\mu \in \mathfrak{X}(G)$. Then $m_\lambda(\mathscr{L}(\chi)) = k_0 + 1$.*

Proof. Every $\eta \in \mathrm{H}^0(O,\mathscr{L}(\chi))^{(B)}_\lambda = \Bbbk[G]^{(B\times H)}_{(\lambda,-\chi)}$ decomposes uniquely as

$$\eta = \sigma_1 \cdots \sigma_{k_0} \eta_1^{k_1} \cdots \eta_s^{k_s} \mu^{-1},$$

where $\sigma_j \in \Bbbk[G]^{(B\times H)}_{(\lambda_0,-\chi_0)}$. Therefore $\dim \Bbbk[G]^{(B\times H)}_{(\lambda,-\chi)} = \dim \Bbbk[G]^{(B\times H)}_{(k_0\lambda_0,-k_0\chi_0)}$, and

$$\Bbbk[G]^{(B\times H)}_{(k_0\lambda_0,-k_0\chi_0)} = \mathrm{S}^{k_0}\Bbbk[G]^{(B\times H)}_{(\lambda_0,-\chi_0)}$$

has dimension $k_0 + 1$. □

Corollary 16.3 ([Pan2, 1.2]). *If $\mathfrak{X}(H) = 0$, then $m_\lambda(O) = k_0 + 1$, where $k_0 = \max\{k \mid \lambda - k\lambda_0 \in \Lambda_+(O)\}$.*

In the one-parametric case, generic B-stable prime divisors are G-stable, whence $\varepsilon_x \in \mathscr{V}_x$ for $x \in \mathring{C}$.

Lemma 16.4. *In the one-parametric case, all colors are central.*

Proof. If D is a non-central color, then $v_D(f) > 0$ for some $f \in K^B = K^G$. Hence D is G-stable, a contradiction. □

Remark 16.5. Since a splitting $\mathbf{f} : \Lambda \to K^{(B)}$ is not uniquely defined, the maps $\mathscr{V} \hookrightarrow \breve{\mathscr{E}}$, $\varkappa : \mathscr{D}^B \to \breve{\mathscr{E}}$ are not canonical. But the change of splitting is easily controlled. If $\mathbf{f}'$ is another splitting, then passing from $\mathbf{f}$ to $\mathbf{f}'$ produces a shift of each $\mathscr{E}_{x,+}$: $h' = h$ and $\ell' = \ell + h\ell_x$, where $\langle \ell_x, \lambda \rangle = v_x(\mathbf{f}'_\lambda/\mathbf{f}_\lambda)$. The shifting vectors $\ell_x \in \Lambda^* \subset \mathscr{E}$ have the property that $\sum_{x\in C}\langle \ell_x,\lambda\rangle x$ is a principal divisor on C for each $\lambda \in \Lambda$; in particular, $\sum_{x\in C}\ell_x = 0$. Conversely, any collection of integral shifting vectors $\ell_x \in \Lambda^*$ such that $\sum_{x\in C}\langle \ell_x,\lambda\rangle x$ is a principal divisor for each λ defines a change of splitting. For $C = \mathbb{P}^1$, it suffices to have $\sum \ell_x = 0$.

Now we describe the dual object to the hyperspace.

Definition 16.6. A *linear functional* on the hyperspace is a function φ on $\check{\mathscr{E}}$ such that $\varphi_x = \varphi|_{\mathscr{E}_{x,+}}$ is a $\mathbb{Q}$-linear functional for any $x \in C$ and $\sum_{x\in C}\langle \varepsilon_x, \varphi_x\rangle = 0$. A linear functional φ is *admissible* if $N\sum_{x\in C}\langle \varepsilon_x, \varphi_x\rangle x$ is a principal divisor on C for some $N \in \mathbb{N}$. Denote by $\check{\mathscr{E}}^*$ the space of linear functionals and by $\check{\mathscr{E}}^*_{\mathrm{ad}}$ the subspace of admissible functionals on $\check{\mathscr{E}}$. The set $\operatorname{Ker}\varphi = \bigcup_{x\in C}\operatorname{Ker}\varphi_x$ is called the *kernel* of $\varphi \in \check{\mathscr{E}}^*$.

If $C = \mathbb{P}^1$, then any linear functional is admissible. Any $f = f_0\mathbf{f}_\lambda \in K^{(B)}$, $f_0 \in K^B$, $\lambda \in \Lambda$, determines an admissible linear functional φ by means of $\langle q, \varphi_x\rangle = hv_x(f_0) + \langle \ell, \lambda\rangle$, $\forall q = (h,\ell) \in \mathscr{E}_{x,+}$, and f is determined by φ uniquely up to a scalar multiple. Conversely, a multiple of any admissible functional is determined by a B-eigenfunction.

Any collection of linear functionals φ_x on $\mathscr{E}_{x,+}$ whose restrictions to $\mathscr{E}$ coincide can be deformed to an admissible functional by a "small variation".

Lemma 16.7. *Let φ_x be linear functionals on $\mathscr{E}_{x,+}$ such that $\varphi_x|_{\mathscr{E}}$ does not depend on $x \in C$, $\langle \varepsilon_x, \varphi_x\rangle = 0$ for all but finitely many x, and $\sum\langle \varepsilon_x, \varphi_x\rangle < 0$. Then for any finite subset $C_0 \subset C$ and any $\varepsilon > 0$ there exists $\psi \in \check{\mathscr{E}}^*_{\mathrm{ad}}$ such that $\psi_x \geq \varphi_x$ on $\mathscr{E}_{x,+}$ for all $x \in C$ with the equality for $x \in C_0$ (in particular, $\psi|_{\mathscr{E}} = \varphi_x|_{\mathscr{E}}$) and $|\langle \varepsilon_x, \psi_x\rangle - \langle \varepsilon_x, \varphi_x\rangle| < \varepsilon$.*

Proof. The divisor $-N\sum\langle \varepsilon_x, \varphi_x\rangle x$ is very ample on C for N sufficiently large. Moving the respective hyperplane section of C, we obtain an equivalent very ample divisor of the form $\sum n_x x$, where $n_x \in \{0,1\}$ and $n_x = 0$ whenever $x \in C_0$. Then the divisor $\sum(N\langle \varepsilon_x, \varphi_x\rangle + n_x)x$ is principal, and $\psi \in \check{\mathscr{E}}^*_{\mathrm{ad}}$ defined by $\psi|_{\mathscr{E}} = \varphi_x|_{\mathscr{E}}$, $\langle \varepsilon_x, \psi_x\rangle = \langle \varepsilon_x, \varphi_x\rangle + n_x/N$, is the desired admissible functional. □

16.3 Hypercones. For reorganizing colored data in a way similar to the spherical case, we need some notions from the geometry of the hyperspace.

Definition 16.8. A *cone* in $\check{\mathscr{E}}$ is a cone in some $\mathscr{E}_{x,+}$.

A *hypercone* in $\check{\mathscr{E}}$ is a union $\mathscr{C} = \bigcup_{x\in C}\mathscr{C}_x$ of finitely generated convex cones $\mathscr{C}_x = \mathscr{C} \cap \mathscr{E}_{x,+}$ such that

(1) $\mathscr{C}_x = \mathscr{K} + \mathbb{Q}_+\varepsilon_x$ for all but finitely many x, where $\mathscr{K} = \mathscr{C} \cap \mathscr{E}$.
(2) Either (A) $\exists x \in C:\ \mathscr{C}_x = \mathscr{K}$,
or (B) $\emptyset \neq \mathscr{B} := \sum \mathscr{B}_x \subseteq \mathscr{K}$, where $\varepsilon_x + \mathscr{B}_x = \mathscr{C}_x \cap (\varepsilon_x + \mathscr{E})$.

The hypercone is *strictly convex* if all $\mathscr{C}_x$ are so and $\mathscr{B} \not\ni 0$.

Remark 16.9. The Minkowski sum $\sum \mathscr{B}_x$ of infinitely many polyhedral domains $\mathscr{B}_x$ is defined as the set of all sums $\sum b_x$, $b_x \in \mathscr{B}_x$, that make sense, i.e., $b_x = 0$ for all but finitely many x. In particular, for a hypercone of type A, there exists $x \in C$ such that $\mathscr{B}_x = \emptyset$, whence $\mathscr{B} = \emptyset$.

Definition 16.10. Suppose that $Q \subseteq \check{\mathscr{E}}$ differs from $\{\varepsilon_x \mid x \in \mathring{C}\}$ by finitely many elements. Let $\varepsilon_x + \mathscr{P}_x$ be the convex hull of the intersection points of $\varepsilon_x + \mathscr{E}$ with the rays $\mathbb{Q}_+q$, $q \in Q$. We say that the hypercone $\mathscr{C} = \mathscr{C}(Q)$, where $\mathscr{C}_x$ are generated by $Q \cap \mathscr{E}_{x,+}$ and by $\mathscr{P} := \sum \mathscr{P}_x$, is *generated* by Q.

Fig. 16.1 Hypercones

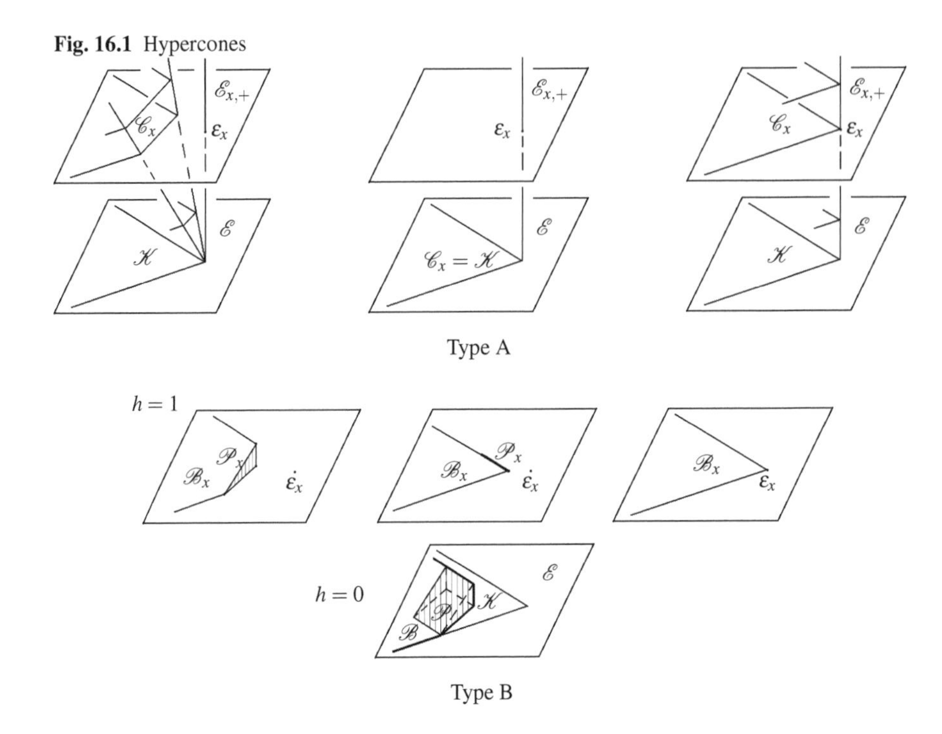

Remark 16.11. We have $\mathscr{B}_x = \mathscr{P}_x + \mathscr{K}$ and $\mathscr{B} = \mathscr{P} + \mathscr{K}$.

Definition 16.12. For a hypercone $\mathscr{C}$ of type B, we define its *interior* $\operatorname{int}\mathscr{C} = \bigcup_{x \in C} \operatorname{int}\mathscr{C}_x \cup \operatorname{int}\mathscr{K}$.

A *face* of a hypercone $\mathscr{C}$ is a face $\mathscr{C}'$ of some $\mathscr{C}_x$ such that $\mathscr{C}' \cap \mathscr{B} = \emptyset$.

A *hyperface* of $\mathscr{C}$ is a hypercone $\mathscr{C}' = \mathscr{C} \cap \operatorname{Ker}\varphi$, where $\varphi \in \check{\mathscr{E}}^*$, $\langle \mathscr{C}, \varphi \rangle \geq 0$. Such φ is called a *supporting functional* of the hyperface $\mathscr{C}'$. The hyperface $\mathscr{C}'$ is *admissible* if φ is admissible.

A hypercone is *admissible* if all supporting functionals of the hyperfaces of type B are admissible.

Remark 16.13. A hyperface $\mathscr{C}' \subseteq \mathscr{C}$ is of type B if and only if $\mathscr{C}' \cap \mathscr{B} \neq \emptyset$. Indeed, $\langle \varepsilon_x + \mathscr{B}_x, \varphi_x \rangle \geq 0$ $(\forall x \in C)$ and $\mathscr{C}'_x \not\subseteq \mathscr{E}$ $(\forall x \in C)$ $\iff$ $\langle \varepsilon_x + \mathscr{B}_x, \varphi_x \rangle \ni 0$ $(\forall x \in C)$ $\iff$ $\sum \langle \varepsilon_x + \mathscr{B}_x, \varphi_x \rangle = \sum \langle \varepsilon_x, \varphi_x \rangle + \sum \langle \mathscr{B}_x, \varphi_x \rangle = \langle \mathscr{B}, \varphi \rangle \ni 0$.

Remark 16.14. If $C = \mathbb{P}^1$, then every hypercone is admissible.

Properties of hypercones are similar to properties of convex polyhedral cones. Let $\mathscr{C}$ be a hypercone. There is a separation property:

Lemma 16.15. (1) $q \notin \mathscr{C} \implies \exists \varphi \in \check{\mathscr{E}}^* : \langle \mathscr{C}, \varphi \rangle \geq 0,\ \langle q, \varphi \rangle < 0.$

(2) *If $\mathscr{C}$ is strictly convex, then one may assume that $\varphi \in \check{\mathscr{E}}^*_{\mathrm{ad}}$ and $\langle \mathscr{C}_x \setminus \{0\}, \varphi_x \rangle > 0$ for any given finite set of $x \in C$.*

Proof. (1) If $q \in \mathscr{E}_{y,+}$, then we construct a collection of functionals φ_x on $\mathscr{E}_{x,+}$ such that $\varphi_x|_{\mathscr{E}} = \varphi_y|_{\mathscr{E}}$, $\langle q, \varphi_y \rangle < 0$, $\langle \mathscr{C}_x, \varphi_x \rangle \geq 0$ for all $x \in C$ and the equality is reached on $\mathscr{C}_x \setminus \mathscr{K}$ whenever $\mathscr{C}_x \neq \mathscr{K}$. If $\mathscr{C}$ is of type A, then we decrease φ_x for $x \in C$ such that $\mathscr{C}_x = \mathscr{K}$ and obtain $\sum \langle \varepsilon_x, \varphi_x \rangle \leq 0$. If $\mathscr{C}$ is of type B, then $\sum \langle \varepsilon_x + \mathscr{B}_x, \varphi_x \rangle = \sum \langle \varepsilon_x, \varphi_x \rangle + \langle \mathscr{B}, \varphi_y \rangle \geq 0$ and the equality is reached. But $\langle \mathscr{B}, \varphi_y \rangle \geq 0 \implies \sum \langle \varepsilon_x, \varphi_x \rangle \leq 0$. It remains to modify φ_x by a small variation (Lemma 16.7) if necessary.
(2) In the proof of (1), we may assume that $\langle \mathscr{K} \setminus \{0\}, \varphi_y \rangle > 0$. If $\mathscr{C}$ is of type A, then we decrease φ_x for $x \in C$ such that $\mathscr{C}_x = \mathscr{K}$ and obtain $\sum \langle \varepsilon_x, \varphi_x \rangle < 0$. If $\mathscr{C}$ is of type B, then $\langle \mathscr{B}, \varphi_y \rangle > 0 \implies \sum \langle \varepsilon_x, \varphi_x \rangle < 0$. Now we may increase finitely many φ_x to have $\langle \mathscr{C}_x \setminus \{0\}, \varphi_x \rangle > 0$. □

This implies a dual characterization of a hypercone:

Lemma 16.16. *For a (strictly convex) hypercone $\mathscr{C} = \mathscr{C}(Q)$, $q \in \mathscr{C}$ if and only if $\langle Q, \varphi \rangle \geq 0 \implies \langle q, \varphi \rangle \geq 0$, $\forall \varphi \in \check{\mathscr{E}}^*$ ($\forall \varphi \in \check{\mathscr{E}}^*_{\mathrm{ad}}$).*

Proof. $\langle Q, \varphi \rangle \geq 0 \implies \forall x \in C\colon \langle \varepsilon_x + \mathscr{P}_x, \varphi_x \rangle \geq 0 \implies \langle \mathscr{P}, \varphi \rangle \geq 0 \implies \langle \mathscr{C}, \varphi \rangle \geq 0$. Lemma 16.15 completes the proof. □

For any $v \in \mathscr{C}$, there is a unique face or hyperface of type B containing v in its interior.

Lemma 16.17. *The face or (admissible) hyperface $\mathscr{C}' \subseteq \mathscr{C}$ such that $v \in \operatorname{int} \mathscr{C}'$ is the intersection of (admissible) hyperfaces of $\mathscr{C}$ containing v.*

Proof. If $\langle \mathscr{C}, \varphi \rangle \geq 0$ and $\langle v, \varphi \rangle = 0$, then $\langle \mathscr{C}', \varphi \rangle = 0$. (If $\mathscr{C}'$ is a hyperface, then $\langle \mathscr{K}', \varphi \rangle = 0 \implies \langle \varepsilon_x + \mathscr{B}'_x, \varphi_x \rangle = 0 \implies \langle \mathscr{C}'_x, \varphi_x \rangle = 0$ for all $x \in C$.) Conversely, if $\mathscr{C}' \subseteq \mathscr{C}_y$ is a face and $q \in \mathscr{C} \setminus \mathscr{C}'$, then we construct an (admissible) functional φ such that $\langle \mathscr{C}, \varphi \rangle \geq 0$, $\langle \mathscr{C}', \varphi \rangle = 0$, $\langle q, \varphi \rangle > 0$ as follows. Take φ_y on $\mathscr{E}_{y,+}$ such that $\langle \mathscr{C}_y, \varphi_y \rangle \geq 0$ and $\mathscr{C}' = \operatorname{Ker} \varphi_y \cap \mathscr{C}$. We may include φ_y in a collection of functionals φ_x on $\mathscr{E}_{x,+}$ such that $\varphi_x|_{\mathscr{E}} = \varphi_y|_{\mathscr{E}}$, $\langle \mathscr{C}_x, \varphi_x \rangle \geq 0$ and the equality is reached on $\mathscr{C}_x \setminus \mathscr{K}$ whenever $\mathscr{C}_x \neq \mathscr{K}$. (To this end, we have to decrease φ_y in case $\mathscr{C}' \subseteq \mathscr{K} \neq \mathscr{C}_y$.) As in the proof of Lemma 16.15(2), we may assume that $\sum \langle \varepsilon_x, \varphi_x \rangle < 0$: if $\mathscr{C}$ is of type A, then we decrease φ_x for $x \in C$ such that $\mathscr{C}_x = \mathscr{K}$; otherwise $\mathscr{C}' \cap \mathscr{B} = \emptyset \implies \langle \mathscr{B}, \varphi_y \rangle > 0 \implies \sum \langle \varepsilon_x, \varphi_x \rangle < 0$. Then we increase some φ_x to obtain $\langle q, \varphi \rangle > 0$ and apply Lemma 16.7. □

16.4 Colored Data. Now let $(\mathscr{W}, \mathscr{R})$ be colored data from **CD** and consider the hypercone $\mathscr{C} = \mathscr{C}(\mathscr{W}, \mathscr{R})$ generated by $\mathscr{W} \cup \varkappa(\mathscr{R})$.

Condition (C) means that $\mathscr{C}$ is strictly convex and $\varkappa(\mathscr{R}) \not\ni 0$ (Lemma 16.15(2)). We assume it in the sequel.

Condition (W) means that the elements of $\mathscr{W}$ are exactly the generators of those edges of $\mathscr{C}$ that do not intersect $\varkappa(\mathscr{R})$. (Indeed, (W) $\iff w \notin \mathscr{C}(\mathscr{W} \setminus \{w\}, \mathscr{R})$ by Lemma 16.16.)

Condition (F) means that $\mathscr{C}$ is admissible. (This is non-trivial, see [Tim2, Pr. 4.1].)

Condition (V) means that $v \in \mathscr{C}$ (Lemma 16.16).

Conditions (V′) and (D′) say that $\mathscr{V}_Y$ and $\mathscr{D}_Y^B$ consist of those elements of $\mathscr{W}$ and $\mathscr{R}$ which lie in the face or hyperface (of type B) $\mathscr{C}_Y = \mathscr{C}(\mathscr{V}_Y, \mathscr{D}_Y^B) \subseteq \mathscr{C}$ such that $v \in \operatorname{int} \mathscr{C}_Y$ (Lemma 16.17).

Condition (S) says that $v \in \mathscr{V} \cap \operatorname{int} \mathscr{C}_Y$ (Lemma 16.17).

Definition 16.18. A *colored hypercone* is a pair $(\mathscr{C}, \mathscr{R})$, where $\mathscr{R} \subseteq \mathscr{D}^B$, $\varkappa(\mathscr{R}) \not\ni 0$, and $\mathscr{C}$ is a strictly convex hypercone generated by $\varkappa(\mathscr{R})$ and $\mathscr{W} \subseteq \mathscr{V}$.

A colored hypercone $(\mathscr{C}, \mathscr{R})$ (of type B) is *supported* if $\operatorname{int} \mathscr{C} \cap \mathscr{V} \neq \emptyset$.

A *(hyper)face* of $(\mathscr{C}, \mathscr{R})$ is a colored (hyper)cone $(\mathscr{C}', \mathscr{R}')$, where $\mathscr{C}'$ is a (hyper)face of $\mathscr{C}$ and $\mathscr{R}' = \mathscr{R} \cap \varkappa^{-1}(\mathscr{C}')$.

A *colored hyperfan* is a set of supported colored cones and hypercones of type B whose interiors are disjoint inside $\mathscr{V}$ and which is obtained as the set of all supported (hyper)faces of finitely many admissible colored hypercones.

The next theorem follows from the above discussion and the results of 14.2.

Theorem 16.19. (1) *B-charts are in bijection with admissible colored hypercones in* $\breve{\mathscr{E}}$.
(2) *G-germs are in bijection with supported colored cones and admissible supported colored hypercones of type B. If* $Y_1, Y_2 \subseteq X$ *are irreducible G-subvarieties in a G-model, then* $Y_1 \subseteq \overline{Y_2}$ *if and only if* $(\mathscr{C}_{Y_2}, \mathscr{D}_{Y_2}^B)$ *is a (hyper)face of* $(\mathscr{C}_{Y_1}, \mathscr{D}_{Y_1}^B)$.
(3) *Normal G-models are in bijection with colored hyperfans.*

Corollaries 15.5–15.7 are easily generalized to the case of complexity 1.

Remark 16.20. Let $\mathring{X}$ be a B-chart defined by a colored hypercone $(\mathscr{C}, \mathscr{R})$. Then $\Bbbk[\mathring{X}]^B = \Bbbk$ if and only if $\mathscr{C}$ is of type B: if $f \in \Bbbk[\mathring{X}]^B$, then the respective $\varphi \in \breve{\mathscr{E}}_{\mathrm{ad}}^{\circ *}$ must be zero on $\mathscr{E}$ and non-negative on $\mathscr{C}$. Thus we have two types of B-charts:

(A) $\Bbbk[\mathring{X}]^B \neq \Bbbk$, or $\mathscr{C}$ is of type A.
(B) $\Bbbk[\mathring{X}]^B = \Bbbk$, or $\mathscr{C}$ is of type B.

There are two types of G-germs:

(A) $\mathscr{C}(\mathscr{V}_Y, \mathscr{D}_Y^B)$ is a colored cone.
(B) $\mathscr{C}(\mathscr{V}_Y, \mathscr{D}_Y^B)$ is a colored hypercone.

A G-germ is of type A if and only if $\mathscr{V}_Y, \mathscr{D}_Y^B$ are finite, and of type B if and only if it has a minimal B-chart.

Example 16.21. Suppose that $G = B = T$ is a torus. We may assume (after factoring out by the kernel of the action) that the stabilizer of general position for any T-model is trivial. The birational type of the action is trivial, i.e., any T-model is birationally isomorphic to $T \times C$, cf. Example 5.4. It follows that $\mathscr{E} = \operatorname{Hom}(\mathfrak{X}(T), \mathbb{Q})$, $\mathscr{V} = \breve{\mathscr{E}}$, $\mathscr{D}^B = \mathscr{D}^B(T) = \emptyset$, cf. Example 20.2.

A T-model is given by a set of cones and admissible hypercones of type B with disjoint interiors which consists of all faces and hyperfaces of type B of finitely many admissible hypercones. (The word "colored" is needless here, since there are no colors.) A T-chart $\mathring{X}$ is of type A (type B) if and only if $\mathring{X}/\!/T \subset C$ is an open subset ($\mathring{X}/\!/T$ is a point).

If all germs of a T-model X are of type A, then all T-charts of X are of type (A) and the quotient morphisms of the T-charts may be glued together into a regular map $\pi : X \to C$ separating T-orbits of general position. Such T-models were classified by Mumford in [KKMS, Ch. IV] in the framework of the theory of toroidal embeddings (for this theory see [KKMS, Ch. II]). The hyperfan of X is a union of fans $\mathscr{F}_x$ in $\mathscr{E}_{x,+}$ having common central part $\mathscr{F} = \{\mathscr{C} \in \mathscr{F}_x \mid \mathscr{C} \subseteq \mathscr{E}\}$ and such that $\mathscr{F}_x$ is a cylinder over $\mathscr{F}$ for $x \neq x_1, \dots, x_s$ (finitely many exceptional points).

It is proved in [KKMS, Ch. IV] that C is covered by open neighborhoods C_i of x_i such that $\pi^{-1}(C_i) \simeq C_i \times_{\mathbb{A}^1} X_i$, where $\nu_i : C_i \to \mathbb{A}^1$ are étale maps such that $\nu_i^{-1}(0) = \{x_i\}$ and X_i are toric $(T \times \Bbbk^\times)$-varieties with fans $\mathscr{F}_{x_i}$ mapping $\Bbbk^\times$-equivariantly onto $\mathbb{A}^1$.

Torus actions of complexity one were studied and classified in [Tim7].

Remark 16.22. The admissibility of a hypercone is essential for condition (F) as the following example [Kn4] shows.

Let C be a smooth projective curve of genus $g > 0$ and let $\delta_i = \sum n_{ix} x$, $i = 1, 2$, be divisors on C having infinite order in $\operatorname{Pic} C$ and such that $\deg \delta_1 = 0$, $\deg \delta_2 \geq g$. Put $\mathscr{L}_i = \mathscr{O}_C(\delta_i)$.

The total space X of $\mathscr{L}_1^* \oplus \mathscr{L}_2^*$ is a $T = (\Bbbk^\times)^2$-model, where the factors $\Bbbk^\times$ act on $\mathscr{L}_i^*$ by homotheties. There are the following T-germs in X: the divisors D_i, D_x ($i = 1, 2$, $x \in C$), where D_i is the total space of $\mathscr{L}_j^*$, $\{i, j\} = \{1, 2\}$, and D_x is the fiber of $X \to C$ over $\{x\}$; $Y_{ix} = D_i \cap D_x$; $C = D_1 \cap D_2$; $\{x\} = D_1 \cap D_2 \cap D_x$.

Let $\mathbf{f}_i$ be a rational section of $\mathscr{L}_i$ such that $\operatorname{div} \mathbf{f}_i = \delta_i$. Then $\Lambda(X)$ is generated by the T-weights ω_1, ω_2 of $\mathbf{f}_1, \mathbf{f}_2$. If w_i, w_x are the T-valuations corresponding to D_i, D_x, then $w_1 = (0, 1, 0)$, $w_2 = (0, 0, 1)$, $w_x = (1, n_{1x}, n_{2x})$ in the basis $\varepsilon_x, \omega_1^\vee, \omega_2^\vee$, where $\omega_i^\vee$ are the dual coweights to ω_i.

The algebra $\Bbbk[X]$ is bigraded by the T-action: $\Bbbk[X]_{m,n} = \mathrm{H}^0(C, \mathscr{L}_1^{\otimes m} \otimes \mathscr{L}_2^{\otimes n}) \neq 0$ if and only if either $m \geq 0$ and $n > 0$ or $m = n = 0$. Hence $\Lambda_+(X)$ is not finitely generated, neither is $\Bbbk[X]$. The reason is that $\Bbbk[X]$ is defined by a non-admissible hypercone $\mathscr{C} = \bigcup \mathscr{C}_{\{x\}}$. Indeed, a hyperface $\mathscr{C}' = \bigcup \mathscr{C}_{Y_{2x}}$ is defined by an inadmissible supporting functional $\varphi \in \check{\mathscr{E}}^*$ such that $\langle \omega_1^\vee, \varphi \rangle = 1$, $\langle \omega_2^\vee, \varphi \rangle = 0$, $\langle \varepsilon_x, \varphi \rangle = -n_{1x}$, and no multiple of $-\delta_1 = \sum \langle \varepsilon_x, \varphi \rangle x$ is a principal divisor. See Fig. 16.2.

Fig. 16.2

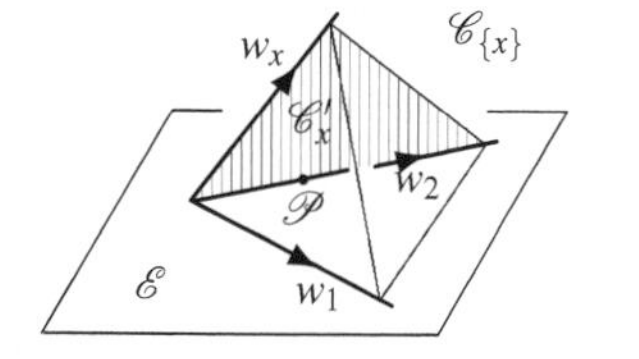

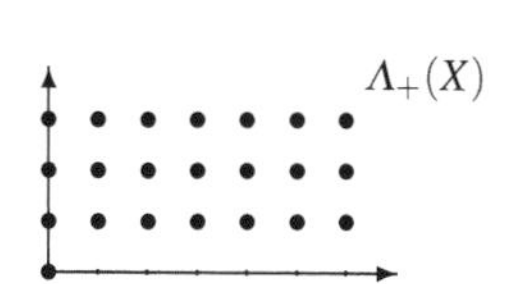

16.5 Examples.

Example 16.23 (SL_2-embeddings). Suppose that $G = \mathrm{SL}_2(\Bbbk)$, $H = \{e\}$. Then $O = \mathrm{SL}_2$ has complexity one. Embeddings of O were described in [LV, §9].

The elements of G are matrices

$$g = \begin{pmatrix} g_{11} & g_{12} \\ g_{21} & g_{22} \end{pmatrix}, \qquad g_{11}g_{22} - g_{21}g_{12} = 1,$$

and B consists of upper-triangular matrices ($g_{21} = 0$). Let ω be the fundamental weight: $\omega(g) = g_{11}$ for $g \in B$.

All colors in O are regular. Their equations are the nonzero elements of the two-dimensional subspace $M = \Bbbk[G]^{(B)}_{\omega}$ generated by $\eta_1(g) = g_{21}$, $\eta_2(g) = g_{22}$. Let $\eta_x = c_1\eta_1 + c_2\eta_2$ be an equation of $x \in \mathbb{P}^1 = \mathbb{P}(M^*)$.

The field $K^B = \Bbbk(\mathbb{P}^1)$ consists of rational functions in η_1, η_2 of degree 0.

The group Λ equals $\mathfrak{X}(B) = \langle\omega\rangle \simeq \mathbb{Z}$. We may take $\mathbf{f}_\omega = \eta_\infty$, where $\infty \in \mathbb{P}^1$ is a certain fixed point.

The set of G-valuations is computed by the method of formal curves (§24). First we determine G-valuations corresponding to divisors with a dense orbit. Up to a multiple, any such valuation is defined by the formula $v_{x(t)}(f) = \mathrm{ord}_t\, f(gx(t))$, where $x(t) \in \mathrm{SL}_2(\Bbbk((t)))$ and g is the generic $\Bbbk(\mathrm{SL}_2)$-point of SL_2. By the Iwasawa decomposition (see 24.2), we may even assume that

$$x(t) = \begin{pmatrix} t^m & u(t) \\ 0 & t^{-m} \end{pmatrix}, \qquad u(t) \in \Bbbk((t)), \quad \mathrm{ord}_t\, u(t) = n \leq -m.$$

The number

$$d = v_{x(t)}(\eta_x) = \mathrm{ord}_t\left((c_1t^m + c_2u(t))g_{21} + c_2t^{-m}g_{22}\right)$$

is constant along $\mathbb{P}^1$ except for one x, where it jumps, so that $v_{x(t)} = (h,\ell) \in \mathscr{V}_x$. The following cases are possible:

$$m \leq n \implies \begin{cases} d = m, & \mathrm{ord}_t(c_1t^m + c_2u(t)) = m \\ d \in (m, -m], & \mathrm{ord}_t(c_1t^m + c_2u(t)) > m \end{cases} \implies \begin{cases} h \in (0, -2m] \\ \ell = m \text{ (or } m+h) \end{cases}$$

$$m > n \implies \begin{cases} d = n, & c_2 \neq 0 \\ d = m, & c_2 = 0 \end{cases} \implies \begin{cases} h = m - n \\ \ell = n \text{ (or } n+h) \end{cases}$$

(Here $\ell = v_{x(t)}(\mathbf{f}_\omega)$ increases by h if the jump occurs at $x = \infty$.)

In both cases, we obtain the subset in $\mathscr{E}_{x,+}$ defined by the inequalities $h > 0$, $2\ell + h \leq 0$ (or $2\ell - h \leq 0$ if $x = \infty$). Thus $\mathscr{V}_x$ is defined by $2\ell + h \leq 0$ (or $2\ell - h \leq 0$) by §24. The colored equipment is represented in Fig. 16.3. (Elements of $\Lambda^* \times \mathbb{Z}_+ \subset \mathscr{E}_{x,+}$ are marked by dots.)

G-germs are given by colored cones or hypercones of type B hatched vertically in Fig. 16.4; their colors are marked by bold dots. The notation for G-germs is taken from [LV, §9]. Up to a "coordinate transform" (Remark 16.5), we may assume that

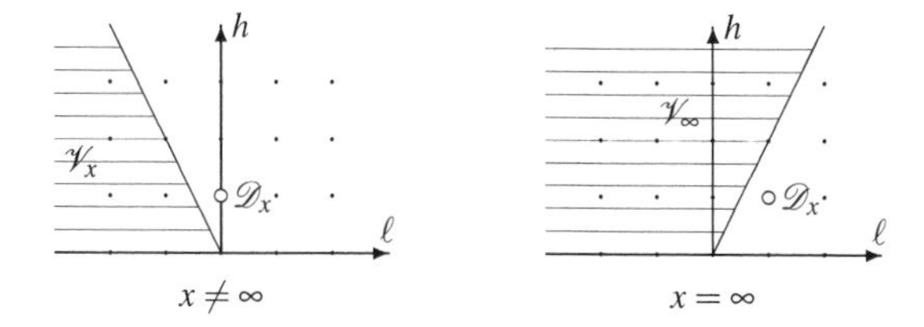

Fig. 16.3

Fig. 16.4 G-germs of SL_2-embeddings

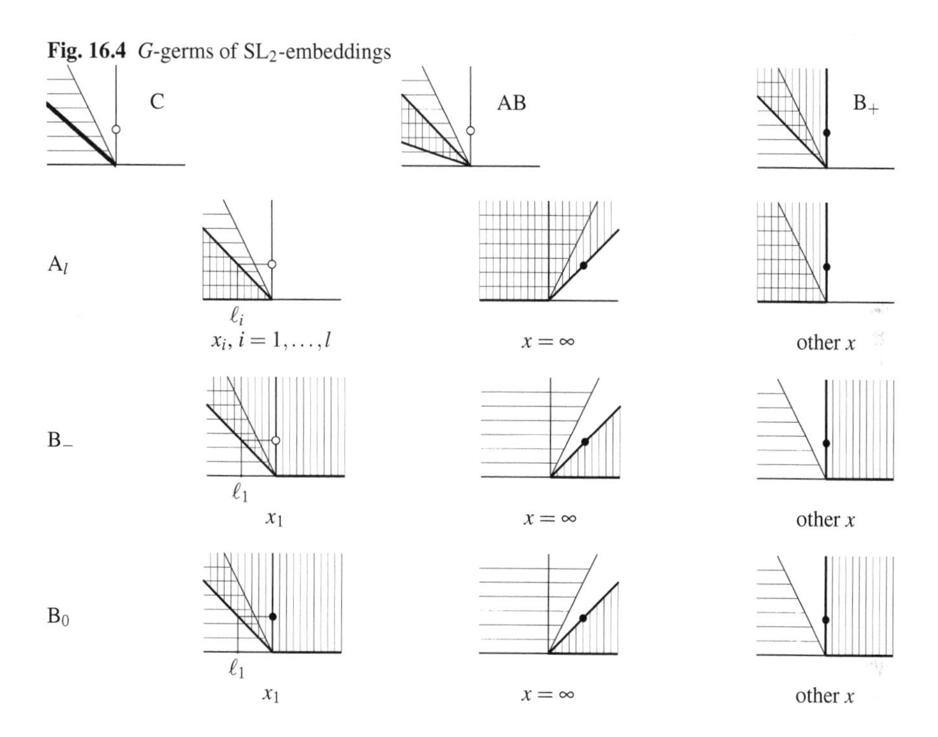

for germs of types $\mathrm{C}, \mathrm{AB}, \mathrm{B}_+$ the colored cone lies in $\mathscr{E}_{x,+}$, $x \neq \infty$, and for germs of types $\mathrm{A}_l, \mathrm{B}_-, \mathrm{B}_0$, $x_i \neq \infty$. Moreover, for the hypercone to be strictly convex, we must have $\sum \ell_i < -1$ for A_l, $l \geq 1$, and $\ell_1 > -1$ for B_- and B_0.

Affine SL_2-embeddings correspond to minimal B-charts of G-germs of type B_0. They were first classified by Popov [Po1]. Embeddings of SL_2/H, where H is finite, were classified in [M-J1]. Embeddings of G/H, where G has semisimple rank 1 and H is finite, were classified in [Tim2, §5]. These results hold in characteristic zero.

Example 16.24 (ordered triangles). Suppose that $G = \mathrm{SL}_3(\Bbbk)$ and $H = T$ is the diagonal torus. Then $O = G/H$ is the space of ordered triangles on a projective plane. The standard Borel subgroup B consists of upper-triangular matrices $g = (g_{ij})$ with $\det g = 1$ and $g_{ij} = 0$ for $i > j$. As usual, $\varepsilon_i(g) = g_{ii}$ are the tautological weights of T, $\omega_1 = \varepsilon_1$, $\omega_2 = \varepsilon_1 + \varepsilon_2$ are the fundamental weights, and $\alpha_1 = \varepsilon_1 - \varepsilon_2$, $\alpha_2 = \varepsilon_2 - \varepsilon_3$ are the simple roots. Denote by $\omega_i^\vee$ the fundamental coweights, and let $\rho = \omega_1 + \omega_2 = \alpha_1 + \alpha_2$.

The subregular colors $D_i, \widetilde{D}_i$ are defined by the $(B \times H)$-eigenfunctions $\eta_i(g) = g_{3i}$, $\widetilde{\eta}_i(g) = \begin{vmatrix} g_{2j} & g_{2k} \\ g_{3j} & g_{3k} \end{vmatrix}$ of biweights $(\omega_2, \varepsilon_i)$, $(\omega_1, -\varepsilon_i)$. $\widetilde{D}_i$ consists of triangles whose i-th side contains the B-fixed point in $\mathbb{P}^2$, and D_i consists of triangles whose i-th vertex lies on the B-fixed line.

The functions $\eta_i \widetilde{\eta}_i$ generate the two-dimensional subspace $M = \Bbbk[G]_{(\rho,0)}^{(B\times H)}$, $\eta_1\widetilde{\eta}_1 + \eta_2\widetilde{\eta}_2 + \eta_3\widetilde{\eta}_3 = 0$. Let $x_i \in \mathbb{P}^1 = \mathbb{P}(M^*)$ be the points corresponding to $\eta_i\widetilde{\eta}_i$. The regular colors D_x, $x \neq x_1, x_2, x_3$, are defined by equations $\eta_x = c_1\eta_1\widetilde{\eta}_1 + c_2\eta_2\widetilde{\eta}_2 + c_3\eta_3\widetilde{\eta}_3$.

The group $\Lambda = \langle \alpha_1, \alpha_2 \rangle$ is the root lattice, $\mathbf{f}_{\alpha_1} = \widetilde{\eta}_1\widetilde{\eta}_2\widetilde{\eta}_3/\eta_\infty$, $\mathbf{f}_{\alpha_2} = \eta_1\eta_2\eta_3/\eta_\infty$, where $\infty \in \mathbb{P}^1$ is a certain fixed point.

By 24.2, any G-valuation corresponding to a divisor with dense G-orbit is proportional to $v_{x(t)}$, where

$$x(t) = \begin{pmatrix} 1 & t^m & u(t) \\ 0 & 1 & t^n \\ 0 & 0 & 1 \end{pmatrix},$$

and we may assume that $m, n, r = \operatorname{ord}_t u(t) \leq 0$. Computing the values $v_{x(t)}(\eta_x)$ as in Example 16.23, one finds that the set of G-valuations $v = (h, \ell) \in \mathscr{E}_{x,+}$ corresponding to divisors with dense G-orbit is determined by the inequalities $a_1, a_2 \leq 0 \leq h$ (if $x = x_i$) or $a_1, a_2 \leq -2h \leq 0$ (if $x = \infty$) or $a_1, a_2 \leq -h \leq 0$ (otherwise), and $h = 0 \implies a_1$ or $a_2 = 0$, where $\ell = a_1\omega_1^\vee + a_2\omega_2^\vee$. Hence $\mathscr{V}_x$ is determined by the same inequalities without any restrictions for $h = 0$.

The colored equipment is represented in Fig. 16.5. (The intersections of $\mathscr{V}$ with the hyperplane sections $\mathscr{E} = \{h = 0\}$ and $\{h = 1\}$ of $\mathscr{E}_{x,+}$ are hatched horizontally.)

The space of ordered triangles has three natural completions: $(\mathbb{P}^2)^3$, $(\mathbb{P}^{2*})^3$, and

$$X = \{(p_1, p_2, p_3, l_1, l_2, l_3) \mid p_j \in \mathbb{P}^2,\ l_i \in \mathbb{P}^{2*},\ p_j \in l_i \text{ whenever } i \neq j\}.$$

If z_{kj} are the homogeneous coordinates of p_j in $\mathbb{P}^2$, and y_{ik} are the dual coordinates of l_i in $\mathbb{P}^{2*}$, then X is determined by 6 equations $(y \cdot z)_{ij} = 0$ $(i \neq j)$ in $(\mathbb{P}^2)^3 \times (\mathbb{P}^{2*})^3$. One verifies that the Jacobian matrix is non-degenerate everywhere on $X \setminus Y$, where $Y \subset X$ is given by the equations $p_1 = p_2 = p_3$, $l_1 = l_2 = l_3$, and $\operatorname{codim}_X Y = 3$. Thence, by Serre's normality criterion, X is a normal complete intersection, smooth outside Y. It contains the following G-subvarieties of degenerate triangles:

W_i: $p_j = p_k$ and $l_j = l_k$, $\{i, j, k\} = \{1, 2, 3\}$. (A divisor.)

$\widetilde{W}$: p_1, p_2, p_3 are collinear and $l_1 = l_2 = l_3$. (The proper pullback of the divisor $\{\det z = 0\}$ in $(\mathbb{P}^2)^3$.)

W: $p_1 = p_2 = p_3$ and l_1, l_2, l_3 pass through this point. (The proper pullback of the divisor $\{\det y = 0\}$ in $(\mathbb{P}^{2*})^3$.)

$\widetilde{Y}_i$: $p_j = p_k$ and $l_1 = l_2 = l_3$ (codim $= 2$).

Fig. 16.5

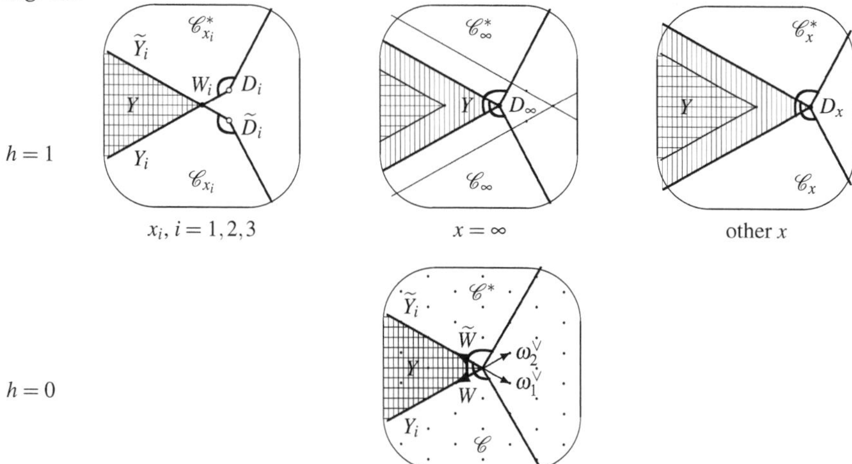

Y_i: $l_j = l_k$ and $p_1 = p_2 = p_3$ (codim $= 2$).
Y: $p_1 = p_2 = p_3$ and $l_1 = l_2 = l_3$ (codim $= 3$).

Note that η_i and $\widetilde{\eta}_i$ may be regarded as certain homogeneous coordinates in the i-th copy of $\mathbb{P}^2$, resp. $\mathbb{P}^{2*}$, restricted to O:

$$\begin{cases} \eta_i = z_{3i}, \\ \widetilde{\eta}_i = \begin{vmatrix} z_{2j} & z_{2k} \\ z_{3j} & z_{3k} \end{vmatrix}, \end{cases} \qquad \text{or dually,} \qquad \begin{cases} \eta_i = \begin{vmatrix} y_{j1} & y_{j2} \\ y_{k1} & y_{k2} \end{vmatrix}, \\ \widetilde{\eta}_i = y_{i1}, \end{cases}$$

and η_∞ is a 3-form in the matrix entries of y or z. Then

$$\mathbf{f}_{\alpha_1} = \frac{\widetilde{\eta}_1(y)\widetilde{\eta}_2(y)\widetilde{\eta}_3(y)}{\eta_\infty(y)} = \frac{\widetilde{\eta}_1(z)\widetilde{\eta}_2(z)\widetilde{\eta}_3(z)}{\eta_\infty(z)\det z}$$
$$\mathbf{f}_{\alpha_2} = \frac{\eta_1(y)\eta_2(y)\eta_3(y)}{\eta_\infty(y)\det y} = \frac{\eta_1(z)\eta_2(z)\eta_3(z)}{\eta_\infty(z)}$$

One easily deduces that $\mathbf{f}_{\alpha_1}, \eta_x$ ($\forall x \in C$) are regular and do not vanish along W, $\mathbf{f}_{\alpha_2}, \eta_x$ along $\widetilde{W}$, and $\mathbf{f}_{\alpha_1}$ (resp. $\mathbf{f}_{\alpha_2}$) has the first order pole along $\widetilde{W}$ (resp. W). Hence the G-valuations of $W, \widetilde{W}, W_i$ are $-\omega_2^\vee, -\omega_1^\vee, \varepsilon_{x_i}$.

Since X is complete and contains the minimal G-germ Y (the closed orbit), we have $X = G\mathring{X}$, where $\mathring{X}$ is the minimal B-chart of Y determined by the colored hypercone $(\mathscr{C}_Y, \mathscr{D}_Y^B)$ of type B such that $\mathscr{C}_Y \supseteq \mathscr{V}$ and $\mathscr{V}_Y = \{-\omega_1^\vee, -\omega_2^\vee, \varepsilon_{x_1}, \varepsilon_{x_2}, \varepsilon_{x_3}\}$. It is easy to see from Fig. 16.5 that there exists a unique such hypercone (indicated by vertical hatching) and $\mathscr{D}_Y^B = \{D_x \mid x \neq x_1, x_2, x_3\}$. The (hyper)faces corresponding to various G-germs of X (including Y) are indicated in Fig. 16.5 by the same letters.

A similar argument shows that $(\mathbb{P}^2)^3$ is defined by the colored hypercone $(\mathscr{C}, \{\widetilde{D}_i, D_x \mid i = 1,2,3,\ x \neq x_1,x_2,x_3\})$ and $(\mathbb{P}^{2^*})^3$ by $(\mathscr{C}^*, \{D_i, D_x \mid i = 1,2,3,\ x \neq x_1,x_2,x_3\})$.

The space $\mathrm{SL}_3(\Bbbk)/N_G(T)$ of unordered triangles and its completion is studied in [Tim3, §9]. The resolution of singularities of X was studied already by Schubert with applications to enumerative geometry, see [CF], [Tim3, §9], and §18.

16.6 Local Properties. We say that a normal G-model X *is of type A* if it contains no G-germs of type B, i.e., any G-orbit in X is contained in finitely many B-stable prime divisors. For any X, there is a proper birational morphism $\nu : \check{X} \to X$ such that $\check{X}$ is of type A.

In characteristic zero, singularities of G-models of type A are good.

Theorem 16.25. *If X is of type A, then all irreducible G-subvarieties $Y \subseteq X$ are normal and have rational singularities.*

Proof. By the local structure theorem, we may assume that X is affine and of type A, i.e., $\Bbbk[X]^B \neq k$. Passing to the categorical quotient by U, we may assume that $G = B = T$. In the notation of Example 16.21, we may replace X by X_i and assume that X is an affine toric $(T \times \Bbbk^\times)$-variety such that $X/\!/T \simeq \mathbb{A}^1$ ($\Bbbk^\times$-equivariantly). Then each T-stable irreducible closed subvariety of X is either $(T \times \Bbbk^\times)$-stable or lying in the fiber of the quotient map $X \to \Bbbk$ over a nonzero point, which is a toric T-variety. Thus the question is reduced to the case of toric varieties. □

17 Divisors

The study of divisors on normal G-models goes back to M. Brion [Bri4] in the spherical case and was extended to arbitrary complexity in [Tim3].

17.1 Reduction to B-stable Divisors. In the study of divisors on a normal G-model X, we may restrict our attention to B-stable ones, by the following result.

Proposition 17.1. *Let a connected solvable algebraic group B act on a normal variety X. Then any Weil divisor on X is rationally equivalent to a B-stable one.*

Proof. Replacing X by X^{reg}, we may assume that X is smooth. Then any Weil divisor δ on X is Cartier. Furthermore, δ is the difference of two effective divisors. Therefore we may assume that δ is effective. The line bundle $\mathscr{O}(\delta)$ is B-linearizable by Theorem C.4, and the B-module $\mathrm{H}^0(X, \mathscr{O}(\delta))$ contains a nonzero B-eigensection σ. The divisor $\operatorname{div}\sigma$ is B-stable and equivalent to δ. □

Remark 17.2. The proposition is true for any algebraic cycle, see Theorem 18.1.

17.2 Cartier Divisors. Our first aim is to describe Cartier divisors. For simplicity, assume that $\operatorname{char}\Bbbk = 0$.

For any Cartier divisor δ on X, we shall always equip the respective line bundle $\mathscr{O}(\delta)$ with a G-linearization assuming that G is of simply connected type (see Appendix C). Recall that δ is said to be *globally generated* if the sheaf of sections of $\mathscr{O}(\delta)$ is generated by global sections.

Lemma 17.3 ([Kn5, 2.2]). *Any prime divisor $D \subset X$ that does not contain a G-orbit of X is globally generated Cartier.*

Proof. Let $\iota : X^{\mathrm{reg}} \hookrightarrow X$ be the inclusion of the subset of smooth points. Then $D \cap X^{\mathrm{reg}}$ is Cartier on X^{reg}, and $D \cap X^{\mathrm{reg}} = \operatorname{div}\eta$ for some $\eta \in \mathrm{H}^0(X^{\mathrm{reg}}, \mathscr{O}(D \cap X^{\mathrm{reg}}))$. As X is normal, $\mathscr{O}(D) = \iota_*\mathscr{O}(D \cap X^{\mathrm{reg}})$ is a trivial line bundle on $X \setminus D$. As G acts on $\mathscr{O}(D)$, the set of points where $\mathscr{O}(D)$ is not invertible is G-stable and contained in D, and hence empty. Therefore $\mathscr{O}(D)$ is a line bundle on X.

If we regard η as an element of $\mathrm{H}^0(X, \mathscr{O}(D))$, then $D = \operatorname{div}\eta$, because the equality holds on X^{reg} and $\operatorname{codim}_X(X \setminus X^{\mathrm{reg}}) > 1$. Furthermore, $\mathscr{O}(D)$ is generated by $g\eta$, $g \in G$, because the set of their common zeroes is $\bigcap_{g\in G} gD = \emptyset$. The assertion follows. □

The following criterion says that a divisor is Cartier if and only if it is determined by a local equation in a neighborhood of a *general* point of each G-subvariety.

Theorem 17.4. *Suppose that δ is a Weil divisor on X. Then δ is Cartier if and only if for any irreducible G-subvariety $Y \subseteq X$ there exists $f_Y \in K$ such that each prime divisor $D \supseteq Y$ occurs in δ with multiplicity $v_D(f_Y)$. If δ is B-stable, then one may choose $f_Y \in K^{(B)}$.*

Proof. Suppose that δ is locally principal in general points of G-subvarieties. Take any G-orbit $Y \subseteq X$. Replacing δ by $\delta - \operatorname{div} f_Y$, we may assume that no component of δ contains Y. Take an affine open chart $\mathring{X} \subseteq X$ intersecting Y, but not intersecting any component of δ. By Lemma 17.3, δ is Cartier on $G\mathring{X}$, whence on X.

Now suppose that δ is Cartier, and let $Y \subseteq X$ be an irreducible G-subvariety. By Sumihiro's Theorem C.7, there is an open G-stable quasiprojective subvariety $X_0 \subseteq X$ intersecting Y. The restriction of δ to X_0 may be represented as the difference of two globally generated divisors, and hence we may replace X by X_0 and assume that δ is globally generated.

It follows that the annihilator of Y in $\mathrm{H}^0(X, \mathscr{O}(\delta))$ is a proper G-submodule, whence there is a section $\sigma \in \mathrm{H}^0(X, \mathscr{O}(\delta))$ such that $\sigma|_Y \neq 0$. In fact, we may assume that σ is a B-eigensection. Therefore δ is principal on the open subset X_σ intersecting Y, and we may take for f_Y the equation of δ on X_σ. □

Remark 17.5. The theorem extends to characteristic $p > 0$, except that for a B-stable Cartier divisor δ one can guarantee the existence of local equations $f_Y \in K^{(B)}$ not for δ itself, but for $q\delta$, where q is a power of p. To prove the existence of an eigensection $\sigma \in \mathrm{H}^0(X, \mathscr{O}(q\delta))^{(B)}$, $\sigma|_Y \neq 0$, one uses Corollary D.2.

By Theorem 17.4, a B-stable Cartier divisor on X is determined by the following data:

(1) a collection of rational B-eigenfunctions f_Y given for each G-germ $Y \in {}_GX$ and such that $w(f_{Y_1}) = w(f_{Y_2})$, $v_D(f_{Y_1}) = v_D(f_{Y_2})$, $\forall w \in \mathscr{V}_{Y_1} \cap \mathscr{V}_{Y_2}, D \in \mathscr{D}^B_{Y_1} \cap \mathscr{D}^B_{Y_2}$;
(2) a collection of integers m_D, $D \in \mathscr{D}^B \setminus \bigcup_{Y \subseteq X} \mathscr{D}^B_Y$, only finitely many of them being nonzero (m_D is the multiplicity of D in the divisor).

Remark 17.6. It suffices to specify the local equations f_Y only for closed G-orbits $Y \subseteq X$: if a G-subvariety $Y \subseteq X$ contains a closed orbit Y_0, then we may put $f_Y = f_{Y_0}$.

When a Cartier divisor is replaced by a rationally equivalent one, the local equations f_Y are replaced by $f_Y f$ for some $f \in K^{(B)}$, and m_D are replaced by $m_D + v_D(f)$.

17.3 Case of Complexity ≤ 1. In this case, the data (1)–(2) are retranslated into the language of convex geometry.

Consider first the spherical case. Each f_Y defines a function ψ_Y on the cone $\mathscr{C}_Y$, which is the restriction of a linear functional $\lambda_Y \in \Lambda$. We may assume that $f_{Y_1} = f_{Y_2}$ if $\mathscr{C}_{Y_1}$ is a face of $\mathscr{C}_{Y_2}$, whence $\psi_{Y_1} = \psi_{Y_2}|_{\mathscr{C}_{Y_1}}$. In particular, ψ_Y paste together in a piecewise linear function on $\bigcup_{Y \subseteq X}(\mathscr{C}_Y \cap \mathscr{V})$. A collection $\psi = (\psi_Y)$ of functions ψ_Y on $\mathscr{C}_Y$ with the above properties is called an *integral piecewise linear function* on the colored fan $\mathscr{F}$ of X.

Note that generally ψ *does not* define a function on $\bigcup_{\mathscr{C} \in \mathscr{F}} \mathscr{C}$, as the following example shows.

Example 17.7 ([Pau3]). Let $G = \mathrm{SL}_3(\Bbbk)$, $H = \mathrm{SL}_2(\Bbbk) =$ the common stabilizer of $e_1 \in \Bbbk^3$, $x_1 \in (\Bbbk^3)^*$. Then $O = G/H$ is defined in $\Bbbk^3 \oplus (\Bbbk^3)^*$ by an equation $\langle v, v^* \rangle = 1$. The colors $D_1, D_2 \in \mathscr{D}^B$ are defined by the restrictions η_1, η_2 of linear B-eigenfunctions on $\Bbbk^3 \oplus (\Bbbk^3)^*$ of B-weights ω_1, ω_2. Here $\eta_1, \eta_2 \in K^{(B)}$ and $\Lambda = \langle \omega_1, \omega_2 \rangle$, whence $\varkappa(D_i) = \alpha_i^\vee$.

A one-dimensional torus acting on the summands of $\Bbbk^3 \oplus (\Bbbk^3)^*$ by the weights ± 1 commutes with G and preserves O. The respective grading of $\Bbbk[O]$ determines two G-valuations $v_\pm(f) = \pm \deg f_\mp$, where $f_\pm$ is the highest/lowest degree term of $f \in \Bbbk[O]$. Since $\deg \eta_i = (-1)^{i-1}$, we have $v_\pm = \pm\alpha_1^\vee \mp \alpha_2^\vee$. It follows easily from Corollary 15.6 that the colored space looks as in Fig. 17.1.

Fig. 17.1

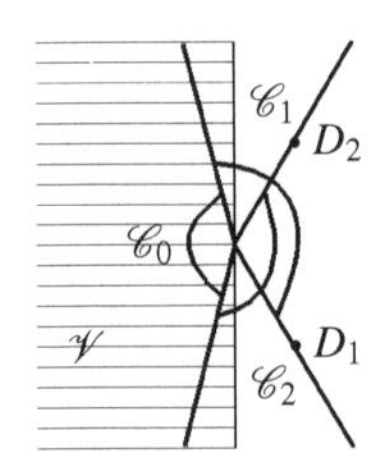

Take a fan $\mathscr{F}$ determined by the cones $\mathscr{C}_0, \mathscr{C}_1, \mathscr{C}_2$ in the figure. Since $\mathscr{C}_1 \cap \mathscr{C}_2$ is a solid cone, a piecewise linear function on $\mathscr{F}$ defines a function on $\mathscr{C}_0 \cup \mathscr{C}_1 \cup \mathscr{C}_2 = \mathscr{E}$ if and only if it is linear.

Let $\mathrm{PL}(\mathscr{F})$ be the group of all integral piecewise linear functions on $\mathscr{F}$, and let $\mathrm{L}(\mathscr{F})$ be its subgroup of linear functions $\psi = (\lambda|_{\mathscr{C}_Y})$, $\lambda \in \Lambda$.

The above discussion yields the following exact sequences:

$$0 \longrightarrow \mathbb{Z}\left(\mathscr{D}^B \setminus \bigcup_{Y \subseteq X} \mathscr{D}_Y^B\right) \longrightarrow \mathrm{CaDiv}(X)^B \longrightarrow \mathrm{PL}(\mathscr{F}) \longrightarrow 0, \tag{17.1}$$

$$\Lambda \cap \mathscr{F}^\perp \longrightarrow \mathrm{PrDiv}(X)^B \longrightarrow \mathrm{L}(\mathscr{F}) \longrightarrow 0, \tag{17.2}$$

where $\mathrm{CaDiv}(\cdot)$ and $\mathrm{PrDiv}(\cdot)$ denote the groups of Cartier and principal divisors, respectively, and $\mathscr{F}^\perp$ is the annihilator of the union of all cones in $\mathscr{F}$.

Theorem 17.8 ([Bri4, 3.1]). *For spherical X there is an exact sequence*

$$\Lambda \cap \mathscr{F}^\perp \longrightarrow \mathbb{Z}\left(\mathscr{D}^B \setminus \bigcup_{Y \subseteq X} \mathscr{D}_Y^B\right) \longrightarrow \operatorname{Pic} X \longrightarrow \mathrm{PL}(\mathscr{F})/\mathrm{L}(\mathscr{F}) \longrightarrow 0.$$

If X contains a complete G-orbit, then $\operatorname{Pic} X$ *is free Abelian of finite rank.*

Proof. The exact sequence is a consequence of (17.1)–(17.2). If $Y \subseteq X$ is a complete G-orbit, then $\mathscr{F}^\perp \subseteq \mathscr{C}_Y^\perp = 0$ by Propositions 15.14 and 10.1. Then it is easy to see that $\mathrm{PL}(\mathscr{F})/\mathrm{L}(\mathscr{F})$ has finite rank and no torsion, whence the second assertion. □

A G-variety X having only one closed orbit $Y \subseteq X$ is called *simple*. If X is spherical and simple, then its fan consists of all supported colored faces of $(\mathscr{C}_Y, \mathscr{D}_Y^B)$.

Corollary 17.9. *If X is spherical and simple with the closed orbit $Y \subseteq X$, then there is an exact sequence*

$$\Lambda \cap \mathscr{C}_Y^\perp \longrightarrow \mathbb{Z}(\mathscr{D}^B \setminus \mathscr{D}_Y^B) \longrightarrow \operatorname{Pic} X \longrightarrow 0.$$

Corollary 17.10. *If X is spherical and simple and the closed orbit $Y \subseteq X$ is complete, then* $\operatorname{Pic} X = \mathbb{Z}(\mathscr{D}^B \setminus \mathscr{D}_Y^B)$ *is free Abelian with basis* $\mathscr{D}^B \setminus \mathscr{D}_Y^B$.

Example 17.11. If $X = G/P$ is a generalized flag variety, then $\operatorname{Pic} X$ is freely generated by the Schubert divisors $D_\alpha = \overline{B(w_G r_\alpha o)}$, $\alpha \in \Pi \setminus I$, where $I \subseteq \Pi$ is the set of simple roots defining the parabolic $P = P_I \supseteq B$ and $r_\alpha \in W$ is the reflection along α.

Example 17.12 ([Bri4]). Let $G = \mathrm{SL}_3(\Bbbk)$, $H = N_G(\mathrm{SO}_3(\Bbbk)) = \mathrm{SO}_3(\Bbbk) \times \mathbf{Z}_3$, where $\mathbf{Z}_3 = Z(\mathrm{SL}_3(\Bbbk))$. Then $O = G/H$ is the space of conics in $\mathbb{P}^2$. The coisotropy representation is the natural representation of $\mathrm{SO}_3(\Bbbk)$ in traceless symmetric matrices, whence

$$H_* = \left\{ \begin{pmatrix} \pm 1 & 0 & 0 \\ 0 & \pm 1 & 0 \\ 0 & 0 & \pm 1 \end{pmatrix} \right\} \times \mathbf{Z}_3$$

and $\Lambda(O) = \langle 2\alpha_1, 2\alpha_2 \rangle$, where α_i are simple roots of $\mathrm{SL}_3(\Bbbk)$. By Example 4.9(2) or (9.1), O is spherical.

We may consider O as the projectivization of the open subset of non-degenerate quadratic forms in $\mathrm{S}^2(\Bbbk^3)^*$. The two $(B \times H)$-eigenfunctions

$$\eta_1(q) = q_{11}, \qquad \eta_2(q) = \begin{vmatrix} q_{11} & q_{12} \\ q_{21} & q_{22} \end{vmatrix} \qquad (q \in \mathrm{S}^2(\Bbbk^3)^*)$$

of biweights $(2\omega_i, i\varepsilon)$ $(i = 1, 2)$, where ω_i are the fundamental weights of G and ε is the weight of $\mathbf{Z}_3$ in $\Bbbk^3$, define the two colors D_1, D_2. They impose the conditions that a conic passes through the B-fixed point, resp. is tangent to the B-fixed line. Since $\mathbf{f}_{2\alpha_1} = \eta_1^2/\eta_2$, $\mathbf{f}_{2\alpha_2} = \eta_2^2/\eta_1 \in K^{(B)}$, and their weights $2\alpha_1, 2\alpha_2$ generate Λ, there are no other colors. (Indeed, if $\eta \in \Bbbk[G]^{(B\times H)}_{(\lambda,\chi)}$, then either $\chi = 0$, $\lambda \in \Lambda$, or $\chi = i\varepsilon$, $\lambda - 2\omega_i \in \Lambda$, hence η is proportional to the product of η_1, η_2 and their inverses.) Furthermore, $\Lambda^* = \langle \omega_1^\vee/2, \omega_2^\vee/2 \rangle$, where $\omega_i^\vee$ denote the fundamental coweights, and $\varkappa(D_i) = \alpha_i^\vee/2$.

The complement to O in $(\mathbb{P}^5)^* = \mathbb{P}(\mathrm{S}^2(\Bbbk^3)^*)$ is a G-stable prime divisor $\{\det q = 0\}$, and the respective G-valuation is $-\omega_2^\vee/2 \in \mathscr{E}$, because in homogeneous coordinates $\mathbf{f}_{2\alpha_1}(q) = \eta_1(q)^2/\eta_2(q)$, $\mathbf{f}_{2\alpha_2}(q) = \eta_2(q)^2/\eta_1(q)\det q$. The unique closed orbit $Y = \{\mathrm{rk}\, q = 1\}$ has the colored data $\mathscr{V}_Y = \{-\omega_2^\vee/2\}$, $\mathscr{D}_Y^B = \{D_2\}$.

Similarly, we embed O in $\mathbb{P}^5 = \mathbb{P}(\mathrm{S}^2\Bbbk^3)$ by the map $q \mapsto q^\vee$ (=the adjoint matrix of q) sending a conic to the dual one. Here the unique closed orbit $Y^\vee = \{\mathrm{rk}\, q^\vee = 1\}$ has the colored data $\mathscr{V}_{Y^\vee} = \{-\omega_1^\vee/2\}$, $\mathscr{D}_{Y^\vee}^B = \{D_1\}$.

Since $\mathbb{P}^5, (\mathbb{P}^5)^*$ are complete, the cones $\mathscr{C}_Y$ and $\mathscr{C}_{Y^\vee}$ contain $\mathscr{V}$, whence $\mathscr{V}$ is generated by $-\omega_1^\vee/2, -\omega_2^\vee/2$. The colored equipment of O is represented in Fig. 17.2.

Fig. 17.2

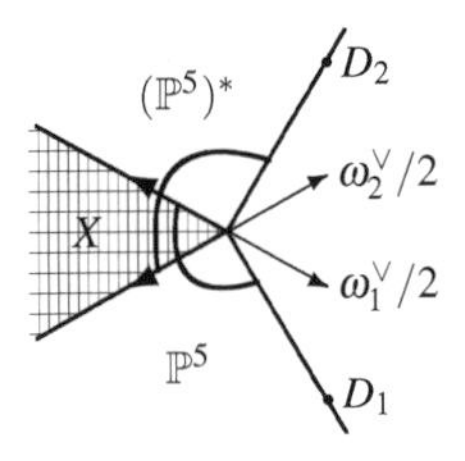

The closure X of the diagonal embedding $O \hookrightarrow \mathbb{P}^5 \times (\mathbb{P}^5)^*$ is called the *space of complete conics*. It is determined in $\mathbb{P}^5 \times (\mathbb{P}^5)^*$ by the equation "$q \cdot q^\vee$ is a scalar matrix", and this implies by direct computations that X is smooth. The unique closed orbit $\check{Y} \subseteq X$ has the colored data $\mathscr{V}_{\check{Y}} = \{-\omega_1^\vee/2, -\omega_2^\vee/2\}$, $\mathscr{D}_{\check{Y}}^B = \emptyset$.

By Corollary 17.10, $\mathrm{Pic}\,\mathbb{P}^5 \simeq \mathrm{Pic}(\mathbb{P}^5)^* \simeq \mathbb{Z}$ are freely generated by D_2, resp. D_1, and $\mathrm{Pic}\, X \simeq \mathbb{Z}^2$ is freely generated by D_1, D_2.

Remark 17.13. In fact, the space of smooth conics is a symmetric variety and the space of complete conics is its "wonderful completion", see §§26, 30.

In the case of complexity 1, the description of Cartier divisors is similar, but one should speak not only of cones, but also of hypercones $\mathscr{C}_Y$, and of admissible functionals λ_Y which are integral on Λ^* and such that $\sum\langle\varepsilon_x,\lambda_Y\rangle x$ is a principal divisor on C.

In particular, if X is simple, then $\operatorname{Pic} X$ is generated by $\mathscr{D}^B\setminus\mathscr{D}^B_Y$ (in fact, by a finite subset), where $Y\subseteq X$ is the closed orbit, and if Y is complete of type B, then $\operatorname{Pic} X$ is free Abelian with finite basis $\mathscr{D}^B\setminus\mathscr{D}^B_Y$.

Example 17.14. If X is the space of complete triangles of Example 16.24, then $\operatorname{Pic} X=\langle D_i,\widetilde{D}_i\mid i=1,2,3\rangle\simeq\mathbb{Z}^6$.

17.4 Global Sections of Line Bundles. In characteristic zero, the G-module structure of the space of global sections of a Cartier divisor is determined by the set of B-eigensections. For any Cartier divisor δ let η_δ denote a rational section of $\mathscr{O}(\delta)$ such that $\operatorname{div}\eta_\delta=\delta$. (It is defined uniquely up to multiplication by an invertible function.) If δ is B-stable, then η_δ is B-semiinvariant, and $\mathrm{H}^0(X,\mathscr{O}(\delta))^{(B)}=\{f\eta_\delta\mid f\in K^{(B)},\ \operatorname{div} f+\delta\geq 0\}$. The B-weight of $\sigma=f\eta_\delta\in\mathrm{H}^0(X,\mathscr{O}(\delta))^{(B)}$ equals $\lambda+\lambda_\delta$, where λ is the B-weight of f and λ_δ is the B-weight of η_δ. The multiplicity of $V(\lambda+\lambda_\delta)$ in $\mathrm{H}^0(X,\mathscr{O}(\delta))$ equals

$$m_\lambda(\delta)=\dim\{f\in K^B\mid\operatorname{div} f+\operatorname{div}\mathbf{f}_\lambda+\delta\geq 0\}. \tag{17.3}$$

The weight λ_δ is defined up to a character of G and may be determined as follows. Consider a general G-orbit $Y\subseteq X$ and let $\widetilde{\delta}$ be the pullback on G of $\delta\cap Y$. As G is a factorial variety, $\widetilde{\delta}$ is defined by an equation $F\in k(G)^{(B)}$. Then λ_δ is the weight of F.

In the case $c(X)\leq 1$, the description of $\mathrm{H}^0(X,\mathscr{O}(\delta))$ is given in the language of convex geometry.

If $c(X)=0$, then the set of highest weights of $\mathrm{H}^0(X,\mathscr{O}(\delta))$ is $\lambda_\delta+\mathscr{P}(\delta)\cap\Lambda$, where

$$\mathscr{P}(\delta)=\left\{\lambda\in\bigcap_{Y\subseteq X}(-\lambda_Y+\mathscr{C}_Y^\vee)\ \middle|\ \forall D\in\mathscr{D}^B\setminus\bigcup_{Y\subseteq X}\mathscr{D}^B_Y:\ \langle\lambda,\varkappa(D)\rangle+m_D\geq 0\right\}, \tag{17.4}$$

and all highest weights occur with multiplicity 1.

Example 17.15. In the notation of Example 17.11, a Schubert divisor $D_{\alpha_i}\subseteq G/P_I$, $\alpha_i\in\Pi\setminus I$, is defined by an equation $\langle v^*,gv\rangle=0$, $v^*\in V(\omega_i^*)^{(B)}$, $v\in V(\omega_i)^{(P_I)}$. Hence $\lambda_{D_{\alpha_i}}=\omega_i^*$. For $\delta=\sum a_iD_{\alpha_i}$, we have $\mathscr{O}(\delta)=\mathscr{L}(-\sum a_i\omega_i)$, $\lambda_\delta=\sum a_i\omega_i^*$, $\mathscr{P}(\delta)=\{0\}$, and $\mathrm{H}^0(G/P_I,\mathscr{O}(\delta))=V(\sum a_i\omega_i^*)$ (the Borel–Weil theorem, cf. 2.6).

Example 17.16. Consider $X=\mathbb{P}^{d-1}\times(\mathbb{P}^{d-1})^*$ as a simple projective embedding of a symmetric space $O=\mathrm{SL}_d/\mathrm{S}(\mathrm{L}_1\times\mathrm{L}_{d-1})$, where $\mathrm{S}(\mathrm{L}_1\times\mathrm{L}_{d-1})$ denotes the group of unimodular block-diagonal matrices with blocks of size 1 and $d-1$. Then $Y = X \setminus O$ is a homogeneous divisor consisting of all pairs (x,y) such that the point x lies in the hyperplane y. It is defined by an equation $\sum x_i y_i = 0$, where $x_1,\dots,x_d$ $(y_1,\dots,y_d)$ are projective coordinates on $\mathbb{P}^{d-1}$ (resp. $(\mathbb{P}^{d-1})^*$). The two colors D, D' are defined by B-eigenfunctions y_1, x_d of biweights $(\omega_1,(d-1)\varepsilon)$, $(\omega_{d-1},(1-d)\varepsilon)$, respectively, where ε generates $\mathfrak{X}(\mathrm{S}(\mathrm{L}_1 \times \mathrm{L}_{d-1}))$. One has $\Lambda = \langle \omega_1 + \omega_{d-1} \rangle \simeq \mathbb{Z}$, $\mathbf{f}_{\omega_1+\omega_{d-1}}(x,y) = x_d y_1 / \sum x_i y_i$. It follows easily that $\mathscr{E} \simeq \mathbb{Q} \supset \mathscr{V} = \mathscr{C}_Y = \mathbb{Q}_-$, $\varkappa(D) = \varkappa(D') = 1$.

By Corollary 17.10, $\mathrm{Pic}(X) = \mathbb{Z}D \oplus \mathbb{Z}D'$, and for $\delta = mD + nD'$ we have $\lambda_\delta = m\omega_1 + n\omega_{d-1}$, $\mathscr{P}(\delta) = \{\lambda = -k(\omega_1 + \omega_{d-1}) \mid k, m-k, n-k \geq 0\}$. On the other hand, $\mathscr{O}(\delta) = \mathscr{O}_{\mathbb{P}^{d-1}}(n) \otimes \mathscr{O}_{(\mathbb{P}^{d-1})^*}(m)$, whence $\mathrm{H}^0(X,\mathscr{O}(\delta)) = \mathrm{S}^m \Bbbk^d \otimes \mathrm{S}^n (\Bbbk^d)^* = V(m\omega_1) \otimes V(n\omega_{d-1})$. We obtain a decomposition formula

$$V(m\omega_1) \otimes V(n\omega_{d-1}) = \bigoplus_{0 \leq k \leq \min(m,n)} V((m-k)\omega_1 + (n-k)\omega_{d-1}).$$

For other applications to computing tensor product decompositions, including Pieri formulæ, see [Bri4, 2.5], cf. 11.4.

Now assume that $c(X) = 1$. Let C be a smooth projective curve with $\Bbbk(C) = K^B$. Put $\delta = \sum m_D D$ (D runs through all B-stable prime divisors on X) and $\varkappa(D) = (h_D, \ell_D) \in \mathscr{E}_{x_D,+}$, $x_D \in C$.

Definition 17.17. A *pseudodivisor* on C is a formal linear combination $\mu = \sum_{x \in C} m_x \cdot x$, where $m_x \in \mathbb{R} \cup \{\pm\infty\}$, and all but finitely many m_x are 0. Put $\mathrm{H}^0(C,\mu) = \{f \in \Bbbk(C) \mid \operatorname{div} f + \mu \geq 0\}$. (Here we assume that $c + (\pm\infty) = \pm\infty$ for any $c \in \mathbb{R}$.)

If all $m_x \neq -\infty$, then $\mathrm{H}^0(C,\mu)$ is just the space of global sections of the divisor $[\mu] = \sum [m_x] \cdot x$ on $C \setminus \{x \mid m_x = +\infty\}$, otherwise $\mathrm{H}^0(C,\mu) = 0$.

Consider the pseudodivisor

$$\mu = \mu(\delta,\lambda) = \sum_{x \in C} \left(\min_{x_D = x} \frac{\langle \lambda, \ell_D \rangle + m_D}{h_D} \right) x.$$

$$\left(\text{Here we assume that } \frac{c}{0} = \begin{cases} +\infty, & c \geq 0, \\ -\infty, & c < 0. \end{cases} \right)$$

Since $\operatorname{div} f = \sum h_D v_{x_D}(f) \cdot D$, $\forall f \in K^B$, and $\operatorname{div} \mathbf{f}_\lambda = \sum \langle \lambda, \ell_D \rangle \cdot D$, it follows from (17.3) that $m_\lambda(\delta) = h^0(\delta,\lambda) := \dim \mathrm{H}^0(C,\mu)$.

We have $h^0(\delta,\lambda) = 0$ outside the polyhedral domain

$$\mathscr{P}(\delta) = \{\lambda \mid \langle \lambda, \ell_D \rangle \geq -m_D \text{ for all } D \text{ such that } h_D = 0\} \subseteq \Lambda \otimes \mathbb{R}.$$

If there is $x \in C$ such that $x_D \neq x$ for all D with $h_D > 0$, then $h^0(\delta,\lambda) = \infty$, $\forall \lambda \in \mathscr{P}(\delta)$, because in this case $h^0(\delta,\lambda)$ is the dimension of the space of sections of a divisor on an affine curve. Otherwise, by the Riemann–Roch theorem,

$$h^0(\delta,\lambda) = \deg[\mu] - g + 1 + h^1(\delta,\lambda)$$
$$= A(\delta,\lambda) - \sigma(\delta,\lambda) - g + 1 + h^1(\delta,\lambda),$$

where g is the genus of C, $h^1(\delta,\lambda) = \dim \mathrm{H}^1(C,[\mu])$,

$$A(\delta,\lambda) = \sum_{x\in C} \left(\min_{x_D=x} \frac{\langle\lambda,\ell_D\rangle + m_D}{h_D} \right)$$

is a piecewise affine concave function of λ, and $\sigma(\delta,\lambda)$ is bounded non-negative for all δ,λ. Furthermore, as $h^0(\delta,\lambda) \le \deg[\mu]+1$ whenever $\deg[\mu] \ge 0$ [Har2, Ex. IV.1.5], we have $h^1(\delta,\lambda) \le g$ if $A(\delta,\lambda) \ge \sigma(\delta,\lambda)$. Note also that $A(n\delta,n\lambda) = nA(\delta,\lambda)$.

It follows that $h^0(\delta,\lambda) = 0$ if $A(\delta,\lambda) < 0$, and $h^0(\delta,\lambda)$ differs from $A(\delta,\lambda)$ by a globally bounded function whenever $A(\delta,\lambda) \ge 0$, $\lambda \in \mathscr{P}(\delta)$. This gives the asymptotic behavior of $h^0(\delta,\lambda)$ as $(\delta,\lambda) \to \infty$ in a fixed direction.

17.5 Ample Divisors. Now we give criteria for a Cartier divisor to be globally generated and ample.

Theorem 17.18. *Suppose that δ is a Cartier divisor on X determined by the data $\{f_Y\}$, $\{m_D\}$.*

(1) *δ is globally generated if and only if local equations f_Y can be chosen in such a way that for any irreducible G-subvariety $Y \subseteq X$ the following two conditions are satisfied:*

(a) *For any other irreducible G-subvariety $Y' \subseteq X$ and each B-stable prime divisor $D \supseteq Y'$, $v_D(f_Y) \le v_D(f_{Y'})$.*
(b) $\forall D \in \mathscr{D}^B \setminus \bigcup_{Y'\subseteq X} \mathscr{D}^B_{Y'} :\ v_D(f_Y) \le m_D$.

(2) *δ is ample if and only if, after replacing δ by a certain multiple, local equations f_Y can be chosen in such a way that, for any irreducible G-subvariety $Y \subseteq X$, there exists a B-chart $\mathring{X}$ of Y such that* (a) *and* (b) *are satisfied and*

(c) *the inequalities therein are strict if and only if $D \cap \mathring{X} = \emptyset$.*

Proof. (1) δ is globally generated if and only if for any irreducible G-subvariety $Y \subseteq X$, there is $\eta \in \mathrm{H}^0(X,\mathscr{O}(\delta))$ such that $\eta|_Y \ne 0$. We may assume η to be a B-eigensection. This means that there exists $f \in K^{(B)}$ such that $\operatorname{div} f + \delta \ge 0$, and no $D \supseteq Y$ occurs in $\operatorname{div} f + \delta$ with positive multiplicity. Replacing f_Y by f^{-1} yields the conditions (a)–(b). Conversely, if (a)–(b) hold then $f = f_Y^{-1}$ yields the desired global section.
(2) Suppose that δ is ample. Replacing δ by a multiple, we may assume that δ is very ample. Consider the G-equivariant projective embedding $X \hookrightarrow \mathbb{P}(M^*)$ defined by a certain finite-dimensional G-submodule $M \subseteq \mathrm{H}^0(X,\mathscr{O}(\delta))$. Take an irreducible G-subvariety $Y \subseteq X$. There exists a homogeneous B-eigenpolynomial in homogeneous coordinates on $\mathbb{P}(M^*)$ (i.e., a section in $\mathrm{H}^0(X,\mathscr{O}(\delta)^{\otimes N})^{(B)}$) that vanishes on $\overline{X} \setminus X$ but not on Y. Replacing δ by $N\delta$, we may assume that there exists $\eta \in \mathrm{H}^0(X,\mathscr{O}(\delta))^{(B)}$ such that $\eta|_{\overline{X}\setminus X} = 0$, $\eta|_Y \ne 0$. Then $\mathring{X} = X_\eta$ is a B-chart of Y,

and there exists $f \in K^{(B)}$ such that $\operatorname{div} f + \delta = \operatorname{div} \eta \geq 0$. It remains to replace f_Y by f^{-1}.

Conversely, assume that the conditions (a)–(c) hold. For any irreducible G-subvariety $Y \subseteq X$, there is a section $\eta \in \mathrm{H}^0(X, \mathscr{O}(\delta))^{(B)}$ determined by f_Y^{-1}, and $\mathring{X} = X_\eta$ is a B-chart of Y. We may pick finitely many B-charts $\mathring{X}_i$ of this kind in such a way that $G\mathring{X}_i$ cover X. Let $\eta_i \in \mathrm{H}^0(X, \mathscr{O}(\delta))^{(B)}$ be the respective global sections. Then

$$\Bbbk[\mathring{X}_i] = \bigcup_{n \geq 0} \eta_i^{-n} \mathrm{H}^0(X, \mathscr{O}(\delta)^{\otimes n}) = \Bbbk\left[\frac{\sigma_{i1}}{\eta_i^{n_i}}, \dots, \frac{\sigma_{i,s_i}}{\eta_i^{n_i}}\right]$$

for some $n_i, s_i \in \mathbb{N}$, $\sigma_{ij} \in \mathrm{H}^0(X, \mathscr{O}(\delta)^{\otimes n_i})$. Replacing δ by a multiple, we may assume that $n_i = 1$.

Take the finite-dimensional G-submodule $M \subseteq \mathrm{H}^0(X, \mathscr{O}(\delta))$ generated by η_i, σ_{ij}, $\forall i, j$. The respective rational map $X \dashrightarrow \mathbb{P}(M^*)$ is G-equivariant and defined on $\mathring{X}_i$, hence everywhere. Moreover, $\varphi^{-1}(\mathbb{P}(M^*)_{\eta_i}) = \mathring{X}_i$ and $\varphi|_{\mathring{X}_i}$ is a closed embedding in $\mathbb{P}(M^*)_{\eta_i}$. Therefore φ is a locally closed embedding and δ is very ample. □

Remark 17.19. If X is complete and δ is very ample, then the conditions (a)–(c) hold for δ itself.

Corollary 17.20. *If $\mathring{X}$ is a B-chart and $X = G\mathring{X}$, then a divisor $\sum_{D \subseteq X \setminus \mathring{X}} m_D D$ is globally generated (ample) whenever all $m_D \geq 0$ ($m_D > 0$). In particular, X is quasiprojective.*

Remark 17.21. Theorem 17.18 extends to characteristic $p > 0$ if one reformulates (1) as a criterion for a certain multiple $q\delta$ to be globally generated, where q is a power of p, in terms of the data defining $q\delta$. Another way to extend the theorem to positive characteristic is to waive B-stability and B-semiinvariance. Corollary 17.20 still holds in positive characteristic.

Corollary 17.22. *If X is spherical and simple and $Y \subseteq X$ is the closed G-orbit, then globally generated (ample) divisor classes in* $\operatorname{Pic} X$ *are those $\delta = \sum_{D \in \mathscr{D}^B \setminus \mathscr{D}_Y^B} m_D D$ with $m_D \geq 0$ ($m_D > 0$). In particular, any simple G-variety is quasiprojective. (This also stems from Sumihiro's Theorem* C.7.*)*

In the case of complexity ≤ 1, conditions (a)–(b) mean that $\lambda_Y \leq \psi_{Y'}$ on $\mathscr{C}_{Y'}$ and $\langle \lambda_Y, \varkappa(D) \rangle \leq m_D$, and (c) means that the inequalities therein are strict exactly outside $\mathscr{C} = \mathscr{C}(\mathscr{W}, \mathscr{R})$ and $\mathscr{R}$.

The description of globally generated and ample divisors on spherical X in terms of piecewise linear functions is more transparent if X is complete (or all closed G-orbits $Y \subseteq X$ are complete). Then maximal cones $\mathscr{C}_Y \in \mathscr{F}$ are solid, and λ_Y are determined by ψ_Y.

Definition 17.23. A function $\psi \in \mathrm{PL}(\mathscr{F})$ is (*strictly*) *convex* if $\lambda_Y \leq \psi_{Y'}$ on $\mathscr{C}_{Y'}$ (resp. $\lambda_Y < \psi_{Y'}$ on $\mathscr{C}_{Y'} \setminus \mathscr{C}_Y$) for any two maximal cones $\mathscr{C}_Y, \mathscr{C}_{Y'} \in \mathscr{F}$.

Corollary 17.24. *If X is complete (or all closed G-orbits in X are complete) and spherical, then δ is globally generated (ample) if and only if ψ is (strictly) convex on $\mathscr{F}$ and $\langle \lambda_Y, \varkappa(D) \rangle \leq m_D$ (resp. $< m_D$) for any closed G-orbit $Y \subseteq X$ and any $D \in \mathscr{D}^B \setminus \bigcup_{Y' \subseteq X} \mathscr{D}^B_{Y'}$.*

Proof. It suffices to note that Y has a unique B-chart $\mathring{X}_Y$ given by the colored cone $(\mathscr{C}_Y, \mathscr{D}^B_Y)$, and conditions (a)–(c) are satisfied for δ if and only if they are satisfied for a multiple of δ. □

Corollary 17.25. *On a complete spherical variety, every ample divisor is globally generated.*

The above results extend to the case of complexity 1, if all closed G-orbits in X are complete and of type B.

Remark 17.26. It follows from the proof of Theorem 17.18(2) that δ is very ample if $\Bbbk[\mathring{X}_i]$ is generated by $\eta_i^{-1}\mathrm{H}^0(X, \mathscr{O}(\delta))$ for each i. This is reduced to the equality on B-semiinvariants: $\Bbbk[\mathring{X}_i]^{(B)} = \bigcup \eta_i^{-n} R_n^{(B)} = \bigcup \eta_i^{-n} S_n^{(B)}$ for $R = \bigoplus_{n \geq 0} \mathrm{H}^0(X, \mathscr{O}(\delta)^{\otimes n})$, $S = \bigoplus_{n \geq 0} \left[\mathrm{H}^0(X, \mathscr{O}(\delta))\right]^n$; and can be effectively verified in some cases. For example, if X is complete and spherical, then it suffices to verify that for each closed G-orbit $Y \subseteq X$ the polyhedral domain $\lambda_Y + \mathscr{P}(\delta)$ contains the generators of the semigroup $\mathscr{C}_Y^\vee \cap \Lambda$. Using this observation, it is easy to show that on a complete smooth toric variety every ample divisor is very ample (Demazure [Oda, Cor. 2.15]).

Example 17.27. On a generalized flag variety $X = G/P$, globally generated (ample) divisors are distinguished in the set of all B-stable divisors $\delta = \sum a_i D_{\alpha_i}$ by the conditions $a_i \geq 0$ (resp. $a_i > 0$), in the notation of Example 17.11. Every ample divisor is very ample.

Example 17.28. The variety X defined by the colored fan $\mathscr{F}$ from Example 17.7 is complete, but not projective. Indeed, since $\mathscr{C}_1 \cap \mathscr{C}_2$ is a solid cone (Fig. 17.1), a convex piecewise linear function on $\mathscr{F}$ is forced to be linear on $\mathscr{C}_1 \cup \mathscr{C}_2$, whence globally on $\mathscr{E}$. Hence there are no non-principal globally generated divisors on X.

Remark 17.29. If a fan $\mathscr{F}$ in a two dimensional colored space has no colors, then the interiors of all cones in $\mathscr{F}$ are disjoint and there exists a strictly convex piecewise linear function on $\mathscr{F}$. Therefore all toroidal spherical varieties (in particular, all toric varieties) of rank 2 are quasiprojective. However, one can construct a complete, but not projective, toric variety of rank 3 [Ful2, p. 71].

Example 17.30. The same reasoning as in Example 17.28 shows that an SL_2-embedding X containing at least two G-germs of types B_-, B_0 (Fig. 16.4) is not quasiprojective. (Here $r(X) = 1$, $\dim X = 3$.) On the other hand, if X contains at most one G-germ of type B_- or B_0, then it is easy to construct a strictly convex piecewise linear function on the hyperfan of X, whence X is quasiprojective. (For smooth X, this was proved in [M-J2, 6.4].)

18 Intersection Theory

Our basic reference in intersection theory is [Ful1].

18.1 Reduction to B-stable Cycles. We begin our study of algebraic cycles on G-models with the following general result reducing everything to B-stable cycles.

Theorem 18.1 ([FMSS]). *Let a connected solvable algebraic group B act on a variety X. Then the Chow group $A_d(X)$ is generated by the B-stable d-cycles with the relations $[\operatorname{div} f] = 0$, where f is a rational B-eigenfunction on an irreducible closed B-stable $(d+1)$-subvariety of X. Here $[z] \in A_*(X)$ denotes the class of a cycle z.*

Proof. Using the normalization of X, equivariant completion [Sum, Th. 3], and the equivariant Chow lemma [Sum, Th. 2], one reduces the assertion to the case of normal projective X by induction on $\dim X$ with the help of the standard technique of exact sequences [FMSS, §2]. The projective case was handled by Hirschowitz [Hir], Vust [M-J2, 6.1], and Brion [Bri10, 1.3]. The idea is to consider the B-action on the Chow variety Z containing a given effective d-cycle z. Applying the Borel fixed point theorem, we find a B-stable cycle $z_0 \in \overline{Bz}$. An easy induction on $\dim B$ shows that z_0 can be connected with z by a sequence of rational curves, whence z_0 is rationally equivalent to z. The assertion on relations is proved by a similar technique, see [Bri10, 1.3] for details. □

This theorem clarifies almost nothing in the structure of Chow groups of general G-varieties, because the set of B-stable cycles is almost as vast as the set of all cycles; however it is very useful for G-varieties of complexity ≤ 1.

Assume that X is a unirational G-variety of complexity ≤ 1 or a B-stable subvariety in it. (The assumption of unirationality is needless in the spherical case, since X has an open B-orbit. If $c(X) = 1$, then unirationality means that $\Bbbk(X)^B \simeq \Bbbk(\mathbb{P}^1)$.)

Corollary 18.2. *$A_*(X)$ is finitely generated. If $U : X$ has finitely many orbits, then $A_*(X)$ is freely generated by U-orbit closures.*

Proof. If $c(X) = 0$, then $B : X$ has finitely many orbits, whence $A_*(X)$ is generated by B-orbit closures. If $c(X) = 1$, then, by Theorem 5.7, each irreducible B-stable subvariety $Y \subseteq X$ is either a B-orbit closure or the closure of a one-parameter family of B-orbits. In the second case, Y is one of finitely many irreducible components of $X_k = \{x \in X \mid \dim Bx \leq k\}$, $0 \leq k < \dim X$, and it follows from Lemma 5.8 that an open B-stable subset $\mathring{Y} \subseteq Y$ admits a geometric quotient $\mathring{Y}/B$ which is a smooth rational curve. Hence all B-orbits in $\mathring{Y}$ are rationally equivalent and each B-orbit, except finitely many of them, lies in one of $\mathring{Y}$. Therefore $A_*(X)$ is generated by finitely many B-orbit closures and irreducible components of X_k. □

Corollary 18.3. *If X is complete, then:*

(1) *The cone of effective cycles $A_d^+(X)_\mathbb{Q} \subseteq A_d(X) \otimes \mathbb{Q}$ is a polyhedral cone generated by the classes of rational subvarieties.*
(2) *Algebraic equivalence coincides with rational equivalence of cycles on X.*

Proof. (1) Similar to the proofs of Theorem 18.1 and Corollary 18.2.
(2) The group of cycles algebraically equivalent to 0 modulo rational equivalence is divisible [Ful1, Ex. 19.1.2]. □

Corollary 18.4. *If X is smooth and complete (projective if $c(X) = 1$), then the cycle map $A_*(X) \to \mathrm{H}_*(X)$ is an isomorphism of free Abelian groups of finite rank. (Here $\Bbbk = \mathbb{C}$, but one may also consider étale homology and Chow groups with corresponding coefficients for arbitrary $\Bbbk$.)*

Proof. If $c(X) = 0$, then it is easy to deduce from Theorem 18.1 that the Künneth map $A_*(X) \otimes A_*(Y) \to A_*(X \times Y)$ is an isomorphism for any Y, and the assertion follows from the fact that $z = \sum(u_i \cdot z)v_i$, $\forall z \in A_*(X)$, where $\sum u_i \otimes v_i$ is the class of the diagonal in $X \times X$ [FMSS, §3]. If X is projective, then one uses the Białynicki-Birula decomposition [B-B1]: X is covered by finitely many B-stable locally closed strata X_i, where each X_i is a vector bundle over a connected component X_i^T of X^T, and either $X_i^T = \mathrm{pt}$ or $X_i^T = \mathbb{P}^1$. This yields a cellular decomposition of X, and we conclude by [Ful1, Ex. 19.1.11]. □

Remark 18.5. The corollaries extend to an arbitrary variety X with an action of a connected solvable group B having finitely many orbits.

Remark 18.6. If X is not unirational, then Corollaries 18.2, 18.3(1) remain valid after replacing $A_*(X)$ by the group $B_*(X)$ of cycles modulo algebraic equivalence.

Example 18.7. If X is a generalized flag variety or a Schubert subvariety, then $A_*(X) \simeq \mathrm{H}_*(X)$ is freely generated by Schubert subvarieties in X.

Now we discuss intersection theory on varieties of complexity ≤ 1. To the end of this section, we assume that $\operatorname{char} \Bbbk = 0$.

18.2 Intersection of Divisors. Let X be a projective normal G-model of complexity ≤ 1. A method for computing intersection numbers of Cartier divisors on X was introduced by Brion [Bri4, §4] in the spherical case and generalized in [Tim3, §8] to the case of complexity 1.

Put $\dim X = d$, $c(X) = c$ $(= 0, 1)$, $r(X) = r$.

The Néron–Severi group $\mathrm{NS}(X)$ of Cartier divisors modulo algebraic equivalence is finitely generated, and the intersection form is a d-linear form on the finite-dimensional vector space $\mathrm{NS}(X) \otimes \mathbb{Q}$. This form is reconstructed via polarization from the form $\delta \mapsto \deg_X \delta^d$ on $\mathrm{NS}(X) \otimes \mathbb{Q}$ of degree d. Moreover, each Cartier divisor on X is a difference of two ample divisors, whence ample divisors form an open solid convex cone in $\mathrm{NS}(X) \otimes \mathbb{Q}$, and the intersection form is determined by values of $\deg \delta^d$ for ample δ.

Retain the notation of 17.5. Also put $A(\delta, \lambda) \equiv 1$, $\forall \lambda \in \mathscr{E}^*$, if $c = 0$, and $\mathscr{P}_+(\delta) = \{\lambda \in \mathscr{P}(\delta) \mid A(\delta, \lambda) \geq 0\}$.

Theorem 18.8. *Suppose that δ is an ample B-stable divisor on X. Then*

$$d = c + r + |\Delta_+^\vee \setminus (\Lambda + \mathbb{Z}\lambda_\delta)^\perp|, \qquad \textit{and} \tag{18.1}$$

$$\deg \delta^d = d! \int\limits_{\lambda_\delta + \mathscr{P}_+(\delta)} A(\delta, \lambda - \lambda_\delta) \prod_{\alpha^\vee \in \Delta_+^\vee \setminus (\Lambda + \mathbb{Z}\lambda_\delta)^\perp} \frac{\langle \lambda, \alpha^\vee \rangle}{\langle \rho, \alpha^\vee \rangle} \, \mathrm{d}\lambda, \tag{18.2}$$

where $\rho = \rho_G$ is half the sum of positive roots of G and the Lebesgue measure on $\Lambda \otimes \mathbb{R}$ is normalized so that a fundamental parallelepiped of Λ has volume 1.

Proof. Two cases are to be considered: $c = 0$ and $c = 1$. We have

$$\begin{aligned}
\dim \mathrm{H}^0(X, \mathscr{O}(\delta)^{\otimes n}) &= \sum_{\lambda \in n\lambda_\delta + n\mathscr{P}(\delta) \cap \Lambda} \dim V(\lambda) \cdot m_{\lambda - n\lambda_\delta}(n\delta) \\
&= \sum_{\lambda \in \lambda_\delta + \mathscr{P}(\delta) \cap \frac{1}{n}\Lambda} \dim V(n\lambda) \cdot m_{n(\lambda - \lambda_\delta)}(n\delta) \\
&= \sum_{\lambda \in \lambda_\delta + \mathscr{P}_+(\delta) \cap \frac{1}{n}\Lambda} \prod_{\alpha^\vee \in \Delta_+^\vee} \left(1 + n\frac{\langle \lambda, \alpha^\vee \rangle}{\langle \rho, \alpha^\vee \rangle}\right) \qquad \text{if } c = 0, \\
\text{or} \quad & \sum_{\lambda \in \lambda_\delta + \mathscr{P}_+(\delta) \cap \frac{1}{n}\Lambda} \prod_{\alpha^\vee \in \Delta_+^\vee} \left(1 + n\frac{\langle \lambda, \alpha^\vee \rangle}{\langle \rho, \alpha^\vee \rangle}\right) \big[nA(\delta, \lambda - \lambda_\delta) \\
&\qquad - \sigma(n\delta, n(\lambda - \lambda_\delta)) - g + 1 + h^1(n\delta, n(\lambda - \lambda_\delta))\big]
\end{aligned}$$

if $c = 1$, using the Weyl dimension formula for $\dim V(\lambda)$ [Bou2, Ch. VIII, §9, n°2]. In both cases,

$$\dim \mathrm{H}^0(X, \mathscr{O}(\delta)^{\otimes n}) \sim n^{c+r} \int\limits_{\lambda_\delta + \mathscr{P}_+(\delta)} A(\delta, \lambda - \lambda_\delta) \prod_{\alpha^\vee \in \Delta_+^\vee \setminus (\lambda_\delta + \mathscr{P}_+(\delta))^\perp} n\frac{\langle \lambda, \alpha^\vee \rangle}{\langle \rho, \alpha^\vee \rangle} \, \mathrm{d}\lambda.$$

On the other hand, the Euler characteristic $\chi(\mathscr{O}(\delta)^{\otimes n}) = \deg(\delta^d) n^d / d! + \cdots$ equals $\dim \mathrm{H}^0(X, \mathscr{O}(\delta)^{\otimes n})$ for $n \gg 0$. It remains to note that $\mathscr{P}_+(\delta)$ generates $\Lambda \otimes \mathbb{R}$, because each rational B-eigenfunction on X is a quotient of two B-eigensections of some $\mathscr{O}(\delta)^{\otimes n}$. Therefore $(\lambda_\delta + \mathscr{P}_+(\delta))^\perp = (\Lambda + \mathbb{Z}\lambda_\delta)^\perp$. □

Remark 18.9. Formula (18.1) may be proved using the local structure theorem (Theorem 4.7).

Remark 18.10. Formula (18.2) is valid for globally generated δ, because globally generated divisor classes lie on the boundary of the cone of ample divisors in $\mathrm{NS}(X) \otimes \mathbb{Q}$ and the r.h.s. of (18.2) depends continuously on δ.

Remark 18.11. The integral in the theorem can be easily computed using a simplicial subdivision of the polyhedral domain $\mathscr{P}_+(\delta)$ and Brion's integration formula [Bri4, 4.2, Rem. (ii)]:

> Suppose that F is a homogeneous polynomial of degree n on $\mathbb{R}^r$, and $[a_0, \ldots, a_r]$ is a simplex with vertices $a_i \in \mathbb{R}^r$. Then

$$\int_{[a_0,\dots,a_r]} F(\lambda)\,\mathrm{d}\lambda = \frac{r!\,\mathrm{vol}[a_0,\dots,a_r]}{(n+1)\cdots(n+r)} \Pi_r F(a_0,\dots,a_r),$$

where

$$\Pi_r F(a_0,\dots,a_r) = \frac{1}{n!} \sum_{n_0+\dots+n_r=n} \frac{\partial^n F(a_0t_0+\dots+a_rt_r)}{\partial t_0^{n_0}\cdots\partial t_r^{n_r}}.$$

Example 18.12. For toric X, $d=r$, $c=0$, and $\deg\delta^r = r!\,\mathrm{vol}\,\mathscr{P}(\delta)$ [Dan, 11.12.2], [Ful2, 5.3].

Example 18.13. If $X = G/P_I$ is a generalized flag variety, then each ample divisor δ defines an embedding $X \hookrightarrow \mathbb{P}(V(\lambda))$, $\lambda = \lambda_\delta^*$, and the degree of this embedding equals

$$|\Delta_+ \setminus (\Delta_I)_+|! \prod_{\alpha\in\Delta_+\setminus(\Delta_I)_+} \frac{\langle\lambda,\alpha^\vee\rangle}{\langle\rho,\alpha^\vee\rangle}.$$

In particular, the degree of the Plücker embedding $\mathrm{Gr}_m(\Bbbk^n) \hookrightarrow \mathbb{P}(\bigwedge^m \Bbbk^n)$ equals

$$[m(n-m)]!\frac{1!\cdots(m-1)!}{(n-m)!\cdots(n-1)!}$$

(Schubert [Sch2]).

Example 18.14. Consider the space X of complete conics of Example 17.12. Here $d=5$, $c=0$, $r=2$. If $\delta = a_1D_1 + a_2D_2$ is an ample divisor, then $\lambda_\delta = 2a_1\omega_1 + 2a_2\omega_2$. Writing $\lambda = -2x_1\alpha_1 - 2x_2\alpha_2$, we have $\mathrm{d}\lambda = \mathrm{d}x_1\,\mathrm{d}x_2$, and $\mathscr{P}(\delta) = \{\lambda \mid x_1,x_2 \geq 0,\ 2x_1 \leq x_2 + a_1,\ 2x_2 \leq x_1 + a_2\}$ is a quadrangle with the set of vertices $\{0, -a_1\alpha_1, -a_2\alpha_2, -\lambda_\delta\}$.

We have $\lambda_\delta + \lambda = (2a_1 - 4x_1 + 2x_2)\omega_1 + (2a_2 - 4x_2 + 2x_1)\omega_2$, and

$$\begin{aligned}\deg\delta^5 &= 5!\int_{\mathscr{P}(\delta)} \tfrac{(2a_1-4x_1+2x_2)(2a_2-4x_2+2x_1)(2a_1+2a_2-2x_1-2x_2)}{2}\,\mathrm{d}x_1\,\mathrm{d}x_2\\ &= a_1^5 + 10a_1^4a_2 + 40a_1^3a_2^2 + 40a_1^2a_2^3 + 10a_1a_2^4 + a_2^5.\end{aligned}$$

Polarizing this 5-form in a_1,a_2, we obtain the intersection form on $\mathrm{NS}(X) = \langle D_1,D_2\rangle$: $\deg D_1^5 = \deg D_2^5 = 1$, $\deg D_1^4D_2 = \deg D_1D_2^4 = 2$, $\deg D_1^3D_2^2 = \deg D_1^2D_2^3 = 4$ (Chasles [Cha]).

This result can be applied to solving various enumerative problems in the space O of plane conics. For example, let us find the number of conics tangent to 5 given conics in general position. The set of conics tangent to a given one is a prime divisor $D \subset O$. It is easy to see that (the closure of) D intersects all G-orbits in X properly. By Kleiman's transversality theorem (see [Har2, III.10.8] and 18.5) five general translates g_iD ($g_i \in G$, $i = 1,\dots,5$) are transversal and intersect only inside O. Thus the number we are looking for equals $\deg_X D^5$.

Using local coordinates, one sees that the degree of (the closure of) D in $\mathbb{P}^5$ or $(\mathbb{P}^5)^*$ equals 6. (Take, e.g., a parabola $\{y = x^2\}$. A conic $\{q(x,y) = 0\}$ is tangent to this parabola if and only if $q(x,x^2) = 0$ and $\begin{vmatrix} 2x & \frac{\partial q}{\partial x}(x,x^2) \\ -1 & \frac{\partial q}{\partial y}(x,x^2) \end{vmatrix} = 0$ for some x. The resultant of these two polynomials has degree 7 in the coefficients of q. Cancelling it by the coefficient at y^2, we obtain the equation of D of degree 6.) Since $\deg_{\mathbb{P}^5} D_1 = \deg_{(\mathbb{P}^5)^*} D_2 = 1$ and $\deg_{\mathbb{P}^5} D_2 = \deg_{(\mathbb{P}^5)^*} D_1 = 2$, one has $D \sim 2D_1 + 2D_2$ (on X) and $\deg_X D^5 = 2^5(1+10+40+40+10+1) = 3264$.

Spaces of conics and of quadrics in higher dimensions were studied intensively from the origin of enumerative geometry [Cha], [Sch1], [Sch3]. For a modern approach, see [Sem1], [Sem2], [Tyr], [CGMP], [Bri4].

Example 18.15. Let X be a completion of the space O of ordered triangles from Example 16.24 with $d = 6$, $c = 1$, $r = 2$. Consider an ample divisor $\delta = a_1\widetilde{D} + a_2 D$, where $D = D_1 + D_2 + D_3$ imposes the condition that one of the vertices of a triangle lies on the B-stable line in $\mathbb{P}^2$ and $\widetilde{D} = \widetilde{D}_1 + \widetilde{D}_2 + \widetilde{D}_3$ imposes the condition that a triangle passes through the B-fixed point.

Writing $\lambda = -x_1\alpha_1 - x_2\alpha_2$, we have $d\lambda = dx_1\,dx_2$, $\mathscr{P}(\delta) = \{\lambda \mid x_1, x_2 \geq 0\}$, $\lambda_\delta = 3a_1\omega_1 + 3a_2\omega_2$, and

$$A(\delta,\lambda) = \begin{cases} A_0(\lambda) = x_1 + x_2, & x_i \leq a_i, \\ A_1(\lambda) = 3a_1 - 2x_1 + x_2, & 0 \leq x_1 - a_1 \geq x_2 - a_2, \\ A_2(\lambda) = 3a_2 - 2x_2 + x_1, & 0 \leq x_2 - a_2 \geq x_1 - a_1. \end{cases}$$

It follows that $\mathscr{P}_+(\delta) = \{\lambda \mid x_1, x_2 \geq 0;\ 2x_1 \leq x_2 + 3a_1;\ 2x_2 \leq x_1 + 3a_2\} = \mathscr{P}_0 \cup \mathscr{P}_1 \cup \mathscr{P}_2$, where $\mathscr{P}_i$ are quadrangles with vertices

$$\begin{gathered} \{0, -a_1\alpha_1, -a_1\alpha_1 - a_2\alpha_2, -a_2\alpha_2\}, \\ \{-a_1\alpha_1, -\tfrac{3a_1}{2}\alpha_1, -\lambda_\delta, -a_1\alpha_1 - a_2\alpha_2\}, \\ \{-a_2\alpha_2, -\tfrac{3a_2}{2}\alpha_2, -\lambda_\delta, -a_1\alpha_1 - a_2\alpha_2\}, \end{gathered}$$

and $A(\delta,\lambda) = A_i(\lambda)$ on $\mathscr{P}_i$. We have $\lambda_\delta + \lambda = (3a_1 - 2x_1 + x_2)\omega_1 + (3a_2 - 2x_2 + x_1)\omega_2$, and

$$\begin{aligned} \deg \delta^6 = 6!\Big(& \int_{\mathscr{P}_0} (x_1 + x_2) \cdot \frac{(3a_1 - 2x_1 + x_2)(3a_2 - 2x_2 + x_1)(3a_1 + 3a_2 - x_1 - x_2)}{2}\, dx_1\, dx_2 \\ & + \int_{\mathscr{P}_1} (3a_1 - 2x_1 + x_2) \cdot \frac{(3a_1 - 2x_1 + x_2)(3a_2 - 2x_2 + x_1)(3a_1 + 3a_2 - x_1 - x_2)}{2}\, dx_1\, dx_2 \\ & + \int_{\mathscr{P}_2} (3a_2 - 2x_2 + x_1) \cdot \frac{(3a_1 - 2x_1 + x_2)(3a_2 - 2x_2 + x_1)(3a_1 + 3a_2 - x_1 - x_2)}{2}\, dx_1\, dx_2 \Big) \\ = {} & 90a_1^6 + 1080a_1^5 a_2 + 4320a_1^4 a_2^2 + 6840a_1^3 a_2^3 + 4320a_1^2 a_2^4 + 1080a_1 a_2^5 + 90a_2^6. \end{aligned}$$

It follows that $\deg D^6 = \deg \widetilde{D}^6 = 90$, $\deg D^5\widetilde{D} = \deg D\widetilde{D}^5 = 180$, $\deg D^4\widetilde{D}^2 = \deg D^2\widetilde{D}^4 = 288$, $\deg D^3\widetilde{D}^3 = 342$.

Since X has finitely many orbits and $D, \widetilde{D}$ intersect all of them properly, it follows from Kleiman's transversality theorem that any 6 general translates δ_i of $D, \widetilde{D}$ are transversal and intersect only inside O. Thus the number of common points of δ_i in O equals $\deg_X(\delta_1 \cdots \delta_6)$. Dividing it by 6 (=the number of ordered triangles corresponding to a given unordered triangle), we obtain the number of triangles satisfying 6 conditions imposed by δ_i. For example, there are $\deg(D^3\widetilde{D}^3)/6 = 57$ triangles passing through 3 given points in general position whose vertices lie on 3 given general lines.

Theorem 18.8 was applied in [Tim3, §10] to computing the degree of a closed 3-dimensional orbit in any SL_2-module.

Brion [Bri9, 4.1] proved a formula similar to (18.2) for the multiplicity of a spherical variety along an orbit in it and deduced a criterion of smoothness for spherical varieties [Bri9, 4.2].

18.3 Divisors and Curves. For any complete variety X, there is a canonical pairing $\operatorname{Pic} X \times A_1(X) \to \mathbb{Z}$ given by the degree of a line bundle restricted to a curve in X (and pulled back to its normalization). The following theorem is essentially due to Brion [Bri10], [Bri12].

Theorem 18.16. (1) *If X is a complete unirational G-model of complexity ≤ 1, then* $\operatorname{Pic} X \hookrightarrow A_1(X)^* = \operatorname{Hom}(A_1(X), \mathbb{Z})$ *via the canonical pairing.*
(2) *If X is complete, normal and spherical, then* $\operatorname{Pic} X \xrightarrow{\sim} A_1(X)^*$.
(3) *If in addition X contains a unique closed G-orbit Y, then $A_1(X)$ is torsion-free, and the basis of $A_1(X)$ dual to the basis $\mathscr{D}^B \setminus \mathscr{D}^B_Y$ of* $\operatorname{Pic} X$ *consists of (classes of) irreducible B-stable curves. Moreover, these basic curves generate the semigroup $A_1^+(X)$ of effective 1-cycles.*

Proof. Using the equivariant Chow lemma [Sum, Th. 2] and resolution of singularities, we construct a proper birational G-morphism $\varphi : \check{X} \to X$, where $\check{X}$ is a smooth projective G-variety. In the commutative diagram

$$\begin{array}{ccc} \operatorname{Pic} X & \longrightarrow & A_1(X)^* \\ \downarrow & & \downarrow \\ \operatorname{Pic} \check{X} & \longrightarrow & A_1(\check{X})^* \end{array}$$

the vertical arrows are injections, and the bottom arrow is an isomorphism by Corollary 18.4 and by Poincaré duality, whence (1). Assertions (2), (3) are proved in [Bri12, §3] using the description of B-stable curves and their equivalences on spherical varieties obtained in [Bri10], [Bri12]. □

Remark 18.17 ([Bri10, 1.6, 2.1]). On a spherical G-model X, any line bundle $\mathscr{L}$ is B-linearized and any B-stable curve C is the closure of a 1-dimensional B-orbit. Let ∞ be a B-fixed point in the normalization $\mathbb{P}^1$ of C and let $0 \in \mathbb{A}^1 = \mathbb{P}^1 \setminus \{\infty\}$ be another T-fixed point. Then T acts on $\mathbb{A}^1 \setminus \{0\}$ via a character $\chi \neq 0$. Let $x, y \in C$ be the images of $0, \infty$ under the normalization map $\nu : \mathbb{P}^1 \to C$, and let χ_x, χ_y be the weights of the T-action on $\mathscr{L}_x, \mathscr{L}_y$. Then $\chi_x - \chi_y$ is a multiple of χ, and $\langle \mathscr{L}, C\rangle = \deg \nu^*\mathscr{L}|_C = (\chi_x - \chi_y)/\chi$.

Example 18.18. For a generalized flag variety $X = G/P_I$, $\operatorname{Pic} X$ is freely generated by Schubert divisors $D_{\alpha_i} = \overline{B(w_G s_i o)}$, and $A_1(X)$ is freely generated by Schubert curves $C_{\alpha_i} = \overline{B(s_i o)} \simeq P_{\alpha_i}/B \simeq \mathbb{P}^1$. We have $\mathscr{O}(D_{\alpha_i}) = G *_{P_I} \Bbbk_{-\omega_i}$ and $\mathscr{O}(D_{\alpha_i})|_{C_{\alpha_j}} = P_{\alpha_j} *_B \Bbbk_{-\omega_i} = \mathscr{O}(1)$ if $i = j$ and $\mathscr{O}(0)$ if $i \neq j$. Here the T-fixed points are $s_j o, o$, $\chi = \alpha_j$, and $\langle D_{\alpha_i}, C_{\alpha_j}\rangle = (\chi_{s_j o} - \chi_o)/\chi = (-s_j\omega_i + \omega_i)/\alpha_j = \delta_{ij}$. Hence the above bases of $\operatorname{Pic} X$ and $A_1(X)$ are dual to each other.

Projective unirational normal varieties of complexity ≤ 1 are well-behaved from the viewpoint of the Mori theory (see Appendix A.2).

Theorem 18.19 ([Bri10], [BKn]). *Suppose that X is a projective unirational G-model of complexity ≤ 1. Then the Mori cone* $\operatorname{NE}(X)$ *is finitely generated by rational B-stable curves and all faces of* $\operatorname{NE}(X)$ *are contractible. If X is $\mathbb{Q}$-factorial, then each contraction of an extremal ray of* $\operatorname{NE}(X)$ *which is an isomorphism in codimension* 1 *can be flipped, and every sequence of directed flips terminates.*

18.4 Chow Rings. Explicit computations of Chow rings for some smooth completions of classical homogeneous spaces were carried on by several authors. Schubert, Pieri, Giambelli, A. Borel, Kostant, Bernstein–Gelfand–Gelfand, Demazure, Lakshmibai–Musili–Seshadri et al contributed to computing Chow (or cohomology) rings of generalized flag varieties.

Here and below we put $A^k(X) = A_{d-k}(X)$, $d = \dim X$.

Without loss of generality, assume that G is semisimple simply connected. Let $\mathfrak{X} = \mathfrak{X}(B)$ be the weight lattice of G. Every $\lambda \in \mathfrak{X}$ defines an induced line bundle $\mathscr{L}(-\lambda) = G *_B \Bbbk_{-\lambda}$ on G/B, and this gives rise to an isomorphism $\mathfrak{X} \xrightarrow{\sim} \operatorname{Pic} G/B \simeq A^1(G/B)$. Put $S = \mathrm{S}^\bullet(\mathfrak{X} \otimes \mathbb{Q})$.

Theorem 18.20. (1) [Bor1], [Dem2] *$A^*(G/B) \otimes \mathbb{Q} \simeq S/SS_+^W$, where $S_+^W \lhd S^W$ is the ideal of W-invariants without constant term.*
(2) [BGG, 5.5] *If $I \subseteq \Pi$, then $A^*(G/P_I) \otimes \mathbb{Q}$ embeds in $A^*(G/B) \otimes \mathbb{Q}$ as $S^{W_I}/S^{W_I}S_+^W$, where $W_I \subseteq W$ is the standard Coxeter subgroup generated by r_α, $\alpha \in I$.*

Bernstein–Gelfand–Gelfand [BGG] and Demazure [Dem2] used divided difference operators to introduce certain functionals D_w on S which represent Schubert cells S_w ($w \in W$) via the Poincaré duality. They also found the basis of S/SS_+^W dual to D_w.

Chow rings of toric varieties were computed by Jurkiewicz and Danilov, cf. [Dan, §10], [Ful2, 5.2]. Namely, if X is a smooth complete toric variety, then $A^*(X) = \mathrm{S}^*(\operatorname{Pic} X)/I$, where the ideal I is generated by monomials $[D_1]\cdots[D_k]$ such that D_i are T-stable prime divisors and $D_1 \cap \dots \cap D_k = \emptyset$.

The above examples are spherical (see also 27.5, 29.5). In the case of complexity 1, Chow rings of complete SL_2-embeddings (cf. Example 16.23) were computed in [M-J2]. The space of complete triangles, which is a desingularization of the space X of Example 16.24, was studied in [CF], and, in particular, its Chow ring was determined there.

18.5 Halphen Ring. Many enumerative problems arise on non-complete homogeneous spaces. Given a homogeneous space $O = G/H$, typically a space of geometric figures or tensors of certain type, one looks for the number of points satisfying a number of conditions in general position. The set of points satisfying a given condition is a closed subvariety $Z \subset O$, and the configuration of conditions $Z_1, \dots, Z_s \subset O$ is put in general position by replacing Z_i with their translates $g_i Z_i$, where $(g_1, \dots, g_s) \in G \times \cdots \times G$ is a general s-tuple. By Kleiman's transversality theorem [Kle], [Har2, III.10.8], the cycles $g_1 Z_1, \dots, g_s Z_s$ intersect transversally in smooth subvarieties of codimension $\sum \operatorname{codim} Z_i$, i.e., $g_1 Z_1 \cap \cdots \cap g_s Z_s$ is empty if $\sum \operatorname{codim} Z_i > \dim O$ and finite if $\sum \operatorname{codim} Z_i = \dim O$, and the cardinality $|g_1 Z_1 \cap \cdots \cap g_s Z_s|$ is stable for general $(g_1, \dots, g_s)$. Thus the natural intersection ring for the enumerative geometry of O is provided by the following

Definition 18.21 ([CP2]). The *intersection number* of irreducible subvarieties $Z_1, \dots, Z_s$ whose codimensions sum up to $\dim O$ is $(Z_1 \cdots Z_s)_O = |g_1 Z_1 \cap \cdots \cap g_s Z_s|$ for all $(g_1, \dots, g_s)$ in a dense open subset of $G \times \cdots \times G$.

This defines a pairing between groups of cycles in O of complementary dimensions. The *group of conditions* $C^*(O)$ is the quotient of the group of all cycles modulo the kernel of this pairing. Write $[[Z]]$ for the image in $C^*(O)$ of a cycle Z.

Theorem 18.22 ([CP2, 6.3]). *If O is spherical, then $C^*(O)$ is a graded ring with respect to the* intersection product $[[Z]] \cdot [[Z']] = [[gZ \cap g'Z']]$, *where $(g, g') \in G \times G$ is a general pair. Furthermore, $C^*(O) = \varinjlim A^*(X)$ over all smooth complete G-embeddings $X \hookleftarrow O$.*

Proof. The proof goes in several steps.

(1) For any subvariety $Z \subset O$ and any smooth complete G-embedding $X \hookleftarrow O$, there is a smooth complete G-embedding $X' \hookleftarrow O$ dominating X such that $\overline{Z}$ intersects all G-orbits in X' properly. First, one constructs a smooth toroidal embedding X' dominating X. Then each G-orbit on X' is a transversal intersection of G-stable prime divisors (Theorem 29.2), and one applies a general result [CP2, 4.7] that a cycle on a complete smooth variety can be put in regular position with respect to a regular configuration of hypersurfaces by blowing up several intersections of pairs of these hypersurfaces.

(2) If the closures of $Z, Z' \subset O$ intersect all G-orbits in $X \setminus O$ properly, then $[\overline{gZ \cap g'Z'}] = [\overline{Z}] \cdot [\overline{Z'}]$ in $A^*(X)$ for general $g, g' \in G$. Indeed, we may apply Kleiman's transversality theorem to intersections of $\overline{Z}, \overline{Z'}$ with each of finitely many G-orbits in X and deduce that $\overline{gZ}$ and $\overline{g'Z'}$ intersect properly with each other and $\overline{gZ} \cap \overline{g'Z'}$ intersects $X \setminus O$ properly.

(3) For any $z \in A^*(X)$, use the Chow moving lemma to represent z as $z = \sum m_i [\overline{Z_i}]$, where $Z_i \subset O$ are closed subvarieties. For any subvariety $Z' \subset O$ of complementary dimension, we may assume by (1) that $\overline{Z_i}, \overline{Z'}$ intersect all orbits in $X \setminus O$ properly and deduce from (2) that $([[Z']] \cdot \sum m_i [[Z_i]])_O = \deg_X([\overline{Z'}] \cdot z)$ depends only on z. Thus we have a well-defined map $A^*(X) \to C^*(O)$, $z \mapsto \sum m_i [[Z_i]]$.

(4) This map gives rise to a homomorphism $\varinjlim A^*(X) \to C^*(O)$ by (2). Its surjectivity is obvious, and injectivity follows from the Poincaré duality on X. □

The ring of conditions $C^*(O)$ is also called the *Halphen ring* of O in honor of G.-H. Halphen, who used it implicitly in the enumerative geometry of conics, see [CX]. If O is a torus, then $C^*(O)$ is McMullen's polytope algebra [FS], [PK], [Bri11, 3.3]. The Halphen ring of the space of plane conics was computed in [CX].

18.6 Generalization of the Bézout Theorem. Theorem 18.22 reflects an idea exploited already by classical geometers that in solving enumerative problems on O one has to consider an appropriate completion $X \hookleftarrow O$ with finitely many orbits such that all "conditions" Z_i under consideration intersect all orbits in $X \setminus O$ properly. (Such a completion always exists in the spherical case.) If $\sum \operatorname{codim} Z_i = \dim O$, then, for general g_i, all intersection points of $\bigcap g_i Z_i$ lie in O, and the intersection number equals $\deg_X \prod [Z_i]$. Applications of this idea can be found in Examples 18.14, 18.15.

Generalizing these examples, we describe a method for computing the intersection number of d divisors on a spherical homogeneous space O of dimension d in characteristic 0. Let δ be an effective divisor on O. Replacing δ by a G-translate, we may assume that δ contains no colors. Let $h \in K$ be an equation of δ on the open B-orbit in O, which is a factorial variety.

Definition 18.23. The *Newton polytope* of δ is the set

$$\mathscr{N}(\delta) = \{\lambda \in \mathscr{E}^* \mid \forall v \in \mathscr{V}: \langle v, \lambda\rangle \geq v(h),\ \forall D \in \mathscr{D}^B: \langle \varkappa(D), \lambda\rangle \geq v_D(h)\}$$

Remark 18.24. We see below that $\mathscr{N}(\delta)$ is indeed a convex polytope in $\mathscr{E}^*$.

Example 18.25. If $G = O = T$ is a torus, then $\mathscr{D}^B = \emptyset$, $h = \sum c_i\lambda_i$, $c_i \in \Bbbk^\times$, $\lambda_i \in \mathfrak{X}(T)$, and $v(h) = \min\langle v, -\lambda_i\rangle$ (cf. Example 20.2). Thus $\mathscr{N}(\delta) = -\operatorname{conv}\{\lambda_1, \dots, \lambda_s\}$ is a usual Newton polytope.

Example 18.26. More generally, suppose that O is quasiaffine and $\delta = \operatorname{div} h$ is principal. We have $h = h_1 + \dots + h_s$, $h_i \in \Bbbk[O]_{(\lambda_i)}$, and $v(h) = \min\langle v, \lambda_i\rangle$, $\forall v \in \mathscr{V}$ (see Remark 19.11 and 21.1). Also $v_D(h) = 0$, $\forall D \in \varkappa(\mathscr{D}^B)$, since δ contains no colors. Hence $\mathscr{N}(\delta) = (\operatorname{conv}\{\lambda_1, \dots, \lambda_s\} + \mathscr{V}^\vee) \cap (\mathbb{Q}_+\varkappa(\mathscr{D}^B))^\vee$.

For any embedding $X \hookleftarrow O$, we have $\operatorname{div} h = \delta - \delta_X$ on X, where $\delta_X = -\sum_i v_i(h) V_i - \sum_{D \in \mathscr{D}^B} v_D(h) D$ is a B-stable divisor and V_i are G-stable prime divisors on X with valuations $v_i \in \mathscr{V}$.

Theorem 18.27. $\mathscr{N}(\delta) = \bigcap_{X \hookleftarrow O} \mathscr{P}(\delta_X)$. *If X is complete and δ intersects all G-orbits in X properly, then $\mathscr{N}(\delta) = \mathscr{P}(\delta_X)$.*

Proof. The first assertion is obvious from (17.4), since every G-valuation corresponds to a divisor on some embedding of X. Suppose that X is complete and δ intersects all orbits properly. Consider a G-linearized line bundle $\mathscr{L} = \mathscr{O}(\delta) \simeq \mathscr{O}(\delta_X)$. Take any $\lambda \in \mathscr{P}(\delta_X)$ and choose $n \in \mathbb{N}$ such that there exists a nonzero section $\eta \in \mathrm{H}^0(X, \mathscr{L}^{\otimes n})^{(B)}_{n\lambda + n\lambda_{\delta_X}}$; then $\eta = c\mathbf{f}_{n\lambda}\sigma_X^n = c\mathbf{f}_{n\lambda}\sigma^n/h^n$, where $c \in \Bbbk^\times$, $\operatorname{div}\sigma_X = \delta_X$, $\operatorname{div}\sigma = \delta$. Take any $v \in \mathscr{V}$. The subdivision of the colored fan of X by the ray $\mathbb{Q}_+ v$ corresponds to a proper birational G-morphism $\nu : \check{X} \to X$ contracting the divisor $D \subset \check{X}$ with $v_D = v$ to the center $Y \subset X$ of v. Since δ intersects Y

properly, D does not occur in the pullback of δ to $\check{X}$, whence $v_D(\nu^*\sigma) = 0$ and $v_D(\nu^*\eta) = v(\mathbf{f}_{n\lambda}/h^n) \geq 0 \implies \langle v, \lambda \rangle \geq v_D(h) \implies \mathscr{P}(\delta_X) \subseteq \mathscr{N}(\delta)$. □

Corollary 18.28 ([Bri11, 4.2]). *For any effective divisor δ on O containing no colors,*

$$(\delta^d)_O = d! \int\limits_{\lambda_\delta + \mathscr{N}(\delta)} \prod_{\alpha^\vee \in \Delta_+^\vee \setminus (\Lambda + \mathbb{Z}\lambda_\delta)^\perp} \frac{\langle \lambda, \alpha^\vee \rangle}{\langle \rho, \alpha^\vee \rangle} \, d\lambda, \tag{18.3}$$

where $\lambda_\delta = -\sum_{D \in \mathscr{D}^B} v_D(h)\lambda_D$

Proof. Follows from Theorems 18.22, 18.27, and 18.8. □

Remark 18.29. In the toric case, (18.3) transforms to $(\delta^d)_O = d! \operatorname{vol} \mathscr{N}(\delta)$. Polarizing this formula, we obtain a theorem of Bernstein [Ber] and Kouchnirenko [Kou]: for any effective divisors $\delta_1, \ldots, \delta_d$ on O, the intersection number $(\delta_1 \cdots \delta_d)_O$ is $d!$ times the mixed volume of $\mathscr{N}(\delta_1), \ldots, \mathscr{N}(\delta_d)$. In the general case, we have a "mixed integral" instead [Bri11, 4.2].

Corollary 18.28 may be considered as a generalization of the classical Bézout theorem.

Example 18.30. Consider $O = G$ as a homogeneous space under the $(G \times G)$-action by left/right multiplication. Here we do not assume that G is of simply connected type. Choose a Borel subgroup $B^- \times B \subseteq G \times G$ and a maximal torus $T \times T \subseteq B^- \times B$. We have $\Lambda(G) = \{(-\lambda, \lambda) \mid \lambda \in \mathfrak{X}(T)\} \simeq \mathfrak{X}(T)$, the colors are $\overline{B^- r_\alpha B}$ ($\alpha \in \Pi$), $\varkappa(\mathscr{D}^{B^- \times B}) = \Pi^\vee$, and $\mathscr{V} = \mathbf{C}(\Delta_-^\vee)$ is the antidominant Weyl chamber, see 27.2. Let δ be an effective divisor on G. In computing the intersection number, there is no essential loss of generality to assume that $\delta = \operatorname{div} h$ for some $h \in \Bbbk[G]$, because a finite cover of G is a factorial variety. Consider the isotypic decomposition $h = h_1 + \cdots + h_s$, $h_i \in \Bbbk[G]_{(\lambda_i)} = \mathrm{M}(V(\lambda_i))$. By Example 18.26,

$$\mathscr{N}(\delta) = (\operatorname{conv}\{\lambda_1, \ldots, \lambda_s\} - \mathbb{Q}_+\Pi) \cap (\mathbb{Q}_+\Pi^\vee)^\vee = \operatorname{conv} W\{\lambda_1, \ldots, \lambda_s\} \cap \mathbf{C}$$

is the dominant part of the weight polytope of $V = V(\lambda_1) \oplus \cdots \oplus V(\lambda_s)$. The positive roots of $G \times G$ with respect to the chosen Borel subgroup $B^- \times B$ are $(-\alpha, 0)$, $(0, \alpha)$ ($\alpha \in \Delta_+$), and half the sum of positive roots equals $(-\rho, \rho)$. Formula (18.3) transforms into

$$(\delta^d)_G = d! \int\limits_{\mathscr{N}(\delta)} \prod_{\alpha^\vee \in \Delta_+^\vee} \frac{\langle \lambda, \alpha^\vee \rangle^2}{\langle \rho, \alpha^\vee \rangle^2} \, d\lambda, \tag{18.4}$$

where $d = \dim G$. This formula was first obtained by Kazarnovskii, who used a different method based on the moment map [Kaz].

Chapter 4
Invariant Valuations

This chapter plays a significant, but auxiliary, rôle in the general context of our survey. We investigate the set of G-invariant valuations of the function field of a G-variety. We have seen in Chap. 3 that G-valuations are of importance in the embedding theory, because they provide a material for constructing combinatorial objects (colored data) that describe equivariant embeddings.

Remarkably, a G-valuation of a given G-field is uniquely determined by its restriction to the multiplicative group of B-eigenfunctions, the latter being a direct product of the weight lattice and of the multiplicative group of B-invariant functions. Thus a G-valuation is essentially a pair composed by a linear functional on the weight lattice and by a valuation of the field of B-invariants. Under these identifications, we prove in §20 that the set of G-valuations is a union of convex polyhedral cones in certain half-spaces.

The common face of these valuation cones is formed by those valuations, called central, that vanish on B-invariant functions. The central valuation cone controls the situation "over the field of B-invariant functions". For instance, its linear part determines the unity component of the group of G-automorphisms acting identically on B-invariants.

This cone has another remarkable property: it is a fundamental chamber of a crystallographic reflection group called the little Weyl group of a G-variety. This group is defined in §22 as the Galois group of a certain symplectic covering of the cotangent bundle constructed in terms of the moment map. The little Weyl group is linked with the central valuation cone via the invariant collective motion on the cotangent variety, which is studied in §23.

For practical applications, we must be able to compute the set of G-valuations. For central valuations, it suffices to know the little Weyl group. In §24 we describe the "method of formal curves" for computing G-valuations on a homogeneous space. Informally, one computes the order of functions at infinity along a formal curve approaching a boundary G-divisor.

Most of the results of this chapter are due to D. Luna and Th. Vust, M. Brion, F. Pauer, and F. Knop. We follow [LV], [Kn3], [Kn5] in our exposition.

D.A. Timashev, *Homogeneous Spaces and Equivariant Embeddings*,
Encyclopaedia of Mathematical Sciences 138, DOI 10.1007/978-3-642-18399-7_4,

19 G-valuations

An algebraic counterpart of a prime divisor on an algebraic variety is the respective valuation of the field of rational functions. Valuations obtained in this way are called geometric (see Appendix B). We consider invariant geometric valuations.

19.1 Basic Properties. Let G be a connected algebraic group and let K be a G-field, i.e., the function field of an irreducible G-variety.

Definition 19.1. A *G-valuation* is a G-invariant geometric valuation of $K/\Bbbk$. The set of G-valuations is denoted by $\mathscr{V} = \mathscr{V}(K)$.

The following approximation result is due to Sumihiro.

Proposition 19.2 ([Sum, §4]). *For any geometric valuation v of K there exists a G-valuation $\overline{v}$ such that for any $f \in K$ one has $\overline{v}(f) = v(gf)$ for general $g \in G$. If $A \subset K$ is a rational G-algebra, then for any $f \in A$ one has $\overline{v}(f) = \min_{g \in G} v(gf)$.*

Proof. We may assume that $v = v_D$ for a prime divisor D on a model X of K. Then $v' = v_{G \times D}$ is a geometric valuation of $\Bbbk(G \times X)$. It is clear that for any $f \in \Bbbk(G \times X)$ one has $v'(f) = v(f(g, \cdot))$ for general $g \in G$. The rational action $G : X$ induces an embedding $\Bbbk(X) \hookrightarrow \Bbbk(G \times X)$. It is easy to see using Proposition B.8(1) that $\overline{v} = v'|_{\Bbbk(X)}$ is the desired G-valuation.

To prove the second assertion, observe that $A^{(d)} = \{f \in A \mid v(f) \geq d\}$ is a filtration of A by linear subspaces, Gf is an algebraic variety and $Gf \cap A^{(d)}$ is a closed subvariety in Gf, $\forall d \in \mathbb{Q}$. □

Remark 19.3. If v is non-geometric, then there still exists a G-invariant valuation $\overline{v}$ satisfying the above properties, but $\overline{v}$ may be no longer geometric. It is constructed in the same way as above, where v' is now *defined* by the formula $v'(f) = v(f(g, \cdot))$ (for general $g \in G$). The results of this section can be extended to the non-geometric case with appropriate modifications.

Remark 19.4. If X is a G-model of K and v has center $Y \subseteq X$, then $\overline{v}$ has center $\overline{GY}$.

Example 19.5. Let $G = \Bbbk$ act rationally on the blow-up X of $\mathbb{A}^2$ at 0 by translations along a fixed axis. In coordinates, $u(x, y) = (x + u, y)$, $\forall u \in G$, $(x, y) \in \mathbb{A}^2$. The valuation v of $\Bbbk(X)$ corresponding to the exceptional divisor is given on $\Bbbk[\mathbb{A}^2] = \Bbbk[x, y]$ by the order of a polynomial in x, y (i.e., the lowest degree of a homogeneous term) and has center 0 on $\mathbb{A}^2$. But $\overline{v}(f) = \min_u v(f(x + u, y))$ is the order of f in y, so that $\overline{v} = v_D$, where $D = \{y = 0\}$ is (the proper pullback of) the x-axis.

Together with Proposition B.8, Sumihiro's approximation immediately implies

Corollary 19.6. *Let $K' \subseteq K$ be a G-subfield. The restriction of a G-valuation of K to K' is a G-valuation, and any G-valuation of K' can be extended to a G-valuation of K.*

The next corollary is useful in applications.

Corollary 19.7. *Let X be a G-model of K and let $\mathscr{L}$ be a G-line bundle on X. Then for any $v \in \mathscr{V}$, any $\sigma, \eta \in \mathrm{H}^0(X, \mathscr{L})$, $\eta \neq 0$, and any $g \in G$ one has $v(\sigma/\eta) = v(g\sigma/\eta)$.*

Proof. Consider a rational G-algebra $R = \bigoplus_{n \geq 0} \mathrm{H}^0(X, \mathscr{L}^{\otimes n})$. Then $\operatorname{Quot} R = K'(\eta)$ is a (purely transcendental) extension of a G-subfield $K' \subseteq K$ (consisting of functions representable as ratio of sections of some $\mathscr{L}^{\otimes n}$). Now apply Corollary 19.6 to extend v to R and conclude by $v(\sigma/\eta) = v(\sigma) - v(\eta) = v(g\sigma) - v(\eta) = v(g\sigma/\eta)$. □

A natural geometric characterization of G-valuations is given by

Proposition 19.8. *Any G-valuation is proportional to v_D for a G-stable prime divisor D on a normal G-model X of K.*

Proof. Let $v \in \mathscr{V}$ and choose $f_1, \dots, f_s \in \mathscr{O}_v$ whose residues generate $\Bbbk(v)$. Take a normal projective G-model X of K and a G-line bundle $\mathscr{L}$ on X such that $f_i = \sigma_i/\sigma_0$ for some $\sigma_0, \dots, \sigma_s \in \mathrm{H}^0(X, \mathscr{L})$. Let $M \subseteq \mathrm{H}^0(X, \mathscr{L})$ be the G-submodule generated by $\sigma_0, \dots, \sigma_s$. The respective rational map $\varphi : X \dashrightarrow \mathbb{P}(M^*)$ is G-equivariant. Replacing X by the normalized closure of the graph of φ, we may assume that φ is a G-morphism. Corollary 19.7 implies $v(M/\sigma_0) \geq 0$, whence the center $Y \subseteq X' = \varphi(X)$ of $v|_{\Bbbk(X')}$ intersects an affine chart X'_{σ_0}, and $f_1, \dots, f_s \in \mathscr{O}_{X',Y}$. Therefore, if D is the center of v on X, then $f_1, \dots, f_s \in \mathscr{O}_{X,D}$, whence D is a divisor. □

Here is a relative version of this proposition.

Proposition 19.9. *Suppose that a G-valuation v has the center Y on a G-model X of K. Then there exists a normal G-model X' and a projective morphism $\varphi : X' \to X$ such that the center of v on X' is a divisor D' and $\varphi(D') = Y$.*

Proof. Take any projective G-model X' such that the center of v is a divisor $D' \subset X'$. The rational map $\varphi : X' \dashrightarrow X$ is defined on an open subset intersecting D', and $\overline{\varphi(D')} = Y$. Now we replace X' by the normalized closure $\widetilde{X}$ of the graph of φ in $X' \times X$. Since $\widetilde{X}$ projects onto X' isomorphically over the domain of definition of φ, we can lift D' to $\widetilde{X}$. □

19.2 Case of a Reductive Group. From now on, G is a connected reductive group.

Lemma 19.10. *If $A \subset K$ is a rational G-algebra, then for any $v \in \mathscr{V}$, $f \in A$ one has $v(f) = \min_{\widetilde{f} \in (M^q)^{(B)}} v(\widetilde{f})/q$, where $M \subset A$ is a G-submodule generated by f, and q is a sufficiently big power of p.*

Proof. As M is generated by f, we have $v(M) \geq v(f)$, whence $v((M^q)^{(B)}) \geq qv(f)$. To prove that the equality is reached, in characteristic zero ($\implies q = 1$) it suffices to note that M is generated by $M^{(B)} \implies v(f) \in v(M) \geq \min v(M^{(B)})$. In the general case, this is not true, and one has to consider powers of M. We organize them in a graded G-algebra $R = \bigoplus_{n \geq 0} M^n$ and consider a graded G-stable ideal $I = \bigoplus I_n \lhd R$, $I_n = \{h \in M^n \mid v(h) > nv(f)\}$.

As $M \not\subseteq I$, there exists $r \in M$ such that $0 \neq r \bmod I \in (R/I)^{(B)}$. By Lemma D.1, $(R/I)^U$ is a purely inseparable finite extension of R^U/I^U. Hence there exists $h \in I_q$ such that $\widetilde{f} = r^q + h \in R_q^{(B)}$, and $v(\widetilde{f}) = v(r^q) = qv(f)$. □

Remark 19.11. In characteristic zero or for $G = T$ Lemma 19.10 yields $v(f) = \min_{\lambda \in \mathfrak{X}_+} v(f_{(\lambda)})$, where $f_{(\lambda)}$ is the projection of f to the isotypic component $A_{(\lambda)}$ of A.

Recall from 13.1 that $\mathscr{D}$ denotes the set of non-G-stable prime divisors on (any) G-model of K and $K_B \subseteq K$ is the subalgebra of rational functions with B-stable divisor of poles.

The following approximation lemma of Knop [Kn3, 3.5] allows us to simplify the study of G-valuations by restricting to B-eigenfunctions.

Lemma 19.12. *For any G-valuation $v \in \mathscr{V}$ and any rational function $f \in K_B$ there exists a rational B-eigenfunction $\widetilde{f} \in K^{(B)}$ such that:*

$$\begin{cases} v(\widetilde{f}) = v(f^q), \\ w(\widetilde{f}) \geq w(f^q), & \forall w \in \mathscr{V}, \\ v_D(\widetilde{f}) \geq v_D(f^q), & \forall D \in \mathscr{D}^B, \end{cases}$$

where q is a sufficiently big power of p.

Proof. Let X be a normal G-model and $\delta = \operatorname{div}_\infty f$, the divisor of poles on X. Then $\sigma = f\eta_\delta \in \mathrm{H}^0(X, \mathscr{O}(\delta))$, where η_δ is the canonical section of $\mathscr{O}(\delta)$. Extend all G-valuations to $R = \bigoplus_{n \geq 0} \mathrm{H}^0(X, \mathscr{O}(n\delta))$ and consider a G-submodule $M \subseteq \mathrm{H}^0(X, \mathscr{O}(\delta))$ generated by σ. For any B-eigensection $\widetilde{\sigma} \in (M^q)^{(B)}$, put $\widetilde{f} = \widetilde{\sigma}/\eta_\delta^q$. Then $v_D(\widetilde{f}) \geq qv_D(f)$, $w(\widetilde{f}) \geq qw(f)$, and $v(\widetilde{f}) = qv(f)$ for some $\widetilde{\sigma} \in (M^q)^{(B)}$ by Lemma 19.10. □

Corollary 19.13. *G-valuations are determined uniquely by their restriction to $K^{(B)}$.*

Proof. As $\operatorname{Quot} K_B = K$, two distinct $v, w \in \mathscr{V}$ differ on some $f \in K_B$, say $v(f) < w(f)$. Lemma 19.12 yields $v(\widetilde{f}) < w(\widetilde{f})$ for some $\widetilde{f} \in K^{(B)}$. □

20 Valuation Cones

We have seen in 19.2 that G-valuations of K are determined by their restriction to $K^{(B)}$. In this section, we give a geometric qualitative description of $\mathscr{V}$ in terms of this restriction. The results of this section go back to Brion and Pauer [BPa] in the spherical case and are due to Knop [Kn3] in full generality.

20.1 Hyperspace. Let $\mathbf{v}$ be a geometric valuation of K^B. Factoring the exact sequence (13.1) by $\mathscr{O}_{\mathbf{v}}^\times$ yields an exact sequence of lattices

$$0 \longrightarrow \mathbb{Z}_{\mathbf{v}} \longrightarrow \Lambda_{\mathbf{v}} \longrightarrow \Lambda \longrightarrow 0, \tag{20.1}$$

where $\mathbb{Z}_{\mathbf{v}} \simeq \mathbb{Z}$ or 0 is the value group of $\mathbf{v}$. Passing to the dual $\mathbb{Q}$-vector spaces, we obtain

$$\begin{array}{ccccccccc} 0 \longleftarrow & \mathbb{Q}_{\mathbf{v}} & \longleftarrow & \mathscr{E}_{\mathbf{v}} & \longleftarrow & \mathscr{E} & \longleftarrow 0 \\ & \cup\!\!\| & & \cup\!\!\| & & \| & \\ 0 \longleftarrow & \mathbb{Q}_{\mathbf{v},+} & \longleftarrow & \mathscr{E}_{\mathbf{v},+} & \longleftarrow & \mathscr{E} & \longleftarrow 0, \end{array} \tag{20.2}$$

where $\mathbb{Q}_{\mathbf{v}} = \mathbb{Q}$ and $\mathscr{E}_{\mathbf{v},+}$ is the preimage of the positive ray $\mathbb{Q}_{\mathbf{v},+}$ for $\mathbf{v} \neq 0$, and $\mathbb{Q}_0 = \mathbb{Q}_{0,+} = 0$, $\mathscr{E}_{0,+} = \mathscr{E}_0 = \mathscr{E}$.

Definition 20.1. The *hyperspace* (of K) is the union $\breve{\mathscr{E}} = \bigcup_{\mathbf{v}} \mathscr{E}_{\mathbf{v},+}$, where $\mathbf{v}$ runs over all geometric valuations of K^B considered up to proportionality. More precisely, $\breve{\mathscr{E}} = \mathscr{E}$ in the spherical case and, if $c(K) > 0$, then $\breve{\mathscr{E}}$ is the union of half-spaces $\mathscr{E}_{\mathbf{v},+}$ (over all $\mathbf{v} \neq 0$) glued together along their common boundary hyperplane $\mathscr{E}$, called the *center* of $\breve{\mathscr{E}}$.

Since Λ is a free Abelian group, the exact sequence (13.1) splits. Any splitting of (13.1) gives rise to simultaneous splittings of (20.1), (20.2), $\forall \mathbf{v}$. From time to time, we will fix such a splitting $\mathbf{f} : \Lambda \to K^{(B)}$, $\lambda \mapsto \mathbf{f}_\lambda$.

If v is a geometric valuation of K dominating $\mathbf{v}$, then $v|_{K^{(B)}}$ factors to a linear functional on $\Lambda_{\mathbf{v}}$ non-negative on $\mathbb{Z}_{\mathbf{v},+}$, i.e., an element of $\mathscr{E}_{\mathbf{v},+}$. Therefore $\mathscr{V} \hookrightarrow \breve{\mathscr{E}}$, and there is a restriction map $\varkappa : \mathscr{D}^B \to \breve{\mathscr{E}}$, which is in general not injective. Put $\mathscr{V}_{\mathbf{v}} = \mathscr{V} \cap \mathscr{E}_{\mathbf{v},+}$ and $\mathscr{D}^B_{\mathbf{v}} = \varkappa^{-1}(\mathscr{E}_{\mathbf{v},+})$. We say that $(\breve{\mathscr{E}}, \mathscr{V}, \mathscr{D}^B, \varkappa)$ is the *colored hyperspace*.

Our aim is to describe $\mathscr{V}_{\mathbf{v}}$.

Example 20.2. Assume that $G = B = T$ is a torus. Since every T-action has trivial birational type, there exists a T-model $X = T/T_0 \times C$, where $T_0 = \operatorname{Ker}(T : K)$ and $C = X/T$. We have $\Lambda = \mathfrak{X}(T/T_0)$, $K^T = \Bbbk(C)$, $K_T = K^T[\Lambda]$. By Remark 19.11, there is only one way to extend $v \in \mathscr{E}$ to a T-valuation of K: put $v(f) = \min_\lambda v(f_\lambda)$, $\forall f = \sum f_\lambda \in K_T$, $f_\lambda \in K^{(T)}_\lambda$, $\lambda \in \Lambda$.

To prove the multiplicative property, for any $f = \sum f_\lambda$, $g = \sum g_\lambda \in K_T$, choose $\gamma \in \mathscr{E}$ such that $\min\langle\gamma,\lambda\rangle$ and $\min\langle\gamma,\mu\rangle$ over all λ with $v(f_\lambda) = \min$, resp. μ with $v(g_\mu) = \min$, are reached at only one point $\lambda = \lambda_0$, resp. $\mu = \mu_0$. Then $fg = \sum f_\lambda g_\mu = f_{\lambda_0} g_{\mu_0} + \sum_{(\lambda,\mu)\neq(\lambda_0,\mu_0)} f_\lambda g_\mu$, and for any term of the second sum we have either $v(f_\lambda g_\mu) > v(f_{\lambda_0} g_{\mu_0})$ or $\langle\gamma,\lambda+\mu\rangle > \langle\gamma,\lambda_0+\mu_0\rangle$. It follows that $v(fg) = v(f_{\lambda_0} g_{\mu_0}) = v(f_{\lambda_0}) + v(g_{\mu_0}) = v(f) + v(g)$. Other properties of a valuation are obvious.

Finally, let $\mathbf{v} = v|_{K^T}$ and consider a short exact subsequence of (13.1):

$$1 \longrightarrow K^T_0 \longrightarrow K^{(T)}_0 \longrightarrow \Lambda_0 \longrightarrow 0, \tag{20.3}$$

where $K^{(T)}_0$ is the kernel of $v : K^{(T)} \to \mathbb{Q}$, $K^T_0 = \mathscr{O}^\times_{\mathbf{v}}$, and $\Lambda_0 \subseteq \Lambda$. Note that any element of K can be written as $f = f_1/f_2$, $f_i \in K_T$, $v(f_2) = 0$. It follows that $\Bbbk(v)$ is the fraction field of $K_T \cap \mathscr{O}_v / K_T \cap \mathfrak{m}_v \simeq \Bbbk(\mathbf{v})[\Lambda_0] \implies \operatorname{tr.deg} K - \operatorname{tr.deg} \Bbbk(v) = \operatorname{tr.deg} K^T + \operatorname{rk}\Lambda - \operatorname{tr.deg}\Bbbk(\mathbf{v}) - \operatorname{rk}\Lambda_0 = \operatorname{rk}(K^T)^\times/K^T_0 + \operatorname{rk}\Lambda/\Lambda_0 = \operatorname{rk} K^{(T)}/K^{(T)}_0 \leq 1$. Hence v is geometric by Proposition B.7.

We conclude that $\mathscr{V} = \breve{\mathscr{E}}$. By the way, we proved that every T-invariant valuation of K is geometric provided that its restriction to K^T is geometric.

20.2 Main Theorem. The main result of this section is

Theorem 20.3. *For any geometric valuation* $\mathbf{v}$ *of* K^B, $\mathscr{V}_{\mathbf{v}}$ *is a finitely generated solid convex cone in* $\mathscr{E}_{\mathbf{v},+}$.

We prove it in several steps.

Lemma 20.4. *For any G-model X, there are only finitely many B-stable prime divisors* $D \subset X$ *such that* v_D *maps to* $\mathscr{E}_{\mathbf{v},+}$.

Proof. Take a sufficiently small B-chart $\mathring{X} \subseteq X$ such that a geometric quotient $\pi : \mathring{X} \to \mathring{X}/B$ exists. Now if v_D maps to $\mathscr{E}_{\mathbf{v},+}$, then either D is an irreducible component of $X \setminus \mathring{X}$ or $D = \pi^{-1}(D_0)$, where D_0 is the center of $\mathbf{v}$ on $\mathring{X}/B$. □

Corollary 20.5. $\mathscr{D}_{\mathbf{v}}^B$ *is finite.*

20.3 A Good G**-model.** In the study of G-valuations, it is helpful to consider their centers on a sufficiently good projective G-model.

Lemma 20.6. *Let P be the common stabilizer of all colors, with a Levi decomposition* $P = P_{\mathrm{u}} \rightthreetimes L$, $L \supseteq T$. *There exists a projective normal G-model X, a P-stable open subset* $\mathring{X} \subseteq X$, *and a T-stable closed subvariety* $Z \subseteq \mathring{X}$ *such that:*

(1) *The action* $P_{\mathrm{u}} : \mathring{X}$ *is proper and has a geometric quotient.*
(2) $\mathring{X} = PZ$ *and the natural maps* $P_{\mathrm{u}} \times Z \to \mathring{X}$, $Z \to \mathring{X}/P_{\mathrm{u}}$ *are finite and surjective.*
(2)′ *In characteristic zero, Z is L-stable, and*

$$P_{\mathrm{u}} \times Z \simeq P *_L Z \xrightarrow{\sim} \mathring{X}.$$

(3) *The* L'*-action on* $\mathring{X}/P_{\mathrm{u}}$ *is trivial.*
(4) *Every G-subvariety* $Y \subset X$ *intersects* $\mathring{X}$, *hence Z.*
(5) *If Y is the center of* $v \in \mathscr{V}_{\mathbf{v}}$, *then* $\mathscr{D}_Y^B = \emptyset$, $\mathscr{V}_Y \subset \mathscr{E}_{\mathbf{v},+}$.

Proof. Take any projective G-model X and choose an ample G-line bundle $\mathscr{L}$ and an eigensection $\sigma \in \mathrm{H}^0(X,\mathscr{L})^{(B)}$ vanishing on sufficiently many colors such that $G_{[\sigma]} = P$. Put $M = \langle G\sigma \rangle$ and take a B^--eigenvector $u \in M^*$, $\langle \sigma, u \rangle \neq 0$, $G_{[u]} = P^-$, so that $G[u] \subseteq \mathbb{P}(M^*)$ is the unique closed orbit.

There is a natural rational G-map $\varphi : X \dashrightarrow \mathbb{P}(M^*)$. Replacing X by the normalized closure of its graph in $X \times \mathbb{P}(M^*)$ makes φ regular. Put $\mathring{X} = X_\sigma = \varphi^{-1}(\mathbb{P}(M^*)_\sigma)$. Then (1), (2), (2)′ follow from the local structure of $\mathbb{P}(M^*)_\sigma$ (Corollary 4.5). Every $(B \cap L)$-stable divisor on $\mathring{X}/P_{\mathrm{u}}$ is L-stable, whence (3). Every closed G-orbit in X maps onto $G[u]$, and hence intersects $\mathring{X}$, which yields (4).

To prove (5), we modify the construction of X. First, we may choose σ vanishing on all $D \in \mathscr{D}_{\mathbf{v}}^B$. Next, consider an affine model C_0 of K^B such that $\mathbf{v}$ has the center $D_0 \subseteq C_0$ which is either a prime divisor or the whole C_0. Let $\Bbbk[C_0] = \Bbbk[f_1, \dots, f_s]$. We may choose $\mathscr{L}$ and σ so that $f_i = \sigma_i/\sigma$, $\sigma_1, \dots, \sigma_s \in \mathrm{H}^0(X,\mathscr{L})^{(B)}$. Let $M' \subseteq \mathrm{H}^0(X,\mathscr{L})$ be the G-submodule generated by $\sigma, \sigma_1, \dots, \sigma_s$.

As above, we may assume that the natural rational map $\varphi' : X \dashrightarrow \mathbb{P}(M'^*)$ is regular. Put $X' = \varphi'(X)$. Consider the composed map $\pi : \mathring{X} \to \mathring{X}' = X'_\sigma \to C_0$. By

Corollary 19.7, $v(M'/\sigma) \geq 0$, whence the center $Y' = \varphi'(Y) \subseteq X'$ of $v|_{\Bbbk(X')}$ intersects the B-chart $\mathring{X}'$. Hence $\mathring{Y} = Y \cap \mathring{X}$ is non-empty and $\overline{\pi(\mathring{Y})} \supseteq D_0$. It follows that $\mathscr{V}_Y \sqcup \mathscr{D}_Y^B$ maps to $\mathscr{E}_{\mathbf{v},+}$. But any $D \in \mathscr{D}_{\mathbf{v}}^B$ is contained in $X \setminus \mathring{X}$, thence $D \not\supseteq Y$, and we are done. □

20.4 Criterion of Geometricity.

Proposition 20.7. *A G-invariant valuation of K is geometric if and only if its restriction to K^B is geometric.*

It suffices to prove the implication "if", because the reverse implication stems from Proposition B.8(1).

First proof [Kn3, 3.9, 4.4]. Let v be a nonzero valuation of K such that $v|_{K^B}$ is geometric. Take a projective G-model X as in Lemma 20.6. Then v has the center $Y \subset X$ and $\mathring{Y} = Y \cap \mathring{X} \neq \emptyset$. By Lemma 20.6(3), $\Bbbk(\mathring{X}/P_{\mathrm{u}}) = K^U$ and $\mathring{Y}/P_{\mathrm{u}}$ is the center of $v|_{K^U}$.

Since K^U is a T-field and $(K^U)^T = K^B$, it follows from Example 20.2 that $v|_{K^U}$ is geometric. Now by Proposition 19.9 there exists a projective birational L-morphism $Z' \to \mathring{X}/P_{\mathrm{u}}$ such that the center of $v|_{K^U}$ on Z' is a divisor $D' \subset Z'$. Consider a Cartesian square

$$\begin{array}{ccc} \mathring{X}' & \longrightarrow & \mathring{X} \\ \downarrow & & \downarrow \\ Z' & \longrightarrow & \mathring{X}/P_{\mathrm{u}}, \end{array}$$

where horizontal arrows are birational projective P-morphisms, and vertical arrows are P_{u}-quotient maps. Therefore v has a center $D \subset \mathring{X}'$, which is P-stable and maps onto D', whence D is the pullback of D', i.e., a divisor. This means that v is geometric. □

Second proof. Here we use the embedding theory of Chap. 3. Assume that $\mathbf{v} = v|_{K^B}$. It is easy to construct an affine model C_0 of K^B containing a principal prime divisor $D_0 = \operatorname{div}(t)$ such that either D_0 is the center of $\mathbf{v}$, or $\mathbf{v} = 0$, $t = 1$, $D_0 = \emptyset$, and $\mathring{C} = C_0 \setminus D_0 = \mathring{X}/B$ for a (sufficiently small) B-chart $\mathring{X}$.

If $\mathscr{R}$ is the set of all B-stable prime divisors in $\mathring{X}$ (= preimages of prime divisors in $\mathring{C}$), then $\{v\} \sqcup \mathscr{R} \in \mathbf{CD}$ defines colored data, and we may consider the respective Krull algebra $\mathscr{A} = \mathscr{A}(v, \mathscr{R})$ (cf. 13.2). Recall that we need not assume a priori that v is geometric (Remark 13.11). Clearly, $\mathscr{A}^U = \Bbbk\big[f \in K^{(B)} \mid \langle v, f\rangle, \langle \mathscr{R}, f\rangle \geq 0\big] \subseteq \Bbbk[\mathring{C}] \otimes \Bbbk[\Lambda]$ is a subalgebra determined by $v(f) \geq 0$, whence $\mathscr{A}^U = \Bbbk[C_0]\big[t^d \mathbf{f}_\lambda \mid (d,\lambda) \in \Lambda_{\mathbf{v}},\ \langle v, (d,\lambda)\rangle \geq 0\big]$.

The generating set of $\mathscr{A}^U$ over $\Bbbk[C_0]$ forms a finitely generated semigroup in $\Lambda_{\mathbf{v}}$ consisting of lattice points in the half-space $\{v \geq 0\}$, whence condition (F) holds for $\mathscr{A}$.

To prove (C), we take $f = t^d \mathbf{f}_\lambda$ such that $\langle v, (d,\lambda)\rangle > 0$ and multiply f by $f_0 \in \Bbbk[C_0]$ vanishing on sufficiently many divisors in $\mathscr{R}$.

Conversely, taking $f = t^d \mathbf{f}_\lambda$ such that $\langle v, (d,\lambda)\rangle < 0$ proves (W) for v.

By Corollary 13.9, $X_0 = \operatorname{Spec}\mathscr{A}$ is a B-chart and v is the valuation of a G-stable divisor intersecting X_0. □

20.5 Proof of the Main Theorem.

Remark 20.8. It is often helpful to assume that $K = \operatorname{Quot} R$, where R is a rational G-algebra. For instance, $R = \Bbbk[X]$, where X is a (quasi)affine G-model of K (if any exists). The general case is reduced to this special one by considering a projectively normal G-model X and taking the affine cone $\widehat{X}$ over X. Then $\widehat{K} = \Bbbk(\widehat{X})$ is a $\widehat{G}$-field, where $\widehat{G} = G \times \Bbbk^\times$ with $\Bbbk^\times$ acting by homotheties. Let us denote various objects related to $\widehat{K}$ in the same way as for K, but equipped with a hat. We have short exact sequences

$$\begin{array}{ccccccccc} 1 & \longrightarrow & (K^B)^\times & \longrightarrow & K^{(B)} & \longrightarrow & \Lambda & \longrightarrow & 0 \\ & & \| & & \cap & & \hookrightarrow & & \\ 1 & \longrightarrow & (\widehat{K}^{\widehat{B}})^\times & \longrightarrow & \widehat{K}^{(\widehat{B})} & \longrightarrow & \widehat{\Lambda} & \longrightarrow & 0 \end{array}$$

and dual sequences

$$\begin{array}{ccccccccc} 0 & \longleftarrow & \mathbb{Q}_{\mathbf{v},+} & \longleftarrow & \mathscr{E}_{\mathbf{v},+} & \longleftarrow & \mathscr{E} & \longleftarrow & 0 \\ & & \| & & \uparrow & & \uparrow & & \\ 0 & \longleftarrow & \mathbb{Q}_{\mathbf{v},+} & \longleftarrow & \widehat{\mathscr{E}}_{\mathbf{v},+} & \longleftarrow & \widehat{\mathscr{E}} & \longleftarrow & 0. \end{array}$$

The set of colors $\widehat{\mathscr{D}}^{\widehat{B}}$ is identified with $\mathscr{D}^B$. The grading of $\Bbbk[\widehat{X}]$ determines two G-valuations $\pm v_0 \in \widehat{\mathscr{E}}$, which generate $\operatorname{Ker}(\widehat{\mathscr{E}} \to \mathscr{E})$. By Corollary 19.6, $\widehat{\mathscr{V}}_{\mathbf{v}}$ surjects onto $\mathscr{V}_{\mathbf{v}}$ and is even the preimage of $\mathscr{V}_{\mathbf{v}}$ by Proposition 20.11 below.

Definition 20.9. Let $f_1, \dots, f_s \in R^{(B)}$ and $f \neq f_1 \cdots f_s$ be any B-eigenvector in $\langle G f_1 \rangle \cdots \langle G f_s \rangle$. Then $f / f_1 \cdots f_s$ is called a *tail vector* of R and its weight is called a *tail weight* or just a *tail*. Note that tails are negative linear combinations of simple roots. In characteristic zero tails are the nonzero differences $\mu - \lambda_1 - \dots - \lambda_s$ over all highest weights μ of simple G-modules occurring in the isotypic decomposition of $R_{(\lambda_1)} \cdots R_{(\lambda_s)}$, $\lambda_1, \dots, \lambda_s \in \Lambda_+$.

Remark 20.10. Knop conjectured that tails of an affine G-variety span a finitely generated semigroup. This conjecture was proved in characteristic zero by Alexeev and Brion [AB3, Pr. 2.13], see Corollary E.15.

Now we proceed in proving Theorem 20.3.

Proposition 20.11. *$\mathscr{V}_{\mathbf{v}}$ is a convex cone in $\mathscr{E}_{\mathbf{v},+}$.*

First proof. We may assume that $K = \operatorname{Quot} R$, where R is a rational G-algebra. The general case is reduced to this one by considering the affine cone over a projectively normal G-model as above. Then we prove for every $v \in \check{\mathscr{E}}$ that $v \in \mathscr{V}$ if and only if

$$v \text{ is non-negative on all tail vectors of } R. \tag{T}$$

Clearly, this condition is necessary, since $v(\langle G f_1 \rangle \cdots \langle G f_s \rangle) \geq v(f_1 \cdots f_s)$, $\forall v \in \mathscr{V}$, $f_1, \dots, f_s \in R$.

Conversely, assume that (T) is satisfied. It can be generalized as follows: let f be any B-eigenvector in $\sum_i \langle Gf_{i1}\rangle \cdots \langle Gf_{is}\rangle$, $f_{ij} \in R^{(B)}$; then $v(f) \geq \min_i \{v(f_{i1}) + \cdots + v(f_{is})\}$. Indeed, if $f = \sum f_i$, where $f_i \in \langle Gf_{i1}\rangle \cdots \langle Gf_{is}\rangle$ are B-eigenvectors of the same weight, then $v(f) \geq \min v(f_i) \geq \min\{v(f_{i1}) + \cdots + v(f_{is})\}$ by (T). The general case is reduced to this one, because a certain power f^q belongs to the image of $\mathrm{S}^q(\bigoplus \langle Gf_{i1}\rangle \cdots \langle Gf_{is}\rangle)^{(B)}$ by Corollary D.2.

Consider a rational B-algebra $A = \Bbbk[gf/h \mid g \in G,\ f,h \in R^{(B)},\ v(f/h) \geq 0]$. The ideal $I = (gf/h \mid v(f/h) > 0) \lhd A$ is proper. Indeed, each $f \in I^{(B)}$ is a linear combination of $(g_{i1}f_{i1}/h_1)\cdots(g_{is}f_{is}/h_s)$, where $v(f_{ij}/h_j) \geq 0$ and $>$ occurs for every i. By the above, $v(fh_1 \cdots h_s) \geq \min\{v(f_{i1}) + \cdots + v(f_{is})\} > v(h_1) + \cdots + v(h_s) \implies v(f) > 0$, whence $v > 0$ on $I^{(B)}$.

Take any valuation v' non-negative on A and positive on I, extend it to K, and take the approximating G-valuation $\overline{v}$ (Proposition 19.2). For any $f \in R^{(B)}$, $g \in G$ we have $v'(gf) \geq v'(f) \implies \overline{v}(f) = v'(f)$. Now for any $f,h \in R^{(B)}$ we have: $v(f/h) \geq 0$ (> 0) $\implies f/h \in A$ ($\in I$) $\implies \overline{v}(f/h) \geq 0$ (> 0). It follows that DVR's of v and $\overline{v}$ on K^B coincide, and hence $\overline{v} \in \mathscr{V}_{\mathbf{v}}$ provided that $v \in \mathscr{E}_{\mathbf{v}}$, and $\overline{v}, v$ determine proportional linear functionals on $\Lambda_{\mathbf{v}}$. Thus $v \in \mathscr{V}$. □

Second proof. This proof relies on the embedding theory. Let $v_1, v_2 \in \mathscr{V}_{\mathbf{v}}$ be two non-proportional vectors. It suffices to prove that $c_1v_1 + c_2v_2 \in \mathscr{V}_{\mathbf{v}}$, $\forall c_1, c_2 \in \mathbb{Z}_+$.

Take an affine model C_0 of K^B as in the second proof of Proposition 20.7 and consider the algebra $\mathscr{A} = \mathscr{A}(v_1, v_2, \mathscr{R})$. Then $\mathscr{A}^U \subseteq \Bbbk[\mathring{C}] \otimes \Bbbk[\Lambda]$ is distinguished by inequalities $v_i(f) \geq 0$, whence $\mathscr{A}^U = \Bbbk[C_0]\big[t^d\mathbf{f}_\lambda \bigm| (d,\lambda) \in \Lambda_{\mathbf{v}},\ \langle v_i, (d,\lambda)\rangle \geq 0,\ i = 1,2\big]$.

The generating set of $\mathscr{A}^U$ over $\Bbbk[C_0]$ forms a finitely generated semigroup of lattice points in the dihedral cone $\{v_1, v_2 \geq 0\} \subseteq \Lambda_{\mathbf{v}} \otimes \mathbb{Q}$, whence condition (F) holds for $\mathscr{A}$. Conditions (C), (W) are verified in the same way as in Proposition 20.7 by taking $f = t^d\mathbf{f}_\lambda$ with $\langle v_i, (d,\lambda)\rangle, \langle v_j, (d,\lambda)\rangle > 0$ or $\langle v_i, (d,\lambda)\rangle < 0 \leq \langle v_j, (d,\lambda)\rangle$, respectively, $\{i,j\} = \{1,2\}$. Thus by Corollary 13.9, $X_0 = \operatorname{Spec} \mathscr{A}$ is a B-chart and v_i correspond to G-stable divisors $D_i \subset X = GX_0$ intersecting X_0.

We blow up the ideal sheaf $\mathscr{O}(-nc_2D_1) + \mathscr{O}(-nc_1D_2)$, $n \gg 0$, and prove that the exceptional divisor corresponds to $c_1v_1 + c_2v_2$.

The local structure of X_0 provided by Proposition 13.12 allows us to replace X_0 by X_0/P_{u} and assume that $G = T$ and $X = X_0$ is an affine T-model of K. We may choose $f_i = t^{d_i}\mathbf{f}_{\lambda_i}$ such that $v_i(f_j) = n\delta_{ij}$, $n \gg 0$. The above ideal sheaf is represented by a proper ideal $I = (f_1^{c_2}, f_2^{c_1}) \lhd \mathscr{A}$. (Indeed, it is easy to see that $v_1 + v_2 > 0$ on $I^{(T)}$.)

The blow-up of I is given in $X \times \mathbb{P}^1$ by the equation $[f_1^{c_2} : f_2^{c_1}] = [t_1 : t_2]$, where t_i are homogeneous coordinates on $\mathbb{P}^1$. The exceptional divisor is given in the open subset $\{t_2 \neq 0\}$ by the equation $f_2 = 0$. Let v_0 be the respective valuation. Up to a power, any $f \in \mathscr{A}^{(T)}$ is represented as $f = f_0f_1^{k_1}f_2^{k_2}$, where $v_i(f_0) = 0$ and $k_i \,\vdots\, c_j$ whenever $\{i,j\} = \{1,2\}$. Then $v_0(f) = v_0(f_0(t_1/t_2)^{k_1/c_2}f_2^{(c_1k_1+c_2k_2)/c_2}) \sim c_1k_1 + c_2k_2$. It follows that $v_0 \sim c_1v_1 + c_2v_2$. □

Proposition 20.12. *The cone $\mathscr{V}_{\mathbf{v}}$ is finitely generated.*

Proof. Take a projective G-model X as in Lemma 20.6. Then any $v \in \mathscr{V}_{\mathbf{v}}$ has the center Y on X and, by Lemma 20.6(5), $\mathscr{D}_Y^B = \emptyset$, $\mathscr{V}_Y \subset \mathscr{V}_{\mathbf{v}}$. Condition (S) yields

$$\forall \lambda \in \Lambda_{\mathbf{v}} : \langle \mathscr{V}_Y, \lambda \rangle \geq 0 \implies \langle v, \lambda \rangle \geq 0.$$

Hence $v \in \mathbb{Q}_+ \mathscr{V}_Y$. It remains to note that $\bigcup_Y \mathscr{V}_Y$ is finite by Lemma 20.4. □

Proof of Theorem 20.3. Due to Propositions 20.11–20.12 and Theorem 21.1, it remains to prove that any geometric valuation $\mathbf{v} \neq 0$ of K^B extends to a G-valuation of K. For this, we modify the first proof of Proposition 20.11.

Namely, in the definition of A we replace v by $\mathbf{v}$ and assume that $f/h \in K^B$. The respective ideal $I \lhd A$ is still proper. For otherwise, 1 is a linear combination of $(g_{i1}f_{i1}/h_1)\cdots(g_{is}f_{is}/h_s)$, where $\mathbf{v}(f_{ij}/h_j) \geq 0$ and $>$ occurs for each i. But the T-weights of all T-eigenvectors in $\langle G f_{ij} \rangle$ except f_{ij} are obtained from the weight of f_{ij} (=the weight of h_j) by subtracting simple roots. Hence we may assume that $g_{ij} = e$, and 1 is a linear combination of $r_i = (f_{i1}/h_1)\cdots(f_{is}/h_s)$, $\mathbf{v}(r_i) > 0$, a contradiction.

Now reproducing the arguments of that proof yields a G-valuation $\overline{v}$ such that $\overline{v}|_{K^B} = \mathbf{v}$. □

20.6 Parabolic Induction. This procedure, which is helpful in various reduction arguments, keeps the colored hyperspace "almost" unchanged. Suppose that K is obtained from a G_0-field K_0 by parabolic induction $G \supseteq Q \twoheadrightarrow G_0$, i.e., a G-model X of K is obtained from a G_0-model X_0 of K_0 by this procedure. There is a natural projection $\pi : X = G *_Q X_0 \to G/Q$. We may assume that $Q \supseteq B^-$. Then the colors of G/Q are the Schubert divisors $D_\alpha = \overline{B(r_\alpha o)}$ ($\alpha \in \Pi$, $r_\alpha \notin Q$). Let us denote various objects related to K_0 in the same way as for K, but with a subscript 0.

Proposition 20.13. *There are natural identifications $\check{\mathscr{E}} = \check{\mathscr{E}}_0$, $\mathscr{V} = \mathscr{V}_0$, and $\mathscr{D}^B = \mathscr{D}_0^{B_0} \sqcup \pi^{-1}(\mathscr{D}^B(G/Q))$ such that $\varkappa = \varkappa_0$ on $\mathscr{D}_0^{B_0}$ and $\varkappa(\pi^{-1}(D_\alpha)) \in \mathscr{E}$ is the restriction of $\alpha^\vee$ to Λ.*

Proof. Since $\pi^{-1}(Bo) \simeq Q_{\mathrm{u}}^- \times X_0$, the restriction to the fiber identifies $K^{(B)}$ with $K_0^{(B_0)}$. Therefore, the hyperspaces are identified and the set of colors is extended by the pullbacks of the Schubert divisors.

The restriction of G-valuations from K to $K^{Q_{\mathrm{u}}^-} \simeq K_0$ yields $\mathscr{V} \subseteq \mathscr{V}_0$. Conversely, each $v \in \mathscr{V}_0$ corresponds to a G_0-stable divisor D on a suitably chosen G_0-model X_0 (Proposition 19.8). The same vector in the hyperspace corresponds to $G *_Q D \subset X$. Thus $\mathscr{V} = \mathscr{V}_0$.

Similar arguments show that $\varkappa = \varkappa_0$ on $\mathscr{D}_0^{B_0}$.

Finally, $\pi^{-1}(D_\alpha)$ is transversal to the subvariety $X_\alpha = L_\alpha *_{B_\alpha^-} X_0 \subseteq X$ along $r_\alpha * X_0$, where L_α is the Levi subgroup of P_α and $B_\alpha^- = B^- \cap L_\alpha$ acts on X_0 via T. The variety X_α is horospherical with respect to L_α and intersects general B-orbits of X. We easily deduce that general B-orbit closures in X intersect $\pi^{-1}(D_\alpha)$ and the intersections cover a dense subset $Br_\alpha * X_0$. Hence all $f \in K^B$ have order 0 along $\pi^{-1}(D_\alpha)$. To determine $\varkappa(\pi^{-1}(D_\alpha))$, it suffices to consider the restriction of $K^{(B)}$ to a general L_α-orbit in X_α, cf. 28.1. □

21 Central Valuations

21.1 Central Valuation Cone. G-valuations of K that vanish on $(K^B)^\times$ are called *central*. They play a distinguished rôle among all G-valuations. By (13.1) a central valuation restricted to $K^{(B)}$ factors through a linear functional on Λ. Thus central valuations form a subset $\mathscr{Z} \subseteq \mathscr{E} = \mathrm{Hom}(\Lambda, \mathbb{Q})$.

If G is linearly reductive (i.e., $\operatorname{char}\Bbbk = 0$ or $G = T$) and $K = \operatorname{Quot} R$ for a rational G-algebra R with the isotypic decomposition $R = \bigoplus_{\lambda \in \Lambda} R_{(\lambda)}$, then any $v \in \mathscr{Z}$ is constant on each isotypic component $R_{(\lambda)}$ (otherwise there would exist two eigenvectors $f_1, f_2 \in R^{(B)}$ of the same weight λ with $v(f_1) \neq v(f_2)$) and $v|_{R_{(\lambda)} \setminus \{0\}} = \langle v, \lambda \rangle$, $\forall \lambda \in \Lambda$.

Theorem 21.1. *$\mathscr{Z} = \mathscr{V} \cap \mathscr{E}$ is a solid convex polyhedral cone in $\mathscr{E}$ containing the image of the antidominant Weyl chamber.*

Proof. By Remark 20.8, we reduce the question to the case $K = \operatorname{Quot} R$, where R is a rational G-algebra. Condition (T) defining the subset $\mathscr{V} \subseteq \check{\mathscr{E}}$ transforms under restriction to $\mathscr{E}$ to the following one: $v \in \mathscr{Z}$ if and only if

$$v \text{ is non-negative on all tails of } R. \tag{T$_0$}$$

Since tails are negative linear combinations of simple roots, we see that (T$_0$) determines a convex cone containing the image of the antidominant Weyl chamber, whence a solid cone. We conclude by Proposition 20.12. □

Example 21.2. Let G act on itself by right translations and $K = \Bbbk(G)$. Here $\Lambda = \mathfrak{X}(T)$. Recall from Proposition 2.14 that $\Bbbk[G]$ is the union of subspaces $\mathrm{M}(V)$ of matrix entries over all G-modules V.

In characteristic zero, $\mathrm{M}(V(\lambda)) = \Bbbk[G]_{(\lambda)}$ are the isotypic components of $\Bbbk[G]$ and $\mathrm{M}(V(\lambda)) \cdot \mathrm{M}(V(\mu)) = \bigoplus \mathrm{M}(V(\nu))$ over all simple submodules $V(\nu)$ occurring in $V(\lambda) \otimes V(\mu)$, by Formula (2.1). Generally, each B-eigenvector $v \in V^{(B)}$ gives rise to B-eigenvectors $f_{\omega,v} \in \mathrm{M}(V)^{(B)}$, $\omega \in V^*$, of the same weight.

If $v = v_\lambda \in V$ is a highest vector of regular highest weight λ, then V contains T-eigenvectors $v_{\lambda-\alpha}$ of weights $\lambda - \alpha$ for all simple roots α, and $w = v_\lambda \otimes v_{\lambda-\alpha} - v_{\lambda-\alpha} \otimes v_\lambda$ are B-eigenvectors of weights $2\lambda - \alpha$ in $V \otimes V$. We get tail vectors $f_{\omega\otimes\omega,w}/f_{\omega,v}^2$ of weights $-\alpha$. It follows that all $-\alpha$ occur among tails of $\Bbbk[G]$, whence $\mathscr{Z}$ is the antidominant Weyl chamber.

In characteristic zero, much more precise information on the structure of $\mathscr{Z}$ can be obtained, as given in Theorem 22.13.

A special case of Theorem 5.7 distinguishes central and non-central G-valuations in terms of the complexity and rank of respective G-stable divisors.

Proposition 21.3. *A G-valuation $v \neq 0$ is central if and only if $c(\Bbbk(v)) = c(K)$, $r(\Bbbk(v)) = r(K) - 1$, and non-central if and only if $c(\Bbbk(v)) = c(K) - 1$, $r(\Bbbk(v)) = r(K)$.*

Proof. Let $Y \subset X$ be a G-stable divisor on a G-model of K corresponding to v. By Lemma 5.8, $\Bbbk(v)^U = \Bbbk(Y)^U$ is a purely inseparable extension of the residue field $\Bbbk(v|_{K^U})$ of $\mathcal{O}^U_{X,Y}$, and similarly for $\Bbbk(v)^B$. Thus by Proposition 5.6, $c(\Bbbk(v)) + r(\Bbbk(v)) = \operatorname{tr.deg}\Bbbk(v)^U = \operatorname{tr.deg} K^U - 1 = c(K) + r(K) - 1$, and $c(\Bbbk(v)) = c(K) \iff \Bbbk(v|_{K^B}) = K^B \iff v \in \mathscr{Z}$. □

21.2 Central Automorphisms. Now we explain the geometric meaning of the linear part of the central valuation cone.

Definition 21.4. A G-automorphism of K acting trivially on K^B is called *central*. Denote by $\operatorname{CAut} K = \operatorname{Ker}(\operatorname{Aut}_G K : K^B)$ the group of central automorphisms.

Theorem 21.5 ([Kn3, 8.1–8.2]). *There exists the largest connected algebraic subgroup $S \subseteq \operatorname{CAut} K$. It has the following properties:*

(1) *S acts on $\mathscr{V}, \mathscr{D}^B, {}_G\mathbb{X}^{\mathrm{norm}}$ trivially.*
(2) *S acts on every normal G-model of K regularly.*
(3) *There is a canonical embedding $S \hookrightarrow A = \operatorname{Hom}(\Lambda, \Bbbk^\times)$ via the action $S : K^{(B)}$, so that $\operatorname{Hom}(\mathfrak{X}(S), \mathbb{Q}) \subseteq \mathscr{Z} \cap -\mathscr{Z}$.*
(4) *There exists a G-subfield $K' \subseteq K$ with $(K')^U = K^U$ and with the same colored hyperspace as for K such that K is purely inseparable over K' and $\operatorname{Hom}(\mathfrak{X}(S'), \mathbb{Q}) = \mathscr{Z} \cap -\mathscr{Z}$ for the largest connected algebraic subgroup $S' \subseteq \operatorname{CAut} K'$.*

Proof. Let $S \subseteq \operatorname{CAut} K$ be any connected algebraic subgroup. Suppose first that $K = \operatorname{Quot} R$, where R is a rational $(G \times S)$-subalgebra. Without loss of generality, we may assume in the reasoning below that $R = \Bbbk[X]$, where X is a normal (quasi)affine G-model of K acted on by S.

If $f \in R^{(B)}$, then $sf \in R^{(B)}$ has the same weight, $\forall s \in S$, whence $sf/f \in K^B \subseteq K^S$. Hence $s\operatorname{div} f - \operatorname{div} f$ is S-stable. (The divisors are considered on a normal completion of X.) But, on the other hand, this divisor has no S-stable components. Hence it is zero, and $sf \in \Bbbk^\times f$. Therefore $R^{(B)} \subseteq R^{(S)} \implies K^{(B)} \subseteq K^{(S)}$.

This yields a homomorphism $S \to A$. Let S_0 be its connected kernel. Then S_0 acts on R^U trivially. As S_0 commutes with G, it acts on R trivially. (In positive characteristic, this stems from Lemma D.4.) Hence $S_0 = \{e\}$ and S is a torus. Then every $s \in S$, $s \neq e$, acts non-trivially on $K^{(B)}$: just take a (G-stable) eigenspace of s in R of eigenvalue $\neq 1$ and choose a B-eigenvector there. Thus $S \hookrightarrow A$.

Since S-action multiplies B-eigenfunctions by scalars and G-valuations are determined by their restriction to $K^{(B)}$, the action $S : \mathscr{V}$ is trivial. As any $D \in \mathscr{D}^B$ is a component of $\operatorname{div} f$, where $f \in K^{(B)} \subseteq K^{(S)}$, S fixes all colors. Then S fixes all G-germs by Proposition 14.1, whence (1). Assertion (2) stems from (1).

Each one-parameter subgroup $\gamma \in \mathfrak{X}^*(S)$ defines a G-stable grading of R, which gives rise to a central valuation (the order of the lowest homogeneous term) represented by γ: $f_\lambda \mapsto \langle \gamma, \lambda \rangle$, $\forall f_\lambda \in R^{(B)}_\lambda$. This finally yields (3).

Furthermore, any $\tau \in \operatorname{CAut} K$ commutes with γ. Indeed, we have two gradings of R defined by $\gamma, \tau\gamma\tau^{-1}$. They coincide on $R^{(B)}$, and hence on the G-subalgebra of R generated by R^U, and therefore on R. (The last implication is easily deduced from

Lemma D.4.) Therefore any two subtori $S_1, S_2 \subseteq \operatorname{CAut} K$ commute, and the natural homomorphism $S_1 \times S_2 \to \operatorname{CAut} K$ provides a larger subtorus. Since the dimensions of subtori are restricted from above by $\dim(\mathscr{Z} \cap -\mathscr{Z})$, there exists a largest subtorus.

We prove (4) in characteristic zero referring to [Kn3, 8.2] for $\operatorname{char} \Bbbk > 0$. Every lattice vector $\gamma \in \mathscr{Z} \cap -\mathscr{Z}$ defines a G-stable grading of R such that $R_{(\lambda)}$ are homogeneous of degree $\langle \gamma, \lambda \rangle$. Since γ vanishes on tails, this grading respects multiplication and defines a 1-subtorus of central automorphisms. These 1-subtori generate a subtorus $S \subseteq A$ which is the connected common kernel of all tails, and $\operatorname{Hom}(\mathfrak{X}(S), \mathbb{Q}) = \mathscr{Z} \cap -\mathscr{Z}$.

Finally, the general case is reduced to the above by taking a projectively normal G-model X acted on by S and considering the affine cone $\widehat{X}$ over X. By Remark 20.8, $\mathscr{Z} = \widehat{\mathscr{Z}}/\mathbb{Q}v_0$, and $S = \widehat{S}/\Bbbk^\times$, where the central valuation v_0 of $\widehat{K}$ is defined by a 1-subtorus $\Bbbk^\times \subseteq \widehat{S}$ acting on $\widehat{X}$ by homotheties. So all assertions on $\widehat{S}$ transfer to S. □

In the generically transitive case, we can say more.

Proposition 21.6 ([Kn3, 8.3]). *If* $K = \Bbbk(G/H)$*, then* $\operatorname{CAut} K$ *is a diagonalizable group extended by a finite p-group and* $\dim \operatorname{CAut} K = \dim(\mathscr{Z} \cap -\mathscr{Z})$.

Proof. Since $\operatorname{Aut}_G K = \operatorname{Aut}_G(G/H) = N_G(H)/H$ is an algebraic group, $\operatorname{CAut} K$ is an algebraic group as well. Central automorphisms preserve general B-orbits in G/H, whence there exist finitely many general points $x_1, \dots, x_s \in G/H$ such that $\operatorname{CAut} K \hookrightarrow \operatorname{Aut}_B Bx_1 \times \dots \times \operatorname{Aut}_B Bx_s$. The latter group is a subquotient of $B \times \dots \times B$, which explains the structure of $\operatorname{CAut} K$ in view of Theorem 21.5(3). By Theorem 21.5(4), there exists a purely inseparable G-map $G/H \twoheadrightarrow O$ such that $\dim \operatorname{CAut} O = \dim(\mathscr{Z} \cap -\mathscr{Z})$. But every G-automorphism of O lifts to G/H, because it is determined by the image no of the base point $o \in O$ with stabilizer $G_o = H$ and $n \in N_G(H)$. Hence $\dim \operatorname{CAut} K$ is the same. □

Remark 21.7. Suppose that $\operatorname{char} \Bbbk = 0$. If G/H is quasiaffine, then $\operatorname{CAut}(G/H)$ is canonically embedded in A as the common kernel of all tails of $\Bbbk[G/H]$ and acts on each $\Bbbk[G/H]_{(\lambda)}$ by the character $\lambda \in \Lambda_+(G/H)$. (Otherwise $\operatorname{CAut}(G/H)$ would act on some $\Bbbk[G/H]^{(B)}_\lambda$ with several distinct eigenweights, whence the action $\operatorname{CAut}(G/H) : \Bbbk(G/H)^B$ would be non-trivial.) For general G/H, $\operatorname{CAut}(G/H)$ acts on each $\Bbbk(G/H)^{(B)}_\lambda$ by homotheties and lies in the center of $\operatorname{Aut}_G(G/H)$. (This is reduced to the quasiaffine case as in the proof of Theorem 21.5.)

Example 21.8. In Example 21.2, $\operatorname{Aut}_G G = G$ (acting by left translations) and $\operatorname{CAut} G = \operatorname{Ker}(G : G/B) = Z(G)$.

Example 21.9. If the orbit map $G \to O$ is not separable, then $\dim \operatorname{CAut} O$ may be smaller than "the proper value". For instance, let $\operatorname{char} \Bbbk = 2$, $G = \operatorname{SL}_2$, $X = \mathbb{P}(\mathfrak{sl}_2)$. Then X has an open orbit O with stabilizer U. By Theorem 21.5(2), the central torus S embeds in $\operatorname{Aut}_G X$, but the latter group is trivial. Indeed, each G-automorphism of X lifts to an intertwining operator of $\mathrm{Ad}\,\mathrm{SL}_2$, but it is easy to see that such an operator has to be scalar. However $\Bbbk(G/U)^B = \Bbbk \implies \mathrm{CAut}(G/U) = \mathrm{Aut}_G(G/U) = T \implies \dim(\mathscr{Z} \cap -\mathscr{Z}) = 1$; cf. Theorem 21.10.

21.3 Valuative Characterization of Horospherical Varieties. We have seen in §7 that horospherical G-varieties play an important rôle in studying general G-varieties. Apparently they can be characterized in terms of central valuation cones.

Theorem 21.10. *A G-variety X is horospherical if and only if $\mathscr{Z}(X) = \mathscr{E}(X)$.*

First proof ($\mathrm{char}\,\Bbbk = 0$) [Po5, §4], [Vin1, §5]. This proof goes back to Popov, who however considered tails of coordinate algebras instead of central valuations; cf. Proposition 7.6. The generically transitive case is due to Pauer [Pau4].

By Remark 20.8, we may assume X to be (quasi)affine. By (T_0), $\mathscr{Z} = \mathscr{E}$ if and only if $R = \Bbbk[X]$ has no tails. However one proves that a rational G-algebra R has no tails if and only if $R \simeq (\Bbbk[G/U^-] \otimes R^U)^T = \bigoplus_{\lambda \in \mathfrak{X}_+} \Bbbk[G/U^-]_{(\lambda)} \otimes R_\lambda^{(B)}$. Here $T = \mathrm{Aut}_G(G/U^-)$ acts on G/U^- by right translations, so that isotypic components $\Bbbk[G/U^-]_{(\lambda)}$ are at the same time T-eigenspaces of weight $-\lambda$ (cf. (2.4)), and the isomorphism is given by $g\mathbf{f}_\lambda \otimes f_\lambda \mapsto gf_\lambda$, $\forall g \in G,\ f_\lambda \in R_\lambda^{(B)}$, where $\mathbf{f}_\lambda \in \Bbbk[G/U^-]_\lambda^{(B)}$, $\mathbf{f}_\lambda(e) = 1$. In our situation, this implies that $X = (G/\!/U^- \times X/\!/U)/\!/T$ is horospherical.

Conversely, if R has tails, then tails do not vanish under restriction of the isotypic decomposition of $R_{(\lambda)} \cdot R_{(\mu)}$ to general G-orbits. But the coordinate algebra of a horospherical homogeneous space has no tails since isotypic components of $\Bbbk[G/U^-]$ are T-eigenspaces. Hence X is not horospherical. □

Second proof [Kn3, 8.5]. If $\mathscr{Z} = \mathscr{E}$, then by Theorem 21.5(4) we may assume that $S = A$ and the geometric quotient X/S exists. Then $r(X/S) = 0$ and, by Propositions 5.11 and 10.1, orbits of $G : X/S$ are projective homogeneous spaces.

Let $x \in X$ and $x \mapsto \bar{x} \in X/S$. We may assume that $G_{\bar{x}} \supseteq U$. Then U preserves Sx, and we have the orbit map $U \to Ux \subseteq Sx$. As U is an affine space (with no non-constant invertible polynomials) and Sx is a torus (whose coordinate algebra is generated by invertibles), this map is constant, whence $Ux = \{x\}$. Thus X is horospherical.

Conversely, for horospherical X put $Z = X^{U^-}$ and consider the natural proper map $X' := G *_{B^-} Z \twoheadrightarrow X$. There are natural maps $\mathscr{E}(X') \twoheadrightarrow \mathscr{E}(X)$, $\mathscr{Z}(X') \to \mathscr{Z}(X)$. Consider the action $T : Z$. By Proposition 20.13 and Example 20.2, $\mathscr{Z}(X') = \mathscr{Z}(Z) = \mathscr{E}(Z) = \mathscr{E}(X')$, whence $\mathscr{Z}(X) = \mathscr{E}(X)$. □

21.4 G-valuations of a Central Divisor. We conclude this section with the description of G-valuations for the residue field of a central valuation.

Proposition 21.11 ([Kn3, 7.4]). *Let X be a smooth G-model of K, let $D \subset X$ be a G-stable prime divisor with $v_D \in \mathscr{Z}$, and let X' be the normal bundle of X along D^{reg}. Then:*

(1) $\breve{\mathscr{E}}(X') = \breve{\mathscr{E}}$ *and* $\mathscr{V}(X') = \mathscr{V} + \mathbb{Q}v_D$*;*
(2) $\breve{\mathscr{E}}(D) = \breve{\mathscr{E}}/\mathbb{Q}v_D$ *and* $\mathscr{V}(D)$ *is the image of* $\mathscr{V}$*.*

Proof. As usual, we may assume that $K = \operatorname{Quot} R$, where R is a rational G-algebra. Then $K' := \Bbbk(X') = \operatorname{Quot} \operatorname{gr} R$, where $\operatorname{gr} R$ is the graded algebra associated with the filtration $R^{(d)} = \{f \in R \mid v_D(f) \geq d\}$ of R. Since v_D is central, it is constant on each B-eigenspace of R^U, whence $R^U \simeq \operatorname{gr}(R^U)$. But $(\operatorname{gr} R)^U$ is a purely inseparable finite extension of $\operatorname{gr}(R^U)$ by Corollary D.3, and hence $(K')^U \supseteq K^U$ is a purely inseparable field extension. This implies that $\breve{\mathscr{E}}(X') = \breve{\mathscr{E}}$.

The G-invariant grading of $\operatorname{gr} R$ is defined by a central 1-torus acting on $\mathscr{V}(X')$ trivially by Theorem 21.5(1). Hence it agrees with all G-valuations. Thus $v \in \mathscr{V}(X')$ if and only if v is non-negative at all tail vectors of the form $\bar{f}_0/\bar{f}_1 \cdots \bar{f}_s$, where $\bar{f}_i \in (\operatorname{gr} R)^{(B)}$ are homogeneous elements represented by $f_i \in R$, and $v_D(f_0) = v_D(f_1) + \cdots + v_D(f_s)$. Replacing $\bar{f}_i$ by suitable powers, we may assume that $f_i \in R^{(B)}$. Thus $\mathscr{V}(X')$ is the set of all $v \in \breve{\mathscr{E}}$ non-negative on tail vectors of R annihilated by v_D, i.e., $\mathscr{V}(X') = \mathscr{V} + \mathbb{Q}v_D$.

By Lemma 5.8, $\Bbbk(D)^U$ is a purely inseparable extension of $\Bbbk(v_D|_{K^U})$, and $\Bbbk(D)^B$ of K^B. Hence $\breve{\mathscr{E}}(D) = \breve{\mathscr{E}}/\mathbb{Q}v_D$. Since X' retracts onto D, $\mathscr{V}(D) = \mathscr{V}(X')/\mathbb{Q}v_D$ by Corollary 19.6. □

22 Little Weyl Group

In Subsection 8.7 we found that important invariants of a G-variety X, such as complexity, rank, and weight lattice, which play an essential rôle in the embedding theory, are closely related to the geometry of the cotangent bundle T^*X. Knop developed these observations further [Kn1], [Kn5] and described the cone of (central) G-valuations as a fundamental chamber for the Galois group of a certain Galois covering of T^*X, called the little Weyl group of X. As this approach requires an infinitesimal technique, we assume that $\operatorname{char} \Bbbk = 0$ in this and the next section. We retain the notation and conventions of §8.

22.1 Normalized Moment Map. The Galois covering of T^*X is defined in terms of the moment map. A disadvantage of the moment map Φ is that its image M_X can be non-normal, and general fibers can be reducible. A remedy is to consider the "Stein factorization" of Φ. Let $\widetilde{M}_X$ be the spectrum of the integral closure of $\Bbbk[M_X]$ (embedded via Φ^*) in $\Bbbk(T^*X)$. We may assume that X is smooth, whence T^*X is smooth and normal, and therefore $\Bbbk[\widetilde{M}_X] \subseteq \Bbbk[T^*X]$. It is easy to see that $\Bbbk(\widetilde{M}_X)$ is algebraically closed in $\Bbbk(T^*X)$. Thus Φ decomposes into the product of a finite morphism $\varphi : \widetilde{M}_X \to M_X$ and the *normalized moment map* $\widetilde{\Phi} : T^*X \to \widetilde{M}_X$ with irreducible general fibers. Set $\widetilde{L}_X = \widetilde{M}_X /\!/ G$. We have the quotient map $\widetilde{\pi}_G : \widetilde{M}_X \to \widetilde{L}_X$ and the natural finite morphism $\varphi/\!/G : \widetilde{L}_X \to L_X$.

The following result illustrates the rôle of the normalized moment map in equivariant symplectic geometry.

Proposition 22.1. *The fields* $\Bbbk(\widetilde{M}_X)$ *and* $\Bbbk(T^*X)^G$ *are the mutual centralizers of each other in* $\Bbbk(T^*X)$ *with respect to the Poisson bracket, and* $\Bbbk(\widetilde{L}_X)$ *is the Poisson center of both* $\Bbbk(\widetilde{M}_X)$ *and* $\Bbbk(T^*X)^G$.

Proof. The field $\Bbbk(M_X)$ is generated by Hamiltonians $H_\xi = \Phi^*\xi$, $\xi \in \mathfrak{g}$. Hence $f \in \Bbbk(T^*X)$ Poisson-commutes with $\Bbbk(M_X)$ if and only if $\{H_\xi, f\} = \xi f = 0, \forall \xi \in \mathfrak{g}$, i.e., f is G-invariant. But then f also commutes with $\Bbbk(\widetilde{M}_X)$. Indeed, let μ_h be the minimal polynomial of $h \in \Bbbk(\widetilde{M}_X)$ over $\Bbbk(M_X)$. Then $\{f, \mu_h(h)\} = \mu_h'(h)\{f,h\} = 0 \implies \{f,h\} = 0$. Therefore $\Bbbk(T^*X)^G$ is the centralizer of $\Bbbk(M_X)$ and $\Bbbk(\widetilde{M}_X)$.

Conversely, as general orbits are separated by invariant functions, $\mathfrak{g}\alpha$ is the common kernel of $\mathrm{d}_\alpha f$, $f \in \Bbbk(T^*X)^G$, for general $\alpha \in T^*X$. Hence $\operatorname{Ker} \mathrm{d}_\alpha \widetilde{\Phi} = \operatorname{Ker} \mathrm{d}_\alpha \Phi = (\mathfrak{g}\alpha)^\angle$ is generated by skew gradients of $f \in \Bbbk(T^*X)^G$. It follows that $h \in \Bbbk(T^*X)$ commutes with $\Bbbk(T^*X)^G$ if and only if $\mathrm{d}h$ vanishes on $\operatorname{Ker} \mathrm{d}_\alpha \widetilde{\Phi} = T_\alpha \widetilde{\Phi}^{-1}\widetilde{\Phi}(\alpha)$, and this holds if and only if h is constant on $\widetilde{\Phi}^{-1}\widetilde{\Phi}(\alpha)$, because general fibers $\widetilde{\Phi}^{-1}\widetilde{\Phi}(\alpha)$ are irreducible. Therefore $\Bbbk(\widetilde{M}_X)$ is the centralizer of $\Bbbk(T^*X)^G$.

Finally, $\Bbbk(\widetilde{L}_X) = \Bbbk(\widetilde{M}_X)^G = \Bbbk(\widetilde{M}_X) \cap \Bbbk(T^*X)^G$, since quotient maps π_G and $\widetilde{\pi}_G$ separate general orbits. □

22.2 Conormal Bundle to General U**-orbits.** Recall the local structure of an open subset of X provided by Theorem 4.7: $\mathring{X} \simeq P_\mathrm{u} \times Z$, where $P = P(X) = P_\mathrm{u} \rtimes L$ is the associated parabolic, and the Levi subgroup L acts on Z with kernel $L_0 \supseteq L'$, so that $Z \simeq A \times C$, $A = L/L_0$.

General U-orbits on X coincide with general P_u-orbits and are of the form $P_\mathrm{u} \times \{z\}$, $z \in Z$. General B-orbits coincide with general P-orbits and are of the form $P_\mathrm{u} \times A \times \{x\}$, $x \in C$. We have $T_xX = \mathfrak{p}_\mathrm{u}x \oplus \mathfrak{a}x \oplus T_xC$, $\forall x \in C$.

General U- and B-orbits on X form two foliations. Consider the respective conormal bundles $\mathscr{U} \supseteq \mathscr{B}$. They are P-vector bundles defined, e.g., over $\mathring{X}$. It follows from the local structure that $\mathscr{U} \simeq P_\mathrm{u} \times T^*Z \simeq P_\mathrm{u} \times A \times \mathfrak{a}^* \times T^*C$ and $\mathscr{B} \simeq P_\mathrm{u} \times A \times T^*C$ over $\mathring{X}$. We have $(\mathscr{U}/\mathscr{B})(x) = (\mathfrak{u}x)^\perp/(\mathfrak{b}x)^\perp = (\mathfrak{b}x/\mathfrak{u}x)^* \simeq (\mathfrak{b}/\mathfrak{u}+\mathfrak{b}_x)^* \subseteq \mathfrak{t}^*$. For $x \in Z$ we have $\mathfrak{b}_x = \mathfrak{b} \cap \mathfrak{l}_0$, whence $\mathscr{U}/\mathscr{B}(x) \simeq \mathfrak{a}^*$. As $\operatorname{Ad} P_\mathrm{u}$ acts on $\mathfrak{b}/\mathfrak{u}$ trivially, there is a canonical isomorphism $\mathscr{U}/\mathscr{B}(x) \simeq \mathfrak{a}^*$, $\forall x \in \mathring{X}$. Therefore $\mathscr{U}/\mathscr{B} \simeq \mathfrak{a}^* \times \mathring{X}$ is a trivial bundle over $\mathring{X}$, and we have a canonical projection $\pi : \mathscr{U} \to \mathfrak{a}^*$, which is nothing else but the moment map for the B-action.

The bundle $\mathscr{U}/\mathscr{B}$ can be lifted (non-canonically) to $\mathscr{U}$ over $\mathring{X}$. Namely consider yet another foliation $\{g(P_\mathrm{u}C) \mid g \in P\}$ and let $\mathscr{A}$ be the respective conormal bundle. By the local structure, $\mathscr{A} \simeq P_\mathrm{u} \times A \times \mathfrak{a}^* \times C$ over $\mathring{X}$, and $\mathscr{U} = \mathscr{A} \oplus \mathscr{B}$.

The isomorphism $\sigma : \mathfrak{a}^* \to \mathscr{A}(x)$, $x \in C$, defined by the formula

$$\sigma(\lambda) = \begin{cases} \lambda & \text{on } \mathfrak{a}x \simeq \mathfrak{a} \\ 0 & \text{on } \mathfrak{p}_\mathrm{u}x \oplus T_xC \end{cases}$$

provides a section for π. It depends on the choice of x and even more—we may replace C by any subvariety in $\mathring{X}^{L_0} = P_\mathrm{u}^{L_0} \times A \times C$ intersecting all P-orbits transversally so that x may be any point in $\mathring{X}$ with $P_x = L_0$ and T_xC may be any L_0-stable complement to $\mathfrak{p}x$ in T_xX.

Recall that $\mathfrak{a}$ embeds in $\mathfrak{l}$ as the orthocomplement to $\mathfrak{l}_0$.

Lemma 22.2. *There is a commutative square*

$$\begin{array}{ccc} \mathscr{U} & \xrightarrow{\Phi} & \mathfrak{a} \oplus \mathfrak{p}_{\mathrm{u}} \subseteq \mathfrak{g} \simeq \mathfrak{g}^* \\ \downarrow \pi & & \downarrow \textit{projection} \\ \mathfrak{a}^* & \xrightarrow{\sim} & \mathfrak{a}. \end{array}$$

Proof. Take $\alpha \in \mathscr{U}(x)$, $x \in \mathring{X}$. Since all maps are P-equivariant, we may assume that $x \in C$, which implies that $\mathscr{U}(x) \simeq (\mathfrak{a}x)^* \oplus T_x^*C$ and $\alpha = \sigma(\lambda) + \beta$ for some $\lambda = \pi(\alpha) \in \mathfrak{a}^*$, $\beta \in T_x^*C = \mathscr{B}(x)$. Hence $\langle \alpha, \xi x \rangle = \langle \lambda, \xi \rangle$, $\forall \xi \in \mathfrak{p} \implies \Phi(\alpha) = \lambda \mod \mathfrak{p}^{\perp} = \mathfrak{p}_{\mathrm{u}}$. □

Corollary 22.3. *There is a commutative square*

$$\begin{array}{ccc} \mathscr{U} & \xrightarrow{\Phi} & M_X \\ \downarrow \pi & & \downarrow \pi_G \\ \mathfrak{a}^* & \xrightarrow{\pi_G} & L_X = \pi_G(\mathfrak{a}^*). \end{array}$$

Lemma 22.4. *There exists a unique morphism $\psi : \mathfrak{a}^* \to \widetilde{L}_X$ making the following square commutative:*

$$\begin{array}{ccc} \mathscr{U} & \xrightarrow{\widetilde{\Phi}} & \widetilde{M}_X \\ \downarrow \pi & & \downarrow \widetilde{\pi}_G \\ \mathfrak{a}^* & \xrightarrow{\psi} & \widetilde{L}_X. \end{array}$$

Proof. The uniqueness is evident. Take $\psi = \widetilde{\pi}_G \widetilde{\Phi} \sigma$. The maps $\psi\pi$ and $\widetilde{\pi}_G \widetilde{\Phi}$ coincide on $\sigma(\mathfrak{a}^*)$, and by Corollary 22.3 they map each $\alpha \in \mathscr{U}$ to one and the same fiber of $\varphi/\!/G$. Thus for every $\lambda \in \mathfrak{a}^*$ the irreducible subvariety $\pi^{-1}(\lambda) \simeq \mathscr{B}$ is mapped by $\widetilde{\pi}_G \widetilde{\Phi}$ to the (finite) fiber of $\varphi/\!/G$ through $\psi(\lambda)$, whence to $\psi(\lambda)$. □

22.3 Little Weyl Group. The normalization of $L_X = \pi_G(\mathfrak{a}^*)$ equals $\mathfrak{a}^*/W(\mathfrak{a}^*)$, where $W(\mathfrak{a}^*) = N_W(\mathfrak{a}^*)/Z_W(\mathfrak{a}^*)$ is the Weyl group of $\mathfrak{a} \subseteq \mathfrak{g}$. By Lemma 22.4, there is a sequence of dominant finite maps of normal varieties $\mathfrak{a}^* \to \widetilde{L}_X \to \mathfrak{a}^*/W(\mathfrak{a}^*)$. It follows from the Galois theory that $\widetilde{L}_X \simeq \mathfrak{a}^*/W_X$ for a certain subgroup $W_X \subseteq W(\mathfrak{a}^*)$ and the left arrow is the quotient map.

Definition 22.5. The group W_X is called the *little Weyl group* of X. It is a subquotient of W.

Remark 22.6. W_X may be a proper subgroup of $W(\mathfrak{a}^*)$, see Example 25.9.

By construction, $\widetilde{M}_X$, $\widetilde{L}_X$, and W_X are G-birational invariants of X. They are related to other invariants such as the horospherical type S.

Proposition 22.7 ([Kn1, 6.4], [Kn6, 7.3]). $\widetilde{M}_{G/S} = \widetilde{M}_X \times_{\widetilde{L}_X} \mathfrak{a}^*$ *and* $\widetilde{M}_X = \widetilde{M}_{G/S}/W_X$.

As well as $M_X, \widetilde{M}_X$ and W_X are determined by a general G-orbit [Kn1, 6.5.4]. For functorial properties of the normalized moment map and the little Weyl group, see [Kn1, 6.5]. For geometric properties of the morphisms $T^*X \to \widetilde{M}_X$, $\widetilde{M}_X \to \widetilde{L}_X$, and $T^*X \to \widetilde{L}_X$, see [Kn1, 7.4, 7.3, 6.6], [Kn6, §§5,7,9]. $\widetilde{M}_X$ has rational singularities [Kn6, 4.3].

Remark 22.8. A non-commutative version of this theory was developed in [Kn6]. Here functions on T^*X are replaced by differential operators on X and $\Bbbk[\mathfrak{g}^*]$ by $U\mathfrak{g}$. The analogue of $\Bbbk[\widetilde{M}_X]$ consists of *completely regular* differential operators generated by velocity fields locally on X and "at infinity". Invariant completely regular operators form a polynomial ring, which coincides with the center of $\mathscr{D}(X)^G$ whenever X is affine. This ring is isomorphic to $\Bbbk[\rho + \mathfrak{a}^*]^{W_X}$ ("Harish-Chandra isomorphism"), where ρ is half the sum of positive roots and W_X is naturally embedded in $N_W(\rho + \mathfrak{a}^*)$ being thus a subgroup, not only a subquotient, of W.

Example 22.9. Take $X = G$ itself, with G acting by left translations. Here $T^*G \simeq G \times \mathfrak{g}^*$, and the moment map Φ is just the coadjoint action map with irreducible fibers isomorphic to G. We have $A = T$, and $\sigma : \mathfrak{t}^* \simeq \mathfrak{t} \hookrightarrow \mathfrak{b} \simeq \mathfrak{u}^\perp = \mathscr{U}(e)$ is the natural inclusion. Therefore $\widetilde{M}_G = M_G = \mathfrak{g}^*$, $\widetilde{L}_G = L_G = \mathfrak{g}^*/\!/G \simeq \mathfrak{t}^*/W$, and $W_G = W$.

Example 22.10. Let $X = G/T$, where G is semisimple. Here $T^*X \simeq G *_T (\mathfrak{u} + \mathfrak{u}^-)$, $A = \mathrm{Ad}_G T$, $M_X = \mathfrak{g}^*$. The subspace $\mathbf{e} + \mathfrak{g}_\mathbf{f} \subset \mathfrak{u} + \mathfrak{u}^-$, where $\mathbf{e} \in \mathfrak{u}$, $\mathbf{f} \in \mathfrak{u}^-$, $\mathbf{h} \in \mathfrak{t}$ form a principal $\mathfrak{sl}_2$-triple, is a cross-section for the fibers of $\pi_G \Phi : T^*X \to \mathfrak{g}^*/\!/G$. Indeed, $\pi_G : \mathbf{e} + \mathfrak{g}_\mathbf{f} \xrightarrow{\sim} \mathfrak{g}^*/\!/G$ [McG, 4.2]. Hence $\widetilde{\pi}_G \widetilde{\Phi}(\mathbf{e} + \mathfrak{g}_\mathbf{f})$ is a cross-section for the fibers of the finite map $\varphi/\!/G$. It follows that $\widetilde{L}_{G/T} = L_{G/T}$, and hence $\widetilde{M}_{G/T} = M_{G/T}$ and $W_{G/T} = W$.

Example 22.11. Consider a horospherical homogeneous space $X = G/S$. We have seen in Theorem 8.7 that the moment map factors as $\Phi = \varphi \pi_A$, where $\varphi : G *_{P^-} \mathfrak{s}^\perp \twoheadrightarrow M_{G/S}$ is generically finite proper and $\pi_A : G *_S \mathfrak{s}^\perp \to G *_{P^-} \mathfrak{s}^\perp$ is the A-quotient map. It immediately follows that $\widetilde{M}_{G/S} = \operatorname{Spec} \Bbbk[G *_{P^-} \mathfrak{s}^\perp]$ and the natural map $G *_{P^-} \mathfrak{s}^\perp \to \widetilde{M}_{G/S}$ is a resolution of singularities. The natural morphisms $G *_{P^-} \mathfrak{s}^\perp \to G *_{P^-} \mathfrak{a}^* = G/P^- \times \mathfrak{a}^* \to \mathfrak{a}^*$ and $\widetilde{\pi}_G : \widetilde{M}_{G/S} \to \widetilde{L}_{G/S}$ are rational G-quotient maps. Indeed, P^- acts on each fiber $\lambda + \mathfrak{p}_\mathrm{u}^-$ of $\mathfrak{s}^\perp \to \mathfrak{a}^*$ generically transitively [McG, 5.5], and fibers of $\widetilde{\pi}_G$ have a dense orbit, because fibers of $\pi_G : M_{G/S} \to L_{G/S}$ do. Passing to rational quotients, we see that $\mathfrak{a}^* \to \widetilde{L}_{G/S}$ is birational, whence an isomorphism. Thus $W_{G/S} = \{e\}$.

The last example has a converse.

Proposition 22.12. *X is horospherical if and only if $W_X = \{e\}$.*

Proof. A horospherical variety of type S is birationally G-isomorphic to $G/S \times C$ by Proposition 7.7. Therefore it suffices to consider $X = G/S$, but then $W_X = \{e\}$ by Example 22.11.

Conversely, suppose that $W_X = \{e\}$ and consider the morphism $\Psi = \widetilde{\pi}_G \widetilde{\Phi} : T^*X \to \widetilde{L}_X = \mathfrak{a}^*$. Then Ψ^* embeds $\mathfrak{a}$ into the space of fiberwise linear G-invariant functions on T^*X, which restrict to linear functions on $\sigma(\mathfrak{a}^*) \simeq \mathfrak{a}^*$.

Geometrically, $\Psi^*\mathfrak{a}$ is an Abelian subalgebra of G-invariant vector fields on X tangent to G-orbits. Furthermore, $\Psi^*\mathfrak{a} = \pi^*\mathfrak{a} = 0$ on $\mathscr{B}$ by Lemma 22.4, and hence $\Psi^*\mathfrak{a}$ is tangent to general P-orbits. It follows that $\Psi^*\mathfrak{a}$ restricts to Px, $x \in C$, as an Abelian subalgebra in the algebra $(\mathfrak{p}/\mathfrak{p}_x)^{P_x} = \mathfrak{a} \oplus \mathfrak{p}_{\mathrm{u}}^{L_0}$ of P-invariant vector fields on Px, and $\Psi^*\xi(x) = \xi x \mod \mathfrak{p}_{\mathrm{u}}^{L_0}x$, $\forall \xi \in \mathfrak{a}$, whence $\Psi^*\mathfrak{a}$ projects onto $\mathfrak{a}$. Since $\mathfrak{z}_{\mathfrak{g}}(\mathfrak{a}) \cap \mathfrak{p}_{\mathrm{u}}^{L_0} = \mathfrak{p}_{\mathrm{u}}^{L} = 0$, $\Psi^*\mathfrak{a}$ is conjugated to $\mathfrak{a}$ by a unique $g_x \in P_{\mathrm{u}}^{L_0}$. Moving each $x \in C$ by g_x, we may assume that $\Psi^*\xi(x) = \xi x$, $\forall \xi \in \mathfrak{a}$, $x \in C$ (or $\forall x \in Z = AC$).

Therefore velocity fields of $A : Z$ extend to G-invariant vector fields on X. These vector fields can be integrated to an A-action on X by central automorphisms, which restricts to the natural A-action on $\mathring{X} = P *_L Z$ provided by $A : Z$. (The induced action $A : T^*X$ integrates the invariant collective motion, cf. §23.) We conclude by Theorems 21.5–21.10 that X is horospherical. □

22.4 Relation to Valuation Cones. Here comes the main result linking the little Weyl group with equivariant embeddings.

Theorem 22.13 ([Kn5, 7.4]). *The little Weyl group W_X acts on $\mathfrak{a}^*$ as a crystallographic reflection group preserving the lattice $\Lambda(X)$, and the central valuation cone $\mathscr{Z}(X)$ is a fundamental chamber of W_X in $\mathscr{E} = \mathrm{Hom}(\Lambda(X), \mathbb{Q})$.*

The proof relies on the integration of the invariant collective motion in T^*X and the study of the asymptotic behavior of its projection to X, see §23. The description of $\mathscr{Z} = \mathscr{Z}(X)$ as a fundamental chamber of a crystallographic reflection group was first obtained by Brion [Bri8] in the spherical case and generalized to arbitrary complexity in [Kn3, §9]. In particular, $\mathscr{Z}$ is a cosimplicial cone.

From this theorem, Knop derived the geometric shape of all valuation cones.

Corollary 22.14 ([Kn3, §9]). *The cones $\mathscr{V}_v$ are cosimplicial.*

In proving Theorem 22.13, we shall use its formal consequence:

Lemma 22.15. *W_X acts trivially on $\mathscr{Z} \cap -\mathscr{Z}$.*

Proof. By Theorem 21.5, there exists a torus E of central automorphisms such that $\mathfrak{e} = (\mathscr{Z} \cap -\mathscr{Z}) \otimes \Bbbk$ with respect to the canonical embedding $\mathfrak{e} \hookrightarrow \mathfrak{a} = \mathscr{E} \otimes \Bbbk$. Consider the action $G^+ = G \times E : X$ and indicate all objects related to this action by the superscript "+". In particular, $A^+ \simeq A$ is the quotient of $A \times E$ modulo the antidiagonal copy of E, and $\mathfrak{X}(A^+) \subset \mathfrak{X}(A) \oplus \mathfrak{X}(E)$ is the graph of the restriction homomorphism $\mathfrak{X}(A) \to \mathfrak{X}(E)$.

Obviously, $\Phi = \tau\Phi^+$, where $\tau : (\mathfrak{g}^+)^* \to \mathfrak{g}^*$ is the canonical projection. It follows that $\widetilde{\Phi} = \widetilde{\tau}\widetilde{\Phi}^+$ for a certain morphism $\widetilde{\tau} : \widetilde{M}_X^+ \to \widetilde{M}_X$. The subalgebra $\Bbbk[\widetilde{M}_X]$ is integrally closed in $\Bbbk[\widetilde{M}_X^+]$, whence $\Bbbk[\widetilde{L}_X]$ is integrally closed in $\Bbbk[\widetilde{L}_X^+]$. On the other hand, we have a commutative square

$$\begin{array}{ccc} (\mathfrak{a}^+)^* & \xrightarrow[\sim]{\tau} & \mathfrak{a}^* \\ \downarrow{\scriptstyle \psi^+} & & \downarrow{\scriptstyle \psi} \\ \widetilde{L}_X^+ & \xrightarrow{\widetilde{\tau}/\!/G} & \widetilde{L}_X, \end{array}$$

where ψ, ψ^+, and hence $\widetilde{\tau}/\!/G$ are finite morphisms. Hence $\widetilde{L}_X^+ \simeq \widetilde{L}_X$ and $W_X^+ = W_X$.

It follows that W_X preserves $(\mathfrak{a}^+)^* \subset \mathfrak{a}^* \oplus \mathfrak{e}^*$ and acts trivially on the second summand $\mathfrak{e}^* \simeq (\mathfrak{a}^+)^*/(\mathfrak{a}^+)^* \cap \mathfrak{a}^* \simeq \mathfrak{a}^*/(\mathfrak{a}^+)^* \cap \mathfrak{a}^*$. Thus W_X acts trivially on $\mathfrak{e}$ embedded in $\mathfrak{a}$. □

23 Invariant Collective Motion

The skew gradients of functions in $\Bbbk[L_X]$ (or $\Bbbk[\widetilde{L}_X]$) pulled back to T^*X generate an Abelian flow of G-automorphisms preserving G-orbits, which is called the invariant collective motion, see 8.6. Restricted to a general orbit $G\alpha \subset T^*X$, the invariant collective motion gives rise to a connected Abelian subgroup $A_\alpha = (G_{\Phi(\alpha)}/G_\alpha)^0 \subseteq \operatorname{Aut}_G G\alpha$. It turns out that $A_\alpha \simeq A$. However, in general, this isomorphism cannot be made canonical in order to produce an A-action on (an open subset of) T^*X integrating the invariant collective motion. This obstruction is overcome by unfolding the cotangent variety by means of a Galois covering with Galois group W_X. This construction is due to Knop [Kn5].

23.1 Polarized Cotangent Bundle.

Definition 23.1. The fiber product $\widehat{T}X = T^*X \times_{\widetilde{L}_X} \mathfrak{a}^*$ is called the *polarized cotangent bundle* of X. Since general fibers of $T^*X \to \widetilde{L}_X$ are irreducible, $\widehat{T}X$ is irreducible. Actually $\widehat{T}X$ is an irreducible component of $T^*X \times_{L_X} \mathfrak{a}^*$, W_X is its stabilizer in $W(\mathfrak{a}^*)$ acting on the set of components, and $\widehat{T}X \to T^*X = \widehat{T}X/W_X$ is a rational Galois cover.

Consider the *principal stratum* $\mathfrak{a}^{\mathrm{pr}} \subseteq \mathfrak{a}^*$ obtained by removing all proper intersections with kernels of coroots and with W-translates of $\mathfrak{a}^*$ in $\mathfrak{t}^*$. The group $W(\mathfrak{a}^*)$ acts on $\mathfrak{a}^{\mathrm{pr}}$ freely. Put $L_X^{\mathrm{pr}} = \pi_G(\mathfrak{a}^{\mathrm{pr}}) = \mathfrak{a}^{\mathrm{pr}}/W(\mathfrak{a}^*)$, the quotient map being an étale finite Galois covering. The preimages of L_X^{pr} in various varieties under consideration will be called *principal strata* and denoted by the superscript "pr".

In particular, $\widehat{T}^{\mathrm{pr}}X \subseteq \widehat{T}X$ is a smooth open subvariety (provided that X is smooth) and the projection $\widehat{T}^{\mathrm{pr}}X \twoheadrightarrow T^{\mathrm{pr}}X \subseteq T^*X$ is an étale finite quotient map by W_X. The G-invariant symplectic structure on T^*X is pulled back to $\widehat{T}^{\mathrm{pr}}X$ so that $\widehat{T}^{\mathrm{pr}}X \to T^*X \to M_X$ is the moment map.

The invariant collective motion on $\widehat{T}^{\mathrm{pr}}X$ is generated by the skew gradients of Poisson-commuting functions from $\pi^*\mathfrak{a}$, where $\pi : \widehat{T}^{\mathrm{pr}}X \to \mathfrak{a}^*$ is the other projection. These skew gradients constitute a commutative r-dimensional subalgebra of Hamiltonian vector fields ($r = r(X) = \dim \mathfrak{a}$). Our aim is to show that these vector fields are the velocity fields of a symplectic A-action so that π is the respective moment map [AG, Ch. 3, 3.1], [Vin3, II.2.3].

Remark 23.2. In particular, it will follow that the W_X-action on $\mathfrak{a}$ lifts to A, so that $\widehat{T}^{\mathrm{pr}}X$ comes equipped with the Hamiltonian $G \times (W_X \leftthreetimes A)$-action.

23.2 Integration of Invariant Collective Motion. Following [Kn5], we shall restrict our considerations to the symplectically stable case (Definition 8.13) for technical reasons.

Proposition 23.3. *$G : T^*X$ is symplectically stable if and only if $T^*X = \overline{G\mathscr{U}}$.*

Proof. Take $x \in C$, $\alpha \in \mathscr{U}^{\mathrm{pr}}(x) \simeq T^{\mathrm{pr}}Z$. By Lemma 22.2, we have $\Phi(\alpha) \in (\mathfrak{a}^{\mathrm{pr}} + \mathfrak{p}_{\mathrm{u}})^{L_0} = \mathfrak{a}^{\mathrm{pr}} + \mathfrak{p}_{\mathrm{u}}^{L_0}$, i.e., $\Phi(\alpha) \in \xi + \mathfrak{p}_{\mathrm{u}}^{L_0}$ for some $\xi \in \mathfrak{a}^{\mathrm{pr}}$. Since $\mathfrak{z}_{\mathfrak{g}}(\xi) = \mathfrak{m}$ and $\mathfrak{m} \cap \mathfrak{p}_{\mathrm{u}}^{L_0} = \mathfrak{p}_{\mathrm{u}}^{L} = 0$, we have $[\mathfrak{p}_{\mathrm{u}}^{L_0}, \xi] = \mathfrak{p}_{\mathrm{u}}^{L_0} \implies \xi + \mathfrak{p}_{\mathrm{u}}^{L_0} = (P_{\mathrm{u}}^{L_0})\xi$ by Lemma 3.4. Therefore

$$\Phi(\mathscr{U}^{\mathrm{pr}}) = P(\mathfrak{a}^{\mathrm{pr}} + \mathfrak{p}_{\mathrm{u}}^{L_0}) = P\mathfrak{a}^{\mathrm{pr}}, \tag{23.1}$$

whence

$$\Phi(G\mathscr{U}^{\mathrm{pr}}) = G\mathfrak{a}^{\mathrm{pr}}. \tag{23.2}$$

Hence density of $G\mathscr{U}$ implies symplectic stability. Conversely, in the symplectically stable case $P_{\mathrm{u}}^{-}\alpha$ is transversal to $\mathscr{U}$ for general $\alpha \in \mathscr{U}$. Indeed, we may assume that $\Phi(\alpha) \in \mathfrak{a}^{\mathrm{pr}}$, but then $[\mathfrak{p}_{\mathrm{u}}^{-}, \Phi(\alpha)] = \mathfrak{p}_{\mathrm{u}}^{-}$ is transversal to $\overline{\Phi(\mathscr{U})} = \mathfrak{a} + \mathfrak{p}_{\mathrm{u}}$. Therefore $\dim \overline{P_{\mathrm{u}}^{-}\mathscr{U}} = \dim P_{\mathrm{u}} + \dim \mathscr{U} = \dim T^*X$. □

Remark 23.4. Similar arguments prove Theorem 8.8 in the case $M = L$, e.g., for quasiaffine X. Indeed, $P_{\mathrm{u}}^{-}\alpha$ is transversal to $\mathscr{U}$ whenever $\Phi(\alpha) \in \mathfrak{a}^{\mathrm{pr}}$, whence $P_{\mathrm{u}}^{-}\mathscr{U}$ is dense in T^*X and $M_X = \overline{\Phi(G\mathscr{U})} = \overline{G\mathfrak{a}} = G(\mathfrak{a} + \mathfrak{p}_{\mathrm{u}})$, because $P\mathfrak{a}^{\mathrm{pr}} = \mathfrak{a}^{\mathrm{pr}} + \mathfrak{p}_{\mathrm{u}}$.

Suppose that the action $G : T^*X$ is symplectically stable. We have observed in 8.5 that $M_X^{\mathrm{pr}} \simeq G *_{N_G(\mathfrak{a})} \mathfrak{a}^{\mathrm{pr}}$. Then $\widetilde{M}_X^{\mathrm{pr}} \simeq M_X^{\mathrm{pr}} \times_{L_X} \widetilde{L}_X \simeq G *_{N_X} \mathfrak{a}^{\mathrm{pr}}$, where $N_X \subseteq N_G(\mathfrak{a})$ is the extension of W_X by $Z_G(\mathfrak{a}) = L$. Hence $T^{\mathrm{pr}}X \simeq G *_{N_X} \Sigma$ has a structure of a homogeneous bundle over G/N_X. Fibers of this bundle are called *cross-sections*. They are smooth and irreducible, because general fibers of $\widetilde{\Phi}$ are irreducible. We may choose a canonical N_X-stable cross-section Σ, namely the unique cross-section in $\Phi^{-1}(\mathfrak{a}^{\mathrm{pr}})$ intersecting $\mathscr{U}$.

Remark 23.5. In fact, $\mathscr{U} \cap \Sigma$ is dense in Σ. Indeed, $\Sigma \cap T^*\mathring{X} \subseteq \mathscr{U}$.

Lemma 23.6. *The kernel of $N_X : \Sigma$ is L_0.*

Proof. By (23.1), $\Phi(\mathscr{U}^{\mathrm{pr}}) = \mathfrak{a}^{\mathrm{pr}} + \mathfrak{p}_{\mathrm{u}} \simeq P *_L \mathfrak{a}^{\mathrm{pr}} \simeq P_{\mathrm{u}} \times \mathfrak{a}^{\mathrm{pr}}$. Hence $\mathscr{U}^{\mathrm{pr}} = P *_L (\mathscr{U} \cap \Sigma) \simeq P_{\mathrm{u}} \times (\mathscr{U} \cap \Sigma)$. On the other hand, $\mathscr{U}|_{\mathring{X}} = P *_L \mathscr{U}|_Z \simeq P_{\mathrm{u}} \times \mathscr{U}|_Z$, and all the stabilizers of $L : \mathscr{U}|_Z \simeq T^*Z$ are equal to L_0. It follows that generic stabilizers of $L : \Sigma$ are P-conjugate to L_0, and hence coincide with L_0. □

Corollary 23.7. *W_X acts on $A = L/L_0$, i.e., preserves the character lattice $\mathfrak{X}(A) = \Lambda(X) \subset \mathfrak{a}^*$.*

Remark 23.8. The lemma implies Theorem 8.18 in the symplectically stable case. This observation was made in [Kn5, §4].

Lemma 23.9. *The A-action integrates the invariant collective motion on Σ.*

Proof. The skew gradient of $\Phi^* f$ ($f \in \Bbbk(M_X)$) at $\alpha \in T^*X$ equals $(\mathrm{d}_{\Phi(\alpha)} f) \cdot \alpha$, where $\mathrm{d}_{\Phi(\alpha)} f$ is considered as an element of $\mathfrak{g}^{**} = \mathfrak{g}$ up to a shift from $(T_{\Phi(\alpha)} M_X)^\perp$. Indeed, the skew gradient of a function is determined by its differential at a point, and for linear functions $f \in \mathfrak{g}$ the assertion holds by the definition of Φ.

If $\alpha \in \Sigma$, then $T_{\Phi(\alpha)} M_X = \mathfrak{a} + [\mathfrak{g}, \Phi(\alpha)] = \mathfrak{a} \oplus \mathfrak{p}_\mathrm{u} \oplus \mathfrak{p}_\mathrm{u}^-$, $(T_{\Phi(\alpha)} M_X)^\perp = \mathfrak{l}_0$. The differentials $\mathrm{d}_{\Phi(\alpha)} f$ of $f \in \Bbbk[L_X^{\mathrm{pr}}]$ generate the conormal space of $G \cdot \Phi(\alpha)$ in M_X at $\Phi(\alpha)$, i.e., $[\mathfrak{g}, \Phi(\alpha)]^\perp / (T_{\Phi(\alpha)} M_X)^\perp = \mathfrak{l}/\mathfrak{l}_0 = \mathfrak{a}$. Thus the invariant collective motion at α is $\mathfrak{a}_\alpha = \mathfrak{l}\alpha = \mathfrak{a}\alpha$. □

Translation by G permutes cross-sections transitively and extends the A-action to each cross-section. These actions integrate the invariant collective motion but, in general, they do not globalize to a regular A-action on the whole cotangent bundle, due to non-trivial monodromy.

However, unfold $T^{\mathrm{pr}}X$ to $\widehat{T}^{\mathrm{pr}}X = T^{\mathrm{pr}}X \times_{\widetilde{L}_X} \mathfrak{a}^{\mathrm{pr}} \simeq G *_L \widehat{\Sigma}$, where $\widehat{\Sigma} = \{\hat{\alpha} = (\alpha, \Phi(\alpha)) \mid \alpha \in \Sigma\}$. We retain the name “cross-sections” for the fibers of this homogeneous bundle, which are isomorphic to the cross-sections in $T^{\mathrm{pr}}X$. Now there is a natural A-action on $\widehat{T}^{\mathrm{pr}}X$ provided by $A : \widehat{\Sigma} \simeq \Sigma$, which integrates the invariant collective motion on $\widehat{T}^{\mathrm{pr}}X$. The W_X-action on $\widehat{T}^{\mathrm{pr}}X$ is induced from the N_X-action on $G \times \widehat{\Sigma}$ given by $n(g, \hat{\alpha}) = (gn^{-1}, n\hat{\alpha})$, $\forall n \in N_X$, $g \in G$, $\hat{\alpha} \in \widehat{\Sigma}$. We sum up in the following

Theorem 23.10 ([Kn5, 4.1–4.2]). *There is a Hamiltonian $G \times (W_X \leftthreetimes A)$-action on $\widehat{T}^{\mathrm{pr}}X$ with the moment map $\Phi \times \pi : \widehat{T}^{\mathrm{pr}}X \to \mathfrak{g}^* \oplus \mathfrak{a}^*$.*

Proof. It remains only to explain why π is the moment map for the A-action. Take any $\hat{\alpha} \in \widehat{\Sigma}$ over $\alpha \in \Sigma$. By (the proof of) Lemma 23.9, for any $\xi \in \mathfrak{a}$ there exists $f \in \Bbbk[L_X^{\mathrm{pr}}]$ such that $\mathrm{d}_{\Phi(\alpha)} f = \xi \mod \mathfrak{l}_0$. The skew gradient of $\pi^*\xi$ at $\hat{\alpha}$ is pulled back from that of $\Phi^* f$ at α, i.e., from $\xi\alpha$, hence it equals $\xi\hat{\alpha}$. We conclude by G-equivariance. □

23.3 Flats and Their Closures. In particular, the orbit of the invariant collective motion through $\hat{\alpha} \in \widehat{T}^{\mathrm{pr}}X$ over $\alpha \in T^{\mathrm{pr}}X$ is $A\hat{\alpha} = G_{\Phi(\alpha)}\hat{\alpha} \simeq G_{\Phi(\alpha)}/G_\alpha$. For the purposes of the embedding theory it is important to study the projections of these orbits to X and their boundaries.

Definition 23.11. A *flat* in X is $F_\alpha = \pi_X(A\hat{\alpha}) = G_{\Phi(\alpha)}x$, where $\alpha \in T_x^{\mathrm{pr}}X$, $\hat{\alpha} \in \widehat{T}^{\mathrm{pr}}X$ lies over α, and $\pi_X : T^*X \to X$ is the canonical projection. The composed map $A \to A\hat{\alpha} \to F_\alpha$ is called the *polarization* of the flat.

For general α the polarization map is an isomorphism: indeed, without loss of generality assume that $\alpha \in \Sigma \cap T_x^*\mathring{X}$, whence $G_{\Phi(\alpha)} = L$ and $G_{\Phi(\alpha)} \cap G_x = G_\alpha = L_0$, which implies $F_\alpha \simeq A$. Generic flats are nothing else but G-translates of L- (or A-) orbits in Z, under appropriate choice of Z. Namely, by Lemma 22.2, there is a commutative diagram

$$\begin{array}{ccc} \mathring{X} \times \mathfrak{a}^* \simeq \mathscr{A} & \xrightarrow{\Phi} & \mathfrak{a} \oplus \mathfrak{p}_u \\ \downarrow{\scriptstyle \pi} & & \downarrow \\ \mathfrak{a}^* & \xrightarrow{\sim} & \mathfrak{a}. \end{array} \tag{23.3}$$

For any $\lambda \in \mathfrak{a}^{\mathrm{pr}}$ we have $\lambda + \mathfrak{p}_u = P \cdot \lambda \simeq P/L$, and hence $\mathring{X} \simeq P *_L Z^\lambda$, where $Z^\lambda \simeq \pi_X(\Phi^{-1}(\lambda) \cap \mathscr{A})$. Clearly, all L-orbits in $Z = Z^\lambda$ are flats. On the other hand, for any $\alpha \in \Sigma \cap T_x^*\mathring{X}$ it is easy to construct a subvariety $C \subset \mathring{X}^{L_0}$ through x intersecting all P-orbits transversally such that $\alpha = 0$ on T_xC, whence $x \in Z^{\Phi(\alpha)}$.

The rigidity of torus actions implies that the closures of generic flats are isomorphic.

Proposition 23.12 ([Kn5, §6]). *The closures $\overline{F}_\alpha$ for general $\alpha \in T^{\mathrm{pr}}X$ are A-isomorphic toric varieties, and the W_X-action on $A \simeq F_\alpha$ extends to $\overline{F}_\alpha$.*

Proof. We explain the affine case, the general case being reduced to this one by standard techniques of invariant quasiprojective open coverings and affine cones. Generic flats are G-translates of generic L-orbits in $\overline{\pi_X(\Sigma)}$. We may assume that X is embedded into a G-module. Since $\overline{\pi_X(\Sigma)}$ is N_X-stable, the set of eigenweights of $A = L/L_0$ in the L-submodule spanned by $\pi_X(\Sigma)$ is W_X-stable. For general $\alpha \in \Sigma$, $\Bbbk[\overline{F}_\alpha]$ is just the semigroup algebra generated by these eigenweights. □

The following result is crucial for interdependence between flats and central valuations. It partially describes the boundary of a generic flat.

Proposition 23.13 ([Kn5, 7.3]). *Let $D \subset X$ be a G-stable divisor with $v = v_D \in \mathscr{Z}$. The closure $\overline{F}_\alpha$ of a generic flat contains A-stable prime divisors $D_{wv} \subseteq D$, $w \in W_X$, which correspond to wv regarded as A-valuations of $\Bbbk(A)$. Furthermore, $\overline{F}_\alpha$ is generically smooth along D_{wv}.*

Proof. Without loss of generality we may assume that $\alpha \in \Sigma$. The W_X-action on $\overline{F}_\alpha$ is given by $w : \overline{F}_\alpha \to \overline{F}_{n\alpha} \to \overline{F}_\alpha$, where the left arrow is the translation by $n \in N_X$ representing $w \in W_X$ and the right arrow is the unique A-isomorphism mapping $n\alpha$ back to α. Since D is N_X-stable, it suffices to prove the assertion for $w = e$.

Shrinking $\mathring{X}$ if necessary, we find a B-chart X_0 intersecting D such that $\mathring{X} = X_0 \setminus D$.

Lemma 23.14. *The morphism $\mathring{X} \times \mathfrak{a}^* \to \mathfrak{a} + \mathfrak{p}_u$ in (23.3) extends to $X_0 \times \mathfrak{a}^*$.*

Proof. Trivializing sections of $\mathscr{A} \simeq \mathring{X} \times \mathfrak{a}^*$ corresponding to $\lambda \in \Lambda$ are $\mathbf{d}\mathbf{f}_\lambda/\mathbf{f}_\lambda$, where $\mathbf{f}_\lambda$ are B-eigenfunctions on $\mathring{X}$ which are constant on P_uC. These sections extend to sections of $T^*X(\log D)$ over X_0, which trivialize the subbundle $\mathscr{A}(\log D) = \overline{\mathscr{A}} \subseteq T^*X_0(\log D)$, because any linear combination of $\mathbf{d}\mathbf{f}_\lambda/\mathbf{f}_\lambda$ maps to the same combination of $-\lambda$ mod $\mathfrak{p}_u$ under the logarithmic moment map. The moment map of $T^*X(\log D)$ restricted to $\mathscr{A}(\log D)$ provides the desired extension. □

Consequently $X_0 \simeq P *_L Z_0$, $Z_0 = \overline{Z^{\Phi(\alpha)}}$, and $F_\alpha = Ax$ is a general A-orbit in Z_0. The proposition stems from

Lemma 23.15. *After possible shrinking of X_0, $Z_0 \simeq F \times C$, where $F = \overline{Ax}$ is the closure of a general A-orbit in Z_0.*

Proof. Since v is central and by Lemma 5.8, the restriction of functions identifies $\Bbbk(D)^B \simeq \Bbbk(Z_0 \cap D)^A$ with $\Bbbk(X)^B \simeq \Bbbk(Z_0)^A \simeq \Bbbk(C)$. Hence removing zeroes/poles of a B-invariant function preserves non-empty intersection with D. In particular, we may assume that $\Bbbk(Z_0 \cap D)^A = \operatorname{Quot}\Bbbk[Z_0 \cap D]^A$ and $\Bbbk[Z_0 \cap D]^A \simeq \Bbbk[Z_0]^A \simeq \Bbbk[C]$ by shrinking X_0. We have $\Bbbk[Z] = \Bbbk[\mathbf{f}_\lambda \mid \lambda \in \Lambda] \otimes \Bbbk[C]$ and $\Bbbk[Z_0] \subseteq \Bbbk[\mathbf{f}_\lambda \mid \lambda \in \Lambda_0] \otimes \Bbbk[C]$, where Λ_0 is the weight semigroup of Z_0. There exist $h_\lambda \in \Bbbk[C]$ such that $h_\lambda \mathbf{f}_\lambda \in \Bbbk[Z_0]$. Shrinking X_0, we may assume that $\mathbf{f}_\lambda \in \Bbbk[Z_0]$, $\forall \lambda \in \Lambda_0$ (because Λ_0 is finitely generated). Hence $Z_0 \simeq F \times C$, where $F = \operatorname{Spec}\Bbbk[\Lambda_0]$. □

23.4 Non-symplectically Stable Case. Now we explain how to deal with non-symplectically stable case.

We may assume X to be quasiprojective. By $\widehat{X}$ denote the cone over X without the origin. In the notation of Remark 20.8, the $\widehat{G}$-action on $T^*\widehat{X}$ is symplectically stable by Proposition 8.14.

The quotient space $T^*\widehat{X}/\Bbbk^\times$ is a vector bundle over X containing T^*X as a subbundle, the quotient bundle being the trivial line bundle. The moment map for $\widehat{G} : T^*\widehat{X}$ factors through $T^*\widehat{X}/\Bbbk^\times$, so that there is a commutative diagram

$$\begin{array}{ccccc} & & T^*\widehat{X} & & \\ & & \downarrow & & \\ T^*X & \subset & T^*\widehat{X}/\Bbbk^\times & \xrightarrow{\Upsilon} & \Bbbk \\ \downarrow \Phi & & \downarrow \widehat{\Phi} & & \| \\ M_X & \subset & M_{\widehat{X}} & \longrightarrow & \Bbbk. \end{array}$$

Here Υ is induced by the evaluation at the Euler vector field in $\widehat{X}$, i.e., by the moment map for the $\Bbbk^\times$-action, and the lower right arrow is the projection of $M_{\widehat{X}} \subseteq \widehat{\mathfrak{g}}^* = \mathfrak{g}^* \oplus \Bbbk$ to $\Bbbk$. T^*X and M_X are the zero-fibers of the respective maps to $\Bbbk$. Also, we have $\mathfrak{a}^* = \widehat{\mathfrak{a}}^* \cap \mathfrak{g}^*$. The morphism $\widehat{\Phi}$ factors through $\widetilde{M}_{\widehat{X}}$, hence Φ factors through the zero-fiber M'_X of $\widetilde{M}_{\widehat{X}} \to \Bbbk$. Since $M'_X \to M_X$ is finite, there is a commutative diagram

$$\begin{array}{ccccc} T^*X & & \subset & & T^*\widehat{X}/\Bbbk^\times \\ \downarrow \widetilde{\Phi} & & & & \downarrow \\ \widetilde{M}_X & \longrightarrow & M'_X & \subset & \widetilde{M}_{\widehat{X}}. \end{array}$$

Passing to quotients, we obtain

$$\begin{array}{ccc} \mathfrak{a}^* & \subset & \widehat{\mathfrak{a}}^* \\ \downarrow & & \downarrow \\ \widetilde{L}_X & \longrightarrow & \widetilde{L}_{\widehat{X}}, \end{array}$$

whence $W_X \subseteq W_{\widehat{X}}$. Actually these groups coincide by Theorem 22.13. By Corollary 23.7, $W_{\widehat{X}}$ preserves $\Lambda(\widehat{X})$, whence W_X preserves $\Lambda(X) = \Lambda(\widehat{X}) \cap \mathfrak{a}^*$.

Instead of flats, one considers *twisted flats* defined as projectivizations of usual flats in $\widehat{X}$. The above results on flats and their closures in $\widehat{X}$ descend to twisted flats in X. If T^*X is symplectically stable, then $T^{\mathrm{pr}}X \subset T^{\mathrm{pr}}\widehat{X}/\Bbbk^\times$, and flats are a particular case of twisted flats.

Example 23.16. Let $G = \mathrm{SL}_2$ and $\widehat{X} \subset V(3) = \Bbbk[x,y]_3$ be the variety of (nonzero) degenerate binary cubic forms (in the variables x,y). Essentially $\widehat{G} = \mathrm{GL}_2$. The form $v = xy^2$ has the open orbit $\widehat{O} \subset \widehat{X}$ and the stabilizer $\widehat{H} = \left\{ \left(\begin{smallmatrix} t^2 & 0 \\ 0 & t^{-1} \end{smallmatrix} \right) \mid t \in \Bbbk^\times \right\}$ in $\widehat{G}$. Passing to projectivizations, we obtain a hypersurface $X \subset \mathbb{P}(V(3))$ with the open orbit $O = G[v]$, and $G_{[v]} =: H = \left\{ \left(\begin{smallmatrix} t & 0 \\ 0 & t^{-1} \end{smallmatrix} \right) \mid t \in \Bbbk^\times \right\}$. The flats through $[v]$ are the orbits of the stabilizers in G of non-degenerate matrices from $\mathfrak{h}^\perp = \left\{ \left(\begin{smallmatrix} 0 & * \\ * & 0 \end{smallmatrix} \right) \right\}$, i.e., the orbits of $L = \left\{ \left(\begin{smallmatrix} a & b \\ b & a \end{smallmatrix} \right) \mid a^2 - b^2 = 1 \right\}$ and of all H-conjugates of L. However the twisted flats are the orbits of the stabilizers in G of matrices from $\widehat{\mathfrak{h}}^\perp = \left\{ \left(\begin{smallmatrix} c & * \\ * & 2c \end{smallmatrix} \right) \mid c \in \Bbbk \right\}$ with non-degenerate projection to $\mathfrak{sl}_2$, i.e., the orbits of arbitrary 1-tori in G.

The boundary of the open orbit is a single orbit $X \setminus O = G[v_0]$, $v_0 = y^3$, with the stabilizer $G_{v_0} = B$. Put $Y = \mathbb{P}(\langle v_0, v \rangle)$, a B-stable subspace in X. The natural bijective morphism $\widetilde{X} = G *_B Y \to X$ is a desingularization. The closures of generic twisted flats are isomorphic to $\mathbb{P}^1$ and intersect the boundary divisor $D = \widetilde{X} \setminus O$ transversally in two points permuted by the (little) Weyl group. Indeed, it suffices to verify it for general T-orbits in O, which is easy.

Remark 23.17. The fibers $T^cX = \Upsilon^{-1}(c)$, $c \in \Bbbk \setminus \{0\}$, are called *twisted cotangent bundles* [BoB, §2]. They carry a structure of affine bundles over X associated with the vector bundle T^*X. Thus each T^cX has a natural symplectic structure. (This is a particular case of symplectic reduction [AG, Ch. 3, 3.2] for the $\Bbbk^\times$-action on $T^*\widehat{X}$.) The action $G : T^cX$ is Hamiltonian and symplectically stable: the moment map Φ^c is the composition of $\widehat{\Phi}$ and the projection $\widehat{\mathfrak{g}}^* \to \mathfrak{g}^*$, so that $\overline{\operatorname{Im} \Phi^c}$ is identified with the fiber of $M_{\widehat{X}} \to \Bbbk$ over c, and the symplectic stability stems from that of $T^*\widehat{X}$.

The whole theory can be developed for arbitrary G-varieties replacing the usual cotangent bundle by its twisted analogue [Kn5, §9]. If X is quasiaffine, i.e., embedded in a G-module V, then $X \subset \mathbb{P}(V \oplus \Bbbk)$, $\widehat{X} \simeq X \times \Bbbk^\times$, and $T^cX \simeq T^*X$. Therefore in the quasiaffine case the classical theory is included in the twisted one.

23.5 Proof of Theorem 22.13. We already know from the above that W_X preserves $\Lambda(X)$ and acts on $\mathscr{E}$. Let $W_X^\# \subset \mathrm{GL}(\mathscr{E})$ be the subgroup generated by reflections at the walls of $\mathscr{Z}$. The first step is to show that $W_X^\# \subseteq W_X$.

Choose a wall of $\mathscr{Z}$ and a vector v in its interior. We may assume that $v = v_D$ for a certain G-stable prime divisor $D \subset X$. Consider the normal bundle X' of X

along D. By Proposition 21.11(1), the central valuation cone of X' is a half-space $\mathscr{Z}' = \mathscr{Z} + \mathbb{Q}v$. By Theorem 21.10, X' is not horospherical, whence $W_{X'} \neq \{e\}$ by Proposition 22.12. By Lemma 22.15, $W_{X'}$ acts trivially on $\mathscr{Z}' \cap -\mathscr{Z}'$, whence $W_{X'}$ is generated by the reflection at the chosen face of $\mathscr{Z}$.

On the other hand, X' is deformed to X, i.e., it is the zero-fiber of the $(G \times \Bbbk^\times)$-equivariant flat family $E \to \mathbb{A}^1$ with the other fibers isomorphic to X [Ful1, 5.1]. Since $E \setminus X' \simeq X \times \Bbbk^\times$ and the moment map of $T^*(E \setminus X')$ factors through the projection onto T^*X, we have $\widetilde{M}_X = \widetilde{M}_E$. There is a commutative diagram

$$\begin{array}{ccc} T^*X' \leftarrow T^*E|_{X'} & \subset & T^*E \\ \downarrow & & \downarrow \\ M_{X'} & \subseteq & M_E = M_X. \end{array}$$

As $\widetilde{M}_E \to M_E$ is finite and $\Bbbk[\widetilde{M}_{X'}]$ is integrally closed in $\Bbbk[T^*E|_{X'}]$, there is a finite morphism $\widetilde{M}_{X'} \to \widetilde{M}_E$, whence $\mathfrak{a}^* \to \widetilde{L}_{X'} \to \widetilde{L}_E = \widetilde{L}_X$. Thus $W_{X'} \subseteq W_E = W_X$.

At this point we may reduce the problem to the symplectically stable case, because $\widehat{\mathscr{Z}} = \mathscr{Z}(\widehat{X})$ is the preimage of $\mathscr{Z}$ and $W^\#_{\widehat{X}} = W^\#_X \subseteq W_X \subseteq W_{\widehat{X}}$.

It follows that $W^\#_X$ is a finite crystallographic reflection group and $\mathscr{Z}$ is a union of some fundamental chambers of $W^\#_X$. To conclude the proof, it remains to show that different vectors $v_1, v_2 \in \mathscr{Z}$ cannot be W_X-equivalent.

Assume the converse, i.e., $v_2 = wv_1$, $w \in W_X$. Without loss of generality X contains two G-stable prime divisors D_1, D_2 corresponding to v_1, v_2. (Replace X by the normalized closure of the graph of the birational map $X_1 \dashrightarrow X_2$, where X_i is a complete G-model of K having a divisor with valuation v_i.) Removing $D_1 \cap D_2$, we may assume that D_1, D_2 are disjoint. By Proposition 23.13, the closure of a generic (twisted) flat contains two A-stable prime divisors D_{v_1}, D_{v_2} both lying in D_1 and in D_2, a contradiction. □

23.6 Sources. Proposition 23.13, together with Theorem 22.13, leads to a description of the whole boundary of a generic flat (to a certain extent). We no longer assume that X is smooth.

Definition 23.18. A *source* $Y \subset X$ is the center of a central valuation.

Proposition 23.19 ([Kn5, 7.6]). *Let $F_\alpha \subseteq X$ be a generic (twisted) flat. A vector $v \in \mathscr{E}$, regarded as an A-valuation of $\Bbbk(F_\alpha)$, has a center F_v in $\overline{F}_\alpha$ if and only if the unique $v' \in W_X v \cap \mathscr{Z}$ has a center $Y \subseteq X$. Furthermore, $F_v \subseteq Y$ and Y is uniquely determined by F_v.*

Proof. Since W_X acts on $\overline{F}_\alpha$, we may assume that $v = v'$. Take a G-equivariant completion $\overline{X} \supseteq X$ [Sum, Th. 3] and construct a proper birational G-morphism $\varphi : X' \to \overline{X}$ such that X' contains a divisor D with $v_D = v$ (Proposition 19.9). By Proposition 23.13, the center of v on the closure $\overline{F}'_\alpha$ of F_α in X' is a divisor $D_v \subseteq D$. Hence $\varphi(D_v) \subseteq \varphi(D)$ is the center of v on the closure $\overline{\overline{F}}_\alpha$ of F_α in $\overline{X}$. It intersects $\overline{F}_\alpha$ (exactly in F_v) if and only if $\varphi(D)$ intersects X (in Y).

Suppose that $F_w = F_v$, but $w' \in W_X w \cap \mathscr{Z}$ has center $Y' \not\subseteq Y$ on X. Then w' has center $Y' \setminus Y$ on $X \setminus Y$, whence w has center on $\overline{F}_\alpha \setminus Y$, a contradiction with $F_w \subseteq Y$. □

Remark 23.20. A-valuations of $\Bbbk(F_\alpha)$ are determined by one-parameter subgroups of A, see Example 24.8. The open orbit in F_v can be reached from F_α by taking the limits of trajectories of the respective one-parameter subgroup. Thus Proposition 23.19 gives a full picture of the asymptotic behavior of the invariant collective motion.

Corollary 23.21. *There are finitely many sources in a G-variety, and the closure of a generic (twisted) flat intersects all of them.*

Corollary 23.22. *Suppose that an affine G-variety X contains a proper source; then* $\Bbbk[X]$ *has a G-invariant non-negative grading induced by a certain central one-parameter subgroup.*

Proof. Corollary 23.21 and the assumptions imply that $\overline{F}_\alpha \neq F_\alpha$ is an affine toric variety acted on by W_X. Its normalization is determined by a strictly convex W_X-stable cone $\mathscr{C} \subset \mathscr{E}$ (Example 15.8). Clearly, $\operatorname{int}\mathscr{C}$ contains a W_X-invariant vector $v \neq 0$. Hence $v \in \mathscr{Z} \cap -\mathscr{Z}$ defines a central one-parameter subgroup γ acting on X by Theorem 21.5, whence a G-invariant grading of $\Bbbk[X]$. Generic γ-orbits in X are contained in generic flats and non-closed therein, because γ contracts F_α to the unique closed orbit in $\overline{F}_\alpha$. Hence the grading is non-negative. □

See [Kn5, §§8–9] and 29.6 for a deeper analysis of sources, flats, and their closures.

23.7 Root System of a G-variety. To any G-variety X one can relate a root system in $\Lambda(X)$, which is a birational invariant of the G-action. Namely, let $\Pi_X^{\min} \subset \Lambda(X)$ be the set of indivisible vectors generating the rays of the simplicial cone $-\mathscr{Z}(X)^\vee$. It is easy to deduce from Theorem 22.13 that $\Delta_X^{\min} = W_X \Pi_X^{\min}$ is a root system with base $\Pi_X^{\min}$ and the Weyl group W_X, called the *minimal root system* of X. It is a generalization of the (reduced) root system of a symmetric variety (see §26 and 30.3). The minimal root system was defined by Brion [Bri8] for a spherical variety and by Knop [Kn8] in the general case.

Remark 23.23. There are several natural root systems related to $G : X$ which generate one and the same Weyl group W_X [Kn8, 6.2, 6.4, 7.5]; $\Delta_X^{\min}$ is the "minimal" one.

Example 23.24. If $X = G$ comes equipped with the G-action by left translations and G' is adjoint, then $\Delta_X^{\min} = \Delta_G$ by Example 21.2. If G' is not adjoint, then $\Delta_X^{\min}$ may differ from Δ_G: this happens if and only if some roots in Δ_G are divisible in $\mathfrak{X}(T)$. For simple G, $\Delta_G^{\min} = \Delta_G$ unless $G = \mathrm{Sp}_{2n}(\Bbbk)$; in the latter case $\Delta_G, \Delta_G^{\min}$ are of types $\mathbf{C}_n, \mathbf{B}_n$, respectively.

An important result of Knop establishes a relation between the minimal root system and central automorphisms.

Theorem 23.25 ([Kn8, 6.4]). *A diagonalizable group* $S_X = \bigcap_{\alpha\in\Delta_X^{\min}} \operatorname{Ker}\alpha \subseteq A$ *is canonically embedded in* $\operatorname{CAut} X$.

Note that S_X^0 is the largest connected algebraic subgroup of $\operatorname{CAut} X$ by Theorem 21.5(3), but Theorem 23.25 is much more subtle.

Synopsis of a proof. Standard reductions allow us to assume that X is quasiaffine. It is clear that $S_X \subseteq A^{W_X}$. The action $A : \widehat{T}^{\mathrm{pr}}X$ descends to $A^{W_X} : T^{\mathrm{pr}}X$.

The most delicate part of the proof is to show that the action of S_X extends to T^*X in codimension one. Knop shows that the A-actions on the orbits of the invariant collective motion patch together in an action on T^*X of a smooth group scheme $\mathscr{S}$ over $\widetilde{L}_X$ with connected fibers, and $A^{W_X} \subset \mathscr{S}(\Bbbk(\widetilde{L}_X))$. Furthermore, $s \in A^{W_X}$ induces a rational section of $\mathscr{S} \to \widetilde{L}_X$ which is defined in codimension one whenever $\alpha(s) = 1$, $\forall\alpha \in \Delta_X^{\min}$, whence the claim.

Now S_X acts on an open subset $R \subseteq T^*X$ whose complement has codimension ≥ 2, and this action commutes with G and with homotheties on the fibers. Hence S_X acts by G-automorphisms on $\mathbb{P}(R) \subseteq \mathbb{P}(T^*X)$. Since X is quasiaffine, and general fibers of $\mathbb{P}(R) \to X$ have no non-constant regular functions, we deduce that S_X permutes the fibers. This yields a birational action $S_X : X$ commuting with G and preserving generic flats. The description of generic flats shows that S_X preserves P-orbits in $\mathring{X}$, whence $S_X \hookrightarrow \operatorname{CAut} X$. □

24 Formal Curves

In the previous sections of this chapter we examined G-valuations on arbitrary G-varieties. However our main interest is in homogeneous varieties. In this section we take a closer look at $\mathscr{V} = \mathscr{V}(O)$, where O is a homogeneous space.

Namely we describe the subset $\mathscr{V}^1 \subseteq \mathscr{V}$ consisting of G-valuations v such that $\Bbbk(v)^G = \Bbbk$. In geometric terms, if v is proportional to v_D for a G-stable divisor D on a G-embedding $X \hookleftarrow O$, then $v \in \mathscr{V}^1$ if and only if D contains a dense G-orbit.

The subset $\mathscr{V}^1$ is big enough. For instance, if $c(O) = 0$, then $\mathscr{V}^1 = \mathscr{V}$ and, if $c(O) = 1$, then $\mathscr{V}^1 \supseteq \mathscr{V} \setminus \mathscr{Z}$, because in this case $c(\Bbbk(v)) = 0$ by Proposition 21.3. In general, any G-valuation can be approximated by $v \in \mathscr{V}^1$ in a sense [LV, 4.11].

24.1 Valuations via Germs of Curves. In [LV] Luna and Vust suggested computing $v(f)$, $v \in \mathscr{V}^1$, $f \in K = \Bbbk(O)$, by restricting f to a (formal) curve in O approaching D, in the above notation. More precisely, take a smooth curve $\Theta \subseteq X$ meeting D transversally in x_0, $\overline{Gx_0} = D$. It is clear that $v_D(f)$ equals the order of $f|_{g\Theta}$ at gx_0 for general $g \in G$. More generally, take a germ of a curve $\chi : \Theta \dashrightarrow O$ that converges to x_0 in X, i.e., χ extends regularly to the base point $\theta_0 \in \Theta$ and $\chi(\theta_0) = x_0$, see Appendix A.3. Then

$$v_D(f) \cdot \langle D, \Theta\rangle_{x_0} = v_{\chi,\theta_0}(f) := v_{\theta_0}(\chi^*(gf)) \qquad \text{for general } g \in G, \tag{24.1}$$

where $\langle D, \Theta\rangle_{x_0}$ is the local intersection number [Ful1, Ch. 7].

Theorem 24.1. *For any germ of a curve* $(\chi : \Theta \dashrightarrow O,\ \theta_0 \in \Theta)$*, Formula* (24.1) *defines a G-valuation* $v_{\chi,\theta_0} \in \mathscr{V}^1$*, and every* $v \in \mathscr{V}^1$ *is proportional to some* v_{χ,θ_0}*. Furthermore, if* $X \supseteq O$ *is a G-model of K and* $Y \subseteq X$ *the center of v, then the germ converges in X to* $x_0 \in Y$ *such that* $\overline{Gx_0} = Y$*.*

Proof. The G-action yields a rational dominant map $\alpha : G \times \Theta \dashrightarrow O$, $(g,\theta) \mapsto g\chi(\theta)$. By construction, v_{χ,θ_0} is the restriction of $v_{G\times\{\theta_0\}} \in \mathscr{V}^1(G\times\Theta)$ to K, whence $v = v_{\chi,\theta_0} \in \mathscr{V}^1$. If v has the center Y on X, then $\alpha : G\times\Theta \dashrightarrow X$ is regular along $G\times\{\theta_0\}$ and $\overline{\alpha(G\times\{\theta_0\})} = Y$, whence χ converges to $x_0 = \alpha(e,\theta_0)$ in the dense G-orbit of Y. □

24.2 Valuations via Formal Curves. For computations, it is more practical to adopt a more algebraic point of view, namely to replace germs of curves by germs of formal curves, i.e., by $\Bbbk((t))$-points of O, see Appendix A.3.

Any germ of a curve $(\chi : \Theta \dashrightarrow O,\ \theta_0 \in \Theta)$ defines a formal germ $x(t) \in O(\Bbbk((t)))$ if we replace Θ by the formal neighborhood of θ_0. We have

$$v_{\chi,\theta_0}(f) = v_{x(t)}(f) := \operatorname{ord}_t f(gx(t)) \qquad \text{for generic } g, \tag{24.2}$$

where "generic" means a sufficiently general point of G (depending on $f \in K$) or the generic $\Bbbk(G)$-point of G (Example A.10).

The counterpart of Theorem 24.1 is

Theorem 24.2. *For any* $x(t) \in O(\Bbbk((t)))$*, Formula* (24.2) *defines a G-valuation* $v_{x(t)} \in \mathscr{V}^1$*, and every* $v \in \mathscr{V}^1$ *is proportional to some* $v_{x(t)}$*. Furthermore, if* $X \supseteq O$ *is a G-model of K and* $Y \subseteq X$ *the center of v, then* $x(t) \in X(\Bbbk[[t]])$ *and* $Y = \overline{Gx(0)}$*.*

To prove this theorem it suffices to show that $v_{x(t)} = v_{\chi,\theta_0}$ for a certain germ of a curve (χ,θ_0). This stems from the two subsequent lemmas.

Lemma 24.3 ([LV, 4.4]). $\forall g(t) \in G(\Bbbk[[t]]),\ x(t) \in O(\Bbbk((t))) :\ v_{g(t)x(t)} = v_{x(t)}$.

Proof. The G-action on $x(t)$ yields $\Bbbk(O) \hookrightarrow \Bbbk(G)((t))$, so that $v_{x(t)}$ coincides with ord_t with respect to this inclusion. The lemma stems from the fact that $G(\Bbbk[[t]])$ acts on $\Bbbk(G)((t))$ "by right translations" preserving ord_t. □

Lemma 24.4 ([LV, 4.5]). *Every germ of a formal curve in O is* $G(\Bbbk[[t]])$*-equivalent to a formal germ induced by a germ of a curve.*

Proof. Since O is homogeneous, $G(\Bbbk[[t]])$-orbits are open in $O(\Bbbk((t)))$ in t-adic topology. Now the lemma stems from Theorem A.16. □

Germs of formal curves in O are more accessible if they come from formal germs in G. Luckily, in characteristic zero this is "almost" always the case. From now on suppose that $\operatorname{char}\Bbbk = 0$.

Proposition 24.5 ([LV, 4.3]). *For any* $x(t) \in O(\Bbbk((t)))$ *there exists* $n \in \mathbb{N}$ *such that* $x(t^n) = g(t)\cdot o$ *for some* $g(t) \in G(\Bbbk((t)))$*.*

Proof. Consider the algebraic closure $\overline{\Bbbk((t))}$ of $\Bbbk((t))$. The set $O(\overline{\Bbbk((t))})$ is equipped with a structure of an algebraic variety over $\overline{\Bbbk((t))}$ with the transitive $G(\overline{\Bbbk((t))})$-action. However $\overline{\Bbbk((t))} = \bigcup_{n=1}^{\infty} \Bbbk((\sqrt[n]{t}))$, whence $x(t) = g(\sqrt[n]{t}) \cdot o$ for some $g(\sqrt[n]{t}) \in G(\Bbbk((\sqrt[n]{t})))$. □

Note that $v_{x(t^n)} = n \cdot v_{x(t)}$. Thus we may describe $\mathscr{V}^1$ in terms of germs of formal curves in G, i.e., points of $G(\Bbbk((t)))$, considered up to left translations by $G(\Bbbk[[t]])$ and right translations by $H(\Bbbk((t)))$. There is a useful structural result shrinking the set of formal germs under consideration:

Iwasawa decomposition [IM]. $G(\Bbbk((t))) = G(\Bbbk[[t]]) \cdot \mathfrak{X}^*(T) \cdot U(\Bbbk((t)))$, *where* $\mathfrak{X}^*(T)$ *is regarded as a subset of* $T(\Bbbk((t)))$.

Corollary 24.6. *Every* $v \in \mathscr{V}^1$ *is proportional to (the restriction of)* $v_{g(t)}$, $g(t) \in \mathfrak{X}^*(T) \cdot U(\Bbbk((t)))$.

Let us mention a related useful result on the structure of $G(\Bbbk((t)))$:

Cartan decomposition [IM]. $G(\Bbbk((t))) = G(\Bbbk[[t]]) \cdot \mathfrak{X}^*(T) \cdot G(\Bbbk[[t]])$.

Example 24.7. Suppose that $O = G/S$ is horospherical. We may assume that $S \supseteq U$; then $N_G(S) \supseteq B$ and $A := \operatorname{Aut}_G O \simeq N_G(S)/S = T/T \cap S$ is a torus. Since O is spherical, $\mathscr{V} = \mathscr{V}^1$. Due to the Iwasawa decomposition, every $v \in \mathscr{V}$ is proportional to some v_γ, $\gamma \in \mathfrak{X}^*(T)$. Let $\overline{\gamma}$ be the image of γ^{-1} in $\mathfrak{X}^*(A)$. By definition, $v_\gamma(f) = \operatorname{ord}_{t=0} f(g\gamma(t)o) = \operatorname{ord}_{t=0} f(\overline{\gamma}(t) \cdot go)$ is the order of f along generic trajectories of $\overline{\gamma}$ as $t \to 0$. In particular, $\mathscr{V} = \mathfrak{X}^*(A) \otimes \mathbb{Q}$, cf. Theorem 21.10.

Example 24.8. Specifically, let $O = G = T$ be a torus. Every T-valuation of $\Bbbk(T)$ is proportional to v_γ, $\gamma \in \mathfrak{X}^*(T)$, where v_γ is the order of a function restricted to $s\gamma(t)$ as $t \to 0$ for general $s \in T$. By Theorem 24.1, v_γ has a center Y on a toric variety $X \supseteq T$ if and only if $\gamma(0) := \lim_{t \to 0} \gamma(t)$ exists and belongs to the dense T-orbit in Y. Thus the lattice points in $\mathscr{V}_Y$ are exactly the one-parameter subgroups of T converging to a point in the dense T-orbit of Y. This is the classical description of the fan of a toric variety [Oda, Pr. 1.6(v)], [Ful2, 2.3].

Example 24.9. Let $O = G$ be acted on by $G \times G$ via left/right multiplication. By the Bruhat decomposition, it is a spherical homogeneous space. Due to the Cartan decomposition, every $v \in \mathscr{V}$ is proportional to v_γ, $\gamma \in \mathfrak{X}^*(T)$. Conjugating by $N_G(T)$, we may assume that γ is antidominant. Choose a Borel subgroup $B^- \times B \subseteq G \times G$. The weight lattice $\Lambda(G)$ is identified with $\mathfrak{X}(T)$ and for $\lambda \in \mathfrak{X}_+$ the respective $(B^- \times B)$-eigenfunction is $\mathbf{f}_\lambda(g) = \langle v_{-\lambda}, gv_\lambda \rangle$, where $v_\lambda \in V$, $v_{-\lambda} \in V^*$ are highest, resp. lowest, vectors in a simple G-module V of highest weight λ and its dual, see 27.2. Then $v_\gamma(\mathbf{f}_\lambda) = \operatorname{ord}_t \mathbf{f}_\lambda(g_1\gamma(t)g_2)$ (for general $g_1, g_2 \in G$) $= \operatorname{ord}_t \langle g_1^{-1} v_{-\lambda}, \gamma(t) g_2 v_\lambda \rangle = \langle \gamma, \lambda \rangle$. Hence $\mathscr{V}$ is the antidominant Weyl chamber in $\mathfrak{X}^*(T) \otimes \mathbb{Q}$, cf. 27.2.

Some other examples can be found in 16.5.

Chapter 5
Spherical Varieties

Although the theory developed in the previous chapters applies to arbitrary homogeneous spaces of reductive groups, and even to more general group actions, it acquires its most complete and elegant form for spherical homogeneous spaces and their equivariant embeddings, called spherical varieties. A justification of the fact that spherical homogeneous spaces are a significant mathematical object is that they arise naturally in various fields, such as embedding theory, representation theory, symplectic geometry, etc. In §25 we collect various characterizations of spherical spaces, the most important being: the existence of an open B-orbit, the "multiplicity-free" property for spaces of global sections of line bundles, commutativity of invariant differential operators and of invariant functions on the cotangent bundle with respect to the Poisson bracket.

Then we examine the most interesting classes of spherical homogeneous spaces and spherical varieties in more detail. Algebraic symmetric spaces are considered in §26. We develop the structure theory and classification of symmetric spaces, compute the colored data required for the description of their equivariant embeddings, and study B-orbits and (co)isotropy representation. §27 is devoted to $(G \times G)$-equivariant embeddings of a reductive group G. A particular interest in this class is explained, for example, by an observation that linear algebraic monoids are nothing else but affine equivariant group embeddings. Horospherical varieties of complexity 0 are classified and studied in §28.

The geometric structure of toroidal varieties, considered in §29, is the best understood among all spherical varieties, since toroidal varieties are "locally toric". They can be defined by several equivalent properties: their fans are "colorless", they are spherical and pseudo-free, and the action sheaf on a toroidal variety is the log-tangent sheaf with respect to a G-stable divisor with normal crossings. An important property of toroidal varieties is that they are rigid as G-varieties. The so-called wonderful varieties are the most remarkable subclass of toroidal varieties. They are canonical completions with nice geometric properties of (certain) spherical homogeneous spaces. The theory of wonderful varieties is developed in §30. Applications include computation of the canonical divisor of a spherical variety and

D.A. Timashev, *Homogeneous Spaces and Equivariant Embeddings*,
Encyclopaedia of Mathematical Sciences 138, DOI 10.1007/978-3-642-18399-7_5,

Luna's conceptual approach to the classification of spherical subgroups through the classification of wonderful varieties.

The concluding §31 is devoted to Frobenius splitting, a technique for proving geometric and algebraic properties (normality, rationality of singularities, cohomology vanishing, etc) in positive characteristic. However, this technique can be applied to zero characteristic using reduction $\bmod p$ provided that reduced varieties are Frobenius split. This works for spherical varieties. As a consequence, one obtains the vanishing of higher cohomology of ample or numerically effective line bundles on complete spherical varieties, normality and rationality of singularities for G-stable subvarieties, etc. Some of these results can be proved by other methods, but Frobenius splitting provides a simple uniform approach.

25 Various Characterizations of Sphericity

25.1 Spherical Spaces. Spherical homogeneous spaces can be considered from diverse viewpoints: orbits and equivariant embeddings, representation theory and multiplicities, symplectic geometry, harmonic analysis, etc. The definition and some other implicit characterizations of this remarkable class of homogeneous spaces are already scattered in the text above. In this section, we review these issues and introduce other important properties of homogeneous spaces which are equivalent or closely related to sphericity.

As usual, G is a connected reductive group, O denotes a homogeneous G-space with the base point o, and $H = G_o$.

Definition–Theorem. *A* spherical homogeneous space O *(resp. a* spherical subgroup $H \subseteq G$, *a* spherical subalgebra $\mathfrak{h} \subseteq \mathfrak{g}$, *a* spherical pair (G,H) *or* $(\mathfrak{g},\mathfrak{h})$*) can be defined by any one of the following equivalent properties:*

(S1) $\Bbbk(O)^B = \Bbbk$.
(S2) *B has an open orbit in O.*
(S3) *H has an open orbit in G/B.*
(S4) $(\operatorname{char}\Bbbk = 0)$ $\exists g \in G:\ \mathfrak{b} + (\operatorname{Ad} g)\mathfrak{h} = \mathfrak{g}$.
(S5) $(\operatorname{char}\Bbbk = 0)$ *There exists a Borel subalgebra $\widetilde{\mathfrak{b}} \subseteq \mathfrak{g}$ such that $\mathfrak{h} + \widetilde{\mathfrak{b}} = \mathfrak{g}$.*
(S6) *H acts on G/B with finitely many orbits.*
(S7) *For any G-variety X and any $x \in X^H$, $\overline{Gx}$ contains finitely many G-orbits.*
(S8) *For any G-variety X and any $x \in X^H$, $\overline{Gx}$ contains finitely many B-orbits.*

The term "spherical homogeneous space" is traced back to Brion, Luna, and Vust [BLV], and "spherical subgroup" to Krämer [Krä], though the notions themselves appeared much earlier.

Proof. (S1) $\Longleftrightarrow$ (S2) B-invariant functions separate general B-orbits [PV, 2.3].
(S2) $\Longleftrightarrow$ (S3) Both conditions are equivalent to requiring that $B \times H : G$ has an open orbit, where B acts by left and H by right translations, or vice versa.

(S4) and (S5) are just reformulations of (S2) and (S3) in terms of tangent spaces. Note that in positive characteristic (S4) and (S5) are stronger than (S2) and (S3), because the orbit map onto the open B- or H-orbit may be inseparable.
(S2) $\Longrightarrow$ (S8) Gx satisfies (S2), too, and we conclude by Corollary 6.5.
(S8) $\Longrightarrow$ (S7) Obvious.
(S7) $\Longrightarrow$ (S2) Stems from Corollary 6.10.
(S8) $\Longrightarrow$ (S6) B acts on G/H with finitely many orbits, which are in bijection with $(B \times H)$-orbits on G and with H-orbits on G/B.
(S6) $\Longrightarrow$ (S3) Obvious. □

In particular, spherical spaces are characterized in the framework of embedding theory as those having finitely many orbits in the boundary of any equivariant embedding. The embedding theory of spherical spaces is considered in §15.

25.2 "Multiplicity-free" Property. Another important characterization of spherical spaces is in terms of representation theory, due to Kimelfeld and Vinberg [VK]. Recall from 2.6 that the multiplicity of a highest weight λ in a G-module M is

$$m_\lambda(M) = \dim \operatorname{Hom}_G(V(\lambda), M) = \dim M_\lambda^{(B)}.$$

In characteristic zero, $m_\lambda(M)$ is the multiplicity of the simple G-module $V(\lambda)$ in the decomposition of M. In positive characteristic, $V(\lambda)$ denotes the respective Weyl module. The module M is said to be multiplicity-free if all multiplicities in M are ≤ 1.

Theorem 25.1. *O is spherical if and only if the following equivalent conditions hold:*

(MF1) $\mathbb{P}(V^*(\lambda))^H$ *is finite for all* $\lambda \in \mathfrak{X}_+$.
(MF2) $\forall \lambda \in \mathfrak{X}_+,\ \chi \in \mathfrak{X}(H):\ \dim V^*(\lambda)_\chi^{(H)} \leq 1$.
(MF3) *For any G-line bundle $\mathscr{L}$ on O, $\mathrm{H}^0(O, \mathscr{L})$ is multiplicity-free.*

If O is quasiaffine, then the last two conditions can be weakened to

(MF4) $\forall \lambda \in \mathfrak{X}_+:\ \dim V^*(\lambda)^H \leq 1$.
(MF5) $\Bbbk[O]$ *is multiplicity-free.*

The spaces satisfying these conditions are called *multiplicity-free*.

Proof. (S1) $\Longleftrightarrow$ (MF3) If $m_\lambda(\mathscr{L}) \geq 2$, then there exist two non-proportional sections $\sigma_0, \sigma_1 \in \mathrm{H}^0(O, \mathscr{L})_\lambda^{(B)}$. Their ratio $f = \sigma_1/\sigma_0$ is a non-constant B-invariant function. Conversely, any $f \in \Bbbk(O)^B$ can be represented in this way: the G-line bundle $\mathscr{L}$ together with the canonical B-eigensection σ_0 is defined by a sufficiently big multiple of $\operatorname{div}_\infty f$ (cf. Corollary C.6).

Finally, if O is quasiaffine, then we may take for $\mathscr{L}$ the trivial bundle: for σ_0 take a sufficiently big power of any B-eigenfunction in $\mathscr{I}(D) \lhd \Bbbk[O]$, where $D \subset O$ is the support of $\operatorname{div}_\infty f$. Hence (S1) $\Longleftrightarrow$ (MF5).
(MF1) $\Longleftrightarrow$ (MF2) Stems from $\mathbb{P}(V^*(\lambda))^H = \mathbb{P}\left(V^*(\lambda)^{(H)}\right) = \bigsqcup_\chi \mathbb{P}\left(V^*(\lambda)_\chi^{(H)}\right)$ (a finite disjoint union).

(MF2) $\Longleftrightarrow$ (MF3) If $O = G/H$ is a quotient space, then this is the Frobenius reciprocity (2.2). Generally, there is a bijective purely inseparable morphism $G/H \to O$ (Remark 1.4), and O is spherical if and only if G/H is so. But we have already seen that the sphericity is equivalent to (MF3).
(MF4) $\Longleftrightarrow$ (MF5) is proved in the same way. □

25.3 Weakly Symmetric Spaces and Gelfand Pairs. The "multiplicity-free" property leads to an interpretation of sphericity in terms of automorphisms and group algebras associated with G. Since the complete reducibility of rational representations is essential here, we assume that $\operatorname{char}\Bbbk = 0$ up to the end of this section.

Recall from 2.5 the algebraic versions of the group algebra $\mathscr{A}(G)$ and the Hecke algebra $\mathscr{A}(O)$.

Theorem 25.2 ([AV], [Vin3]). *An affine homogeneous space O is spherical if and only if any of the following four equivalent conditions is satisfied:*

(GP1) $\mathscr{A}(O) = \mathscr{A}(G)^{H\times H}$ *is commutative.*
(GP2) $\mathscr{A}(V)^{H\times H}$ *is commutative for all G-modules V.*
(WS1) (Selberg condition) *The G-action on O extends to a cyclic extension $\widehat{G} = \langle G, s\rangle$ of G so that (sx, sy) is G-equivalent to (y, x) for general $x, y \in O$.*
(WS2) (Gelfand condition) *There exists $\theta \in \operatorname{Aut} G$ such that $\theta(H) = H$ and $\theta(g) \in Hg^{-1}H$ for general $g \in G$.*

The condition (GP1) is an algebraization of a similar commutativity condition for the group algebra of a Lie group, see [Gel], [Vin3, I.2], and 25.5. The condition (WS2) appeared in [Gel], and (WS1) was first introduced by Selberg in the seminal paper on the trace formula [Sel], and by Akhiezer and Vinberg [AV] in the context of algebraic geometry. The spaces satisfying (WS1)–(WS2) are called *weakly symmetric* and (G, H) is said to be a *Gelfand pair* if (GP1)–(GP2) hold.

Proof. (MF5) $\Longleftrightarrow$ (GP1) Stems from Schur's lemma.
(GP1) $\Longleftrightarrow$ (GP2) Obvious.
(MF5) $\Longrightarrow$ (WS1) There exists a Weyl involution $\theta \in \operatorname{Aut} G$, $\theta(H) = H$ [AV]. There is a conceptual argument for symmetric spaces and in general a case-by-case verification using the classification from 10.2. Define $s \in \operatorname{Aut} O$ by $s(go) = \theta(g)o$ and $\widehat{G} = G \leftthreetimes \langle s\rangle$ by $sgs^{-1} = \theta(g)$. The G-action on $O \times O$ is extended to $\widehat{G}$ by $s(x, y) = (sy, sx)$.

Consider the $(G \times G)$-isotypic decomposition

$$\Bbbk[O\times O] = \bigoplus_{\lambda,\mu\in\Lambda_+(O)} \Bbbk[O\times O]_{(\lambda,\mu)},$$
$$\text{where}\quad \Bbbk[O\times O]_{(\lambda,\mu)} = \Bbbk[O]_{(\lambda)} \otimes \Bbbk[O]_{(\mu)} \simeq V(\lambda)\otimes V(\mu).$$

Clearly, s twists the G-action by θ, and hence maps $\Bbbk[O\times O]_{(\lambda,\mu)}$ to $\Bbbk[O\times O]_{(\mu^*,\lambda^*)}$ and preserves each summand of

$$\Bbbk[O\times O]^G = \bigoplus_{\lambda\in\Lambda_+(O)} \Bbbk[O\times O]^G_{(\lambda,\lambda^*)}.$$

A $\widehat{G}$-invariant inner product $(f_1,f_2) \mapsto (f_1f_2)^\natural$ on $\Bbbk[O]$ is non-degenerate. (Otherwise its kernel would be a non-trivial $\widehat{G}$-stable ideal in $\Bbbk[O]$.) Hence it induces a nonzero pairing between simple G-modules $\Bbbk[O]_{(\lambda^*)}$ and $\Bbbk[O]_{(\lambda)}$, whence by duality a $\widehat{G}$-invariant function in $\Bbbk[O\times O]_{(\lambda,\lambda^*)}$, which spans $\Bbbk[O\times O]^G_{(\lambda,\lambda^*)}$. It follows that s acts trivially on $\Bbbk[O\times O]^G$.

But general G-orbits in $O\times O$ are closed (Theorem 8.27), whence s preserves general orbits, which is exactly the Selberg condition.

(WS1) $\Longrightarrow$ (WS2) Multiplying s by $g\in G$ preserves the Selberg condition. Also, if $(sx,sy)\sim(y,x)$, then the same is true for any G-equivalent pair. Hence, without loss of generality, $so=o=x$. Define $\theta\in\operatorname{Aut}G$ by $\theta(g)=sgs^{-1}$; then $(so,sgo)=(o,\theta(g)o)\sim(go,o)$ for general $g\in G$. Hence $g'go=o$, $g'o=\theta(g)o$, i.e., $g'g=h\in H$, $\theta(g)=g'h'=hg^{-1}h'$ for some $h'\in H$.

(WS2) $\Longrightarrow$ (WS1) Since $\theta(H)=H$, there is a well-defined automorphism $s\in\operatorname{Aut}O$, $s(go)=\theta(g)o$. Put $\widehat{G}=G\rtimes\langle s\rangle$, $sgs^{-1}=\theta(g)$. The Selberg condition is verified by reversing the previous arguments.

(WS2) $\Longrightarrow$ (GP1) Both θ and the inversion map $g\mapsto g^{-1}$ on G give rise to automorphisms of $\Bbbk[G]$ preserving $\Bbbk[G]^{H\times H}$, whose restrictions to $\Bbbk[G]^{H\times H}$ coincide. By the complete reducibility of $(H\times H)$-modules, $\mathscr{A}(G)^{H\times H}\simeq(\Bbbk[G]^{H\times H})^*$. Hence the antiautomorphism of $\mathscr{A}(G)^{H\times H}$ induced by the inversion coincides with the automorphism induced by θ. Therefore $\mathscr{A}(G)^{H\times H}$ is commutative. □

Remark 25.3. Already in the quasiaffine case the classes of weakly symmetric and spherical spaces are not contained in each other [Zor].

25.4 Commutativity. Now we characterize sphericity in terms of symplectic geometry.

Recall from 8.2 that the action $G:T^*O$ is Hamiltonian with respect to the natural symplectic structure. Thus we have a G-invariant Poisson bracket of functions on T^*O. Homogeneous functions on T^*O are locally the symbols of differential operators on O, and the Poisson bracket is induced by the commutator of differential operators.

The functions pulled back under the moment map $\Phi: T^*O\to\mathfrak{g}^*$ are called *collective*. They Poisson-commute with G-invariant functions on T^*O (Proposition 22.1).

Theorem 25.4. *O is spherical if and only if the following equivalent conditions hold:*

(WC1) *General orbits of $G:T^*O$ are coisotropic, i.e., $\mathfrak{g}\alpha\supseteq(\mathfrak{g}\alpha)^\angle$ for general $\alpha\in T^*O$.*

(WC2) *$\Bbbk(T^*O)^G$ is commutative with respect to the Poisson bracket.*

(CI) *There exists a complete system of collective functions in involution on T^*O.*

If O is affine, then these conditions are equivalent to

(Com) *$\mathscr{D}(O)^G$ is commutative.*

The theorem goes back to Guillemin, Sternberg [GS], and Mikityuk [Mik]. The spaces satisfying (WC1)–(WC2) are called *weakly commutative* and those satisfying (Com) are said to be *commutative*.

Proof. (S2) $\iff$ (WC1) By Theorem 8.17, $\operatorname{cork} T^*O = 2c(O)$ is zero if and only if O is spherical, and this means exactly that general orbits are coisotropic.
(WC1) $\iff$ (WC2) Skew gradients of $f \in \Bbbk(T^*O)^G$ at a point α of general position span $(\mathfrak{g}\alpha)^{\angle}$. All G-invariant functions Poisson-commute if and only if their skew gradients are skew-orthogonal to each other, i.e., if and only if $(\mathfrak{g}\alpha)^{\angle}$ is isotropic.
(GP1) $\iff$ (Com) If O is quasiaffine, then $\mathscr{D}(O)$ acts faithfully on $\Bbbk[O]$ by linear endomorphisms. Hence $\mathscr{D}(O)^G$ is a subalgebra in $\mathscr{A}(O)$. It remains to utilize the approximation of linear endomorphisms by differential operators.

Lemma 25.5. *Let X be a smooth affine G-variety.*

(1) *For any linear operator $\varphi : \Bbbk[X] \to \Bbbk[X]$ and any finite-dimensional subspace $M \subset \Bbbk[X]$ there exists $\partial \in \mathscr{D}(X)$ such that $\partial|_M = \varphi|_M$.*
(2) *If φ is G-equivariant, then one may assume that $\partial \in \mathscr{D}(X)^G$.*
(3) *Put $\mathscr{I} = \operatorname{Ann} M \lhd \mathscr{D}(X)$; then $\forall f \in \Bbbk[X] : \mathscr{I} f = 0 \implies f \in M$.*

We conclude by Lemma 25.5(2) that $\mathscr{A}(O)$ is commutative if and only if $\mathscr{D}(O)^G$ is so.

Proof of Lemma 25.5. (1) We deduce it from (3). Choose a basis $f_1, \dots, f_n$ of M. It suffices to construct $\partial \in \mathscr{D}(X)$ such that $\partial f_i = 0$, $\forall i < n$, $\partial f_n = 1$. By (3) there exists $\partial' \in \operatorname{Ann}(f_1, \dots, f_{n-1})$, $\partial' f_n \neq 0$. As $\Bbbk[X]$ is a simple $\mathscr{D}(X)$-module [MRo, 15.3.8] we may find $\partial'' \in \mathscr{D}(X)$, $\partial''(\partial' f_n) = 1$ and put $\partial = \partial''\partial'$.
(2) Without loss of generality, M is G-stable. Assertion (1) yields an epimorphism of G-$\Bbbk[X]$-modules $\mathscr{D}(X) \twoheadrightarrow \operatorname{Hom}(M, \Bbbk[X])$ given by restriction to M. But taking G-invariants is an exact functor.
(3) The assertion is trivial for $M = 0$ and we proceed by induction on $\dim M$. In the above notation, put $\mathscr{I}' = \operatorname{Ann}(f_1, \dots, f_{n-1})$. For any $\partial, \partial' \in \mathscr{I}'$ we have $(\partial f_n)\partial' - (\partial' f_n)\partial \in \mathscr{I}$, whence

$$(\partial f_n)(\partial' f) = (\partial' f_n)(\partial f). \tag{25.1}$$

Taking $\partial' = \xi\partial$, $\xi \in \mathrm{H}^0(X, \mathscr{T}_X)$, yields $\xi(\partial f / \partial f_n) = 0 \implies \partial f / \partial f_n = c_\partial = \text{const}$. Substituting this in (25.1) yields $c_{\partial'} = c_\partial = c$ (independent of ∂). Thus $\partial(f - cf_n) = 0 \implies f - cf_n \in \langle f_1, \dots, f_{n-1} \rangle \implies f \in M$. □

(Com) $\implies$ (WC2) If O is affine, then $\operatorname{gr} \mathscr{D}(O) = \Bbbk[T^*O]$. By complete reducibility, $\Bbbk[T^*O]^G = \operatorname{gr} \mathscr{D}(O)^G$ is Poisson-commutative. But general G-orbits in T^*O are closed (Remark 8.15), whence $\Bbbk(T^*O)^G = \operatorname{Quot} \Bbbk[T^*O]^G$ is Poisson-commutative as well.
(CI) $\iff$ (S2) This equivalence is due to Mikityuk [Mik] (for affine O).

A complete system of Poisson-commuting functions on M_O can be constructed by the method of argument shift [MF1]: choose a regular semisimple element $\xi \in \mathfrak{g}^*$ and consider the derivatives $\partial_\xi^n f$ of all $f \in \Bbbk[\mathfrak{g}^*]^G$. (Here ∂_ξ denotes the derivative in direction ξ.) The functions $\partial_\xi^n f$ Poisson-commute and produce a complete involutive system on $Gx \subset \mathfrak{g}^*$ for general ξ whenever $\operatorname{ind}\mathfrak{g}_x = \operatorname{ind}\mathfrak{g}$, where $\operatorname{ind}\mathfrak{h} = d_H(\mathfrak{h}^*)$ [Bol]. In the symplectically stable case, general points $x \in M_O$ are semisimple and $\operatorname{ind}\mathfrak{g}_x = \operatorname{ind}\mathfrak{g} = \operatorname{rk}\mathfrak{g}$. Generally, the equality $\operatorname{ind}\mathfrak{g}_x = \operatorname{ind}\mathfrak{g}$ for all $x \in \mathfrak{g}^*$ was conjectured by Elashvili. It is easily reduced to the case of simple $\mathfrak{g}$ and nilpotent x. Elashvili's conjecture was proved by Yakimova for classical $\mathfrak{g}$ [Yak3] and verified by de Graaf for exceptional $\mathfrak{g}$ using computer calculations [Gra]. Recently a general proof (almost avoiding case-by-case considerations) was given by Charbonnel and Moreau [CM].

Since symplectic leaves of the Poisson structure on M_O are G-orbits, there are $(d_G(M_O) + \dim M_O)/2 = \dim O - c(O)$ independent Poisson-commuting collective functions. Thus a complete involutive system of collective functions exists if and only if $c(O) = 0$. □

Since $T^*O = G *_H \mathfrak{h}^\perp$, weak commutativity is readily reformulated in terms of the coadjoint representation [Mik], [Pan1], [Vin3, II.4.1].

Theorem 25.6. (G,H) *is a spherical pair if and only if general points* $\alpha \in \mathfrak{h}^\perp$ *satisfy any of the equivalent conditions:*

(Ad1) $\dim G\alpha = 2\dim H\alpha$.
(Ad2) $H\alpha$ *is a Lagrangian subvariety in* $G\alpha$ *with respect to the Kirillov form.*
(Ad3) (Richardson condition) $\mathfrak{g}\alpha \cap \mathfrak{h}^\perp = \mathfrak{h}\alpha$.

The Richardson condition means that $G\alpha \cap \mathfrak{h}^\perp$ is a finite union of open H-orbits [PV, 1.5].

Proof. (WC1) $\iff$ (Ad1) Recall that the moment map $\Phi : G *_H \mathfrak{h}^\perp \to \mathfrak{g}^*$ is defined via replacing the $*$-action by the coadjoint action (Example 8.1). We have

$$d_G(T^*O) = \dim O - \dim H\alpha \qquad \text{and}$$
$$\operatorname{def} T^*O = \dim G_{\Phi(e*\alpha)}/G_{e*\alpha} = \dim G_\alpha/H_\alpha.$$

Hence

$$\operatorname{cork} T^*O = d_G(T^*O) - \operatorname{def} T^*O = \dim G\alpha - 2\dim H\alpha.$$

(Ad1) $\iff$ (Ad2) The Kirillov form vanishes on $\mathfrak{h}\alpha$.
(Ad2) $\iff$ (Ad3) Stems from $(\mathfrak{g}\alpha) \cap \mathfrak{h}^\perp = (\mathfrak{h}\alpha)^\angle$, the skew-orthocomplement with respect to the Kirillov form. □

Invariant functions on cotangent bundles of spherical homogeneous spaces have a nice structure.

Proposition 25.7 ([Kn1, 7.2]). *If* $O = G/H$ *is spherical, then* $\Bbbk[T^*O]^G \simeq \Bbbk[\widetilde{L}_O] \simeq \Bbbk[\mathfrak{a}^*]^{W_O}$ *is a polynomial algebra; there are similar isomorphisms for fields of rational functions.*

Proof. By Proposition 22.1 and (WC2), $\Bbbk(T^*O)^G \simeq \Bbbk(\widetilde{L}_O) \simeq \Bbbk(\mathfrak{a}^*)^{W_O}$. By Lemma 22.4, $\widetilde{\pi}_G\widetilde{\Phi} : T^*O \to \widetilde{L}_O$ is a surjective morphism of normal varieties. Therefore any $f \in \Bbbk(T^*O)^G$ having poles on $\widetilde{L}_O$ must have poles on T^*O, whence $\Bbbk[T^*O]^G = \Bbbk[\widetilde{L}_O]$. The latter algebra is polynomial for W_O is generated by reflections (Theorem 22.13). □

In other words, invariants of the coisotropy representation form a polynomial algebra $\Bbbk[\mathfrak{h}^\perp]^H \simeq \Bbbk[\mathfrak{a}^*]^{W_{G/H}}$ for any spherical pair (G,H).

Remark 25.8. A similar assertion in the non-commutative setup was proved in [Kn6]. Namely, all invariant differential operators on a spherical space O are completely regular, whence $\mathscr{D}(O)^G$ is a polynomial ring isomorphic to $\Bbbk[\rho + \mathfrak{a}^*]^{W_O}$ (see Remark 22.8). In particular, every spherical homogeneous space is commutative.

Example 25.9 ([BPa, Ex. 4.3.3]). Let $G = \mathrm{Sp}_4$, $H = \Bbbk^\times \times \mathrm{Sp}_2 \subset \mathrm{Sp}_2 \times \mathrm{Sp}_2 \subset G$. The coisotropy representation of H is $\mathfrak{h}^\perp = \Bbbk^1 \oplus (\Bbbk^1)^* \oplus \Bbbk^2 \oplus (\Bbbk^2)^*$, where $\Bbbk^1, \Bbbk^2$ are the trivial and the tautological Sp_2-module acted on by $\Bbbk^\times$ via the characters $2\varepsilon, \varepsilon$, respectively, where $\mathfrak{X}(\Bbbk^\times) = \langle\varepsilon\rangle$. In the notation of Theorem 9.1, $H_* = Z(G)$, whence $A = \mathrm{Ad}_G T$. The algebra $\Bbbk[\widetilde{L}_O] \simeq \Bbbk[\mathfrak{h}^\perp]^H$ is freely generated by two quadratic invariants ab, $\langle x,y\rangle$, where $(a,b,x,y) \in \mathfrak{h}^\perp$. Hence $W_O \simeq \mathbf{Z}_2^2$ is a subgroup of W generated by the reflections along two orthogonal roots. This example shows that generally $W_X \neq W(\mathfrak{a}^*)$.

25.5 Generalizations. In our considerations G was always assumed to be reductive. However some of the concepts introduced above are reasonable even for non-reductive G assuming H be reductive instead. Some of the above results remain valid:

(1) If $O = G/H$ is weakly symmetric, then (G,H) is a Gelfand pair.
(2) O is commutative if and only if (G,H) is a Gelfand pair.
(3) A commutative space O is weakly commutative provided that $\Bbbk[\mathfrak{h}^\perp]^H$ separates general H-orbits in $\mathfrak{h}^\perp$.

The above proofs work in this case: O is affine, the functor $(\cdot)^G$ is exact on global sections of G-sheaves on O since $(\cdot)^H$ is exact on rational H-modules, and orbit separation in (3) guarantees $\Bbbk(T^*O)^G = \mathrm{Quot}\,\Bbbk[T^*O]^G$. The converse implication in (1) fails, the simplest counterexample being:

Example 25.10 ([Lau]). Put $H = \mathrm{Sp}_{2n}(\Bbbk)$, $G = H \rightthreetimes N$, where $N = \exp\mathfrak{n}$ is a unipotent group associated with the Heisenberg type Lie algebra $\mathfrak{n} = (\Bbbk^{2n} \oplus \Bbbk^{2n}) \oplus \Bbbk^3$, the commutator in $\mathfrak{n}$ being defined by the identification $\bigwedge^2(\Bbbk^{2n} \oplus \Bbbk^{2n})^{\mathrm{Sp}_{2n}(\Bbbk)} \simeq \Bbbk^3 = \mathfrak{z}(\mathfrak{n})$. Then (G,H) is a Gelfand pair, but O is not weakly symmetric.

Also, the implication (3) fails if the orbit separation is violated. The reason is that there may be too few invariant differential operators. For instance, in the previous example, replace H by $\Bbbk^\times$ acting on $\Bbbk^{2n}$ via a character $\chi \neq 0$ and on $\Bbbk^3$ via 2χ. Then O is not weakly commutative while $\mathscr{D}(O)^G = \Bbbk$.

The classes of weakly symmetric and (weakly) commutative homogeneous spaces were first introduced and examined in Riemannian geometry and harmonic analysis, see the survey [Vin3]. We shall review the analytic viewpoint now.

Quitting a somewhat restrictive framework of algebraic varieties, one may consider the above properties of homogeneous spaces in the category of Lie group actions, making appropriate modifications in formulations. For instance, instead of regular or rational functions one considers arbitrary analytic or differentiable functions. Some of these properties receive a new interpretation in terms of differential geometry, e.g., (CI) means that invariant Hamiltonian dynamic systems on T^*O are completely integrable in the class of Noether integrals [MF2], [Mik].

The situation where H is a compact subgroup of a real Lie group G, i.e., $O = G/H$ is a Riemannian homogeneous space, has attracted the main attention of researchers. Most of the above results were originally obtained in this setting.

The properties (MF4), (MF5) are naturally reformulated here in the category of unitary representations of G replacing $\Bbbk[O]$ by $L^2(O)$. In (GP1) one considers the algebra $\mathscr{A}(G)$ of complex measures with compact support on G. The conditions (WS1), (WS2) are formulated for *all* (not only general) points (which is equivalent for compact H); there is also an infinitesimal characterization of weak symmetry [Vin3, I.1.2].

There are the following implications:

$$\begin{array}{c}\text{weakly symmetric space}\\ \Downarrow\\ \text{Gelfand pair}\\ \Updownarrow\\ \text{multiplicity-free space}\\ \Updownarrow\\ \text{commutative space}\\ \Updownarrow\\ \text{weakly commutative space}\end{array}$$

The implication (WS2) $\Longrightarrow$ (GP1) is due to Gelfand [Gel] and (GP1) $\Longleftrightarrow$ (MF5) was proved in [BGGN]. The equivalence (GP1) $\Longleftrightarrow$ (Com) is due to Helgason [Hel2, Ch. IV, B13] and Thomas [Tho], for a proof see [Vin3, I.2.5]. The implication (Com) $\Longrightarrow$ (WC2) is easy [Vin3, I.4.2] and the converse was proved by Rybnikov [Ryb].

A classification of commutative Riemannian homogeneous spaces was obtained by Yakimova [Yak1], [Yak2] using partial results of Vinberg [Vin4] and the classification of affine spherical spaces from 10.2.

An algebraic homogeneous space $O = G/H$ over $\Bbbk = \mathbb{C}$ may be considered as a homogeneous manifold in the category of complex or real Lie group actions. At the same time, if (G,H) is defined over $\mathbb{R}$, then O has a real form $O(\mathbb{R})$ containing $G(\mathbb{R})/H(\mathbb{R})$ as an open orbit (in classical topology). Thus G/H may be regarded as the complexification of $G(\mathbb{R})/H(\mathbb{R})$, a homogeneous space of a real Lie group $G(\mathbb{R})$.

It is easy to see that G/H is commutative (resp. weakly commutative, multiplicity-free, weakly symmetric, satisfies (GP1), (CI)) if and only if $G(\mathbb{R})/H(\mathbb{R})$ is so. In other words, the above listed properties are stable under complexification and passing to a real form.

This observation leads to the following criterion of sphericity, which is a "real form" of Theorem 25.4.

By Chevalley's theorem, there exists a projective embedding $O \subseteq \mathbb{P}(V)$ for some G-module V. Assume that G is reductive and $K \subset G$ is a compact real form. Then V can be endowed with a K-invariant Hermitian inner product $(\cdot|\cdot)$, which induces a Kählerian metric on $\mathbb{P}(V)$ and on O (the *Fubini–Study metric*). The imaginary part of this metric is a real symplectic form. The action $K : \mathbb{P}(V)$ is Hamiltonian, the moment map $\Phi : \mathbb{P}(V) \to \mathfrak{k}^*$ being defined by the formula

$$\langle \Phi([v]), \xi \rangle = \frac{1}{2\mathrm{i}} \cdot \frac{(\xi v|v)}{(v|v)}, \qquad \forall v \in V,\ \xi \in \mathfrak{k}.$$

Theorem 25.11 ([Bri3], [HW], [Akh4, §13]). *O is spherical if and only if general K-orbits in O are coisotropic with respect to the Fubini–Study form or, equivalently, the algebra $C^\infty(O)^K$ of smooth K-invariant functions on O is Poisson-commutative.*

Proof. First note that general K-orbits in O are coisotropic if and only if

$$d_K(O) = \operatorname{def} O = \operatorname{rk} K - \operatorname{rk} K_*, \tag{25.2}$$

where K_* is the stabilizer of general position for $K : O$ [Vin3, II.3.1, 2.6]. The condition (25.2) does not depend on the symplectic structure.

If O is affine, then the assertion can be directly reduced to Theorem 25.4 by complexification. Without loss of generality $K \cap H$ is a compact real form of H. Using the Cartan decompositions $G = K \cdot \exp \mathrm{i}\mathfrak{k}$, $H = (K \cap H) \cdot \exp \mathrm{i}(\mathfrak{k} \cap \mathfrak{h})$, one obtains a K-diffeomorphism

$$O \simeq K *_{K \cap H} \mathrm{i}\mathfrak{k}/\mathrm{i}(\mathfrak{k} \cap \mathfrak{h}) \simeq T^*(K/K \cap H)$$

(see, e.g., [Kob, 2.7]). Complexifying the r.h.s. we obtain T^*O.

In the general case, it is more convenient to apply the theory of doubled actions (see 8.8).

There exists a Weyl involution θ of G commuting with the Hermitian conjugation $g \mapsto g^*$. The mapping $g \mapsto \overline{g} := \theta(g^*)^{-1}$ is a complex conjugation on G defining a split real form $G(\mathbb{R})$. There exists a $G(\mathbb{R})$-stable real form $V(\mathbb{R}) \subset V$ such that $(\cdot|\cdot)$ takes real values on $V(\mathbb{R})$. The complex conjugation on V, $\mathbb{P}(V)$, or G is defined by conjugating the coordinates or matrix entries with respect to an orthonormal basis in $V(\mathbb{R})$.

It follows that the complex conjugate variety $\overline{O}$ is naturally embedded in $\mathbb{P}(V)$ as a G-orbit. Complexifying the action $K : O$ we obtain the diagonal action $G : O \times \overline{O}$, $g(x,\overline{y}) = (gx, \theta(g)\overline{y})$, $\forall g \in G,\ x,y \in O$. This action differs slightly from the doubled action, but Theorems 8.24–8.25 remain valid, together with the proofs. Now it follows from (8.3)–(8.4) that O is spherical if and only if

$$d_G(O \times \overline{O}) = \operatorname{rk} G - \operatorname{rk} G_*,$$

where $G_* = K_*(\mathbb{C})$ is the stabilizer of general position for $G : O \times \overline{O}$. The latter condition coincides with (25.2). □

26 Symmetric Spaces

26.1 Algebraic Symmetric Spaces. The concept of a Riemannian symmetric space was introduced by É. Cartan [Car1], [Car2]. A (globally) symmetric space is defined as a connected Riemannian manifold O such that for any $x \in O$ there exists an isometry s_x of O inverting the geodesics passing through x. Symmetric spaces form a very important class of Riemannian spaces including all classical geometries. The theory of Riemannian symmetric spaces is well developed, see [Hel1].

In particular, it is easy to see that a symmetric space O is homogeneous with respect to the unity component G of the full isometry group, so that $O = G/H$, where $H = G_o$ is the stabilizer of a fixed base point. The geodesic symmetry $s = s_o$ is an involutive automorphism of O normalizing G. It defines an involution $\theta \in \operatorname{Aut} G$ by $\theta(g) = sgs^{-1}$. From the definition of a geodesic symmetry one deduces that $(G^\theta)^0 \subseteq H \subseteq G^\theta$. Furthermore, reducing G to a smaller transitive isometry group if necessary, one may assume that $\mathfrak{g}$ is a reductive Lie algebra. This leads to the following algebraic definition of a symmetric space, which we accept in our treatment.

Definition 26.1. An (algebraic) *symmetric space* is a homogeneous algebraic variety $O = G/H$, where G is a connected reductive group equipped with a non-identical involution $\theta \in \operatorname{Aut} G$, and $(G^\theta)^0 \subseteq H \subseteq G^\theta$.

Riemannian symmetric spaces are locally isomorphic to real forms (with compact isotropy subgroups) of algebraic symmetric spaces over $\mathbb{C}$.

It is reasonable to impose a restriction $\operatorname{char} \Bbbk \neq 2$ on the ground field.

Remark 26.2. If G is semisimple simply connected, then G^θ is connected [St, 8.2], whence $H = G^\theta$. On the other hand, if G is adjoint, then $G^\theta = N_G(H)$ [Vu2, 2.2].

The differential of θ, denoted by the same letter by abuse of notation, induces a $\mathbf{Z}_2$-grading

$$\mathfrak{g} = \mathfrak{h} \oplus \mathfrak{m}, \tag{26.1}$$

where $\mathfrak{h}, \mathfrak{m}$ are the (± 1)-eigenspaces of θ.

The subgroup H is reductive [St, §8], and hence O is an affine algebraic variety. More specifically, consider a morphism $\tau : G \to G$, $\tau(g) = \theta(g)g^{-1}$. Observe that τ is the orbit map at e for the G-action on G by *twisted conjugation*: $g \circ x = \theta(g)xg^{-1}$. It is not hard to prove the following result.

Proposition 26.3 ([Sp1, 2.2]). *$\tau(G) \simeq G/G^\theta$ is a connected component of $\{x \in G \mid \theta(x) = x^{-1}\}$.*

Example 26.4. Let $G = \mathrm{GL}_n(\Bbbk)$ and let θ be defined by $\theta(x) = (x^\top)^{-1}$. Then $G^\theta = \mathrm{O}_n(\Bbbk)$ and $\tau(G) = \{x \in G \mid \theta(x) = x^{-1}\}$ is the set of non-degenerate symmetric matrices, which is isomorphic to $\mathrm{GL}_n(\Bbbk)/\mathrm{O}_n(\Bbbk)$.

However, if θ is an inner involution, i.e., the conjugation by a matrix of order 2, then the set of matrices x such that $\theta(x) = x^{-1}$ is disconnected. The connected components are determined by the collection of eigenvalues of x, which are ± 1.

26.2 θ-stable Tori. The local and global structure of symmetric spaces is examined in [KoR], [Hel1] (transcendental methods), [Vu1], [Vu2] ($\operatorname{char}\Bbbk = 0$), [Ri2], [Sp1]. We follow these sources in our exposition. The starting point is an analysis of θ-stable tori.

Lemma 26.5. *Every Borel subgroup $B \subseteq G$ contains a θ-stable maximal torus T.*

Proof. The group $B \cap \theta(B)$ is connected, solvable, and θ-stable. By [St, 7.6] it contains a θ-stable maximal torus T, which is a maximal torus in G, too. □

Corollary 26.6. *Every θ-stable torus $S \subseteq G$ is contained in a θ-stable maximal torus T.*

Proof. Take for T any θ-stable maximal torus in $Z_G(S)$. □

A θ-stable torus T decomposes into an almost direct product $T = T_0 \cdot T_1$, where $T_0 \subseteq H$ and T_1 is *θ-split*, which means that θ acts on T_1 as the inversion.

Let Δ denote the root system of G with respect to T and $\mathfrak{g}^\alpha \subset \mathfrak{g}$ the root subspace corresponding to $\alpha \in \Delta$. One may choose root vectors $e_\alpha \in \mathfrak{g}^\alpha$ in such a way that $e_\alpha, e_{-\alpha}, h_\alpha = [e_\alpha, e_{-\alpha}]$ form an $\mathfrak{sl}_2$-triple for all $\alpha \in \Delta$. Clearly, θ acts on $\mathfrak{X}(T)$ leaving Δ stable. Choosing e_α in a compatible way allows us to subdivide all roots into *complex*, *real*, and *imaginary* (*compact* or *non-compact*) ones, according to Table 26.1.

Table 26.1 Root types with respect to an involution

α	complex	real	imaginary	
			compact	non-compact
$\theta(\alpha)$	$\neq \pm\alpha$	$-\alpha$	α	α
$\theta(e_\alpha)$	$e_{\theta(\alpha)}$	$e_{-\alpha}$	e_α	$-e_\alpha$

We fix an inner product on $\mathfrak{X}(T) \otimes \mathbb{Q}$ invariant under the Weyl group $W = N_G(T)/T$ and θ. Then $\mathfrak{X}(T) \otimes \mathbb{Q}$ is identified with $\mathfrak{X}^*(T) \otimes \mathbb{Q}$ and with the orthogonal sum of $\mathfrak{X}(T_0) \otimes \mathbb{Q}$ and $\mathfrak{X}(T_1) \otimes \mathbb{Q}$. The coroots $\alpha^\vee \in \Delta^\vee$ (for $\alpha \in \Delta$) are identified with $2\alpha/(\alpha,\alpha)$. Let $\langle\alpha|\beta\rangle = \langle\alpha^\vee,\beta\rangle = 2(\alpha,\beta)/(\alpha,\alpha)$ denote the Cartan pairing on $\mathfrak{X}(T)$ and let $r_\alpha(\beta) = \beta - \langle\alpha|\beta\rangle\alpha$ be the reflection of β along α.

Two opposite classes of θ-stable maximal tori are of particular importance in the theory of symmetric spaces.

26.3 Maximal θ-fixed Tori.

Lemma 26.7. *If* $\dim T_0$ *is maximal possible, then* T_0 *is a maximal torus in* H *and* $Z_G(T_0) = T$. *Moreover,* T *is contained in a* θ*-stable Borel subgroup* $B \subseteq G$ *such that* $(B^\theta)^0$ *is a Borel subgroup in* H.

Proof. If $Z_G(T_0) \neq T$, then the commutator subgroup $Z_G(T_0)'$ and $(Z_G(T_0)')^\theta$ have positive dimension. Hence T_0 can be extended by a subtorus in $(Z_G(T_0)')^\theta$, a contradiction. Now choose a Borel subgroup of H containing T_0 and extend it to a Borel subgroup B of G. Then $B \supseteq T$. If B were not θ-stable, then there would exist a root $\alpha \in \Delta_+$ such that $\theta(\alpha) \in \Delta_-$. Then $e_{\pm\alpha} + \theta(e_{\pm\alpha})$ are opposite root vectors in $\mathfrak{h}$ outside the Borel subalgebra $\mathfrak{b}^\theta$, a contradiction. □

In particular, if T_0 is maximal, then there are no real roots, and θ preserves the set Δ_+ of positive roots (with respect to B) and induces a diagram involution $\overline{\theta}$ of the set $\Pi \subseteq \Delta_+$ of simple roots. If G is of simply connected type, then $\overline{\theta}$ extends to an automorphism of G so that $\theta = \overline{\theta} \cdot \theta_0$, where θ_0 is an inner automorphism defined by an element of T_0.

Consider the set $\overline{\Delta} = \{\overline{\alpha} = \alpha|_{T_0} \mid \alpha \in \Delta\} \subset \mathfrak{X}(T_0)$. Clearly, $\overline{\Delta}$ consists of the roots of H with respect to T_0 and the nonzero weights of $T_0 : \mathfrak{m}$. The restrictions of complex roots belong to both subsets, the eigenvectors being $e_\alpha + \theta(e_\alpha) \in \mathfrak{h}$, $e_\alpha - \theta(e_\alpha) \in \mathfrak{m}$, whereas (non-)compact roots restrict to roots of H (resp. weights of $\mathfrak{m}$).

Lemma 26.8. $\overline{\Delta}$ *is a (possibly non-reduced) root system with base* $\overline{\Pi} = \{\overline{\alpha} \mid \alpha \in \Pi\}$. *The simple roots of* H *and the (nonzero) lowest weights of* $H : \mathfrak{m}$ *form an affine simple root system* $\widetilde{\Pi}$, *i.e.,* $\langle\overline{\alpha}|\overline{\beta}\rangle \in \mathbb{Z}_-$ *for distinct* $\overline{\alpha}, \overline{\beta} \in \widetilde{\Pi}$.

Proof. Note that the restriction of $\alpha \in \Delta$ to T_0 is the orthogonal projection to $\mathfrak{X}(T_0) \otimes \mathbb{Q}$, so that $\overline{\alpha} = (\alpha + \theta(\alpha))/2$. If α is complex, then $\langle\alpha|\theta(\alpha)\rangle = 0$ or -1 (otherwise $\alpha - \theta(\alpha)$ would be a real root), In the second case, $2\overline{\alpha} = \alpha + \theta(\alpha)$ is a non-compact root with a root vector $e_{\alpha+\theta(\alpha)} = [e_\alpha, \theta(e_\alpha)]$.

A direct computation shows that $\langle\overline{\alpha}|\overline{\beta}\rangle \in \mathbb{Z}$, $\forall \alpha, \beta \in \Delta$, and the reflections $r_{\overline{\alpha}}$ preserve $\overline{\Delta}$, see Table 26.2. Hence $\overline{\Delta}$ is a root system. The subset $\overline{\Pi}$ is linearly

Table 26.2 Cartan numbers and reflections for restricted roots

Case	$\langle\overline{\alpha}\|\overline{\beta}\rangle$	$r_{\overline{\alpha}}(\overline{\beta})$
$\alpha = \theta(\alpha)$	$\langle\alpha\|\beta\rangle$	$\overline{r_\alpha(\beta)}$
$\langle\alpha\|\theta(\alpha)\rangle = 0$	$\langle\alpha\|\beta\rangle + \langle\theta(\alpha)\|\beta\rangle$	$\overline{r_\alpha r_{\theta(\alpha)}(\beta)}$
$\langle\alpha\|\theta(\alpha)\rangle = -1$	$2\langle\alpha\|\beta\rangle + 2\langle\theta(\alpha)\|\beta\rangle$	$r_{2\overline{\alpha}}(\overline{\beta}) = \overline{r_{\alpha+\theta(\alpha)}(\beta)}$

independent. (Otherwise there would be a linear dependence between $2\overline{\alpha} = \alpha + \theta(\alpha)$, where $\alpha, \theta(\alpha) \in \Pi$, i.e., between roots in Π.) Restricting to T_0 the expression of $\alpha \in \Delta$ as a linear combination of Π with integer coefficients of the same sign yields a similar expression of $\overline{\alpha}$ in terms of $\overline{\Pi}$. Thus $\overline{\Pi}$ is a base of $\overline{\Delta}$.

Note that $\overline{\alpha}=\overline{\beta}$ if and only if $\alpha=\beta$ or $\theta(\alpha)=\beta$. (Otherwise $\alpha-\beta$ or $\theta(\alpha)-\beta$ would be a real root, depending on whether $\langle\alpha|\beta\rangle>0$ or $\langle\theta(\alpha)|\beta\rangle>0$.) Therefore the nonzero weights occur in $\mathfrak{m}$ with multiplicity 1.

To prove the second assertion, it suffices to consider the Cartan numbers $\langle\overline{\alpha}|\overline{\beta}\rangle$ of lowest weights of $\mathfrak{m}$. The assumption $\langle\overline{\alpha}|\overline{\beta}\rangle>0$ yields without loss of generality $\langle\alpha|\beta\rangle>0$, whence $\gamma=\alpha-\beta\in\Delta$, $e_\alpha=[e_\beta,e_\gamma]$. If β is non-compact, then $[e_\beta,e_\gamma+\theta(e_\gamma)]=e_\alpha-\theta(e_\alpha)$. If β is complex, then γ is also complex. (This is clear if α is non-compact; otherwise $(\overline{\gamma},\overline{\gamma})=\min\{(\overline{\alpha},\overline{\alpha}),(\overline{\beta},\overline{\beta})\}<\min\{(\alpha,\alpha),(\beta,\beta)\}=(\gamma,\gamma)$.) Then $\beta+\theta(\gamma),\theta(\beta)+\gamma\notin\Delta$, whence $[e_\beta-\theta(e_\beta),e_\gamma+\theta(e_\gamma)]=e_\alpha-\theta(e_\alpha)$. In both cases, either $\overline{\beta}$ or $\overline{\alpha}$ is not a lowest weight, a contradiction. □

Remark 26.9. If some of the Cartan numbers of Δ vanish in $\Bbbk$, then the previous argument concerning lowest weights does not work. The assertion on $\widetilde{\Pi}$ is true only if one interprets lowest weights in the combinatorial sense as those weights of $\mathfrak{m}$ which cannot be obtained from other weights by adding simple roots of H. However this happens only for $G=\mathbf{G}_2$ (if $\operatorname{char}\Bbbk=3$), where the unique (up to conjugation) involution is easy to describe by hand.

In a usual way, the system $\widetilde{\Pi}$ together with the respective Cartan numbers is encoded by an (affine) Dynkin diagram. Marking the nodes corresponding to the simple roots of H by black, and those corresponding to the lowest weights of $\mathfrak{m}$ by white, one obtains the so-called *Kac diagram* of the involution θ, or of the symmetric space O. From the Kac diagram one easily recovers $\mathfrak{h}$ and (at least in characteristic zero) the (co)isotropy representation $H^0:\mathfrak{m}$.

Example 26.10. Let H be diagonally embedded in $G=H\times H$, where θ permutes the factors. Here the Kac diagram is the affine Dynkin diagram of H with the white nodes corresponding to the lowest roots, e.g.:

26.4 Maximal θ-split Tori. Now consider an opposite class of θ-stable maximal tori.

Lemma 26.11. *There exist non-trivial θ-split tori.*

Proof. In the converse case θ acts identically on every θ-stable torus. Lemma 26.5 implies that all Borel subgroups are θ-stable. Then all maximal tori are θ-stable and even pointwise fixed, whence θ is identical. □

Lemma 26.12. *If T_1 is a maximal θ-split torus, then $L=Z_G(T_1)$ decomposes into an almost direct product $L=L_0\cdot T_1$, where $L_0=L\cap H$.*

Proof. Clearly, L and the commutator subgroup L' are θ-stable. If $L'\nsubseteq H$, then T_1 could be extended by a non-trivial θ-split torus in L' by Lemma 26.11, a contradiction. The assertion follows from $L'\subseteq H$. □

Corollary 26.13. *Every maximal torus $T\supseteq T_1$ is θ-stable.*

Choose a general one-parameter subgroup $\gamma \in \mathfrak{X}^*(T_1)$ and consider the associated parabolic subgroup $P = P(\gamma)$ with the Lie algebra $\mathfrak{p} = \mathfrak{t} \oplus \bigoplus_{\langle\alpha,\gamma\rangle\geq 0} \mathfrak{g}^\alpha$. Clearly, $L \subseteq P$ is a Levi subgroup and $\mathfrak{p}_u = \bigoplus_{\langle\alpha,\gamma\rangle>0} \mathfrak{g}^\alpha$. Note that $\theta(P) = P^-$ (since $\langle\theta(\alpha),\gamma\rangle = -\langle\alpha,\gamma\rangle$, $\forall\alpha \in \Delta$). In fact, all minimal parabolics having this property are obtained as above [Vu1, 1.2]. It follows that $\mathfrak{h}$ is spanned by $\mathfrak{l}_0$ and $e_\alpha + \theta(e_\alpha)$ over all $\alpha \in \Delta$ such that $\langle\alpha,\gamma\rangle > 0$. This yields:

Iwasawa decomposition. $\mathfrak{g} = \mathfrak{h} \oplus \mathfrak{t}_1 \oplus \mathfrak{p}_u$.

As a consequence, we obtain

Theorem 26.14. *Symmetric spaces are spherical.*

Indeed, choosing a Borel subgroup $B \subseteq P$, $B \supseteq T$ yields (S5). There are many other ways to verify this fact. For instance, it is easy to verify the Richardson condition (Ad3): for any $\xi \in \mathfrak{m} \simeq \mathfrak{h}^\perp$ one has $[\mathfrak{g},\xi] \cap \mathfrak{m} = [\mathfrak{h},\xi]$, because $[\mathfrak{m},\xi] \subseteq \mathfrak{h}$. One can also check the Gelfand condition (WS2) for elements in a dense subset $\tau(G)H \subseteq G$: $g = xh$, $x \in \tau(G)$, $h \in H \implies \theta(g) = x^{-1}h = hg^{-1}h$. The multiplicity-free property (for compact Riemannian symmetric spaces and unitary representations) was established already by É. Cartan [Car3, n°17].

The Iwasawa decomposition clarifies the local structure of a symmetric space. Namely, O contains a dense orbit $P \cdot o \simeq P/L_0 \simeq P_u \times A$, where $A = T/T \cap H$ is the quotient of T_1 by an elementary Abelian 2-group $T_1 \cap H$. We have $\mathfrak{a} \simeq \mathfrak{t}_1$, $\Lambda(O) = \mathfrak{X}(A)$, $r(O) = \dim\mathfrak{a}$. The notation here agrees with Theorem 4.7 and Subsection 7.2.

Lemma 26.15. *All maximal θ-split tori are H^0-conjugate.*

Proof. In the above notation, PH is open in G, whence the H^0-orbit of P is open in G/P. Since P coincides with the normalizer of the open B-orbit in O, all such parabolics are G-conjugate and therefore H^0-conjugate. Hence the Levi subgroups $L = P \cap \theta(P)$ and finally the maximal θ-split tori $T_1 = (Z(L)^0)_1$ are H^0-conjugate. □

If T_1 is maximal, then every imaginary root is compact and θ maps positive complex or real roots to negative ones. Compact (simple) roots form (the base of) the root system of L.

The endomorphism $\iota = -w_L\theta$ of $\mathfrak{X}(T)$ preserves Δ_+ and induces a diagram involution of the set Π of simple roots. (Here w_L is the longest element in the Weyl group of L.) Since $w_G w_L \theta$ preserves Δ_+ and differs from θ by an inner automorphism, it coincides with the diagram automorphism $\overline{\theta}$, whence $\iota(\lambda) = \overline{\theta}(\lambda)^*$, $\forall\lambda \in \mathfrak{X}(T)$.

Consider the set $\Delta_O \subset \mathfrak{X}(T_1)$ and the subset $\Pi_O \subset \Delta_O$ consisting of the restrictions $\overline{\alpha} = \alpha|_{T_1}$ of complex and real roots $\alpha \in \Delta$ (resp. $\alpha \in \Pi$) to T_1.

Lemma 26.16. *Δ_O is a (possibly non-reduced) root system with base Π_O, called the* (little) root system *of the symmetric space O.*

Proof. The proof is similar to that of Lemma 26.8. The restriction of $\alpha \in \Delta$ to T_1 is identified with the orthogonal projection to $\mathfrak{X}(T_1) \otimes \mathbb{Q}$ given by $\overline{\alpha} = (\alpha - \theta(\alpha))/2$. We have $\alpha + \theta(\alpha) \notin \Delta$, $\forall\alpha \in \Delta$. (Otherwise $\alpha + \theta(\alpha)$ would be a non-compact root.) The involution ι coincides with $-\theta$ modulo the root lattice of L. One easily deduces that $\overline{\alpha} = \overline{\beta}$ if and only if $\alpha = \beta$ or $\iota(\alpha) = \beta$, $\forall \alpha, \beta \in \Pi$, and that Π_O is linearly independent. Taking these remarks into account, the proof repeats that of Lemma 26.8 with θ replaced by $-\theta$. □

The Dynkin diagram of Π with the "compact" nodes marked by black and the remaining nodes by white, where the white nodes transposed by ι are joined by two-headed arrows, is called the *Satake diagram* of the involution θ, or of the symmetric space O. The Satake diagram encodes the embedding of $\mathfrak{h}$ into $\mathfrak{g}$. Besides, it contains information on the weight lattice (semigroup) of the symmetric space (see Propositions 26.22, 26.24).

Example 26.17. The Satake diagram of the symmetric space $O = H \times H/\operatorname{diag} H$ of Example 26.10 consists of two Dynkin diagrams of H, so that all nodes are white and each node of the first diagram is joined with the respective node of the second diagram, e.g.:

26.5 Classification. The classification of symmetric spaces goes back to Cartan. To describe it, first note that θ preserves the connected center and either preserves or transposes the simple factors of G. Hence every symmetric space is locally isomorphic to a product of a torus $Z/Z \cap H$, of symmetric spaces $H \times H/\operatorname{diag} H$ with H simple, and of symmetric spaces of simple groups.

Thus the classification reduces to simple G. It can be obtained using either Kac diagrams [Hel1, X.5], [GOV, Ch. 3, §3] or Satake diagrams [Sp2], [GOV, Ch. 4, §4]. For simple G both Kac and Satake diagrams are connected.

Further analysis shows that the underlying affine Dynkin diagram for the Kac diagram of θ depends only on the diagram involution $\overline{\theta}$. This diagram is easily recovered from the Dynkin diagram of Π and from $\overline{\theta}$ using Table 26.2. Since the weight system of $T_0 : \mathfrak{m}$ is symmetric, for each "white" root $\overline{\alpha} \in \widetilde{\Pi}$ there exists a "white" root $\overline{\alpha}_0$ and "black" roots $\overline{\alpha}_1, \dots, \overline{\alpha}_r$ such that $-\overline{\alpha} = \overline{\alpha}_0 + \overline{\alpha}_1 + \dots + \overline{\alpha}_r$. As $\widetilde{\Pi}$ is bound by a unique linear dependence, the coefficients being positive integers, there exists either a unique "white" root, with the coefficient 1 or 2, or exactly two "white" roots, with the coefficients 1. The first possibility occurs exactly for outer involutions, because in this case the weight system contains the zero weight, while the other two possibilities correspond to inner involutions. Using these observations, it is easy to write down all possible Kac diagrams, see Table 26.3.

On the other hand, all a priori possible Satake diagrams can also be classified. One verifies that a Satake diagram cannot be one of the following:

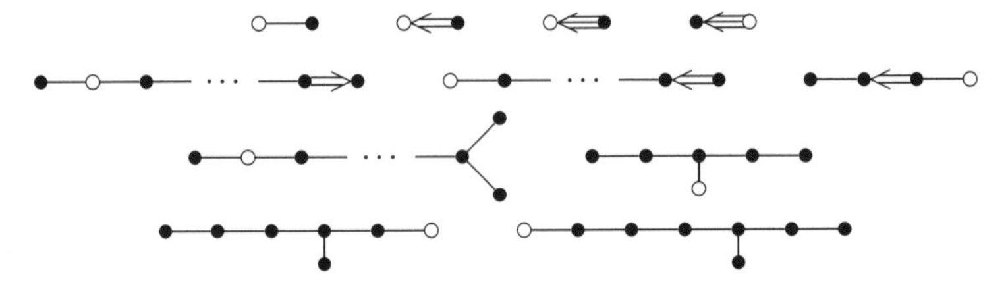

In the first seven cases, the sum of all simple roots would be a complex root α such that $\alpha+\theta(\alpha)\in\Delta$ is a non-compact root, a contradiction. In the remaining four cases, θ would be an inner involution represented by an element $s\in S=Z_G(L_0)^0$. The group S is a simple SL_2-subgroup corresponding to the highest root $\delta\in\Delta$ and T_1 is a maximal torus in S. Replacing T_1 by another maximal torus containing s, one obtains $\delta(s)=-1$. However the unique $\alpha\in\Pi$ such that $\alpha(s)=-1$ occurs in the decomposition of δ with coefficient 2, a contradiction.

By a *fragment* of a Satake diagram we mean an ι-stable subdiagram such that no one of its nodes is joined with a black node outside the fragment. A fragment is the Satake diagram of a Levi subgroup in G. It follows that a Satake diagram cannot contain the above listed fragments. Also, if a Satake diagram contains a fragment ●— ⋯ —● of length >1, then there are no other black nodes and ι is non-trivial. Having this in mind, it is easy to write down all possible Satake diagrams, see Table 26.3.

Both Kac and Satake diagrams uniquely determine the involution θ, up to conjugation. All a priori possible diagrams are realized for simply connected G. It follows that symmetric spaces of simple groups are classified, up to a local isomorphism, by Kac or Satake diagrams.

The classification is presented in Table 26.3. $\mathrm{S}(\mathrm{L}_m\times\mathrm{L}_{n-m})$ in the column "H" denotes the group of unimodular block-diagonal matrices with blocks of size m and $n-m$. The column "θ" describes the involution for classical G in matrix terms. Here

$$I_{n,m}=\begin{pmatrix}-E_m & 0\\ 0 & E_{n-m}\end{pmatrix},\quad K_{n,m}=\begin{pmatrix}I_{n,m} & 0\\ 0 & I_{n,m}\end{pmatrix},\quad \text{and } \Omega_n=\begin{pmatrix}0 & E_n\\ -E_n & 0\end{pmatrix}$$

is the matrix of a standard symplectic form fixed by $\mathrm{Sp}_{2n}(\Bbbk)$, where E_k is the unit $k\times k$ matrix.

Example 26.18. Let us describe the symmetric spaces of $G=\mathrm{SL}_n(\Bbbk)$. Take the standard Borel subgroup of upper-triangular matrices $B\subset G$ and the standard diagonal torus $T\subset B$. By $\varepsilon_1,\dots,\varepsilon_n$ denote the weights of the tautological representation in $\Bbbk^n$ (i.e., the diagonal entries of T).

If θ is inner, then $\overline{\Delta}=\Delta$ and the Dynkin diagram of $\widetilde{\Pi}$ is the following one:

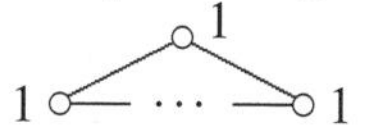

The coefficients of the unique linear dependence on $\widetilde{\Pi}$ are indicated at the diagram. It follows that there are exactly two white nodes in the Kac diagram. The involution ι is non-trivial, whence there is at most one black fragment in the Satake diagram, which is located in the middle. Thus we obtain No. 1 of Table 26.3.

The involution θ is the conjugation by an element of order 2 in $\mathrm{GL}_n(\Bbbk)$. In a certain basis, $\theta(g)=I_{n,m}\cdot g\cdot I_{n,m}$. Then $T_0=T$, $H=\mathrm{S}(\mathrm{L}_m\times\mathrm{L}_{n-m})$ is embedded in G by the two diagonal blocks, the simple roots being $\varepsilon_i-\varepsilon_{i+1}$, $1\le i<n$, $i\ne m$, and $\mathfrak{m}=\Bbbk^m\otimes(\Bbbk^{n-m})^*\oplus(\Bbbk^m)^*\otimes\Bbbk^{n-m}$ is embedded in $\mathfrak{g}$ by the two antidiagonal blocks, the lowest weights of the summands being $\varepsilon_m-\varepsilon_{m+1}$, $\varepsilon_n-\varepsilon_1$, in accordance with the Kac diagram.

Table 26.3 Symmetric spaces of simple groups

No.	G	H	θ	Kac diagram	Satake diagram	$\Delta_{G/H}$
1	SL_n	$\mathrm{S}(\mathrm{L}_m \times \mathrm{L}_{n-m})$ ($m \le n/2$)	$g \mapsto I_{n,m} g I_{n,m}$	m	m	$\mathbf{BC}_m$
				($n = 2$)	($m = n/2$)	$\mathbf{C}_{n/2}$
2	SL_{2n}	Sp_{2n}	$g \mapsto \Omega_n (g^\top)^{-1} \Omega_n^\top$			$\mathbf{A}_{n-1}$
3	SL_n	SO_n	$g \mapsto (g^\top)^{-1}$	(n even)		$\mathbf{A}_{n-1}$
				(n odd)		
				($n = 3$)		
4	Sp_{2n}	$\mathrm{Sp}_{2m} \times \mathrm{Sp}_{2(n-m)}$ ($m \le n/2$)	$g \mapsto K_{n,m} g K_{n,m}$	m	$2m$	$\mathbf{BC}_m$
					($m = n/2$)	$\mathbf{C}_{n/2}$
5	Sp_{2n}	GL_n	$g \mapsto I_{2n,n} g I_{2n,n}$			$\mathbf{C}_n$
6	SO_n	$\mathrm{SO}_m \times \mathrm{SO}_{n-m}$ ($m \le n/2$)	$g \mapsto I_{n,m} g I_{n,m}$	$m/2$ or $(n-m)/2$ (n odd)	(n odd)	$\mathbf{B}_m$
				(n odd, $m = 2$)		
				$m/2$ (n even)	m (n even)	$\mathbf{B}_m$
				$(m+1)/2$ (n even)	(n even, $m = n/2 - 1$)	$\mathbf{B}_{n/2-1}$
				(n even, $m = 2$)	(n even, $m = n/2$)	$\mathbf{D}_{n/2}$
7	SO_{2n}	GL_n	$g \mapsto \Omega_n g \Omega_n^\top$		(n odd)	$\mathbf{BC}_{\lfloor n/2 \rfloor}$
					(n even)	$\mathbf{C}_{n/2}$
8	$\mathbf{E}_6$	$\mathbf{A}_5 \times \mathbf{A}_1$				$\mathbf{F}_4$
9	$\mathbf{E}_6$	$\mathbf{D}_5 \times \Bbbk^\times$				$\mathbf{BC}_2$
10	$\mathbf{E}_6$	$\mathbf{C}_4$				$\mathbf{E}_6$
11	$\mathbf{E}_6$	$\mathbf{F}_4$				$\mathbf{A}_2$
12	$\mathbf{E}_7$	$\mathbf{A}_7$				$\mathbf{E}_7$
13	$\mathbf{E}_7$	$\mathbf{D}_6 \times \mathbf{A}_1$				$\mathbf{F}_4$
14	$\mathbf{E}_7$	$\mathbf{E}_6 \times \Bbbk^\times$				$\mathbf{C}_3$
15	$\mathbf{E}_8$	$\mathbf{D}_8$				$\mathbf{E}_8$
16	$\mathbf{E}_8$	$\mathbf{E}_7 \times \mathbf{A}_1$				$\mathbf{F}_4$
17	$\mathbf{F}_4$	$\mathbf{B}_4$				$\mathbf{BC}_1$
18	$\mathbf{F}_4$	$\mathbf{C}_3 \times \mathbf{A}_1$				$\mathbf{F}_4$
19	$\mathbf{G}_2$	$\mathbf{A}_1 \times \mathbf{A}_1$				$\mathbf{G}_2$

In another basis, $\theta(g) = J_{n,m} \cdot g \cdot J_{n,m}$, where

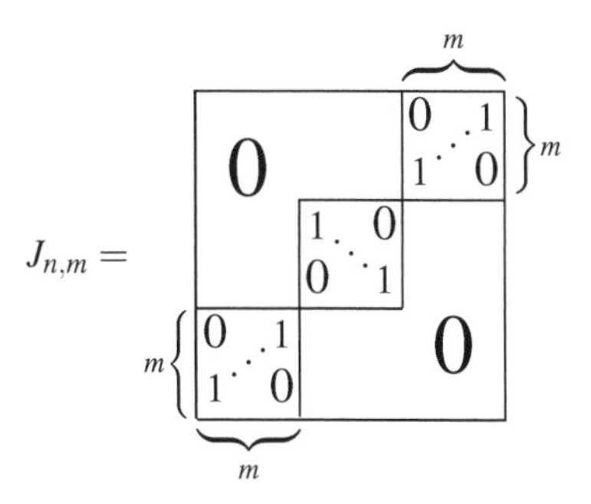

Now $T_1 = \{t = \mathrm{diag}(t_1, \dots, t_m, 1, \dots, 1, t_m^{-1}, \dots, t_1^{-1})\}$ is a maximal θ-split torus and the (compact) imaginary roots are $\varepsilon_i - \varepsilon_j$, $m < i \neq j \leq n-m$, in accordance with the Satake diagram. The little root system Δ_O consists of the nonzero restrictions $\overline{\varepsilon}_i - \overline{\varepsilon}_j$, $1 \leq i, j \leq n$, i.e., of $\pm\overline{\varepsilon}_i \pm \overline{\varepsilon}_j$, $\pm 2\overline{\varepsilon}_i$, and $\pm\overline{\varepsilon}_i$ unless $m = n/2$, $1 \leq i \neq j \leq m$. Thus Δ_O is of type $\mathbf{BC}_m$ or $\mathbf{C}_{n/2}$.

If θ is outer, then $\overline{\theta}(\varepsilon_i) = -\varepsilon_{n+1-i}$ and $T^{\overline{\theta}} = \{t = \mathrm{diag}(t_1, t_2, \dots, t_2^{-1}, t_1^{-1})\}$. Restricting the roots to this subtorus, we see that $\overline{\Delta}$ consists of $\pm\varepsilon_i' \pm \varepsilon_j'$, $\pm 2\varepsilon_i'$, and $\pm\varepsilon_i'$ for odd n, where ε_i' are the restrictions of ε_i, $1 \leq i \leq n/2$. The Dynkin diagram of $\widetilde{\Pi}$ has one of the following forms:

1 2 … 2 2   1 2 … 2 1
1

depending on whether n is odd or even. Therefore the Kac diagram has a unique white node, namely an extreme one.

The involution ι is trivial, whence either all nodes of the Satake diagram are white or the black nodes are isolated from each other and alternate with the white ones, the extreme nodes being black. (Otherwise, there would exist an inadmissible fragment ○—● .) Thus we obtain Nos. 2–3 of Table 26.3.

Any outer involution has the form $\theta(g) = (g^*)^{-1}$, where $*$ denotes the conjugation with respect to a non-degenerate (skew-)symmetric bilinear form on $\Bbbk^n$. In the symmetric case, choosing an orthonormal basis yields $\theta(g) = (g^\top)^{-1}$, whence $T_1 = T$ is a maximal θ-split torus and $\Delta_O = \Delta$. In a hyperbolic basis, $\theta(g) = (g^\dagger)^{-1}$, where $\dagger$ denotes the transposition with respect to the secondary diagonal. Then $T_0 = T^{\overline{\theta}}$ is a maximal torus in $H = \mathrm{SO}_n(\Bbbk)$. The roots of H are $\pm\varepsilon_i' \pm \varepsilon_j'$, $1 \leq i \neq j \leq n/2$, and $\pm\varepsilon_i'$ for odd n. The space $\mathfrak{m}$ consists of traceless symmetric matrices, and the lowest weight is $-2\varepsilon_1'$.

In the skew-symmetric case, choosing an appropriately ordered symplectic basis yields $\theta(g) = I_{n,n/2}(g^\dagger)^{-1} I_{n,n/2}$. Here T_0 is a maximal torus in $H = \mathrm{Sp}_n(\Bbbk)$ and $T_1 = \{t = \mathrm{diag}(t_1, t_2, \dots, t_2, t_1) \mid t_1 \cdots t_{n/2} = 1\}$ is a maximal θ-split torus. The roots of H are $\pm\varepsilon_i' \pm \varepsilon_j'$, $\pm 2\varepsilon_i'$, $1 \leq i \neq j \leq n/2$, and the lowest weight of $\mathfrak{m}$ is $-\varepsilon_1' - \varepsilon_2'$. The compact roots are $\varepsilon_i - \varepsilon_{n+1-i}$ $(1 \leq i \leq n)$, and Δ_O consists of $\overline{\varepsilon}_i - \overline{\varepsilon}_j$, $1 \leq i \neq j \leq n/2$, thus having the type $\mathbf{A}_{n/2-1}$.

From now on we assume that T_1 is a maximal θ-split torus.

26.6 Weyl Group. Consider the Weyl group W_O of the little root system Δ_O.

Proposition 26.19. $W_O \simeq N_{H^0}(T_1)/Z_{H^0}(T_1) \simeq N_G(T_1)/Z_G(T_1)$.

Proof. First we prove that each element of W_O is induced by an element of $N_{H^0}(T_1)$. It suffices to consider a root reflection $r_{\overline{\alpha}}$. Let $T_1^{\overline{\alpha}} \subseteq T_1$ be the connected kernel of $\overline{\alpha}$. Replacing G by $Z_G(T_1^{\overline{\alpha}})$ we may assume that $W_O = \{e, r_{\overline{\alpha}}\}$. The same argument as in Lemma 26.15 shows that $P^- = \theta(P) = hPh^{-1}$ for some $h \in H^0$. It follows that $h \in N_{H^0}(L) = N_{H^0}(T_1)$ acts on $\mathfrak{X}(T_1)$ as $r_{\overline{\alpha}}$.

On the other hand, $N_G(T_1)$ acts on T_1 as a subgroup of the "big" Weyl group $W = N_G(T)/T$. Indeed, any $g \in N_G(T_1)$ normalizes $L = Z_G(T_1)$ and may be replaced by another element in gL normalizing T. Since the Weyl chambers of W_O in $\mathfrak{X}(T_1) \otimes \mathbb{Q}$ are the intersections of Weyl chambers of W with $\mathfrak{X}(T_1) \otimes \mathbb{Q}$, the orbits of $N_G(T_1)/Z_G(T_1)$ intersect them in single points. Thus $N_G(T_1)/Z_G(T_1)$ cannot be bigger than W_O. This concludes the proof. □

26.7 *B*-orbits. Since O is spherical, there are finitely many B-orbits in O (Corollary 6.5). Their structure plays an important rôle in some geometric problems and, for $\Bbbk = \mathbb{C}$, in the representation theory of the real reductive Lie group $G(\mathbb{R})$ acting on the Riemannian symmetric space $O(\mathbb{R})$, the non-compact real form of O [Vog]. The classification and the adherence relation for B-orbits were described in [Sp1], [RS1], [RS2] (cf. Example 6.7). We explain the basic classification result under the assumption $H = G^{\theta}$. This is not an essential restriction [RS2, 1.1(b)].

By Proposition 26.3, O is identified with $\tau(G)$, where G (and B) acts by twisted conjugation.

Proposition 26.20. *The (twisted) B-orbits in $\tau(G) \simeq O$ intersect $N_G(T)$ in T-orbits. Thus $\mathfrak{B}(O)$ is in bijective correspondence with the set of twisted T-orbits in $N_G(T) \cap \tau(G)$.*

Proof. Consider a B-orbit $Bgo \subseteq O$. By Lemma 26.5, replacing g by bg, $b \in B$, one may assume that $g^{-1}Tg$ is a θ-stable maximal torus in $g^{-1}Bg$. This holds if and only if $\tau(g) \in N_G(T)$. One the other hand, taking another point $g'o \in Bgo$, $g' = bgh$, $b \in B$, $h \in H$, we have $\tau(g') = \theta(b)\tau(g)b^{-1} \in N_G(T)$ if and only if $\tau(g') = \theta(t)\tau(g)t^{-1}$, where $b = tu$, $t \in T$, $u \in U$, by standard properties of the Bruhat decomposition [Hum, 28.4]. □

There is a natural map $\mathfrak{B}(O) \to W$, $Bgo \mapsto w$, where $\theta(B)wB$ is the unique Bruhat double coset containing the respective B-orbit $\tau(BgH)$. By Proposition 26.20, $\tau(BgH) \cap N_G(T) \subseteq wT$. This map plays an important rôle in the study of B-orbits [RS1], [RS2]. Its image is contained in the set of *twisted involutions* $\{w \in W \mid \theta(w) = w^{-1}\}$, but in general is neither injective nor surjective onto this set.

Example 26.21. Let $G = \mathrm{GL}_n(\Bbbk)$, $\theta(g) = (g^{\top})^{-1}$, $H = \mathrm{O}_n(\Bbbk)$. Then $\tau(G)$ is the set of non-degenerate symmetric matrices, viewed as quadratic forms on $\Bbbk^n$. The group B of upper-triangular matrices acts on $\tau(G)$ by base changes preserving the standard flag in $\Bbbk^n$. It is an easy exercise in linear algebra that, for any inner product on $\Bbbk^n$, one can choose a basis $e_1, \dots, e_n$ compatible with a given flag and having the property that for any i there is a unique j such that $(e_i, e_j) = 1$ and $(e_i, e_k) = 0$, $\forall k \neq j$. The matrix of the quadratic form in this basis is the permutation matrix of the involution transposing i and j. It lies in $N_G(T)$ (where T is the diagonal torus) and is uniquely determined by the B-orbit of the quadratic form. Thus $\mathfrak{B}(O)$ is in bijection with the set of involutions in $W = \mathrm{S}_n$.

26.8 Colored Equipment. Now we describe the colored equipment of a symmetric space, according to [Vu2].

The weight lattice of a symmetric space is read off the Satake diagram, at least up to a finite extension. Let $Z = Z(G)^0$ and ω_i be the fundamental weights corresponding to the simple roots $\alpha_i \in \Pi$.

Proposition 26.22. *If G is of simply connected type, then*

$$\Lambda(O) = \mathfrak{X}(Z/Z\cap H) \oplus \left\langle\, \widehat{\omega}_j,\ \omega_k + \omega_{\iota(k)} \;\middle|\; j,k \right\rangle, \tag{26.2}$$

where j,k run over all ι-fixed, resp. ι-unstable, white nodes of the Satake diagram, and $\widehat{\omega}_j = \omega_j$ or $2\omega_j$, depending on whether the j-th node is adjacent to a black one or not. In the general case, $\Lambda(O)$ is a sublattice of finite index in the r.h.s. of (26.2).

Remark 26.23. The weight lattice $\Lambda(O) = \mathfrak{X}(T/T\cap H) = \mathfrak{X}(T_1/T_1\cap H)$ injects into $\mathfrak{X}(T_1)$ via restriction of characters from T to T_1. The space $\mathscr{E} = \mathrm{Hom}(\Lambda(O),\mathbb{Q})$ is then identified with $\mathfrak{X}^*(T_1)\otimes\mathbb{Q}$. The second direct summand in the r.h.s. of (26.2) is nothing else but the doubled weight lattice $2(\mathbb{Z}\Delta_O^\vee)^*$ of the little root system Δ_O. Indeed, $\widehat{\omega}_j/2$ and $(\omega_k + \omega_{\iota(k)})/2$ restrict to the fundamental weights dual to the simple coroots $\overline{\alpha}_j^\vee = \alpha_j^\vee - \theta(\alpha_j^\vee)$ or $\alpha_j^\vee$ and $\overline{\alpha}_k^\vee = \alpha_k^\vee - \theta(\alpha_k^\vee)$.

Proof. Without loss of generality we may assume that G is semisimple simply connected, whence $H = G^\theta$. The sublattice $\Lambda(O) \subseteq \mathfrak{X}(T)$ consisting of the weights vanishing on T^θ, i.e., of $\mu - \theta(\mu)$, $\mu \in \mathfrak{X}(T)$, is contained in $\mathfrak{X}(T/T_0) = \{\lambda \in \mathfrak{X}(T) \mid \theta(\lambda) = -\lambda\} = \langle \omega_j,\ \omega_k + \omega_{\iota(k)} \mid j,k\rangle$. The latter lattice injects into $\mathfrak{X}(T_1)$ so that $\Lambda(O)$ is identified with $2\mathfrak{X}(T_1)$. It remains to prove that $\mathfrak{X}(T_1) = (\mathbb{Z}\Delta_O^\vee)^*$ or, equivalently, that $\mathfrak{X}^*(T_1) = \mathbb{Z}\Delta_O^\vee$ is the coroot lattice of the little root system.

We have $\mathfrak{X}^*(T) = \mathbb{Z}\Delta^\vee$ and $\mathfrak{X}^*(T_1) = \mathbb{Z}\Delta^\vee \cap \mathscr{E} \supseteq \mathbb{Z}\Delta_O^\vee$. The *alcoves* (= fundamental polyhedra, see [Bou1, Ch. VI, §2, n°1]) of the affine Weyl group $W_{\mathrm{aff}}(\Delta_O)$ are the intersections of $\mathscr{E}$ with alcoves of $W_{\mathrm{aff}}(\Delta_O)$. Hence each alcove of $W_{\mathrm{aff}}(\Delta_O)$ contains a unique point from $\mathfrak{X}^*(T_1)$. It follows that $\mathfrak{X}^*(T_1)$ coincides with $\mathbb{Z}\Delta_O^\vee$. □

Let $\mathbf{C} = \mathbf{C}(\Delta_+)$ denote the dominant Weyl chamber of a root system Δ (with respect to a chosen subset of positive roots Δ_+). The weight semigroup $\Lambda_+(O)$ is contained both in $\Lambda(O)$ and in $\mathbf{C}$. Note that $\mathbf{C}\cap\mathscr{E} = \mathbf{C}(\Delta_O^+)$.

Proposition 26.24. $\Lambda_+(O) = \Lambda(O)\cap\mathbf{C}(\Delta_O^+)$.

Proof. Since $\Lambda_+(O)$ is the semigroup of all lattice points in a cone (see 15.1), it suffices to prove that $\mathbb{Q}_+\Lambda_+(O) = \mathbf{C}(\Delta_O^+)$. Take any dominant $\lambda \in \Lambda(O)$. We prove that $2\lambda \in \Lambda_+(O)$.

First note that $\lambda = -\theta(\lambda)$ is orthogonal to compact roots, whence λ is extended to P and $V^*(\lambda) = \mathrm{Ind}_P^G \Bbbk_{-\lambda}$. Consider another dual Weyl module obtained by twisting the G-action by θ: $V^*(\lambda)^\theta = \mathrm{Ind}_{\theta(P)}^G \Bbbk_{-\theta(\lambda)} \simeq V^*(\lambda^*)$. We have the canonical H-equivariant linear isomorphism $\omega : V^*(\lambda)^\theta \xrightarrow{\sim} V^*(\lambda)$. (If the dual Weyl modules are realized in $\Bbbk[G]$ as in Example 2.10, then ω is just the restriction of θ acting on $\Bbbk[G]$.) In other words, $\omega \in (V^*(\lambda) \otimes V(\lambda^*))^H$. Note that ω maps a T-eigenvector of weight μ to an eigenvector of weight $\theta(\mu)$. Hence

$$\omega = v_{-\lambda} \otimes v'_{-\lambda} + \sum_{\mu \neq \lambda} v_{\theta(\mu)} \otimes v'_{-\mu}, \tag{26.3}$$

where v_χ, v'_χ denote basic eigenvectors of weight χ in $V^*(\lambda)$ and $V(\lambda^*)$, respectively. Applying the homomorphisms $V(\lambda^*) \to V^*(\lambda)$, $v'_{-\lambda} \mapsto v_{-\lambda}$, and $V^*(\lambda) \otimes V^*(\lambda) \to V^*(2\lambda)$ (induced by multiplication in $\Bbbk[G]$), we obtain a nonzero element $\overline{\omega} \in V^*(2\lambda)^H$, whence $2\lambda \in \Lambda_+(O)$ by (2.2). □

Now we are ready to describe the colors and G-valuations of a symmetric space.

Theorem 26.25. *The colors of a symmetric space O are represented by the vectors from $\frac{1}{2}\Pi_O^\vee \subset \mathscr{E}$ (where $\Pi_O^\vee$ is the base of $\Delta_O^\vee \subset \mathfrak{X}^*(T_1)$). The valuation cone $\mathscr{V}$ is the antidominant Weyl chamber of $\Delta_O^\vee$ in $\mathscr{E}$.*

Corollary 26.26. *W_O is the little Weyl group of O in the sense of* 22.3.

Proof. Without loss of generality G is assumed to be of simply connected type. In the notation of Remarks 13.4 and 15.1, each $f \in \Bbbk[O]_\lambda^{(B)}$ is represented as $f = \eta_1^{d_1} \cdots \eta_s^{d_s}$, where the η_i are equations of the colors $D_i \in \mathscr{D}^B$, $d_i \in \mathbb{Z}_+$, and $\lambda = \sum d_i\lambda_i$, $\sum d_i\chi_i = 0$, where (λ_i, χ_i) are the biweights of η_i.

In the notation of Proposition 26.22, if $\lambda = \widehat{\omega}_j$ or $\omega_k + \omega_{\iota(k)}$, then $f = \eta_j$, or $\eta'_j\eta''_j$, or η_k, or $\eta'_k\eta''_k$, where the biweights of $\eta_j, \eta'_j, \eta''_j, \eta_k, \eta'_k, \eta''_k$ are $(\widehat{\omega}_j, 0)$, (ω_j, χ_j), $(\omega_j, -\chi_j)$, $(\omega_k + \omega_{\iota(k)}, 0)$, (ω_k, χ_k), $(\omega_{\iota(k)}, -\chi_k)$, respectively, for some nonzero $\chi_j, \chi_k \in \mathfrak{X}(H)$. In particular, the respective colors $D_j, D'_j, D''_j, D_k, D'_k, D''_k$ are pairwise distinct, and all colors occur among them since these f's span the multiplicative semigroup $\Bbbk[O]^{(B)}/\Bbbk[O]^\times$ by Proposition 26.24 and Remark 26.23. The assertion on colors stems now from Remarks 15.1 and 26.23.

Now we treat G-valuations. Take any $v = v_D \in \mathscr{V}$, where D is a G-stable prime divisor on a G-model X of $\Bbbk(O)$. It follows from the local structure theorem that $F = \overline{T_1 o}$ is an $N_H(T_1)$-stable subvariety of X intersecting D in the union of T_1-stable prime divisors D_{wv}, $w \in W_O$, that correspond to wv regarded as T_1-valuations of $\Bbbk(T_1 o)$ (cf. Proposition 23.13). By Theorem 21.1 $\mathscr{V}$ contains the antidominant Weyl chamber. It remains to show as in the proof of Theorem 22.13 that different vectors from $\mathscr{V}$ cannot be W_O-equivalent. □

The proof of Theorem 26.25 shows that the map $\varkappa : \mathscr{D}^B \to \mathscr{E}$ may be non-injective if H is not semisimple. There is a more precise description of colors in the spirit of Proposition 26.20 [Sp1, 5.4], [CS, §4].

It suffices to consider simple G. Assume first that H is connected. For any $\overline{\alpha}^\vee \in \Pi_O^\vee$ there exist either a unique or exactly two colors mapping to $\overline{\alpha}^\vee/2$. They correspond to the twisted T-orbits in $\tau(G) \cap r_\alpha T$ (for real α) or in $\tau(G) \cap (r_{\theta(\alpha)} r_\alpha T \cup r_{\theta(\iota(\alpha))} r_{\iota(\alpha)} T)$ (for complex α).

If H is semisimple, then such an orbit (and the respective color D_α) is always unique. In particular, $\iota(\alpha) = \alpha$ or $\iota(\alpha) = -\theta(\alpha) \perp \alpha$.

If H is not semisimple (*Hermitian case*), then inspection of Table 26.3 shows that $\dim Z(H) = 1$ and Δ_O is of type $\mathbf{BC}_n$ or $\mathbf{C}_n$. The color mapped to $\overline{\alpha}^\vee/2$ is unique except for the case where $\overline{\alpha}^\vee$ is the short simple coroot.

In the latter case, if α is complex, then $\tau(G) \cap r_{\theta(\alpha)} r_\alpha T$ and $\tau(G) \cap r_{\theta(\iota(\alpha))} r_{\iota(\alpha)} T$ are the twisted T-orbits corresponding to the two colors $D_\alpha, D_{\iota(\alpha)}$ mapped to $\overline{\alpha}^\vee/2$. Here $\Delta_O = \mathbf{BC}_n$ and $c(G/H') = 0$.

If α is real, then $\tau(G) \cap r_\alpha T$ consists of two twisted T-orbits corresponding to the two colors $D_\alpha^\pm$ mapped to $\alpha^\vee/2$ and swapped by $\operatorname{Aut}_G O \simeq \mathbf{Z}_2$. Here $\Delta_O = \mathbf{C}_n$ and $c(G/H') = 1$.

For disconnected H the divisors $D_\alpha^\pm \in \mathscr{D}^B(G/H^0)$ patch together into a single divisor $D_\alpha \in \mathscr{D}^B(G/H)$.

26.9 Coisotropy Representation. The (co)isotropy representation $H : \mathfrak{m}$ has nice invariant-theoretic properties in characteristic zero. They were examined by Kostant and Rallis [KoR]. From now on assume that $\operatorname{char} \Bbbk = 0$.

Semisimple elements in $\mathfrak{m}$ are exactly those having closed H-orbits, and the unique closed H-orbit in $\overline{H\xi}$ ($\xi \in \mathfrak{m}$) is $H\xi_s$. General elements of $\mathfrak{m}$ are semisimple. One may deduce it from the fact that T^*O is symplectically stable (Proposition 8.14) or prove directly: $\mathfrak{g} = \mathfrak{l} \oplus [\mathfrak{g}, \mathfrak{t}_1] \implies \mathfrak{m} = \mathfrak{t}_1 \oplus [\mathfrak{h}, \mathfrak{t}_1] \implies \mathfrak{m} = \overline{H\mathfrak{t}_1}$. This argument also shows that H-invariant functions on $\mathfrak{m}$ are uniquely determined by their restrictions to $\mathfrak{t}_1$. A more precise result was obtained by Kostant and Rallis.

Proposition 26.27 ([KoR]). *Every semisimple H-orbit in $\mathfrak{m}$ intersects $\mathfrak{t}_1$ in a W_O-orbit. Restriction of functions yields an isomorphism $\Bbbk[\mathfrak{m}]^H \simeq \Bbbk[\mathfrak{t}_1]^{W_O}$.*

Proof. Every semisimple element $\xi \in \mathfrak{m}$ is contained in the Lie algebra of a maximal θ-split torus. Hence by Lemma 26.15, $\xi' = (\operatorname{Ad} h)\xi \in \mathfrak{t}_1$ for some $h \in H$. If $\xi \in \mathfrak{t}_1$, then $T_1, h^{-1}T_1h$ are two maximal θ-split tori in $Z_G(\xi)$. Again by Lemma 26.15, $zT_1z^{-1} = h^{-1}T_1h$ for some $z \in Z_G(\xi) \cap H$, whence $h' = hz \in N_H(\mathfrak{t}_1)$, $\xi' = (\operatorname{Ad} h')\xi \in W_O\xi$.

The second assertion is a particular case of Proposition 25.7. It suffices to observe that the surjective birational morphism $\mathfrak{m}/\!/H \to \mathfrak{t}_1/W_O$ of two normal affine varieties has to be an isomorphism. □

Global analogues of these results for the H-action on O (in any characteristic) were obtained by Richardson [Ri2].

26.10 Flats. It is not incidental that the description of the valuation cone of a symmetric space was obtained by the same reasoning as in 23.5.

Proposition 26.28 ([Kn5, §6]). *Flats in O are exactly the G-translates of $T_1 \cdot o$.*

Proof. It suffices to consider flats F_α, $\alpha \in T_o^{\mathrm{pr}}O$. We have $T^*O = G *_H \mathfrak{m}$, $\alpha = e * \xi$, $\xi = \Phi(\alpha) \in \mathfrak{m}^{\mathrm{pr}}$. By Proposition 26.27, $\xi \in (\mathrm{Ad}\,h)\mathfrak{t}_1^{\mathrm{pr}}$, $h \in H^0$. It follows that $G_\xi = hLh^{-1}$, whence $F_\alpha = hLo = hT_1o$. □

The W_O-action on the flat $T_1 \cdot o$ comes from $N_H(T_1)$.

In the case $\Bbbk = \mathbb{C}$, flats in O are (G-translates of) the complexifications of maximal totally geodesic flat submanifolds in a Riemannian symmetric space $O(\mathbb{R})$ which is a real form of O [Hel1].

27 Algebraic Monoids and Group Embeddings

27.1 Algebraic Monoids. Similarly to algebraic groups, defined by superposing the concepts of an abstract group and an algebraic variety, it is quite natural to consider *algebraic semigroups*, i.e., algebraic varieties equipped with an associative multiplication law which is a regular map.

Example 27.1. All linear operators on a finite-dimensional vector space V form an algebraic semigroup $\mathrm{L}(V) \simeq \mathrm{L}_n(\Bbbk)$ ($n = \dim V$). The operators (matrices) of rank $\leq r$ form a closed subsemigroup $\mathrm{L}^{(r)}(V)$ ($\mathrm{L}_n^{(r)}(\Bbbk)$), a particular example of a determinantal variety.

However the category of *all* algebraic semigroups is immense. (For instance, every algebraic variety X turns into an algebraic semigroup being equipped with the "zero" multiplication $X \times X \to \{0\}$, where $0 \in X$ is a fixed element.) In order to make the theory really substantive, one has to restrict the attention to algebraic semigroups not too far from algebraic groups.

Definition 27.2. An *algebraic monoid* is an algebraic semigroup with unit, i.e., an algebraic variety X equipped with a morphism $\mu : X \times X \to X$, $\mu(x,y) =: x \cdot y$ (the *multiplication law*), and with a distinguished *unity element* $e \in X$ such that $(x \cdot y) \cdot z = x \cdot (y \cdot z)$, $e \cdot x = x \cdot e = x$, $\forall x,y,z \in X$.

Let $G = G(X)$ denote the group of invertible elements in X. The following elementary result can be found, e.g., in [Rit1, §2].

Proposition 27.3. *G is open in X.*

Proof. Since the left translation $x \mapsto g \cdot x$ by an element $g \in G$ is an automorphism of X, it suffices to prove that G contains an open subset of an irreducible component of X. Without loss of generality we may assume that X is irreducible. Let p_1, p_2 be the two projections of $\mu^{-1}(e) \subset X \times X$ to X. By the fiber dimension theorem, every component of $\mu^{-1}(e)$ has dimension $\geq \dim X$, and $p_i^{-1}(e) = (e,e)$. Hence p_i are dominant maps and $G = p_1(\mu^{-1}(e)) \cap p_2(\mu^{-1}(e))$ is a dense constructible set containing an open subset of X. □

Corollary 27.4. *G is an algebraic group.*

Those irreducible components of X which do not intersect G do not "feel the presence" of G and their behavior is beyond control. Therefore it is reasonable to restrict oneself to algebraic monoids X such that $G = G(X)$ is dense in X. In this case, left translations by G permute the components of X transitively, and many questions are reduced to the case where X is irreducible.

Monoids of this kind form an interesting category of algebraic structures closely related to algebraic groups (e.g., they arise as the closures of linear algebraic groups in the spaces of linear operators). The theory of algebraic monoids was created in major part during the last 30 years by M. S. Putcha, L. E. Renner, E. B. Vinberg, A. Rittatore, et al. The interested reader may consult a detailed survey [Ren3] of the theory from the origin up to the latest developments. In this section, we discuss algebraic monoids from the viewpoint of equivariant embeddings. A link between these two theories is provided by the following result.

Theorem 27.5 ([Rit1, §2]).

(1) *Any algebraic monoid X is a $(G\times G)$-equivariant embedding of $G = G(X)$, where the factors of $G\times G$ act by left/right multiplication, having a unique closed $(G\times G)$-orbit.*
(2) *Conversely, any* affine *$(G\times G)$-equivariant embedding $X \hookleftarrow G$ carries a structure of algebraic monoid with $G(X) = G$.*

Proof. (1) One has only to prove the uniqueness of a closed orbit $Y \subseteq X$. Note that $X\cdot Y\cdot X = \overline{G\cdot Y\cdot G} = Y$, i.e., Y is a (two-sided) *ideal* in X. For any other ideal $Y' \subseteq X$ we have $Y\cdot Y' \subseteq Y \implies Y = Y\cdot Y' \subseteq Y'$. Thus Y is the smallest ideal, called the *kernel* of X.
(2) The actions of the left and right copy of $G\times G$ on X define coactions $\Bbbk[X] \to \Bbbk[G]\otimes\Bbbk[X]$ and $\Bbbk[X] \to \Bbbk[X]\otimes\Bbbk[G]$, which are the restrictions to $\Bbbk[X] \subseteq \Bbbk[G]$ of the comultiplication $\Bbbk[G] \to \Bbbk[G]\otimes\Bbbk[G]$. Hence the image of $\Bbbk[X]$ lies in $(\Bbbk[G]\otimes\Bbbk[X])\cap(\Bbbk[X]\otimes\Bbbk[G]) = \Bbbk[X]\otimes\Bbbk[X]$, and we have a comultiplication in $\Bbbk[X]$. Now G is open in $X = \overline{G}$ and consists of invertibles. For any invertible $x \in X$, we have $xG\cap G \neq \emptyset$, and hence $x \in G$. □

Remark 27.6. Assertion (2) was first proved for reductive G by Vinberg [Vin2] in a different way.

Among general algebraic groups, affine (= linear) ones occupy a privileged position due to their most rich and interesting structure. The same holds for algebraic monoids. We provide two results confirming this observation.

Theorem 27.7 ([Mum, §4]). *Complete irreducible algebraic monoids are just Abelian varieties.*

Theorem 27.8 ([Rit3]). *An algebraic monoid X is affine provided that $G(X)$ is affine.*

This theorem was proved by Renner [Ren1] for quasiaffine X using some structure theory, and Rittatore [Rit3] reduced the general case to the quasiaffine one by considering total spaces of certain line bundles over X.

A theorem of Barsotti [Bar] and Rosenlicht [Ros] says that every connected algebraic group G has a unique affine normal connected subgroup G_{aff} such that the quotient group $G_{\mathrm{ab}} = G/G_{\mathrm{aff}}$ is an Abelian variety. An analogous structure result for algebraic monoids was obtained by Brion and Rittatore [BR]: for any normal irreducible algebraic monoid X the quotient map $G = G(X) \to G_{\mathrm{ab}}$ extends to a homomorphism of algebraic monoids $X \to G_{\mathrm{ab}}$, whose fiber at unity $X_{\mathrm{aff}} = \overline{G_{\mathrm{aff}}}$ is an affine algebraic monoid.

Theorem 27.9. *Any affine algebraic monoid X admits a closed homomorphic embedding $X \hookrightarrow \mathrm{L}(V)$. Furthermore, $G(X) = X \cap \mathrm{GL}(V)$.*

The proof is essentially the same as that of a similar result for algebraic groups [Hum, 8.6]. Thus the adjectives "affine" and "linear" are synonyms for algebraic monoids, in the same way as for algebraic groups.

In the notation of Theorem 27.9, the space of matrix entries $\mathrm{M}(V)$ generates $\Bbbk[X] \subseteq \Bbbk[G]$. Generally, $\Bbbk[X] \supset \mathrm{M}(V)$ if and only if the representation $G : V$ is extendible to X. It follows from Theorem 27.9 and (2.1) that

$$\Bbbk[X] = \bigcup \mathrm{M}(V) \tag{27.1}$$

over all G-modules V that are X-modules (cf. Proposition 2.14).

Example 27.10. By Theorem 27.5(2), every affine toric variety X carries a natural structure of algebraic monoid extending the multiplication in the open torus T. By Theorem 27.9, X is the closure of T in $\mathrm{L}(V)$ for some faithful representation $T : V$, i.e., a closed submonoid in the monoid of all diagonal matrices in some $\mathrm{L}_n(\Bbbk)$. The coordinate algebra $\Bbbk[X]$ is the semigroup algebra of the semigroup $\Sigma \subseteq \mathfrak{X}(T)$ consisting of all characters $T \to \Bbbk^\times$ extendible to X. Conversely, every finitely generated semigroup $\Sigma \ni 0$ such that $\mathbb{Z}\Sigma = \mathfrak{X}(T)$ defines a toric monoid $X \supseteq T$.

27.2 Reductive Monoids. The classification and structure theory for algebraic monoids is most well developed in the case where the group of invertibles is reductive.

Definition 27.11. An irreducible algebraic monoid X is called *reductive* if $G = G(X)$ is a reductive group.

In the sequel we consider only reductive monoids, thus returning to the general convention of our survey that G is a connected reductive group. By Theorems 27.5 and 27.8, reductive monoids are nothing else but $(G \times G)$-equivariant affine embeddings of G. They were classified by Vinberg [Vin2] in characteristic zero. Rittatore [Rit1] extended this classification to arbitrary characteristic using the embedding theory of spherical homogeneous spaces.

Considered as a homogeneous space under $G \times G$ acting by left/right multiplication, G is a symmetric space (Example 26.10). All θ-stable maximal tori of $G \times G$ are of the form $T \times T$, where T is a maximal torus in G. The maximal θ-split tori are $(T \times T)_1 = \{(t^{-1}, t) \mid t \in T\}$. Choose a Borel subgroup $B \supseteq T$ of G. Then $B^- \times B$ is a

Borel subgroup in $G\times G$ containing $T\times T$ and $\theta(B^-\times B)=B\times B^-$ is the opposite Borel subgroup.

The weight lattice $\Lambda=\mathfrak{X}(T\times T/\operatorname{diag}T)=\{(-\lambda,\lambda)\mid\lambda\in\mathfrak{X}(T)\}$ is identified with $\mathfrak{X}(T)$ and the little root system with $\frac{1}{2}\Delta$. The eigenfunctions $\mathbf{f}_\lambda\in\Bbbk(G)^{(B^-\times B)}$ ($\lambda\in\mathfrak{X}(T)$) are defined on the "big" open cell $U^-\times T\times U\subseteq G$ by the formula $\mathbf{f}_\lambda(u^-tu)=\lambda(t)$. For $\lambda\in\mathfrak{X}_+$ they are matrix entries: $\mathbf{f}_\lambda(g)=\langle v_{-\lambda},gv_\lambda\rangle$, where $v_\lambda\in V$, $v_{-\lambda}\in V^*$ are $B^\pm$-eigenvectors of weights $\pm\lambda$.

By Theorem 26.25, the valuation cone $\mathscr{V}$ is identified with the antidominant Weyl chamber in $\mathfrak{X}^*(T)\otimes\mathbb{Q}$ (cf. Example 24.9) and the colors are represented by the simple coroots $\alpha_1^\vee,\dots,\alpha_l^\vee\in\Pi^\vee$. In fact, the respective colors are $D_i=\overline{B^-r_{\alpha_i}B}$. Indeed, the equation of D_i in $\Bbbk[\widetilde{G}]$ is $\mathbf{f}_{\omega_i}$, where ω_i denote the fundamental weights.

By Example 24.9, every $(G\times G)$-valuation is proportional to $v=v_\gamma$, where $\gamma\in\mathfrak{X}^*(T)$ is antidominant, and $v(\mathbf{f}_\lambda)=\langle\gamma,\lambda\rangle$, $\forall\lambda\in\mathfrak{X}_+$, whence v is identified with γ (as a vector in the valuation cone).

Now Corollary 15.5 yields

Theorem 27.12. *Normal reductive monoids X are in bijection with strictly convex cones $\mathscr{C}=\mathscr{C}(X)\subset\mathfrak{X}^*(T)\otimes\mathbb{Q}$ generated by all simple coroots and finitely many antidominant vectors.*

Remark 27.13. The normality assumption is not so restrictive, because the multiplication on X lifts to its normalization $\widetilde{X}$ turning it into a monoid with the same group of invertibles.

Corollary 27.14. *There are no non-trivial monoids with semisimple group of invertibles.*

Corollary 27.15 ([Put], [Rit1, Pr. 9]). *Every normal reductive monoid has the structure $X=(X_0\times G_1)/Z$, where X_0 is a monoid with zero, and Z is a finite central subgroup in $G(X_0)\times G_1$ not intersecting the factors.*

Proof. Identify $\mathfrak{X}(T)\otimes\mathbb{Q}$ with $\mathscr{E}=\mathfrak{X}^*(T)\otimes\mathbb{Q}$ via a W-invariant inner product. Consider an orthogonal decomposition $\mathscr{E}=\mathscr{E}_0\oplus\mathscr{E}_1$, where $\mathscr{E}_0=\langle\mathscr{C}\cap\mathscr{V}\rangle$, $\mathscr{E}_1=(\mathscr{C}\cap\mathscr{V})^\perp$. It is easy to see that each root is contained in one of the $\mathscr{E}_i$. Then $G=G_0\cdot G_1=(G_0\times G_1)/Z$, where G_i are the connected normal subgroups with $\mathfrak{X}^*(T\cap G_i)=\mathscr{E}_i\cap\mathfrak{X}^*(T)$. Take a reductive monoid $X_0\supseteq G_0$ defined by $\mathscr{C}_0=\mathscr{C}\cap\mathscr{E}_0$. Since $\operatorname{int}\mathscr{C}_0$ intersects $\mathscr{V}(G_0)=\mathscr{V}\cap\mathscr{E}_0$, the kernel of X_0 is a complete variety, and hence a single point 0, the zero element with respect to the multiplication on X_0. Now X coincides with $(X_0\times G_1)/Z$, because both monoids have the same colored data. □

This classification can be made more transparent via coordinate algebras and representations. Recall from 15.1 that $\Bbbk[X]^{U^-\times U}=\Bbbk[\mathscr{C}^\vee\cap\mathfrak{X}(T)]$. The algebra $\Bbbk[X]$ itself is given by (27.1). It remains to determine which representations of G extend to X.

Proposition 27.16. *The following conditions are equivalent:*

(1) *The representation $G:V$ is extendible to X.*

(2) *The highest weights of the simple factors of* V *are in* $\mathscr{C}^\vee$.
(3) *All dominant* T*-weights of* V *are in* $\mathscr{C}^\vee$.

Proof. (1) $\Longrightarrow$ (2) Choose a G-stable filtration of V with simple factors and consider the associated graded G-module $\operatorname{gr} V$. If $G : V$ extends to X, then $\operatorname{gr} V$ is an X-module. Hence $\mathbf{f}_\lambda \in \Bbbk[X]$ whenever λ is a highest weight of a simple factor of $\operatorname{gr} V$.
(2) $\Longleftrightarrow$ (3) All T-weights of V are obtained from the highest weights of simple factors by subtracting positive roots. The structure of $\mathscr{C}$ implies that all dominant vectors obtained this way from $\lambda \in \mathscr{C}^\vee$ belong to $\mathscr{C}^\vee$.
(3) $\Longrightarrow$ (1) It follows from Example 24.9 that $\overline{T} \subseteq X$ intersects all $(G \times G)$-orbits, cf. Proposition 27.18 below. Thus it suffices to prove that $G : V$ extends to $\overline{T}$. Choose a closed embedding $X \hookrightarrow \mathrm{L}(V_0)$ such that the dominant T-weights of V_0 generate $\mathscr{C}^\vee \cap \mathfrak{X}(T)$. Then, clearly, $\Bbbk[\overline{T}] = \Bbbk[W\mathscr{C}^\vee \cap \mathfrak{X}(T)]$. Since all T-weights of V are in $W\mathscr{C}^\vee$, they are well defined on $\overline{T}$. □

Corollary 27.17. *If* $X \subseteq \mathrm{L}(V)$ *is a closed submonoid, then* $\mathscr{C}^\vee = \mathscr{K}(V) \cap \mathbf{C}$, *where* $\mathscr{K}(V)$ *denotes the convex cone spanned by the* T*-weights of* V.

Proof. The proposition implies that $\mathscr{C}^\vee \supseteq \mathscr{K}(V) \cap \mathbf{C}$. On the other hand, all $(T \times T)$-weights of $\Bbbk[X]$ are of the form $(-\lambda, \mu)$, $\lambda, \mu \in \mathscr{K}(V)$, whence $\mathscr{C}^\vee \subseteq \mathscr{K}(V)$. □

In characteristic zero, Proposition 27.16 together with (27.1) yields

$$\Bbbk[X] = \bigoplus_{\lambda \in \mathscr{C}^\vee \cap \mathfrak{X}(T)} \mathrm{M}(V(\lambda)) \tag{27.2}$$

(cf. Theorem 2.15 and (2.3)). In positive characteristic, $\Bbbk[X]$ has a "good" filtration with factors $V^*(\lambda) \otimes V^*(\lambda^*)$ [Do], [Rit2, §4], [Ren3, Cor. 9.9].

27.3 Orbits. The embedding theory provides a combinatorial encoding for $(G \times G)$-orbits in X, which reflects the adherence relation. This description can be made more explicit using the following

Proposition 27.18. *Suppose that* $X \hookleftarrow G$ *is an equivariant normal embedding. Then* $F = \overline{T}$ *intersects each* $(G \times G)$*-orbit* $Y \subset X$ *in finitely many* T*-orbits permuted transitively by* W. *Exactly one of these orbits* $F_Y \subseteq F \cap Y$ *satisfies* $\operatorname{int} \mathscr{C}_{F_Y} \cap \mathscr{V} \neq \emptyset$; *then* $\mathscr{C}_{F_Y} = \bigcup w(\mathscr{C}_Y \cap \mathscr{V})$ *over all* $w \in W$ *such that* $w(F_Y) = F_Y$.

Remark 27.19. Since T is a flat of G (Proposition 26.28), some of the assertions stem from the results of §23. However, the proposition here is more precise. In particular, it completely determines the fan of F.

Proof. Take any $v \in \mathscr{S}_Y$; then $v = v_\gamma$, $\gamma \in \mathfrak{X}^*(T) \cap \mathscr{V}$, and $\exists \lim_{t \to 0} \gamma(t) = \gamma(0) \in Y$. The associated parabolic subgroup $P = P(\gamma)$ contains B^-. Consider the Levi decomposition $P = P_\mathrm{u} \rtimes L$, $L \supseteq T$. One verifies that $(G \times G)_{\gamma(0)} \supseteq (P_\mathrm{u}^- \times P_\mathrm{u}) \cdot \operatorname{diag} L$. It easily follows that $(B^- \times B)\gamma(0) = \mathring{Y}$ is the open $(B^- \times B)$-orbit in Y and $F_Y := T\gamma(0) = \mathring{Y}^{\operatorname{diag} T}$ is the unique T-orbit in F intersecting $\mathring{Y}$.

In view of Example 24.8, this implies $\operatorname{int}\mathscr{C}_{F_Y} \supseteq (\operatorname{int}\mathscr{C}_Y) \cap \mathscr{V}$. On the other hand, each T-orbit in $F \cap Y$ is accessed by a one-parameter subgroup $\gamma \in \mathfrak{X}^*(T)$, $\gamma(0) \in Y$. Taking $w \in W$ such that $w\gamma \in \mathscr{V}$ yields $w(T\gamma(0)) = F_Y$. All assertions of the proposition are deduced from these observations. □

Now suppose that $X \subseteq \mathrm{L}(V)$ is a closed normal submonoid and denote $\mathscr{K} = \mathscr{K}(V)$.

Theorem 27.20. *The $(G \times G)$-orbits in X are in bijection with the faces of $\mathscr{K}$ whose interiors intersect $\mathbf{C}$. The orbit Y corresponding to a face $\mathscr{F}$ is represented by the T-equivariant projector $e_{\mathscr{F}}$ of V onto the sum of T-eigenspaces of weights in $\mathscr{F}$. The cone $\mathscr{C}_Y$ is dual to the barrier cone $\mathscr{K} \cap \mathbf{C} - \mathscr{F} \cap \mathbf{C}$ of $\mathscr{K} \cap \mathbf{C}$ at the face $\mathscr{F} \cap \mathbf{C}$, and $\mathscr{D}_Y^B$ consists of the simple coroots orthogonal to $\mathscr{F}$.*

Proof. A complete set of T-orbit representatives in $F = \overline{T}$ is formed by the limits of one-parameter subgroups, i.e., by the $e_{\mathscr{F}}$ over *all* faces $\mathscr{F}$ of $\mathscr{K}$. The respective cones in the fan of F are the dual faces $\mathscr{F}^* = \mathscr{K}^\vee \cap \mathscr{F}^\perp$ of $\mathscr{K}^\vee = W(\mathscr{C} \cap \mathscr{V})$. By Proposition 27.18, the orbits Y are bijectively represented by those $e_{\mathscr{F}}$ which satisfy $\operatorname{int}\mathscr{F}^* \cap \mathscr{V} \neq \emptyset$. This happens if and only if $\mathscr{F}^*$ lies on a face of $\mathscr{C}$ of the same dimension (namely on $\mathscr{C}_Y$) or, equivalently, $\mathscr{F}$ contains a face of $\mathscr{C}^\vee = \mathscr{K} \cap \mathbf{C}$ of the same dimension (namely $\mathscr{C}_Y^* = \mathscr{F} \cap \mathbf{C}$), i.e., $\operatorname{int}\mathscr{F} \cap \mathbf{C} \neq \emptyset$. The assertion on $(\mathscr{C}_Y, \mathscr{D}_Y)$ stems from the description of a dual face. □

Example 27.21. Let $G = \mathrm{GL}_n(\Bbbk)$ and $X = \mathrm{L}_n(\Bbbk)$. For B and T take the standard Borel subgroup of upper-triangular matrices and diagonal torus, respectively. We have $\mathfrak{X}(T) = \langle \varepsilon_1, \dots, \varepsilon_n \rangle$, where the ε_i are the diagonal matrix entries of T. We identify $\mathfrak{X}(T)$ with $\mathfrak{X}^*(T)$ via the inner product such that the ε_i form an orthonormal basis. Let $(k_1, \dots, k_n)$ denote the coordinates on $\mathfrak{X}(T) \otimes \mathbb{Q}$ with respect to this basis. The Weyl group $W = S_n$ permutes them.

The weights $\lambda_i = \varepsilon_1 + \dots + \varepsilon_i$ span $\mathfrak{X}(T)$ and $\mathbf{f}_{\lambda_i} \in \Bbbk[X]$ are the upper-left corner i-minors of a matrix. Put $D_i = \{x \in X \mid \mathbf{f}_{\lambda_i}(x) = 0\}$. Then $\mathscr{D}^B = \{D_1, \dots, D_{n-1}\}$, D_i are represented by $\alpha_i = \varepsilon_i - \varepsilon_{i+1}$, $\forall i < n$, and D_n is the unique G-stable prime divisor, $v_{D_n} = \varepsilon_n$.

Therefore $\mathscr{C} = \{k_1 + \dots + k_i \geq 0,\ i = 1, \dots, n\}$ is the cone spanned by $\varepsilon_i - \varepsilon_{i+1}, \varepsilon_n$, and $\mathscr{C}^\vee = \{k_1 \geq \dots \geq k_n \geq 0\}$ is spanned by λ_i. The lattice vectors of $\mathscr{C}^\vee$ are exactly the dominant weights of polynomial representations (cf. Proposition 27.16). The lattice vectors of $\mathscr{K} = W\mathscr{C}^\vee = \{k_1, \dots, k_n \geq 0\}$ are all polynomial weights of T.

The $(G \times G)$-orbits in X are $Y_r = \{x \in X \mid \operatorname{rk} x = r\}$. Clearly, $\mathscr{D}_{Y_r}^B = \{D_i \mid r < i < n\}$ and $\mathscr{C}_{Y_r}$ is a face of $\mathscr{C}$ cut off by the equations $k_1 = \dots = k_r = 0$. The dual face $\mathscr{C}_{Y_r}^*$ of $\mathscr{C}^\vee$ is the dominant part of the face $\mathscr{F}_r = \{k_i \geq 0 = k_j \mid i \leq r < j\} \subseteq \mathscr{K}$, and all faces of $\mathscr{K}$ whose interiors intersect $\mathbf{C} = \{k_1 \geq \dots \geq k_n\}$ are obtained this way. Clearly, the respective projectors $e_{\mathscr{F}_r} = \operatorname{diag}(1, \dots, 1, 0, \dots, 0)$ are the $(G \times G)$-orbit representatives, and the representatives of all T-orbits in $\overline{T}$ are obtained from $e_{\mathscr{F}_r}$ by the W-action.

27.4 Normality and Smoothness. In characteristic zero, it is possible to classify (to a certain extent) arbitrary (not necessarily normal) reductive monoids [Vin2] via their coordinate algebras similarly to (27.2). The question is to describe finitely generated $(G \times G)$-stable subalgebras of $\Bbbk[G]$ with the quotient field $\Bbbk(G)$. They are of the form

$$\Bbbk[X] = \bigoplus_{\lambda \in \Sigma} \mathrm{M}(V(\lambda)), \tag{27.3}$$

where Σ is a finitely generated subsemigroup of $\mathfrak{X}_+$ such that $\mathbb{Z}\Sigma = \mathfrak{X}(T)$ and the r.h.s. of (27.3) remains closed under multiplication, i.e., all highest weights of $V(\lambda) \otimes V(\mu)$ belong to Σ whenever $\lambda, \mu \in \Sigma$. Such a semigroup Σ is called *perfect*.

Definition 27.22. We say that $\lambda_1, \dots, \lambda_m$ *G-generate* Σ if Σ consists of all highest weights of G-modules $V(\lambda_1)^{\otimes k_1} \otimes \cdots \otimes V(\lambda_m)^{\otimes k_m}$, $k_1, \dots, k_m \in \mathbb{Z}_+$. (In particular any generating set G-generates Σ.) All weights in Σ are of the form $\sum k_i \lambda_i - \sum l_j \alpha_j$, $k_i, l_j \in \mathbb{Z}_+$.

Example 27.23. In Example 27.21, $\Sigma = \mathscr{C}^\vee \cap \mathfrak{X}(T)$ is generated by $\lambda_1, \dots, \lambda_n$ and G-generated by λ_1.

It is easy to see that $X \hookrightarrow \mathrm{L}(V)$ if and only if the highest weights $\lambda_1, \dots, \lambda_m$ of the simple summands of V G-generate Σ. The highest weight theory implies that $\mathscr{K} = \mathscr{K}(V)$ is the W-span of

$$\mathscr{K} \cap \mathbf{C} = (\mathbb{Q}_+\{\lambda_1, \dots, \lambda_m, -\alpha_1, \dots, -\alpha_l\}) \cap \mathbf{C}. \tag{27.4}$$

Theorem 27.20 generalizes to this context.

By Theorem D.5(3), X is normal if and only if $\Bbbk[X]^{U^- \times U} = \Bbbk[\Sigma]$ is integrally closed, i.e., if and only if Σ is the semigroup of all lattice vectors in a polyhedral cone. In general, taking the integral closure yields

$$\mathbb{Q}_+\Sigma = \mathscr{C}^\vee = \mathscr{K} \cap \mathbf{C},$$

where $\mathscr{C}$ is the cone associated with the normalization of X. Indeed, the inclusion $\mathbb{Q}_+\Sigma \subseteq \mathscr{K} \cap \mathbf{C}$ stems from the structure of Σ, $\mathscr{K} \cap \mathbf{C} \subseteq \mathscr{C}^\vee$ stems from the structure of $\mathscr{C}$, and $\mathbb{Q}_+\Sigma = \mathscr{C}^\vee$ is due to Lemma D.6. Here is a representation-theoretic interpretation: a multiple of each dominant vector in $\mathscr{K}$ eventually occurs as a highest weight in a tensor power of V, see [Tim4, §2] for a direct proof.

Given $G : V$, the above normality condition for $X \subseteq \mathrm{L}(V)$ is generally not easy to verify, because the reconstruction of Σ from $\{\lambda_1, \dots, \lambda_m\}$ requires decomposing tensor products of arbitrary G-modules. Of course, there is no problem if λ_i already generate $\mathscr{K} \cap \mathfrak{X}_+$—a sufficient condition for normality. Here is an effective necessary condition:

Proposition 27.24 ([Ren2], [Ren3, Th. 5.4(b)]). *If X is normal, then $F = \overline{T}$ is normal, i.e., the T-weights of V generate $\mathscr{K} \cap \mathfrak{X}(T)$.*

Proof. We can increase V by adding new highest weights λ_i so that $\lambda_1, \dots, \lambda_m$ will generate $\Sigma = \mathscr{K} \cap \mathfrak{X}_+$. (This operation does not change X and F.) Then

$W\{\lambda_1,\dots,\lambda_m\}$ generates $\mathscr{K}\cap\mathfrak{X}(T)$, i.e., $\Bbbk[F]=\Bbbk[\mathscr{K}\cap\mathfrak{X}(T)]$ is integrally closed. □

If $V=V(\lambda)$ is irreducible, then the center of G acts by homotheties, whence $G=\Bbbk^\times\cdot G_0$, where G_0 is semisimple, $\mathfrak{X}(T)\subseteq\mathbb{Z}\oplus\mathfrak{X}(T\cap G_0)$ is a cofinite sublattice, and $\lambda=(1,\lambda_0)$. De Concini showed that $\mathscr{K}(V(\lambda))\cap\mathfrak{X}_+$ is G-generated by the T-dominant weights of $V(\lambda)$ [Con]. However Σ contains no T-weights of $V(\lambda)$ except λ. It follows that X is normal if and only if λ_0 is a minuscule weight for G_0 [Con], [Tim4, §12].

It turns out that Example 27.21 is essentially the unique non-trivial example of a smooth reductive monoid.

Theorem 27.25 (cf. [Ren2], [Tim4, §11]). *Smooth reductive monoids are of the form $X=(G_0\times\mathrm{L}_{n_1}(\Bbbk)\times\cdots\times\mathrm{L}_{n_s}(\Bbbk))/Z$, where $Z\subset G_0\times\mathrm{GL}_{n_1}(\Bbbk)\times\cdots\times\mathrm{GL}_{n_s}$ is a finite central subgroup not intersecting $\mathrm{GL}_{n_1}(\Bbbk)\times\cdots\times\mathrm{GL}_{n_s}(\Bbbk)$.*

Proof. By Corollary 27.15, $X=G_0*_Z X_0$, where X_0 has the zero element. Thus it suffices to consider monoids with zero. We explain how to handle this case in characteristic zero.

Assume that $X\subseteq\mathrm{L}(V)$. There exists a coweight $\gamma\in\operatorname{int}\mathscr{C}\cap\mathscr{V}$, $\gamma\perp\Delta$. It defines a one-parameter subgroup $\gamma(t)\in Z(G)$ contracting V to 0 (as $t\to0$). The algebra $\mathscr{A}=\mathscr{A}(V)$ spanned by X in $\mathrm{L}(V)$ is semisimple, i.e., a product of matrix algebras, and T_0X is an ideal in $\mathscr{A}$. As X is smooth and the multiplication by $\gamma(t)$ contracts X to 0, the equivariant projection $X\to T_0X$ is an isomorphism. □

27.5 Group Embeddings. We conclude this section with a discussion of arbitrary (not necessarily affine) equivariant embeddings of G. For simplicity, we assume that $\operatorname{char}\Bbbk=0$.

In the same way as a faithful linear representation $G:V$ defines a reductive monoid $\overline{G}\subseteq\mathrm{L}(V)$, a faithful projective representation $G:\mathbb{P}(V)$ (arising from a linear representation of a finite cover of G in V) defines a projective completion $X=\overline{G}\subseteq\mathbb{P}(\mathrm{L}(V))$. These group completions are studied in [Tim4]. There are two main tools to reduce their study to reductive monoids.

First, the cone $\widehat{X}\subseteq\mathrm{L}(V)$ over X is a reductive monoid whose group of invertibles $\widehat{G}$ is the extension of G by homotheties. Conversely, any such monoid gives rise to a projective completion. This allows the transfer of some of the above results to projective group completions. For instance, Theorem 27.20 transfers verbatim if we only replace the weight cone $\mathscr{K}(V)$ by the weight polytope $\mathscr{P}=\mathscr{P}(V)$ (= the convex hull of the T-weights of V), see [Tim4, §9] for details.

Another approach, suitable for local study, is to use the local structure theorem. By the above, closed $(G\times G)$-orbits $Y\subset X$ correspond to the dominant vertices $\lambda\in\mathscr{P}$, and the representatives are $y=[v_\lambda\otimes v_{-\lambda}]$, where $v_\lambda\in V$, $v_{-\lambda}\in V^*$ are $B^\pm$-eigenvectors of weights $\pm\lambda$, $\langle v_\lambda,v_{-\lambda}\rangle\neq0$. Consider the parabolic $P=P(\lambda)$ and its Levi decomposition $P=P_\mathrm{u}\rightthreetimes L$, $L\supseteq T$. Then $V_0=\langle v_{-\lambda}\rangle^\perp$ is an L-stable complement to $\langle v_\lambda\rangle$ in V. Put $\mathring{X}=X_{\mathrm{f}_\lambda}$.

Lemma 27.26. $\mathring{X}\simeq P_\mathrm{u}^-\times Z\times P_\mathrm{u}$, *where $Z\simeq\overline{L}\subseteq\mathrm{L}(V_0\otimes\Bbbk_{-\lambda})$ is a reductive monoid with the zero element y.*

Proof. Applying Corollary 4.5 to the projectivization of $G \times G : \mathrm{L}(V) = V \otimes V^*$ and intersecting with X, we obtain a neighborhood of the desired structure with $Z = X \cap \mathbb{P}\big(\Bbbk^\times(v_\lambda \otimes v_{-\lambda}) + E_0\big)$, where

$$E_0 = \big((\mathfrak{g} \times \mathfrak{g})(v_{-\lambda} \otimes v_\lambda)\big)^\perp = (\mathfrak{g}v_{-\lambda} \otimes v_\lambda + v_{-\lambda} \otimes \mathfrak{g}v_\lambda)^\perp \supseteq V_0 \otimes V_0^* = \mathrm{L}(V_0).$$

Hence $Z = \overline{L} \subseteq \mathbb{P}\big(\Bbbk^\times(v_\lambda \otimes v_{-\lambda}) \oplus \mathrm{L}(V_0)\big) \simeq \mathrm{L}(V_0 \otimes \Bbbk_{-\lambda})$. □

The monoids Z are transversal slices to the closed orbits in X. They can be used to study the local geometry of X. For instance, one can derive criteria for normality and smoothness [Tim4, §§10,11].

Example 27.27. Take $G = \mathrm{Sp}_4(\Bbbk)$, with the simple roots $\alpha_1 = \varepsilon_1 - \varepsilon_2$, $\alpha_2 = 2\varepsilon_2$, and the fundamental weights $\omega_1 = \varepsilon_1$, $\omega_2 = \varepsilon_1 + \varepsilon_2$, $\pm\varepsilon_i$ being the weights of the tautological representation $\mathrm{Sp}_4(\Bbbk) : \Bbbk^4$. Let $\lambda_1 = 3\omega_1$, $\lambda_2 = 2\omega_2$ be the highest weights of the simple summands of V. The weight polytope $\mathscr{P}$ is depicted in Fig. 27.1(a), the highest weights are indicated by bold dots. There are two closed orbits $Y_1, Y_2 \subset X$.

Fig. 27.1 A projective completion of $\mathrm{Sp}_4(\Bbbk)$

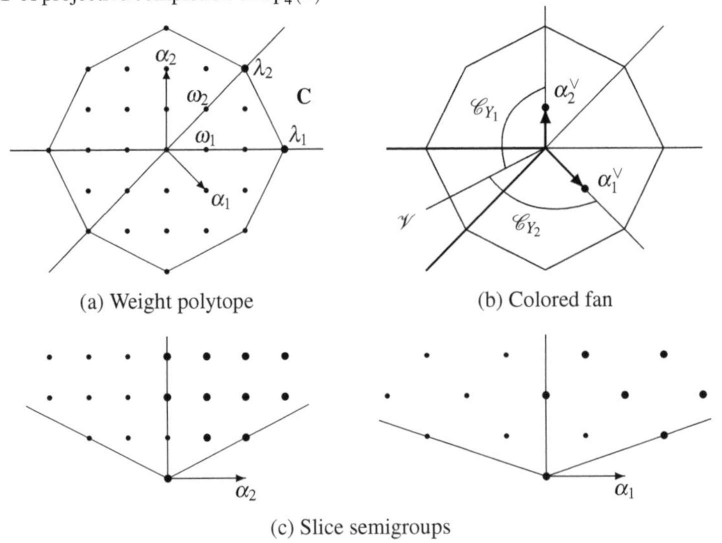

The respective Levi subgroups are $L_1 = \mathrm{SL}_2(\Bbbk) \times \Bbbk^\times$ and $L_2 = \mathrm{GL}_2(\Bbbk)$, with the simple roots α_2 and α_1, respectively.

Consider the slice monoids Z_i for Y_i. The weight semigroups of $F_i = \overline{T}$ (the closure in Z_i) are plotted by dots in Fig. 27.1(c), the bold dots corresponding to the weight semigroups Σ_i of Z_i. (They are easily computed using the Clebsch–Gordan formula.) We can now see that F_i are normal, but Z_i are not, i.e., X is non-normal along Y_1, Y_2. However, if we increase V by adding two highest weights $\lambda_3 = 2\omega_1$, $\lambda_4 = \omega_1 + \omega_2$, then X becomes normal. Its colored fan is depicted in Fig. 27.1(b).

The projective completions of adjoint simple groups in projective linear operators on fundamental and adjoint representation spaces were studied in detail in [Tim4, §12]. In particular, the orbital decomposition was described, and normal and smooth completions were identified.

Example 27.28. Suppose that $G = \mathrm{SO}_{2l+1}(\Bbbk)$ and $V = V(\omega_i)$ is a fundamental representation. We have a unique closed orbit $Y \subset X$. If $i < l$, then $L \not\simeq \mathrm{GL}_{n_1}(\Bbbk) \times \cdots \times \mathrm{GL}_{n_s}(\Bbbk)$. Hence Z and X are singular. But for $i = l$ (the spinor representation), $L \simeq \mathrm{GL}_l(\Bbbk)$ and $V(\omega_l) \otimes \Bbbk_{-\omega_l}$ is L-isomorphic to $\bigwedge^\bullet \Bbbk^l$. It follows that $Z \simeq \mathrm{L}_l(\Bbbk)$, whence X is smooth.

Example 27.29. Suppose that all vertices of $\mathscr{P}$ are regular weights. Then the slice monoids Z are toric and their weight semigroups Σ are generated by the weights $\mu - \lambda$, where μ runs over all T-weights of V. The variety X is toroidal, and normal (smooth) if and only if each Σ consists of all lattice vectors in the barrier cone of $\mathscr{P}$ at λ (resp. Σ is generated by linearly independent weights).

In particular, if $V = V(\lambda)$ is a simple module of regular highest weight, then $\Sigma = \mathbb{Z}_+(-\Pi)$, whence X is smooth. This is a particular case of a wonderful completion, see §30.

A interesting model for the wonderful completion of G in terms of Hilbert schemes was proposed by Brion [Bri17]. Namely, given a generalized flag variety $M = G/Q$, he proves that the closure $X = \overline{(G \times G)[\operatorname{diag} M]}$ in the Hilbert scheme (or the Chow variety) of $M \times M$ is isomorphic to the wonderful completion. If $G = (\operatorname{Aut} M)^0$ (e.g., if $Q = B$), then X is an irreducible component of the Hilbert scheme (the Chow variety). All fibers of the universal family over X are reduced and Cohen–Macaulay (even Gorenstein if $Q = B$).

Toroidal and wonderful group completions were studied intensively in the framework of the general theory of toroidal and wonderful varieties (see §29–§30) and on their own. De Concini and Procesi [CP3] and Strickland [Str2] computed ordinary and equivariant rational cohomology of smooth toroidal completions over $\Bbbk = \mathbb{C}$ (see also [BCP], [LP]). Brion [Bri14] carried out a purely algebraic treatment of these results replacing cohomology by (equivariant) Chow rings.

The basis of the Chow ring $A(X)$ of a smooth toroidal completion $X = \overline{G}$ is given by the closures of the Białynicki-Birula cells [B-B1], which are isomorphic to affine spaces and intersect $(G \times G)$-orbits in $(B^- \times B)$-orbits [BL, 2.3]. The latter were described in [Bri14, 2.1]. The $(B^- \times B)$-orbit closures in X are smooth in codimension 1, but singular in codimension 2 (apart from trivial exceptions arising from $G = \mathrm{PSL}_2(\Bbbk)$) [Bri14, §2]. For wonderful X, the Białynicki-Birula cells are described in [Bri14, 3.3], and the closures of the cells intersecting G (= the closures in X of $(B^- \times B)$-orbits in G) are normal and Cohen–Macaulay [BPo]. The geometry of $(B^- \times B)$-orbit closures in X was studied in [Sp4], [Ka].

The class of reductive group embeddings is not closed under degenerations. Alexeev and Brion [AB1], [AB2] introduced a more general class of *(stable) reductive varieties* closed under flat degenerations with irreducible (resp. reduced) fibers. Affine (stable) reductive varieties may be defined as normal affine spherical $(G\times G)$-varieties X such that $\Lambda(X)=\Lambda(G)\cap\mathscr{S}$ for some subspace $\mathscr{S}\subseteq\Lambda(G)\otimes\mathbb{Q}$ (resp. as seminormal connected unions of reductive varieties); projective (stable) reductive varieties are the projectivizations of affine ones. Affine reductive varieties provide examples of algebraic semigroups without unit.

Alexeev and Brion gave a combinatorial classification and described the orbital decomposition for stable reductive varieties in the spirit of Theorems 27.12, 27.20. They constructed moduli spaces for affine stable reductive varieties embedded in a $(G\times G)$-module and for stable reductive pairs, i.e., projective stable reductive varieties with a distinguished effective ample divisor containing no $(G\times G)$-orbit.

27.6 Enveloping and Asymptotic Semigroups. An interesting family of reductive varieties was introduced by Vinberg [Vin2]. Consider the group $\widehat{G}=(G\times T)/Z$, where $Z=\{(t^{-1},t)\mid t\in Z(G)\}$. The cone $\mathscr{C}\subset\mathscr{E}(\widehat{G})$ spanned by (the projections to $\mathscr{E}$ of) $(\alpha_i^\vee,0)$ and $(-\omega_j^\vee,\omega_j^\vee)$, where $\omega_j^\vee$ are the fundamental coweights, defines a normal reductive monoid $\operatorname{Env}G$, called the *enveloping semigroup* of G, with group of invertibles $\widehat{G}$. The projection $\mathscr{E}(\widehat{G})\to\mathscr{E}(T/Z(G))$ maps $\mathscr{C}$ onto $\mathbf{C}$. Hence by Theorem 15.10 we have an equivariant map $\pi_G:\operatorname{Env}G\to\mathbb{A}^l$, where $G\times G$ acts on $\mathbb{A}^l$ trivially and T acts with the weights $-\alpha_1,\dots,-\alpha_l$.

The algebra $\Bbbk[\operatorname{Env}G]=\bigoplus_{\chi\in\lambda+\mathbb{Z}_+\Pi}\mathrm{M}(V(\lambda))\otimes\Bbbk\chi$ is a free module over $\Bbbk[\mathbb{A}^l]=\Bbbk[\mathbb{Z}_+\Pi]$ and $\Bbbk[\operatorname{Env}G]^{U^-\times U}=\Bbbk[\mathfrak{X}_+]\otimes\Bbbk[\mathbb{Z}_+\Pi]$, i.e., all schematic fibers of π_G have the same algebra of $(U^-\times U)$-invariants $\Bbbk[\mathfrak{X}_+]$. Hence π_G is flat and all its fibers are reduced and irreducible by Theorem D.5(1), i.e., $\operatorname{Env}G$ is the total space of a family of reductive varieties, in the sense of Appendix E.3. (In fact, $\mathbb{A}^l=(\operatorname{Env}G)/\!/(G\times G)$ and π_G is the categorical quotient map.)

It is easy to see that the fibers of π_G over points with nonzero coordinates are isomorphic to G. Degenerate fibers are obtained from G by a deformation of the multiplication law in $\Bbbk[G]$. In particular, the "most degenerate" fiber $\operatorname{As}G:=\pi_G^{-1}(0)$, called the *asymptotic semigroup* of G, is just the horospherical contraction of G (see 7.3). In a sense, the asymptotic semigroup reflects the behavior of G at infinity.

The enveloping semigroup is used in [AB1, 7.5] to construct families of affine reductive varieties with given general fiber X: $\operatorname{Env}X=(\operatorname{Env}G\times X)/\!/G$, where G acts as $\{e\}\times\operatorname{diag}G\times\{e\}\subset G\times G\times G\times G$, so that $\Bbbk[\operatorname{Env}X]=\bigoplus_{\chi\in\lambda+\mathbb{Z}_+\Pi}\Bbbk[X]_{(\lambda)}\otimes\Bbbk\chi\subseteq\Bbbk[X\times T]$. The map π_G induces a flat morphism $\pi_X:\operatorname{Env}X\to\mathbb{A}^l$ with reduced and irreducible fibers.

It was proved in [AB1, 7.6] that π_X is a locally universal family of reductive varieties with general fiber X, i.e., every flat family of affine reductive varieties with reduced fibers over an irreducible base is locally a pullback of π_X. The universal property for enveloping semigroups was already noticed in [Vin2].

Example 27.30. Let us describe the enveloping and asymptotic semigroups of $G=\mathrm{SL}_n(\Bbbk)$ for small n, using the notation of Example 27.21. Here $\Lambda_+(\operatorname{Env}G)$ is generated by (ω_i,ω_i), $(0,\alpha_i)$, $i=1,\dots,n-1$. Recall that $\omega_i=\varepsilon_1+\dots+\varepsilon_i$ is the highest weight of $\bigwedge^i\Bbbk^n$. Thus $\operatorname{Env}\mathrm{SL}_n(\Bbbk)$ is the closure in $\mathrm{L}(V)$ of the image of $\mathrm{SL}_n(\Bbbk)\times T$ acting on $V=\bigwedge^\bullet\Bbbk^n\oplus\Bbbk^{n-1}$, where $\mathrm{SL}_n(\Bbbk)$ acts on $\bigwedge^\bullet\Bbbk^n$ in a natural way, and T

acts on $\bigwedge^k \Bbbk^n$ by the weight $\varepsilon_1 + \cdots + \varepsilon_k$ and on $\Bbbk^{n-1}$ by the weights $\varepsilon_i - \varepsilon_{i+1}$. In other words, the image of $\mathrm{SL}_n(\Bbbk) \times T$ consists of tuples of the form

$$(t_1 g, \dots, t_1 \cdots t_k \textstyle\bigwedge^k g, \dots, t_1/t_2, \dots, t_{n-1}/t_n),$$

where $g \in \mathrm{SL}_n(\Bbbk)$, $t = \mathrm{diag}(t_1, \dots, t_n) \in T$ $(t_1 \cdots t_n = 1)$. It follows that $\mathrm{Env}\,\mathrm{SL}_2(\Bbbk) = \{(a,z) \mid a \in \mathrm{L}_2(\Bbbk),\ z \in \Bbbk,\ \det a = z\} \simeq \mathrm{L}_2(\Bbbk)$ and $\mathrm{As}\,\mathrm{SL}_2(\Bbbk)$ is the subsemigroup of degenerate matrices. Under the identification $\bigwedge^2 \Bbbk^3 \simeq (\Bbbk^3)^* \simeq \Bbbk^3$,

$$\mathrm{Env}\,\mathrm{SL}_3(\Bbbk) = \{(a_1, a_2, z_1, z_2) \mid a_i \in \mathrm{L}_3(\Bbbk),\ z_i \in \Bbbk, \\ \textstyle\bigwedge^2 a_i = z_i a_j\ (i \neq j),\ a_1^\top a_2 = a_1 a_2^\top = z_1 z_2 e\}$$

and $\mathrm{As}\,\mathrm{SL}_3(\Bbbk) = \{(a_1, a_2) \mid a_i \in \mathrm{L}_3(\Bbbk),\ \mathrm{rk}\, a_i \leq 1,\ a_1^\top a_2 = a_1 a_2^\top = 0\}$.

28 S-varieties

28.1 General S-varieties. Horospherical varieties of complexity 0 form another class of spherical varieties whose structure and embedding theory is understood better than in the general case.

Definition 28.1. An *S-variety* is an equivariant embedding of a horospherical homogeneous space $O = G/S$.

This terminology is due to Popov and Vinberg [VP], though they considered only the affine case. General S-varieties were studied by Pauer [Pau1], [Pau2] in the case where S is a maximal unipotent subgroup of G.

S-varieties are spherical. We shall examine them from the viewpoint of the Luna–Vust theory. In order to apply it, we have to describe the colored space $\mathscr{E} = \mathscr{E}(O)$.

It is convenient to assume that $S \supseteq U^-$; then $S = P_{\mathrm{u}}^- \rtimes L_0$ for a certain parabolic $P \supseteq B$ with the Levi subgroup $L \supseteq L_0 \supseteq L'$ and the unipotent radical P_{u} (Lemma 7.4). We may assume that $L \supseteq T$. Put $T_0 = T \cap L_0$.

We have $\Lambda(O) = \mathfrak{X}(A)$, where $A = P^-/S \simeq L/L_0 \simeq T/T_0$. By Theorem 21.10, $\mathscr{V}(O) = \mathscr{E}$. The space $\mathscr{E} = \mathfrak{X}^*(A) \otimes \mathbb{Q}$ may be identified with the orthocomplement of $\mathfrak{X}^*(T_0) \otimes \mathbb{Q}$ in $\mathfrak{X}^*(T) \otimes \mathbb{Q}$. It follows from the Bruhat decomposition that the colors on O are of the form $D_\alpha = \overline{Br_\alpha o}$, $\alpha \in \Pi \setminus \Pi_0$, where $\Pi_0 \subseteq \Pi$ is the simple root set of L. An argument similar to that in 27.2 shows that D_α maps to $\overline{\alpha^\vee}$, the image of $\alpha^\vee$ under the projection $\mathfrak{X}^*(T) \to \mathfrak{X}^*(A)$.

Theorem 15.4(3) says that normal S-varieties are classified by colored fans in $\mathscr{E}$, each fan consisting of finitely many colored cones $(\mathscr{C}_i, \mathscr{R}_i)$, so that the cones $\mathscr{C}_i$ form a polyhedral fan in $\mathscr{E}$, $\mathscr{R}_i \subseteq \mathscr{D}^B$, and each $\mathscr{C}_i \setminus \{0\}$ contains all $\overline{\alpha^\vee}$ such that $D_\alpha \in \mathscr{R}_i$. The colored cones in a fan correspond to the G-orbits Y_i in the respective S-variety X, and X is covered by simple open S-subvarieties $X_i = \{x \in X \mid \overline{Gx} \supseteq Y_i\}$.

The following result "globalizing" Theorem 15.17 is a nice example of how the combinatorial embedding theory of §15 helps to clarify the geometric structure of

S-varieties. For any G-orbit $Y \subseteq X$ let $P(Y) = P[\mathcal{D}^B \setminus \mathcal{D}^B_Y]$ be the normalizer of the open B-orbit in Y and let $S(Y) \subseteq P(Y)$ be the normalizer of general U-orbits, so that $S(Y)^-$ is the stabilizer of $G : Y$ (see 7.2). The Levi subgroup $L(Y) \subseteq P(Y)$ containing T has the simple root set $\Pi_0 \cup \{\alpha \in \Pi \mid D_\alpha \in \mathcal{D}^B_Y\}$, and $S(Y) = P(Y)_{\mathrm{u}} \leftthreetimes L(Y)_0$, where the Levi subgroup $L(Y)_0$ is intermediate between $L(Y)$ and $L(Y)'$ and is in fact the common kernel of all characters in $\Lambda(Y) = \mathfrak{X}(A) \cap \mathscr{C}_Y^\perp$.

Theorem 28.2 (cf. [Paul, 5.4]). *Let X be a simple normal S-variety with the unique closed G-orbit $Y \subseteq X$.*

(1) *There exists a $P(Y)^-$-stable affine closed subvariety $Z \subseteq X$ such that $P(Y)^-_{\mathrm{u}}$ acts on Z trivially and $X \simeq G *_{P(Y)^-} Z$.*
(2) *There exists an $S(Y)^-$-stable closed subvariety $Z_0 \subseteq Z$ with a fixed point such that $Z \simeq P(Y)^- *_{S(Y)^-} Z_0 \simeq L(Y) *_{L(Y)_0} Z_0$ and $X \simeq G *_{S(Y)^-} Z_0$.*
(3) *The varieties Z and Z_0 are equivariant affine embeddings of $L(Y)/L(Y) \cap S$ and $L(Y)_0/L(Y)_0 \cap S$ whose weight lattices are $\mathfrak{X}(A)$ and $\mathfrak{X}(A)/\mathfrak{X}(A) \cap \mathscr{C}_Y^\perp$, colored spaces are $\mathscr{E}$ and $\mathscr{E}_0 := \langle \mathscr{C}_Y \rangle$, and colored cones coincide with $(\mathscr{C}_Y, \mathcal{D}^B_Y)$.*

Proof. The idea of the proof is to construct normal affine S-varieties Z and Z_0 with the colored data as in (3) and then to verify that the colored data of $L(Y) *_{L(Y)_0} Z_0$ coincide with those of Z and the colored data of $G *_{P(Y)^-} Z$ with those of X. In each case both varieties under consideration are simple normal embeddings of one and the same homogeneous space. The restriction of B-eigenfunctions to the fiber of each homogeneous bundle above preserves the orders along B-stable divisors. It follows that the colored cones of both varieties coincide with the colored cone of the fiber, whence the varieties are isomorphic. Note that Z_0 contains a fixed point since it is determined by a colored cone of full dimension. □

28.2 Affine Case. The theorem shows that the local geometry of (normal) S-varieties is completely reduced to the affine case (even to affine S-varieties with a fixed point). Affine S-varieties were studied in [VP] in characteristic 0 and in [Gr2, §17] in arbitrary characteristic.

First note that O is quasiaffine if and only if all $\overline{\alpha^\vee}$ are nonzero (whenever $\alpha \in \Pi \setminus \Pi_0$) and generate a strictly convex cone in $\mathscr{E}$ (Corollary 15.6). This holds if and only if there exists a dominant weight λ such that $\langle \lambda, \Pi^\vee \setminus \Pi_0^\vee \rangle > 0$ and $\lambda|_{T_0} = 1$, i.e, if and only if S is regularly embedded in the stabilizer of a lowest vector of weight $-\lambda$ (cf. Theorem 3.13).

Theorem 28.3. *Let X be the normal affine S-variety determined by a colored cone $(\mathscr{C}, \mathcal{D}^B)$. Then*

$$\Bbbk[X] \simeq \bigoplus_{\lambda \in \mathfrak{X}(A) \cap \mathscr{C}^\vee} V^*(\lambda^*) \subseteq \Bbbk[G/S] = \bigoplus_{\lambda \in \mathfrak{X}(A) \cap \mathbf{C}} V^*(\lambda^*).$$

Here $V^*(\lambda^*)$ is identified with $\Bbbk[G]^{(B^-)}_{-\lambda}$ via right translation by w_G on G. If the semigroup $\mathfrak{X}(A) \cap \mathscr{C}^\vee$ is generated by dominant weights $\lambda_1, \dots, \lambda_m$, then $X \simeq \overline{Gv} \subseteq V(\lambda_1^*) \oplus \cdots \oplus V(\lambda_m^*)$, where $v = v_{-\lambda_1} + \cdots + v_{-\lambda_m}$ is the sum of respective lowest vectors.

Proof. Observe that $R = \bigoplus_{\lambda \in \mathfrak{X}(A) \cap \mathscr{C}^\vee} V^*(\lambda^*)$ is the largest subalgebra of $\Bbbk[G/S]$ with the given algebra of U-invariants $R^U = \Bbbk[X]^U \simeq \Bbbk[\mathfrak{X}(A) \cap \mathscr{C}^\vee]$. Hence $R \supseteq \Bbbk[X] \supseteq \langle G \cdot R^U \rangle$, and the extension is integral by Lemma D.4. Now $R = \Bbbk[X]$ since $\Bbbk[X]$ is integrally closed.

It is easy to see that $\overline{Gv}$ is an affine embedding of O such that $\Bbbk[\overline{Gv}]$ is generated by $V^*(\lambda_1^*) \oplus \cdots \oplus V^*(\lambda_m^*) \subset \Bbbk[O]$. By Lemma 2.23, $\Bbbk[\overline{Gv}] = R$. □

Every (even non-normal) affine S-variety with the open orbit O is realized in a G-module V as $X = \overline{Gv}$, $v \in V^S$. We may assume that $V = \langle Gv \rangle$ and decompose $v = v_{-\lambda_1} + \cdots + v_{-\lambda_m}$, where $v_{-\lambda_i}$ are B^--eigenvectors of certain antidominant weights $-\lambda_i$.

In characteristic zero, $V \simeq V(\lambda_1^*) \oplus \cdots \oplus V(\lambda_m^*)$ and the same arguments as in the proof of Theorem 28.3 show that $\Bbbk[X] = \bigoplus_{\lambda \in \Sigma} V^*(\lambda^*)$, where Σ is the semigroup generated by $\lambda_1, \dots, \lambda_m$, and the dual Weyl modules $V^*(\lambda^*) \simeq V(\lambda)$ are the (simple) G-isotypic components of $\Bbbk[X]$. It is easy to see that $Gv \simeq O$ if and only if $\lambda_1, \dots, \lambda_m$ span $\mathfrak{X}(A)$. Thus we obtain the following

Proposition 28.4 ([VP, 3.1, 3.4]). *In characteristic zero, affine S-varieties X with the open orbit O bijectively correspond to finitely generated semigroups Σ of dominant weights spanning $\mathfrak{X}(A)$, via $\Sigma = \Lambda_+(X)$. The variety X is normal if and only if the semigroup Σ is saturated, i.e., $\Sigma = \mathbb{Q}_+\Sigma \cap \mathfrak{X}(A)$. Moreover, the saturation $\widetilde{\Sigma} = \mathbb{Q}_+\Sigma \cap \mathfrak{X}(A)$ of Σ corresponds to the normalization $\widetilde{X}$ of X.*

G-orbits in an affine S-variety $X = \overline{Gv} \subseteq V$ have a transparent description "dual" to that in Theorem 15.4(4).

Proposition 28.5 ([VP, Th. 8]). *The orbits in X are in bijection with the faces of $\mathscr{C}^\vee = \mathbb{Q}_+\lambda_1 + \cdots + \mathbb{Q}_+\lambda_m$. The orbit corresponding to a face $\mathscr{F}$ is represented by $v_{\mathscr{F}} = \sum_{\lambda_i \in \mathscr{F}} v_{-\lambda_i}$. The adherence of orbits agrees with the inclusion of faces.*

Proof. We have $X = G\overline{Tv}$ since $\overline{Tv}$ is B^--stable (Proposition 2.7). The T-orbits in $\overline{Tv}$ are represented by $v_{\mathscr{F}}$ over all faces $\mathscr{F} \subseteq \mathscr{C}^\vee$, and the adherence of orbits agrees with the inclusion of faces. On the other hand, it is easy to see that the U^--fixed point set in each G-orbit of X is a T-orbit. Hence distinct $v_{\mathscr{F}}$ represent distinct G-orbits. □

In characteristic zero, one can describe the defining equations of X in V. Let $c = \sum \xi_i \xi_i^* \in U\mathfrak{g}$ be the Casimir element with respect to a G-invariant inner product on $\mathfrak{g}$, ξ_i and ξ_i^* being mutually dual bases. It is well known that c acts on $V(\lambda^*)$ by a scalar $c(\lambda) = (\lambda + 2\rho, \lambda)$. Note that $c(\lambda)$ depends on λ monotonously with respect to the partial order induced by positive roots: if $\lambda = \mu + \sum k_i \alpha_i$, $k_i \geq 0$, then $c(\lambda) = c(\mu) + \sum k_i((\lambda + 2\rho, \alpha_i) + (\alpha_i, \mu)) \geq c(\mu)$, and the inequality is strict, except for $\lambda = \mu$. The following result is due to Kostant:

Proposition 28.6 ([LT]). *If $\operatorname{char} \Bbbk = 0$ and $\lambda_1, \dots, \lambda_m$ are linearly independent, then $\mathscr{I}(X) \lhd \Bbbk[V]$ is generated by the relations*

$$c(x_i \otimes x_j) = (\lambda_i + \lambda_j + 2\rho, \lambda_i + \lambda_j)(x_i \otimes x_j), \qquad i, j = 1, \dots, m,$$

where x_k denotes the projection of $x \in V$ to $V(\lambda_k^*)$.

Remark 28.7. There is a characteristic-free version of this result, due to Kempf and Ramanathan [KeR], asserting that $\mathscr{I}(X)$ is generated by quadratic relations, see also [BKu, Ex. 3.5.E(1)].

Proof. The algebra $\Bbbk[V] = \bigoplus_{k_1,\dots,k_m} \mathrm{S}^{k_1}V(\lambda_1) \otimes \cdots \otimes \mathrm{S}^{k_m}V(\lambda_m)$ is multigraded and $\mathscr{I}(X)$ is a multihomogeneous ideal. The structure of $\Bbbk[X]$ implies that each homogeneous component $\mathscr{I}(X)_{k_1,\dots,k_m}$ is the kernel of the natural map $\mathrm{S}^{k_1}V(\lambda_1) \otimes \cdots \otimes \mathrm{S}^{k_m}V(\lambda_m) \to V(k_1\lambda_1 + \cdots + k_m\lambda_m)$.

Consider a series of linear endomorphisms $\pi = c - c(\sum k_i\lambda_i)\mathbf{1}$ of the subspaces $\mathrm{S}^{k_1,\dots,k_m}V = \mathrm{S}^{k_1}V(\lambda_1^*) \otimes \cdots \otimes \mathrm{S}^{k_m}V(\lambda_m^*) \subset \mathrm{S}^\bullet V$. Note that $\operatorname{Ker}\pi \simeq V(\sum k_i\lambda_i^*)$ is the highest irreducible component of $\mathrm{S}^{k_1,\dots,k_m}V$, annihilated by $\mathscr{I}(X)_{k_1,\dots,k_m}$, and $\operatorname{Im}\pi \simeq \mathscr{I}(X)^*_{k_1,\dots,k_m}$ is the complementary G-module.

It follows that $\mathscr{I}(X)$ is spanned by the coordinate functions of all $\pi(x_1^{k_1}\cdots x_m^{k_m})$. An easy calculation shows that

$$\pi(x_1^{k_1}\cdots x_m^{k_m}) = \sum_i \frac{k_i(k_i-1)}{2}\pi(x_i^2)x_1^{k_1}\cdots x_i^{k_i-2}\cdots x_m^{k_m} + \sum_{i<j} k_ik_j\pi(x_ix_j)x_1^{k_1}\cdots x_i^{k_i-1}\cdots x_j^{k_j-1}\cdots x_m^{k_m}.$$

Thus $\mathscr{I}(X)$ is generated by the relations $\pi(x_ix_j) = 0$, $i,j = 1,\dots,m$. □

If the generators of $\Lambda_+(X)$ are not linearly independent, one has to extend the defining equations of X by those arising from the linear dependencies between the λ_i, see [Sm-E].

Proposition 17.1 allows the divisor class group of a normal affine S-variety X to be computed. Every Weil divisor is rationally equivalent to a B-stable one $\delta = \sum m_\alpha D_\alpha + \sum m_i Y_i$, where Y_i are the G-stable prime divisors corresponding to the generators v_i of the rays of $\mathscr{C}$ containing no colors. The divisor δ is principal if and only if $m_\alpha = \langle\lambda, \alpha^\vee\rangle$ and $m_i = \langle\lambda, v_i\rangle$ for a certain $\lambda \in \mathfrak{X}(A)$. This yields a finite presentation for $\operatorname{Cl}X$. In particular, we have

Proposition 28.8. *A normal affine S-variety X is factorial if and only if $\Lambda_+(X)$ is generated by weights $\lambda_1,\dots,\lambda_s,\pm\lambda_{s+1},\dots,\pm\lambda_r$ ($s \le r$), where the λ_i are linearly independent and the projection $\mathfrak{X}(T) \to \mathfrak{X}(T\cap G')$ maps them to distinct fundamental weights or to* 0.

For semisimple G, we conclude that factorial S-varieties are those corresponding to weight semigroups Σ generated by some of the fundamental weights [VP, Th. 11].

The simplest class of affine S-varieties is formed by *HV-varieties*, i.e., cones of highest (or lowest) vectors $X = \overline{Gv_{-\lambda}}$, $v_{-\lambda} \in V(\lambda^*)^{(B^-)}$, see 11.1. Particular examples are quadratic cones or Grassmann cones of decomposable polyvectors. The above results on affine S-varieties imply Proposition 11.2, which describes basic properties of HV-varieties. It follows from Proposition 28.6 that an HV-cone is defined by quadratic equations in the ambient simple G-module. For a Grassmann cone we recover the Plücker relations between the coordinates of a polyvector.

28.3 Smoothness. Now we describe smooth S-varieties in characteristic zero. By Theorem 28.2, the problem is reduced to affine S-varieties with a fixed point, which are nothing else but G-modules with a dense orbit of a U-fixed vector.

Lemma 28.9. *If a G-module V is an S-variety, then $V = V_0 \oplus V_1 \oplus \dots \oplus V_s$ so that $Z = Z(G)^0$ acts on V_0 with linearly independent weights of multiplicity 1 and each V_i ($i > 0$) is a simple submodule acted on non-trivially by a unique simple factor $G_i \subseteq G$, $G_i \simeq \mathrm{SL}(V_i)$ or $\mathrm{Sp}(V_i)$.*

Proof. Since Z has a dense orbit in $V_0 = V^{G'}$, it acts with linearly independent weights of multiplicity 1. If G_i acts non-trivially on two simple submodules V_i, V_j, and $v_i \in V_i^{(B)}$, $v_j \in V_j^{(B^-)}$, then the stabilizer of $v_i + v_j$ is not horospherical, i.e., V is not an S-variety. Therefore we may assume that V is irreducible and each simple factor of G acts non-trivially.

Then G acts transitively on $\mathbb{P}(V)$, which implies that $G' \simeq \mathrm{SL}(V)$ or $\mathrm{Sp}(V)$ [Oni2]. Indeed, we have $V = \mathfrak{b}v_{-\lambda}$, where $v_{-\lambda} \in V$ is a lowest vector. Hence there exists a unique root δ such that $e_\delta v_{-\lambda} = v_{\lambda^*}$ is a highest vector. One easily deduces that the root system of G is indecomposable and δ is the highest root, so that $\delta = \lambda + \lambda^*$ is the sum of two dominant weights, whence the assertion. □

The colored data of such a G-module V are easy to write down. Namely $\Pi \setminus \Pi_0 = \{\alpha_1, \dots, \alpha_s\}$, where α_i are the first simple roots in some components of Π having the type $\mathbf{A}_l$ or $\mathbf{C}_l$. The weight lattice $\mathfrak{X}(A)$ is spanned by linearly independent weights $\lambda_1, \dots, \lambda_r$, where $\lambda_1, \dots, \lambda_s$ are the highest weights of V_i^*, which project to the fundamental weights ω_i corresponding to α_i, and $\lambda_{s+1}, \dots, \lambda_r$ are the weights of V_0^*, which are orthogonal to Π. The cone $\mathscr{C}$ is spanned by the basis $\overline{\alpha_1^\vee}, \dots, \overline{\alpha_s^\vee}, v_{s+1}, \dots, v_r$ of $\mathfrak{X}^*(A)$ dual to $\lambda_1, \dots, \lambda_r$. Using Theorem 28.2 we derive the description of colored data of arbitrary smooth S-varieties:

Theorem 28.10 (cf. [Pau2, 3.5]). *An S-variety X is smooth if and only if all colored cones $(\mathscr{C}_Y, \mathscr{D}_Y^B)$ in the colored fan of X satisfy the following properties:*

(1) *$\mathscr{C}_Y$ is generated by a part of a basis of $\mathfrak{X}^*(A)$, and all $\overline{\alpha^\vee}$ such that $D_\alpha \in \mathscr{D}_Y^B$ are among the generators.*
(2) *The simple roots α such that $D_\alpha \in \mathscr{D}_Y^B$ are isolated from each other in the Dynkin diagram of G, and each α is connected with at most one component Π_α of Π_0; moreover, $\{\alpha\} \cup \Pi_\alpha$ has the type $\mathbf{A}_l$ or $\mathbf{C}_l$, α being the first simple root therein.*

The condition (1) is equivalent to the local factoriality of X.

29 Toroidal Embeddings

In this section we assume that $\mathrm{char}\,\Bbbk = 0$.

29.1 Toroidal Versus Toric Varieties. Recall that a G-equivariant normal embedding X of a spherical homogeneous space $O = G/H$ is said to be *toroidal* if $\mathscr{D}_Y^B = \emptyset$ for each G-orbit $Y \subseteq X$. Toroidal embeddings are defined by fans in $\mathscr{V}$, and G-morphisms between them correspond to subdivisions of these fans in the same way as in toric geometry [Ful2]. There is a more direct relation between toroidal and toric varieties. Put $P = P(O)$, with the Levi decomposition $P = P_u \rightthreetimes L$ and other notation from 4.2 and 7.2.

Theorem 29.1 ([BPa, 3.4], [Bri13, 2.4]). *A toroidal embedding $X \hookleftarrow O$ is covered by G-translates of an open P-stable subset*

$$\mathring{X} = X \setminus \bigcup_{D \in \mathscr{D}^B} D \simeq P *_L Z \simeq P_u \times Z,$$

where Z is a locally closed L-stable subvariety pointwise fixed by L_0. The variety Z is a toric embedding of $A = L/L_0$ defined by the same fan as X, and the G-orbits in X intersect Z in A-orbits.

Proof. The problem is easily reduced to the case where X contains a unique closed orbit Y with $\mathscr{C}_Y = \mathscr{V}$, $\mathscr{D}_Y^B = \emptyset$. Such toroidal embeddings, called standard, are discussed in 30.1. Indeed, consider another spherical homogeneous space $\overline{O} = G/N_G(H)$. Then $\overline{\mathscr{V}} = \mathscr{V}(\overline{O}) = \mathscr{V}/(\mathscr{V} \cap -\mathscr{V})$ is strictly convex, whence there exists a standard embedding $\overline{X} \hookleftarrow \overline{O}$. The canonical map $\varphi : O \to \overline{O}$ extends to $X \to \overline{X}$ by Theorem 15.10. We have $P(\overline{O}) = P$, and $\mathring{X}, Z$ are the preimages of the respective subvarieties defined for $\overline{X}$.

For standard X one applies the local structure theorem in a neighborhood of Y: by Theorem 15.17, $\mathring{X} = \mathring{X}_Y \simeq P *_L Z$, where Z is toric since $Z \cap O$ is a single A-orbit.

The G-stable divisors in X intersect $\mathring{X}$ in the P-stable divisors and Z in the A-stable divisors. Each G-(A-)orbit in X (in Z) meets $\mathring{X}$ and is an intersection of G-(A-)stable divisors, whence the assertion on orbits. The assertion on fans is easy, cf. Remark 15.19. □

It follows that toroidal varieties are locally toric. They inherit many nice geometric properties from toric varieties. On the other hand, each spherical variety is the image of a toroidal one by a proper birational equivariant map: to obtain this toroidal covering variety, just remove all colors from the fan. This universality of toroidal varieties can be used to derive some properties of spherical varieties from the toroidal case.

29.2 Smooth Toroidal Varieties. A toroidal variety is smooth if and only if all cones of its fan are simplicial and generated by a part of a basis of $\Lambda(O)^*$: for toric varieties this is deduced from the description of the coordinate algebra [Ful2, 2.1] (cf. Example 15.8) and the general case follows by Theorem 29.1. For a singular toroidal variety one may construct an equivariant desingularization by subdividing its fan, cf. [Ful2, 2.6].

Every (smooth) toroidal variety admits an equivariant (smooth) completion, which is defined by adding new cones to the fan in order to cover all of $\mathscr{V}(O)$. Smooth complete toroidal varieties have other interesting characterizations.

Theorem 29.2 ([BiB]). *For a smooth G-variety X consider the following conditions:*

(1) *X is toroidal.*
(2) *There is a dense open orbit $O \subseteq X$ such that $\partial X = X \setminus O$ is a divisor with normal crossings, each orbit $Gx \subset X$ is locally the intersection of several components of ∂X, and G_x has a dense orbit in $T_x X/\mathfrak{g}x$.*
(3) *There is a G-stable divisor $D \subset X$ with normal crossings such that $\mathscr{G}_X = \mathscr{T}_X(-\log D)$.*
(4) *X is spherical and pseudo-free.*

Then (4) $\Longrightarrow$ (1) $\Longrightarrow$ (2) $\Longleftrightarrow$ (3)*. If X is complete or spherical, then all conditions are equivalent.*

G-varieties satisfying the condition (2), resp. (3), are known as *regular* in the sense of Bifet–de Concini–Procesi [BCP], resp. of Ginzburg [Gin].

Proof. (1) $\Longrightarrow$ (2)&(3) Theorem 29.1 reduces the problem to smooth toric varieties. The latter are covered by invariant affine open charts of the form $X = \mathbb{A}^m \times (\mathbb{A}^1 \setminus 0)^{n-m}$, where $(\Bbbk^\times)^n$ acts in the natural way, so that $D = \partial X$ is the union of coordinate hyperplanes $\{x_i = 0\}$, X is isomorphic to the normal bundle of the closed orbit, and $\mathscr{T}_X(-\log D)$ is a free sheaf spanned by velocity fields $x_1\partial_1, \dots, x_n\partial_n$ ($\partial_i := \partial/\partial x_i$).
(2) $\Longleftrightarrow$ (3) First observe that $O = X \setminus D$ is a single G-orbit if and only if $\mathscr{G}_{X\setminus D} = \mathscr{T}_{X\setminus D}$. Now consider a neighborhood of any $x \in D$. Due to local nature of the conditions (2) and (3), we may assume that all components $D_1, \dots, D_k$ of D contain x. Choose local parameters $x_1, \dots, x_n$ at x such that D_i are locally defined by the equations $x_i = 0$. Let $\partial_1, \dots, \partial_n$ denote the vector fields dual to $\mathrm{d}x_1, \dots, \mathrm{d}x_n$. Then $\mathscr{T}_X(-\log D)$ is locally generated by $x_1\partial_1, \dots, x_k\partial_k, \partial_{k+1}, \dots, \partial_n$.

Let $Y = D_1 \cap \dots \cap D_k$ and $\pi : N = \operatorname{Spec} \mathrm{S}^\bullet(\mathscr{I}_Y/\mathscr{I}_Y^2) \to Y$ be the normal bundle. There is a natural embedding $\pi^*\mathscr{T}_X(-\log D)|_Y \hookrightarrow \mathscr{T}_N$: each vector field in $\mathscr{T}_X(-\log D)$ preserves $\mathscr{I}_Y$, whence induces a derivation of $\mathrm{S}^\bullet(\mathscr{I}_Y/\mathscr{I}_Y^2)$. The image of $\pi^*\mathscr{T}_X(-\log D)|_Y$ is $\mathscr{T}_N(-\log \bigcup N_i)$, where N_i are the normal bundles to Y in D_i. Indeed, $\bar{x}_i = x_i \bmod \mathscr{I}_Y^2$ ($i \le k$), $\bar{x}_j = \pi^*x_j|_Y$ ($j > k$) are local parameters on N and $x_i\partial_i, \partial_j$ induce the derivations $\bar{x}_i\bar{\partial}_i, \bar{\partial}_j$. Note that $N = \bigoplus L_i$, where $L_i = \bigcap_{j\neq i} N_j$ are G-stable line subbundles. Hence the G_x-action on $T_x X/T_x Y = N(x) = \bigoplus L_i(x)$ is diagonalizable.

Condition (2) implies that Gx is open in Y and the weights of $G_x : L_i(x)$ are linearly independent. This yields velocity fields $\bar{x}_i\bar{\partial}_i$ on $N(x)$ and in transversal directions, which locally generate $\mathscr{T}_N(-\log \bigcup N_i)$. Therefore $\mathscr{T}_X(-\log D)|_Y$ is generated by velocity fields. By Nakayama's lemma, $\mathscr{T}_X(-\log D) = \mathscr{G}_X$ in a neighborhood of x.

Conversely, (3) implies that $\mathscr{G}_Y = \mathscr{T}_Y$ and $\mathscr{G}_N = \mathscr{T}_N(-\log \bigcup N_i)$. Hence Gx is open in Y and $N|_{Gx} = G *_{G_x} T_x X/\mathfrak{g}x$ has an open G-orbit. Thus $T_x X/\mathfrak{g}x$ contains an open G_x-orbit.
(2)&(3) $\Longrightarrow$ (4) Since $\mathscr{T}_X(-\log \partial X)$ is locally free, the implication is trivial provided that X is spherical. It remains to prove that X is spherical if it is complete.

A closed orbit $Y \subseteq X$ intersects a B-chart $\mathring{X} \simeq P *_L Z$, where $L \subseteq P = P(Y)$ is the Levi subgroup and Z is an L-stable affine subvariety intersecting Y in a single point z. Since the maximal torus $T \subseteq L \subseteq G_z = P^-$ acts on $T_z Z \simeq T_z X/\mathfrak{g}z$ with linearly independent weights, $Z \simeq T_z Z$ contains an open T-orbit, whence $\mathring{X}$ has an open B-orbit.
(4) $\Longrightarrow$ (1) There is a morphism $X \to \mathrm{Gr}_k(\mathfrak{g})$, $x \mapsto [\mathfrak{h}_x]$, extending the map $x \mapsto [\mathfrak{g}_x]$ on O. If X is not toroidal, then there exists a G-orbit $Y \subset X$ contained in a color $D \subset X$. Then we have $\mathfrak{b} + \mathfrak{h}_{gy} = \mathfrak{b} + (\mathrm{Ad}\, g)\mathfrak{h}_y \neq \mathfrak{g}$, $\forall y \in Y$, $g \in G$, i.e., $\mathfrak{h}_y$ is not spherical.

To obtain a contradiction, it suffices to prove that all $\mathfrak{h}_x$ are spherical subalgebras. Passing to a toroidal variety mapping onto X, one may assume that X itself is toroidal. Consider the normal bundle N to $Y = Gx$. Since $\mathscr{G}_N = \pi^* \mathscr{G}_X|_Y$, $\mathfrak{h}_x$ is the stabilizer subalgebra of general position for $G : N$. But N is spherical, because the minimal B-chart $\mathring{X}$ of Y is P-isomorphic to $N|_{Y \cap \mathring{X}}$. □

29.3 Cohomology Vanishing. Toric varieties and generalized flag varieties form two "extreme" classes of toroidal varieties. A number of geometric and cohomological results generalize from these particular cases to general toroidal varieties. A powerful vanishing theorem was proved by Bien and Brion (1991) under some restrictions and refined by Knop (1992).

Theorem 29.3. *If X is a smooth complete toroidal variety, then*

$$\mathrm{H}^i(X, \mathrm{S}^\bullet \mathscr{T}_X(-\log \partial X)) = 0, \quad \forall i > 0.$$

For flag varieties, this result is due to Elkik (vanishing of higher cohomology of the tangent sheaf was proved already by Bott in 1957). In fact, Bien and Brion proved a twisted version of Theorem 29.3 [BiB, 3.2]: $\mathrm{H}^i(X, \mathscr{L} \otimes \mathrm{S}^\bullet \mathscr{T}_X(-\log \partial X)) = 0$ for all $i > 0$ and any globally generated line bundle $\mathscr{L}$ on X, under a technical condition that the stabilizer H of O is parabolic in a reductive subgroup of G. (Generally, higher cohomology of globally generated line bundles vanishes on every complete spherical variety, see Corollary 31.7.)

In view of Theorem 29.2, Theorem 29.3 stems from a more general vanishing result of Knop:

Theorem 29.4 ([Kn6, 4.1]). *If X is a pseudo-free smooth equivariant completion of a homogeneous space O, then* $\mathrm{H}^i(X, \mathscr{U}_X^{(m)}) = \mathrm{H}^i(X, \mathrm{S}^m \mathscr{G}_X) = 0$, $\forall i > 0$, $m \geq 0$.

Synopsis of a proof. The assertions on $\mathscr{U}_X$ are reduced to those on $\mathrm{S}^\bullet \mathscr{G}_X = \mathrm{gr}\, \mathscr{U}_X$. Since $\pi_X : T^{\mathfrak{g}}X \to X$ is an affine morphism, the Leray spectral sequence reduces the question to proving $\mathrm{H}^i(T^{\mathfrak{g}}X, \mathscr{O}_{T^{\mathfrak{g}}X}) = 0$. The localized moment map $\overline{\Phi} : T^{\mathfrak{g}}X \to M_X$ factors through $\widetilde{\Phi} : T^{\mathfrak{g}}X \to \widetilde{M}_X$. As $\widetilde{M}_X$ is affine, $\mathrm{H}^i(T^{\mathfrak{g}}X, \mathscr{O}_{T^{\mathfrak{g}}X}) = \mathrm{H}^0(\widetilde{M}_X, \mathrm{R}^i \widetilde{\Phi}_* \mathscr{O}_{T^{\mathfrak{g}}X})$, and it remains to prove that $\mathrm{R}^i \widetilde{\Phi}_* \mathscr{O}_{T^{\mathfrak{g}}X} = 0$. Here one applies to $\widetilde{\Phi}$ a version of Kollár's vanishing theorem [Kn6, 4.2]:

> If Y is smooth, Z has rational singularities, and $\varphi : Y \to Z$ is a proper morphism with connected general fibers F, which satisfy $\mathrm{H}^i(F, \mathscr{O}_F) = 0$, $\forall i > 0$, then $\mathrm{R}^i \varphi_* \mathscr{O}_Y = 0$ for all $i > 0$.

It remains to verify the conditions. The morphism $\widetilde{\Phi}$ is proper by Example 8.2. The variety $\widetilde{M}_X$ has rational singularities by [Kn6, 4.3]. To show vanishing of the higher cohomology of $\mathscr{O}_F$, it suffices to prove that F is unirational [Se2]. Here one may assume that $X = O$, $\widetilde{\Phi} : T^*O \to \widetilde{M}_O$. Unirationality of the fibers of the moment map is the heart of the proof [Kn6, §5]. □

There is a relative version of Theorem 29.4 asserting that $\mathrm{R}^i\psi_*\mathscr{U}_X^{(m)} = \mathrm{R}^i\psi_*\mathrm{S}^m\mathscr{G}_X = 0$, $\forall i > 0$, for a proper G-invariant morphism $\psi : X \to Y$ separating general orbits, where X is smooth pseudo-free and Y has rational singularities.

29.4 Rigidity. The vanishing theorems of Bien–Brion and Knop have a number of important consequences. For instance, on a pseudo-free smooth completion X of O the symbol map $\mathrm{gr}\,\mathrm{H}^0(X, \mathscr{U}_X) \to \Bbbk[T^{\mathfrak{g}}X]$ is surjective. In the toroidal case, $\mathrm{H}^1(X, \mathscr{T}_X(-\log \partial X)) = 0$ implies that the pair $(X, \partial X)$ is locally rigid, by the deformation theory of Kodaira–Spencer [Ser]. Using this observation, Alexeev and Brion proved Luna's conjecture on rigidity of spherical subgroups.

Theorem 29.5 ([AB3, §3]). *For any (irreducible) G-variety with spherical (general) orbits, the stabilizers of points in general position are conjugate.*

Proof. Let $\mathscr{X}$ be a G-variety with spherical orbits. Passing to an open subset, we may assume that $\mathscr{X}$ is smooth quasiprojective and there exists a smooth G-invariant morphism $\pi : \mathscr{X} \to Z$ whose fibers contain dense orbits. Regarding $\mathscr{X}$ as a family of spherical G-orbit closures, we may replace $\mathscr{X}$ by a birationally isomorphic family of smooth projective toroidal varieties.

Indeed, there is a locally closed G-embedding of $\mathscr{X}$ into $\mathbb{P}(V)$, and therefore into $\mathbb{P}(V) \times Z$, for some G-module V. Replacing $\mathscr{X}$ by its closure and taking a pseudo-free desingularization, we may assume that $\mathscr{X}$ is pseudo-free and π is a projective morphism. By Theorem 29.2, the fibers of π are smooth projective toroidal varieties. Shrinking Z if necessary, we obtain that the G-orbits of non-maximal dimension in $\mathscr{X}$ form a divisor with normal crossings $\partial\mathscr{X} = \mathscr{D}_1 \cup \dots \cup \mathscr{D}_k$ whose components $\mathscr{D}_i$ are smooth over Z.

Morally, an equivariant version of Kodaira–Spencer theory should imply that all fibers of π are G-isomorphic, which should complete the proof. An alternative argument uses nested Hilbert schemes (see Appendix E.2).

Let X be any fiber of π, with $\partial X = D_1 \cup \dots \cup D_k$, $D_i = \mathscr{D}_i \cap X$. Applying a suitable Veronese map, we satisfy a technical condition that the restriction map $V^* \to \mathrm{H}^0(X, \mathscr{O}(1))$ is surjective.

The nested Hilbert scheme Hilb parameterizes tuples $(Y, Y_1, \dots, Y_k)$ of projective subvarieties $Y_i \subseteq Y \subseteq \mathbb{P}(V)$ having the same Hilbert polynomials as $X, D_1, \dots, D_k$. The varieties $\mathscr{X}, \mathscr{D}_1, \dots, \mathscr{D}_k$ are obtained as the pullbacks under $Z \to \mathrm{Hilb}$ of the universal families $\mathscr{Y}, \mathscr{Y}_1, \dots, \mathscr{Y}_k \to \mathrm{Hilb}$. The groups $\mathrm{GL}(V)$ and G act on Hilb in a natural way, so that Hilb^G parameterizes tuples of G-subvarieties. Since the centralizer $\mathrm{GL}(V)^G$ of G maps G-subvarieties to G-isomorphic ones, it suffices to prove that the $\mathrm{GL}(V)^G$-orbit of $(X, D_1, \dots, D_k)$ is open in Hilb^G.

This is done by considering tangent spaces. Let $\mathscr{N}_Z, \mathscr{N}_{Z_i/Z}$ denote the normal bundles to Z in $\mathbb{P}(V)$, resp. to Z_i in Z. By Proposition E.7, $T_{(X,D_1,\dots,D_k)}\mathrm{Hilb} =$

$\mathrm{H}^0(X,\mathscr{N})$, where $\mathscr{N} \subset \mathscr{N}_X \oplus \mathscr{N}_{D_1} \oplus \cdots \oplus \mathscr{N}_{D_k}$ is formed by tuples $(\xi,\xi_1,\ldots,\xi_k)$ of normal vector fields such that $\xi|_{D_i} = \xi_i \bmod \mathscr{N}_{D_i/X}$, $i = 1,\ldots,k$. (These vector fields define infinitesimal deformations of $X, D_1,\ldots,D_k$, so that the deformation of D_i is determined by the deformation of X modulo a deformation inside X.) There are exact sequences

$$0 \longrightarrow \mathscr{T}_X(-\log\partial X) \longrightarrow \mathscr{T}_{\mathbb{P}(V)}|_X \longrightarrow \mathscr{N} \longrightarrow 0,$$
$$0 \longrightarrow \mathscr{O}_X \longrightarrow V \otimes \mathscr{O}_X(1) \longrightarrow \mathscr{T}_{\mathbb{P}(V)}|_X \longrightarrow 0.$$

Taking cohomology yields

$$\mathrm{H}^0\left(X,\mathscr{T}_{\mathbb{P}(V)}\right) \longrightarrow T_{(X,D_1,\ldots,D_k)}\mathrm{Hilb} \longrightarrow \mathrm{H}^1(X,\mathscr{T}_X(-\log\partial X)) = 0,$$
$$V \otimes \mathrm{H}^0(X,\mathscr{O}(1)) \longrightarrow \mathrm{H}^0\left(X,\mathscr{T}_{\mathbb{P}(V)}\right) \longrightarrow \mathrm{H}^1(X,\mathscr{O}_X) = 0.$$

(The first cohomologies vanish by Theorem 29.3 and [Se2], since X is a smooth projective rational variety.) Hence the differential of the orbit map

$$\mathfrak{gl}(V) \simeq V \otimes V^* \longrightarrow V \otimes \mathrm{H}^0(X,\mathscr{O}(1)) \longrightarrow \mathrm{H}^0\left(X,\mathscr{T}_{\mathbb{P}(V)}\right) \longrightarrow T_{(X,D_1,\ldots,D_k)}\mathrm{Hilb}$$

is surjective. By linear reductivity of G, the composite map

$$\mathfrak{gl}(V)^G \longrightarrow T_{(X,D_1,\ldots,D_k)}(\mathrm{Hilb}^G) \subseteq \left(T_{(X,D_1,\ldots,D_k)}\mathrm{Hilb}\right)^G$$

is surjective as well. Hence $(X,D_1,\ldots,D_k)$ is a smooth point of Hilb^G and the orbit $\mathrm{GL}(V)^G(X,D_1,\ldots,D_k)$ is open. □

29.5 Chow Rings. Cohomology rings of smooth complete toroidal varieties (over $\Bbbk = \mathbb{C}$) were computed by Bifet–de Concini–Procesi [BCP], see also [LP] for toroidal completions of symmetric spaces. By Corollary 18.4, cohomology coincides with the Chow ring in this situation. The most powerful approach is through equivariant cohomology or the equivariant intersection theory of Edidin–Graham, see [Bri15]. In particular, Chow (or cohomology) rings of smooth (complete) toric varieties and flag varieties are easily computed in this way [BCP, I.4], [Bri11, 2, 3], [Bri15], cf. 18.4.

29.6 Closures of Flats. The local structure of toroidal varieties can be refined in order to obtain a full description for the closures of generic flats.

Proposition 29.6 ([Kn5, 8.3]). *The closure of a generic twisted flat in a toroidal variety X is a normal toric variety whose fan is the W_X-span of the fan of X.*

Proof. It suffices to choose the toric slice Z in Theorem 29.1 in such a way that the open A-orbit in Z is a generic (twisted) flat F_α. Then $\overline{Z} = \overline{F}_\alpha$ (the closure in X), so that Theorem 29.1 and Proposition 23.19 imply the claim.

If X is smooth and T^*X is symplectically stable, then the conormal bundle to general U-orbits extends to a trivial subbundle $\mathring{X} \times \mathfrak{a}^* \hookrightarrow T^*\mathring{X}(\log\partial X)$, the trivializing sections being $\mathbf{d}\mathbf{f}_\lambda/\mathbf{f}_\lambda$, $\lambda \in \Lambda$. The logarithmic moment map restricts to $\Phi : \mathring{X} \times \mathfrak{a}^* \to \mathfrak{a} \oplus \mathfrak{p}_u$, cf. Lemma 23.14. It follows that $\mathring{X} \simeq P *_L Z$, where $Z = \pi_X \Phi^{-1}(\lambda)$, $\lambda \in \mathfrak{a}^{\mathrm{pr}}$, and F_α is the open L-orbit in Z for any $\alpha \in \Phi^{-1}(\lambda) \cap T^*O$.

If X is singular, then it admits a toroidal resolution of singularities $\nu : X' \to X$. Then $\mathring{X}' := \nu^{-1}(\mathring{X}) = X' \setminus \bigcup_{D \in \mathscr{D}^B} D \simeq P *_L Z'$ and $Z' \supseteq F_\alpha$. The map $\Phi : \mathring{X}' \times \mathfrak{a}^* \to \mathfrak{a} \oplus \mathfrak{p}_u$ descends to $\mathring{X}$, because $\Bbbk[\mathring{X}'] = \Bbbk[\mathring{X}]$. Thus one may put $Z = \nu(Z')$.

If T^*X is not symplectically stable, then passing to affine cones and back to projectivizations yields Z such that the open L-orbit in Z is a twisted flat. □

Example 29.7. If X is a toroidal $(G \times G)$-embedding of G, then T is a flat and $F = \overline{T}$ is a toric variety whose fan is the W-span of the fan of X (in the antidominant Weyl chamber), cf. Proposition 27.18. For instance, if $X = \overline{G} \subseteq \mathbb{P}(\mathrm{L}(V))$ for a faithful projective representation $G : \mathbb{P}(V)$ with regular highest weights, then the fan of F is formed by the duals to the barrier cones of the weight polytope $\mathscr{P}(V)$, and the fan of X is its antidominant part (see 27.3).

Example 29.8. Consider the variety of complete conics $X \subset \mathbb{P}(\mathrm{S}^2(\Bbbk^3)^*) \times \mathbb{P}(\mathrm{S}^2\Bbbk^3)$ from Example 17.12. The set $F = \{([q],[q^\vee]) \mid q \text{ diagonal}, \det q \neq 0\}$ is a flat. Using the Segre embedding $\mathbb{P}(\mathrm{S}^2(\Bbbk^3)^*) \times \mathbb{P}(\mathrm{S}^2\Bbbk^3) \hookrightarrow \mathbb{P}(\mathrm{S}^2(\Bbbk^3)^* \otimes \mathrm{S}^2\Bbbk^3)$ and observing that the T-weights occurring in the weight decomposition of $q \otimes q^\vee$ are $2(\varepsilon_i - \varepsilon_j)$, we conclude that the fan of $\overline{F}$ is the set of all Weyl chambers of $G = \mathrm{SL}_3(\Bbbk)$ together with their faces, while the fan of X consists of the antidominant Weyl chamber and its faces.

30 Wonderful Varieties

30.1 Standard Completions. In the study of a homogeneous space O it is useful to consider its equivariant completions. The reason is that properties of O and of related objects (subvarieties and their intersection, functions, line bundles and their sections, etc) often become apparent "at infinity", and equivariant completions of O take into account the points at infinity. Also, complete varieties behave better than non-complete ones from various points of view (e.g., in intersection theory).

Among all equivariant completions of a spherical homogeneous space O one distinguishes two opposite classes. Toroidal completions have nice geometry (see §29) and a universal property: each equivariant completion of O is dominated by a toroidal one. On the other hand, simple completions of O (i.e., those having a unique closed orbit) are the most "economical" ones: their boundaries are "small". Simple completions exist if and only if the valuation cone $\mathscr{V}$ is strictly convex.

These two classes intersect in a unique element, called the standard completion.

Definition 30.1. A spherical subgroup $H \subseteq G$ is called *sober* if $N_G(H)/H$ is finite or, equivalently, if $\mathscr{V}(G/H)$ is strictly convex.

The *standard embedding* of $O = G/H$ is the unique toroidal simple complete G-embedding $X \hookleftarrow O$, defined by the colored cone $(\mathscr{V}, \emptyset)$, provided that H is sober. A smooth standard embedding is called *wonderful*.

The standard embedding has a universal property: for any toroidal completion $X' \hookleftarrow O$ and any simple completion $X'' \hookleftarrow O$, there exist unique proper birational G-morphisms $X' \to X \to X''$ extending the identity map on O.

Wonderful embeddings were first introduced by de Concini and Procesi [CP1] for symmetric spaces. Their remarkable properties were studied by many researchers (see below) mainly in characteristic zero, though some results in special cases, e.g., for symmetric spaces [CS], are obtained in arbitrary characteristic. For simplicity, we assume that $\operatorname{char}\Bbbk = 0$ from now on.

Every spherical subgroup $H \subseteq G$ is contained in the smallest sober overgroup $H \cdot N_G(H)^0$ normalizing H. This stems, e.g., from the following useful lemma.

Lemma 30.2. *If $H \subseteq G$ is a spherical subgroup, then $N_G(H) = N_G(\overline{H})$ for any intermediate subgroup $\overline{H}$ between H and $N_G(H)$.*

Proof. As $N_G(H)/H$ is Abelian, we have $N_G(H) \subseteq N_G(\overline{H})$. In particular, $N_G(H) = N_G(H^0)$. To prove the converse inclusion, we may assume without loss of generality that H is connected and $\mathfrak{b} + \mathfrak{h} = \mathfrak{g}$. Then the right multiplication by $N_G(\overline{H})$ preserves $BH = B\overline{H}$, the unique open $(B \times H)$-orbit in G. Hence the $N_G(\overline{H})$-action on $\Bbbk(G)$ by right translations of an argument preserves $\Bbbk[G]^{(B\times H)}$ (= the set of regular functions on G invertible on BH). Since this action commutes with the G-action by left translations, it preserves $\Bbbk[G]^{(H)}$, whence $\Bbbk(G/H)$, too. Hence $N_G(\overline{H})$ acts on G/H by G-automorphisms, i.e., is contained in $N_G(H)$. □

Now let $H \subseteq G$ be a sober subgroup and X be the standard embedding of $O = G/H$. The local structure theorem reveals the orbit structure and local geometry of X: by Theorem 29.1 there are an affine open chart $\mathring{X} = X \setminus \bigcup_{D\in\mathscr{D}^B} D$ and a closed subvariety $Z \subset \mathring{X}$ such that $\mathring{X}$ is stable under $P = P(O)$, the Levi subgroup $L \subset P$ leaves Z stable and acts on it via the quotient torus $A = L/L_0$, $\mathring{X} \simeq P *_L Z \simeq P_{\mathrm{u}} \times Z$, and each G-orbit of X intersects Z in an A-orbit. Actually $\mathring{X}$ is the unique B-chart of X intersecting all G-orbits.

The affine toric variety Z is defined by the cone $\mathscr{V}$, so that $\Bbbk[Z] = \Bbbk[\mathscr{V}^\vee \cap \Lambda]$, where $\Lambda = \Lambda(O) = \mathfrak{X}(A)$. The orbits (of $A : Z$ or of $G : X$) are in an order-reversing bijection with the faces of $\mathscr{V}$, and each orbit closure is the intersection of invariant divisors containing the orbit. If $\mathscr{V}$ is generated by a basis of Λ^*, then $Z \simeq \mathbb{A}^r$ with the natural action of $A \simeq (\Bbbk^\times)^r$; the eigenweight set for $A : Z$ is $\Pi_O^{\min}$. Generally, since $\mathscr{V}$ is simplicial (Theorem 22.13), one deduces that $Z \simeq \mathbb{A}^r/\Gamma$ with the natural action of $A \simeq (\Bbbk^\times)^r/\Gamma$, where $\Gamma \simeq \Lambda^*/N$ is the common kernel of all $\lambda \in \Lambda$ in $(\Bbbk^\times)^r = N \otimes \Bbbk^\times$, the sublattice $N \subseteq \Lambda^*$ being spanned by the indivisible generators of the rays of $\mathscr{V}$.

In particular, X is smooth if and only if $\mathscr{V}$ is generated by a basis of Λ^*, i.e., if and only if $\Lambda = \mathbb{Z}\Delta_O^{\min}$. It is a delicate problem to characterize the (sober) spherical subgroups $H \subseteq G$ such that the standard embedding $X \hookleftarrow O = G/H$ is smooth.

Note that $N_G(H)/H = \operatorname{Aut}_G O$ acts on a finite set $\mathscr{D}^B$.

Definition 30.3. A spherical subgroup $H \subseteq G$ is called *very sober* if $N_G(H)/H$ acts on $\mathscr{D}^B$ effectively. (In particular, H is sober, because $(N_G(H)/H)^0$ leaves $\mathscr{D}^B$ pointwise fixed.) The *very sober hull* of H is the kernel $\overline{H}$ of $N_G(H) : \mathscr{D}^B$. An alternative terminology is: *spherically closed* subgroup, *spherical closure*.

Remark 30.4. It is easy to deduce from Lemma 30.2 that $\overline{H}$ is the smallest very sober subgroup of G containing H as a normal subgroup. The colored space $\overline{\mathscr{E}} = \mathscr{E}(G/\overline{H})$ is identified with $\mathscr{E}/(\mathscr{V}\cap-\mathscr{V})$, the valuation cone is $\overline{\mathscr{V}} = \mathscr{V}/(\mathscr{V}\cap-\mathscr{V})$, and the set of colors $\overline{\mathscr{D}}^B$ is identified with $\mathscr{D}^B$ via pullback.

Observe that $\overline{H}$ is the kernel of $N_G(H) : \mathfrak{X}(H)$ [Kn8, 7.4]. Indeed, (a multiple of) each B-stable divisor δ on O is defined by an equation $\eta \in \Bbbk(G)^{(B\times H)}_{(\lambda,\chi)}$, and each $\chi \in \mathfrak{X}(H)$ arises in this way (because every G-line bundle $\mathscr{L}_{G/H}(\chi)$ has a rational B-eigensection). The right multiplication by $n \in N_G(H)$ maps η to $\eta' \in \Bbbk(G)^{(B\times H)}_{(\lambda,\chi')}$, the equation of $\delta' = n(\delta)$, where $\chi'(h) = \chi(n^{-1}hn)$. Since $\Bbbk(O)^B = \Bbbk$, we have $\chi' = \chi \iff \eta'/\eta = \mathrm{const} \iff \delta' = \delta$.

In particular, $\overline{H} \supseteq Z_G(H)$.

Theorem 30.5 ([Kn8, 7.6, 7.2]). *If H is very sober, then the standard embedding $X \hookleftarrow G/H$ is smooth. In particular, X is smooth if $N_G(H) = H$.*

Remark 30.6. If all simple factors of G are isomorphic to PSL_{n_i}, then very soberness is also a necessary condition for X be smooth [Lu6, 7.1]. This is not true in general: $S^{n-1} = \mathrm{SO}_n/\mathrm{SO}_{n-1}$ and $\mathrm{SL}_4/\mathrm{Sp}_4$ are symmetric spaces of rank 1, and hence their standard embeddings are smooth (Proposition 30.18), while $\overline{\mathrm{SO}_{n-1}} = \mathrm{S}(\mathrm{O}_1\times\mathrm{O}_{n-1})$, $\overline{\mathrm{Sp}_4} = \mathrm{Sp}_4\cdot Z(\mathrm{SL}_4)$.

Proof. By Theorem 23.25, $S_O = \bigcap_{\alpha\in\Delta_O^{\min}} \operatorname{Ker}\alpha \hookrightarrow \operatorname{Aut}_G O = N_G(H)/H$. It suffices to show that S_O fixes all colors; then $S_O = \{e\}$, i.e., $\Delta_O^{\min}$ spans Λ.

Take any $D \in \mathscr{D}^B$. Replacing D by a multiple, we may assume that $\mathscr{O}(D)$ is G-linearized. Consider the total space $\widehat{O} = \widehat{G}/\widehat{H}$ of $\mathscr{O}(-D)^\times$, where $\widehat{G} = G\times\Bbbk^\times$, cf. Remark 20.8. Using the notation of Remark 20.8, we have

$$0 \longrightarrow \Lambda \longrightarrow \widehat{\Lambda} \longrightarrow \mathbb{Z} \longrightarrow 0,$$

$\mathscr{V} = \widehat{\mathscr{V}}/(\widehat{\mathscr{V}}\cap-\widehat{\mathscr{V}})$, and $\Delta_{\widehat{O}}^{\min} = \Delta_O^{\min}$. Therefore $S_O = S_{\widehat{O}}/\Bbbk^\times$.

However the pullback $\widehat{D} \subset \widehat{O}$ of D is principal. Since $S_{\widehat{O}}$ multiplies the equation of $\widehat{D}$ by scalars, it leaves $\widehat{D}$ stable, whence S_O leaves D stable. □

30.2 Demazure Embedding. If $N_G(H) = H$, then $O \simeq G[\mathfrak{h}]$, the orbit of $\mathfrak{h}$ in $\mathrm{Gr}_k(\mathfrak{g})$, $k = \dim\mathfrak{h}$. The closure $X(\mathfrak{h}) = \overline{G[\mathfrak{h}]} \subseteq \mathrm{Gr}_k(\mathfrak{g})$ is called the *Demazure embedding*.

Proposition 30.7 ([Los1]). *If $N_G(H) = H$, then $X(\mathfrak{h})$ is the wonderful embedding of O.*

Proof. The standard embedding $X \hookleftarrow O$ is wonderful by Theorem 30.5. Brion proved that X is the normalization of $X(\mathfrak{h})$ [Bri8, 1.4]. Finally, Losev proved that the normalization map $X \to X(\mathfrak{h})$ is an isomorphism [Los1]. We give a proof of Brion's result referring to [Los1] for the rest.

The decomposition $\mathfrak{g} = \mathfrak{p}_\mathrm{u}\oplus\mathfrak{a}\oplus\mathfrak{h}$ yields $\mathfrak{h} = \mathfrak{l}_0\oplus\langle e_{-\alpha}+\xi_\alpha \mid \alpha\in\Delta^+\setminus\Delta_L^+\rangle$, where $\xi_\alpha \in \mathfrak{p}_\mathrm{u}\oplus\mathfrak{a}$ is the projection of $-e_{-\alpha}$ along $\mathfrak{h}$. Hence

$$\widehat{\mathfrak{h}} = \widehat{\mathfrak{l}_0} \wedge \bigwedge_{\alpha \in \Delta^+ \setminus \Delta_L^+} (e_{-\alpha} + \xi_\alpha) = \widehat{\mathfrak{s}} + \text{terms of higher } T\text{-weights},$$

where $\widehat{\mathfrak{q}} \in \bigwedge^\bullet \mathfrak{g}$ denotes a generator of $[\mathfrak{q}] \in \mathrm{Gr}(\mathfrak{g})$, $\mathfrak{s} = \mathfrak{l}_0 \oplus \mathfrak{p}_\mathrm{u}^-$, and the weights of other terms differ from that of $\widehat{\mathfrak{s}}$ by $\sum(\alpha_i + \beta_i)$, $\alpha_i, \beta_i \in \Delta^+ \setminus \Delta_L^+$ or $\beta_i = 0$.

Let $Z(\mathfrak{h})$ be the closure of $T[\mathfrak{h}]$ in the affine chart defined by non-vanishing of the covector dual to $\widehat{\mathfrak{s}}$. It is an affine toric variety with the fixed point $[\mathfrak{s}]$. Thus $Y = G[\mathfrak{s}] \subset X(\mathfrak{h})$ is a closed orbit. The local structure theorem in a neighborhood of $[\mathfrak{s}]$ provides a B-chart $\mathring{X}(\mathfrak{h}) \subset X(\mathfrak{h})$, $\mathring{X}(\mathfrak{h}) \simeq P_\mathrm{u} \times Z(\mathfrak{h})$. Note that for any $[\mathfrak{q}] \in Z(\mathfrak{h}) \setminus T[\mathfrak{h}]$ the subalgebra $\mathfrak{q}$ is transversal to $\mathfrak{p}_\mathrm{u} \oplus \mathfrak{a}$ while $\mathfrak{n}_\mathfrak{g}(\mathfrak{q}) \cap \mathfrak{a} \neq 0$, whence $\dim G_{[\mathfrak{q}]} > \dim H$. It follows that $\mathring{X}(\mathfrak{h})$ intersects no colors, i.e., $X(\mathfrak{h})$ is toroidal in a neighborhood of Y.

On the other hand, every smooth toroidal embedding of O maps to $X(\mathfrak{h})$ by Theorem 29.2. It follows that the normalization of $X(\mathfrak{h})$ is simple, and hence wonderful. □

If $N_G(H) \neq H$, then $X(\mathfrak{h})$ is the wonderful embedding of $G/N_G(H)$.

30.3 Case of a Symmetric Space. Let G be an adjoint semisimple group and let $H = G^\theta$ be a symmetric subgroup. Here $N_G(H) = H$. We have $\Lambda(O) = \mathfrak{X}(T/T^\theta) = \{\mu - \theta(\mu) \mid \mu \in \mathfrak{X}(T)\}$, where T is a θ-stable maximal torus such that T_1 is a maximal θ-split torus. Hence $\Lambda(O)$ is the root lattice of $2\Delta_O$. Since $\mathscr{V}(O)$ is the antidominant Weyl chamber of $\Delta_O^\vee$ in $\Lambda(O)^* \otimes \mathbb{Q}$ (by Theorem 26.25), $\Delta_O^{\min}$ is the reduced root system associated with $2\Delta_O$. It follows that the standard completion X is smooth in this case.

Wonderful completions of symmetric spaces were studied in [CP1], [CS]. In particular, a geometric realization for a wonderful completion as an embedded projective variety was constructed. Take any $\lambda \in \Lambda(\widetilde{G}/\widetilde{G}^\theta) \cap \operatorname{int} \mathbf{C}(\Delta_O^+)$. There exists a unique (up to proportionality) nonzero $\widetilde{G}^\theta$-fixed vector $v' \in V^*(\lambda)$. Then $X' = \overline{G[v']} \subseteq \mathbb{P}(V^*(\lambda))$ is the wonderful embedding of $G[v'] \simeq O$.

Indeed, a natural closed embedding $\mathbb{P}(V^*(\lambda)) \hookrightarrow \mathbb{P}(V^*(2\lambda))$ (given by the multiplication $V^*(\lambda) \otimes V^*(\lambda) \to V^*(2\lambda)$ in $\Bbbk[\widetilde{G}]$) identifies X' with $X'' = \overline{G[v'']}$, where $v'' \in V^*(2\lambda)$ is a unique $\widetilde{G}^\theta$-fixed vector. As X'' is a simple projective embedding of $G[v'']$, the natural map $O \to G[v'']$ extends to $X \to X''$. On the other hand, the homomorphism $V^*(\lambda) \otimes V^*(\lambda) \to V^*(2\lambda)$ maps ω to v'', where ω is defined by (26.3). Let Z'' be the closure of $T[v'']$ in the affine chart of $\mathbb{P}(V^*(2\lambda))$ defined by non-vanishing of the highest covector of weight 2λ. From (26.3) it is easy to deduce that $Z'' \simeq \mathbb{A}^r$ is acted on by T via the eigenweight set $\Pi_O^{\min}$ and the closed orbit $G[v_{-2\lambda}]$ is transversal to Z'' at $[v_{-2\lambda}]$. Hence $Z \xrightarrow{\sim} Z''$, $\mathring{X} \xrightarrow{\sim} PZ'' \simeq P_\mathrm{u} \times Z''$, and finally $X \xrightarrow{\sim} X'' \simeq X'$. (A similar reasoning shows $X \simeq \overline{G[\omega]}$. A slight refinement carries over the construction to positive characteristic [CS].)

Another model for the wonderful completion is the Demazure embedding. First note that $\mathfrak{h} = \mathfrak{l}_0 \oplus \langle e_\alpha + e_{\theta(\alpha)} \mid \alpha \in \Delta^+ \setminus \Delta_L^+ \rangle$. Arguing as in the proof of Proposition 30.7, we see that $Z(\mathfrak{h}) = \overline{T[\mathfrak{h}]} \simeq \mathbb{A}^r$ is acted on by T with the eigenweights $\alpha - \theta(\alpha)$, $\alpha \in \Pi \setminus \Delta_L$, and $Y = G[\mathfrak{s}]$ is transversal to $Z(\mathfrak{h})$ at $[\mathfrak{s}]$. This yields $\mathscr{C}_Y = \mathscr{V}$.

Now the Luna–Vust theory together with the description of the colored data for symmetric spaces implies that $X(\mathfrak{h})$ is wonderful [CP1]. The varieties $X(\mathfrak{h})$ were first considered by Demazure in the case, where $G = \mathrm{PSL}_n(\Bbbk)$ and H is the projective orthogonal or symplectic group [Dem4].

30.4 Canonical Class. Using the Demazure embedding, Brion computed the canonical class of any spherical variety.

Proposition 30.8 ([Bri8, 1.6]). *Suppose that X is a spherical variety with the open orbit $O \simeq G/H$. Consider the G-morphism $\varphi : O \to \mathrm{Gr}_k(\mathfrak{g})$, $\varphi(o) = [\mathfrak{h}]$, $k = \dim H$. Then a canonical divisor of X is*

$$K_X = -\sum_i D_i - \overline{\varphi^*\mathscr{H}} = -\sum_i D_i - \sum_{D\in\mathscr{D}^B} m_D D,$$

where D_i runs over all G-stable prime divisors in X, $\mathscr{H}$ is a hyperplane section of $X(\mathfrak{h})$ in $\mathbb{P}(\bigwedge^k \mathfrak{g})$, and $m_D \in \mathbb{N}$.

Explicit formulæ for m_D are given in [Bri12, 4.2], [Lu5, 3.6]. In the notation of 30.10, $m_D = 1$ unless $D = D_\alpha \in \mathscr{D}^b$, in which case $m_D = 2\langle \rho_G - \rho_L, \alpha^\vee \rangle \geq 2$.

Proof. Removing all G-orbits of codimension > 1, we may assume that X is smooth and toroidal. Then by Theorem 29.2, φ extends to X, and we have an exact sequence

$$0 \longrightarrow \varphi^*\mathscr{E} \longrightarrow \mathscr{O}_X \otimes \mathfrak{g} \longrightarrow \mathscr{T}_X(-\log \partial X) \longrightarrow 0,$$

where $\mathscr{E}$ is the tautological vector bundle on $\mathrm{Gr}_k(\mathfrak{g})$. Taking the top exterior powers yields $\omega_X \otimes \mathscr{O}_X(\partial X) \simeq \bigwedge^k \varphi^*\mathscr{E} = \mathscr{O}_X(-\varphi^*\mathscr{H})$, whence the first expression for K_X. If $\mathscr{H}$ is defined by a covector in $(\bigwedge^k \mathfrak{g}^*)^{(B)}$ dual to $\widehat{\mathfrak{s}}$, then $\mathring{X}(\mathfrak{h}) = X(\mathfrak{h}) \setminus \mathscr{H}$ intersects all G-orbits in open B-orbits, and hence $\varphi^*\mathscr{H} = \sum m_D D$ with $m_D > 0$ for all $D \in \mathscr{D}^B$. □

Using the characterization of ample divisors on complete spherical varieties (Corollary 17.24), one deduces that certain wonderful embeddings (e.g., flag varieties, most wonderful completions of symmetric spaces, primitive wonderful varieties of rank 1, see 30.8) are Fano varieties (i.e., the anticanonical divisor is ample).

30.5 Cox Ring. In the study of projective varieties it is very helpful to use homogeneous coordinates. Polynomials in homogeneous coordinates are not functions, but sections of a very ample line bundle and its powers. Instead of taking one line bundle, one may consider all line bundles and their sections. Thus one arrives at the notion of a total coordinate ring or a Cox ring of an algebraic variety [BH]. For simplicity, we define it under some technical restrictions.

Definition 30.9. Suppose that X is a locally factorial irreducible algebraic variety such that $\mathrm{Pic}\,X$ is free and finitely generated. The *Cox sheaf* of X is a sheaf of graded $\mathscr{O}_X$-algebras $\mathscr{R}_X = \bigoplus_\delta \mathscr{O}_X(\delta)$, where δ runs over a subgroup of $\mathrm{CaDiv}\,X$ mapped isomorphically onto $\mathrm{Pic}\,X$. The multiplication in $\mathscr{R}_X$ is given by choosing compatible identifications $\mathscr{O}(\delta) \otimes \mathscr{O}(\delta') \simeq \mathscr{O}(\delta + \delta')$, which is possible since $\mathrm{Pic}\,X$ is free. The *Cox ring* of X is $\mathscr{R}(X) = \mathrm{H}^0(X, \mathscr{R}_X) = \bigoplus_\delta \mathrm{H}^0(X, \mathscr{O}(\delta))$.

$\mathscr{R}_X$ and $\mathscr{R}(X)$ do not depend (up to isomorphism) on the lifting $\operatorname{Pic} X \hookrightarrow \operatorname{CaDiv} X$. The ring $\mathscr{R}(X)$ was first introduced by Cox for smooth complete toric X, in which case $\mathscr{R}(X)$ is a polynomial algebra [Cox].

Example 30.10. For $X = \mathbb{P}(V)$ one has $\mathscr{R}(X) = \Bbbk[V]$.

Put $\mathring{\mathscr{X}} = \operatorname{Spec}_{\mathscr{O}_X} \mathscr{R}_X$. It is a quasiaffine variety [BH, Pr. 3.10] and a principal T_X-bundle over X, where $T_X = \operatorname{Hom}(\operatorname{Pic} X, \Bbbk^\times)$ is the *Néron–Severi torus*. If $\mathscr{R}(X)$ is finitely generated, then $\mathscr{X} = \operatorname{Spec} \mathscr{R}(X)$ is a factorial affine variety containing $\mathring{\mathscr{X}}$ as an open subset with the complement of codimension > 1 [BH]. Indeed, $\mathring{\mathscr{X}}$ is covered by affine open subsets $\mathring{\mathscr{X}}_\eta \simeq \mathscr{X}_\eta$, where $\eta \in \mathrm{H}^0(X, \mathscr{O}(\delta))$, $\Bbbk[\mathring{\mathscr{X}}] = \Bbbk[\mathscr{X}]$, and each T_X-stable divisor on $\mathscr{X}$ is pulled back from a divisor on X, and hence is principal.

The Cox ring of a wonderful variety was investigated by Brion in [Bri18]. Here we describe his results.

Let $X \hookleftarrow O = G/H$ be a wonderful embedding, where G may and will be assumed semisimple and simply connected. By Corollary 17.10, $\operatorname{Pic} X$ is freely generated by the classes of colors $D \in \mathscr{D}^B$. The canonical B-eigensection η_D of $\mathscr{O}(D)$ with $\operatorname{div} \eta_D = D$ may be regarded as a function in $\Bbbk[G]^{(B \times H)}$ (= the equation of the preimage of D in G) of eigenweight $\widehat{\lambda}_D = (\lambda_D, \chi_D)$, so that λ_D is the eigenweight of the section and $\mathscr{O}_{G/H}(D) \simeq \mathscr{L}(-\chi_D)$, cf. Remark 13.4. The biweights $\widehat{\lambda}_D$ are linearly independent by Remark 15.1. Consider also the canonical G-invariant sections $\eta_1, \dots, \eta_r$ corresponding to the components $D_1, \dots, D_r$ of $X \setminus O$.

Proposition 30.11. (1) *$\mathscr{R}(X)^U = \Bbbk[\eta_1, \dots, \eta_r, \eta_D \mid D \in \mathscr{D}^B]$ and $\mathscr{R}(X)^G = \Bbbk[\eta_1, \dots, \eta_r]$ are polynomial algebras.*
(2) *$\mathscr{R}(X)$ is a free module over $\mathscr{R}(X)^G$ and a finitely generated algebra.*
(3) *The categorical quotient map $\pi_G : \mathscr{X} \to \mathscr{X} /\!\!/ G \simeq \mathbb{A}^r$ is flat with reduced and normal fibers.*

Proof. (1) stems from the fact that the divisor of each $(B \times T_X)$-eigensection in $\mathscr{R}(X)^U$ is uniquely expressed as a non-negative integral linear combination of $D_1, \dots, D_r, D \in \mathscr{D}^B$. Since $\mathscr{R}(X)^U$ is free over $\mathscr{R}(X)^G$, this implies (2) in view of Theorem D.5(1). By (2), π_G is flat, and all fibers have one and the same algebra of U-invariants isomorphic to $\Bbbk[\eta_D \mid D \in \mathscr{D}^B]$, which yields (3) by Theorem D.5. □

Let H_0 denote the common kernel of all characters in $\mathfrak{X}(H)$. Then $T_O = H/H_0$ is a diagonalizable group with $\mathfrak{X}(T_O) = \mathfrak{X}(H)$. There is a commutative diagram

$$\begin{array}{ccccc}
\operatorname{Pic} O & \longleftarrow & \operatorname{Pic} X & \longrightarrow & \operatorname{Pic} Y \\
\| & & \| & & \| \\
\mathfrak{X}(T_O) & \longleftarrow & \mathbb{Z}\mathscr{D}^B & \longrightarrow & \mathfrak{X}(P) \\
\chi_D & \longleftarrow & D & \longrightarrow & \lambda_D,
\end{array} \tag{30.1}$$

where $Y \subseteq X$ is the closed G-orbit and $P = P(O)$. The left arrows are surjective, with kernels consisting of $\sum m_D D$ such that $\sum m_D D = \operatorname{div} \mathbf{f}_\lambda$ on O for some $\lambda \in \Lambda(O) =$

$\mathbb{Z}\Pi_O^{\min}$, i.e., $m_D = \langle D, \lambda \rangle$. In particular, $T_O \hookrightarrow T_X \simeq (\Bbbk^\times)^{\mathscr{D}^B}$, $h \mapsto (\chi_D(h))_{D\in\mathscr{D}^B}$. A homomorphism $T \to T_X$, $t \mapsto (\lambda_D(t))_{D\in\mathscr{D}^B}$, induces an isomorphism $A = T/T_0 \xrightarrow{\sim} T_X/T_O$. Indeed, $\mathfrak{X}(T_X/T_O)$ consists of the elements $\sum\langle D,\lambda\rangle D$, which are mapped bijectively to $\sum\langle D,\lambda\rangle\lambda_D = \lambda \in \Lambda(O) = \mathfrak{X}(A)$.

The group $G\times T$ acts on $\mathscr{X}$ via the above homomorphism $T \to T_X$. The preimage $\widehat{O} \subseteq \mathring{\mathscr{X}}$ of $O \subseteq X$ is a single $(G\times T_X)$- or $(G\times T)$-orbit consisting of G-orbits isomorphic to G/H_0 and transitively permuted by T_X, so that $\widehat{O} \simeq T_X *_{T_O} G/H_0 \simeq T *_{T_0} G/H_0$. It follows that

$$\Bbbk[\widehat{O}] \simeq \bigoplus_{\substack{\chi=\sum m_D\chi_D\\ \lambda=\sum m_D\lambda_D\\ m_D\in\mathbb{Z}}} \Bbbk[G/H_0]^{(H)}_{\chi} \otimes \Bbbk\lambda^{-1} = \bigoplus_{\substack{\lambda\in\Lambda_+(G/H_0)=\sum\mathbb{Z}_+\lambda_D\\ \mu\in\lambda+\Lambda(G/H)}} \Bbbk[G/H_0]_{(\lambda)} \otimes \Bbbk\mu^{-1} \subseteq \Bbbk[G/H_0\times T].$$

Note that $D_i \sim \sum\langle D,\lambda_i\rangle D$, where $\lambda_i \in \Pi_X^{\min}$, $\langle v_{D_i},\lambda_i\rangle = -1$, whence $\eta_D, \eta_i \in \Bbbk[\mathscr{X}] \subseteq \Bbbk[\widehat{O}]$ are $(B\times T)$-eigenfunctions of biweights (λ_D,λ_D) and $(0,\lambda_i)$, respectively. Since they generate $\Bbbk[\mathscr{X}]^U$, we deduce the following

Theorem 30.12. (1) $\mathscr{R}(X) \simeq \bigoplus_{\substack{\lambda\in\Lambda_+(G/H_0)\\ \mu\in\lambda+\mathbb{Z}_+\Pi_X^{\min}}} \Bbbk[G/H_0]_{(\lambda)} \otimes \Bbbk\mu^{-1}.$

(2) *π_G is also the categorical quotient map for the action $\widehat{G} = G\times T_O : \mathscr{X}$.*
(3) *$\pi_G : \mathscr{X} \to \mathbb{A}^r$ is a T_X-equivariant flat family of affine spherical $\widehat{G}$-varieties with categorical quotient by U isomorphic to $\mathbb{A}^{|\mathscr{D}^B|}$, where $\widehat{B} = B\times T_O$ and $\widehat{T} = T\times T_O$ act by weights $-\widehat{\lambda}_D$.*
(4) *General fibers of π_G are isomorphic to $G/\!/H_0$ and $\pi_G^{-1}(0)$ is the horospherical contraction of $G/\!/H_0$.*

Proof. (1) follows by choosing in $\Bbbk[\widehat{O}]$ those isotypic components which correspond to $(B\times T)$-eigenweights of $\mathscr{R}(X)^U$. Assertions (2) and (3) are easily derived from the structure of $\mathscr{R}(X)^U$. To deduce (4) from (1), observe that $\mathscr{V}(G/H_0)$ is the preimage of $\mathscr{V}(X)$, whence $\Pi_{G/H_0}^{\min}$ is proportional to $\Pi_X^{\min}$ and, by (T_0), $-\Pi_{G/H_0}^{\min}$ generates the cone spanned by tails of $G/\!/H_0$. □

Example 30.13. If $O = G/Z(G)$ is the adjoint group of G considered as a symmetric space (Example 26.10), then $\mathscr{X} = \operatorname{Env} G$ is the enveloping semigroup.

Generators and relations for $\mathscr{R}(X)$ are described in [Bri18, 3.3].

The Cox sheaf and ring may be defined in a more general setup than above [EKW], [Hau]. Namely let X be any normal variety such that $\operatorname{Cl} X$ is finitely generated. The Cox sheaf $\mathscr{R}_X$ and the Cox ring $\mathscr{R}(X)$ are defined by the formulæ of Definition 30.9, where δ runs over a set of representatives for the divisor classes in $\operatorname{Cl} X$, and $\mathscr{O}_X(\delta)$ is the corresponding reflexive sheaf. If $\operatorname{Cl} X$ is free or $\Bbbk[X]^\times = \Bbbk^\times$, then one may choose multiplication maps $\mathscr{O}(\delta)\otimes\mathscr{O}(\delta')\to\mathscr{O}(\delta+\delta')$ (which are isomorphisms over X^{reg}) making $\mathscr{R}_X$ into a sheaf of graded commutative associative $\mathscr{O}_X$-algebras in a canonical way independent on the choice of representatives of divisor classes, so that $\mathscr{R}(X)$ is indeed a ring. Note that $\mathscr{R}_X=\iota_*\mathscr{R}_{X^{\mathrm{reg}}}$, where $\iota : X^{\mathrm{reg}}\hookrightarrow X$ and $\mathscr{R}(X)=\mathscr{R}(X^{\mathrm{reg}})$. Here X^{reg} may be replaced by any other open subset of X with the complement of codimension >1. So (at least for a free divisor class group) the concepts of Cox sheaf and Cox ring in general can be reduced to Definition 30.9. An equivariant version of these concepts is obtained by replacing $\mathrm{Cl}(X)$ with the group $\mathrm{Cl}_G(X)$ of G-linearized divisor classes (i.e., isomorphism classes of G-linearized reflexive sheaves of rank 1) on a G-variety X. (If $\mathrm{Cl}_G(X)$ has torsion, then one has to require that there are no non-constant G-invariant invertible functions on X.)

If $\mathscr{R}_X$ is a sheaf of finitely generated $\mathscr{O}_X$-algebras (which holds, e.g., whenever X is $\mathbb{Q}$-factorial or $\mathscr{R}(X)$ is finitely generated), then $\mathring{\mathscr{X}}=\mathrm{Spec}_{\mathscr{O}_X}\mathscr{R}_X$ is a quasi-affine normal variety equipped with a natural action of a (possibly disconnected) diagonalizable group T_X such that $\mathfrak{X}(T_X)=\mathrm{Cl}(X)$ (or $\mathrm{Cl}_G(X)$), and the natural map $\mathring{\mathscr{X}}\to X$ is a good quotient for the T_X-action. If $\mathscr{R}(X)$ is finitely generated, then $\mathscr{X}=\mathrm{Spec}\,\mathscr{R}(X)$ is a normal affine variety containing $\mathring{\mathscr{X}}$ as an open subset with complement of codimension >1.

Cox rings of arbitrary spherical varieties were computed by Brion in [Bri18, §4]. Let X be a spherical variety with the open orbit $O\simeq G/H$, where G may and will be assumed to be of simply connected type. For simplicity, we impose a (not very restrictive) condition $\Bbbk[X]^\times=\Bbbk^\times$.

Note that $\mathscr{R}(X)=\mathscr{R}(X')$, where $X'\subseteq X$ is a smooth open subset obtained by removing all G-orbits of codimension >1. Therefore X may be assumed smooth and toroidal whenever necessary.

As in the wonderful case, we use the notation η_D and $\eta_1,\dots,\eta_k$ for the canonical sections corresponding to the colors D and the G-stable prime divisors $D_1,\dots,D_k$, respectively. Since G' is simply connected semisimple, $\mathscr{R}_X$ admits a unique G'-linearization, so that η_D are $(B\cap G')$-semiinvariant of eigenweights λ_D and η_j are G'-invariant.

Proposition 30.11 extends to the general spherical case, except that G is replaced by G'.

Let $\overline{X}$ denote the wonderful embedding of $\overline{O}=G/\overline{H}$. Various objects related to $\overline{X}$ (divisors, sections, ...) will be denoted in the same way as for X, but equipped with a bar. The natural rational map $\varphi : X \dashrightarrow \overline{X}$ (which is regular if and only if X is toroidal) gives rise to a homomorphism $\varphi^* : \mathscr{R}(\overline{X})\to\mathscr{R}(X)$. It is easy to see that $\varphi^*\eta_{\overline{D}}=\eta_D$, where $D=\varphi^{-1}(\overline{D})$, and $\varphi^*\overline{\eta}_i=\prod_j\eta_j^{\langle v_{D_j},-\lambda_i\rangle}$.

Theorem 30.14 ([Bri18, 4.3]). $\mathscr{R}(X)\simeq\mathscr{R}(\overline{X})\otimes_{\mathscr{R}(\overline{X})^{G'}}\mathscr{R}(X)^{G'}$. *In geometric terms,* $\mathscr{X}\simeq\overline{\mathscr{X}}\times_{\overline{\mathscr{X}}/\!/G'}\mathscr{X}/\!/G'$.

A G-linearization of $\mathscr{R}_X$ may not exist and even if it exists, it may be not unique. In order to take into account the G-action, it is more convenient to use the G-equivariant version of the Cox sheaf and ring. The above results generalize to this setup: $\mathscr{O}(D)$ and $\mathscr{O}(D_j)$ are equipped with canonical G-linearizations so that η_D and η_j

are $Z(G)^0$-invariant, $\mathscr{R}(X)^U$ is freely generated by η_D, η_j as a $\Bbbk[Z(G)^0]$-algebra, and Theorem 30.14 extends with G' replaced by G, see [Bri18, §4].

30.6 Wonderful Varieties. Wonderful embeddings can be characterized intrinsically by the configuration of G-orbits.

Theorem 30.15 ([Lu4]). *A smooth complete G-variety X is a wonderful embedding of a spherical homogeneous space if and only if it satisfies the following conditions:*

(1) *X contains a dense open orbit O.*
(2) *$X \setminus O$ is a divisor with normal crossings, i.e., its components $D_1, \dots, D_r$ are smooth and intersect transversally.*
(3) *For each tuple $1 \le i_1 < \cdots < i_k \le r$, the set $D_{i_1} \cap \cdots \cap D_{i_k} \setminus \bigcup_{i \neq i_1,\dots,i_k} D_i$ is a single G-orbit. (In particular, it is non-empty.)*

G-varieties satisfying the conditions of the theorem are called *wonderful varieties.*

Sketch of a proof. Wonderful embeddings obviously satisfy the conditions (1)–(3), as a particular case of Theorem 29.2: the toric slice $Z \simeq \mathbb{A}^r$ is transversal to all orbits and the G-stable prime divisors intersect it in the coordinate hyperplanes.

To prove the converse, consider the local structure of X in a neighborhood of the closed orbit Y which is provided by an embedding of X into a projective space. Let $P = P_{\mathrm{u}} \rightthreetimes L$ be a Levi decomposition of $P = P(Y)$. There is a B-chart $\mathring{X} \simeq P_{\mathrm{u}} \times Z$ such that Z is a smooth L-stable locally closed subvariety intersecting Y transversally at the unique P^--fixed point z. It is easy to see that a general dominant one-parameter subgroup $\gamma \in \mathfrak{X}^*(Z(L))$ contracts $\mathring{X}$ to z. Hence Z is L-isomorphic to T_zZ.

Consider the wonderful subvarieties $X_i = \bigcap_{j \neq i} D_j$ and let λ_i be the T-weights of $T_z(Z \cap X_i)$, $i = 1, \dots, r$. Since $T_zZ = \bigoplus T_z(Z \cap X_i)$, it suffices to prove that $\lambda_1, \dots, \lambda_r$ are linearly independent.

The latter is reduced to the cases $r = 1$ or 2. Indeed, if we already know that X_i and $X_{ij} = \bigcap_{k \neq i,j} D_k$ are wonderful embeddings of spherical spaces, then $\Pi_{X_i}^{\min} = \{\lambda_i\}$ and $\Pi_{X_{ij}}^{\min} = \{\lambda_i, \lambda_j\}$. Thus the λ_i are positive linear combinations of positive roots located at obtuse angles to each other. This implies the linear independence.

The case $r = 1$ stems from Proposition 30.17.

The case $r = 2$ can be reduced to $G = \mathrm{SL}_2$. Indeed, assuming that λ_1, λ_2 are proportional, we see that $c(X) = r(X) = 1$. By Proposition 10.3, O is obtained from a 3-dimensional homogeneous SL_2-space by parabolic induction. Let us describe the colored hypercone $(\mathscr{C}_Y, \mathscr{D}_Y^B)$.

Since T_zZ is contracted to 0 by γ, we have $\Lambda(X) = \mathbb{Z}\lambda$ and $\lambda_i = h_i\lambda$, where $\langle \lambda, \gamma \rangle > 0$ and h_1, h_2 are coprime positive integers. Without loss of generality $\ell_1 h_1 - \ell_2 h_2 = 1$ for some $\ell_1, \ell_2 \in \mathbb{N}$. Consider $T_z(Z \cap X_i)$ as coordinate axes in $T_zZ \simeq Z$ and extend the respective coordinates to $f_1, f_2 \in \Bbbk(X)^{(B)}$. Then we may put $\mathbf{f}_\lambda = f_2^{\ell_2}/f_1^{\ell_1}$, and $\Bbbk(X)^B = \Bbbk(f_2^{h_1}/f_1^{h_2})$. We have the following picture for $(\mathscr{C}_Y, \mathscr{D}_Y^B)$ ($f_2^{h_1}/f_1^{h_2}$ is regarded as an affine coordinate on $\mathbb{P}^1$, colors in $\mathscr{D}_Y^B$ are marked by bold dots, and the rays corresponding to the G-stable divisors X_1, X_2 are marked, too):

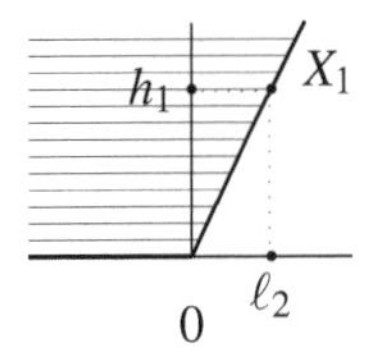

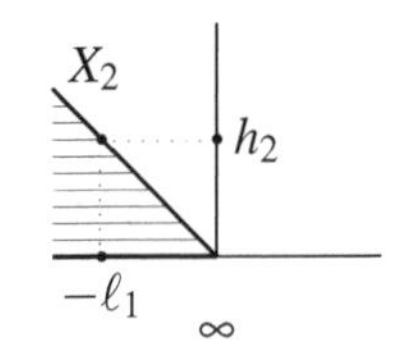

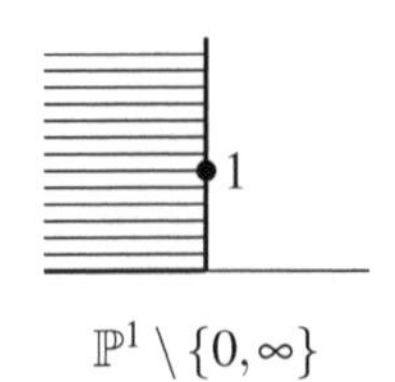

By Corollary 12.14, $\mathscr{C}_Y \supseteq \mathscr{V}$. Since $\mathscr{D}_Y^B$ does not contain central colors, Propositions 20.13 and 14.4 imply that X is induced from a wonderful SL_2-variety. However it is easy to see (e.g., from the classification in [Tim2, §5]) that there exist no SL_2-germs with the colored data as above. (Luna uses different arguments in [Lu4].) □

Wonderful varieties play a distinguished rôle in the study of spherical homogeneous spaces, because they are the canonical completions of these spaces having nice geometric properties. To a certain extent this rôle is analogous to that of (generalized) flag varieties in the theory of reductive groups. For symmetric spaces this was already observed by de Concini and Procesi [CP1]. For general spherical spaces this principle was developed by Brion, Knop, Luna, et al [BPa], [Bri8], [Kn8], [Lu3], [Lu5], [Lu6].

30.7 How to Classify Spherical Subgroups. In particular, wonderful varieties are applied to classification of spherical subgroups. The strategy, proposed by Luna, is to reduce the classification to very sober subgroups, which are stabilizers of general position for wonderful varieties, and then to classify the wonderful varieties.

By Theorem 29.5, there are no continuous families of non-conjugate spherical subgroups, and even more:

Proposition 30.16 ([AB3, Cor. 3.2]). *There are finitely many conjugacy classes of sober spherical subgroups $H \subseteq G$.*

Proof. Sober spherical subalgebras of dimension k form a locally closed G-subvariety in $\mathrm{Gr}_k(\mathfrak{g})$. Indeed, the set of spherical subalgebras is open in the variety of k-dimensional Lie subalgebras, and sober subalgebras are those having orbits of maximal dimension. Theorem 29.5 implies that this variety is a finite union of locally closed strata such that all orbits in each stratum have the same stabilizer. But the isotropy subalgebras are nothing else but the points of the strata. Hence each stratum is a single orbit, i.e., there are finitely many sober subalgebras, up to conjugation. As for subgroups, there are finitely many ways to extend H^0 by a (finite) subgroup in $N_G(H^0)/H^0$. □

Note that finiteness fails for non-sober spherical subgroups: H can be extended by countably many diagonalizable subgroups in $N_G(H)/H$.

These results create evidence that spherical subgroups should be classified by some discrete invariants. Such invariants were suggested by Luna, under the names of *spherical systems* and *spherical homogeneous data* (Definition 30.21). They are defined in terms of roots and weights of G and wonderful G-varieties of rank 1.

30.8 Spherical Spaces of Rank 1. For spherical homogeneous spaces of rank 1, standard embeddings are always smooth. Indeed, they are normal G-varieties consisting of two G-orbits—a dense one and another of codimension 1. Furthermore, spherical homogeneous spaces of rank 1 are characterized by existence of a completion by homogeneous divisors.

Proposition 30.17 ([Akh1], [Bri5]). *The following conditions are equivalent:*

(1) *$O = G/H$ is a spherical homogeneous space of rank* 1.
(2) *There exists a smooth complete embedding $X \hookleftarrow O$ such that $X \setminus O$ is a union of G-orbits of codimension* 1.

Moreover, if O is horospherical, then $X \setminus O$ consists of two orbits and $X \simeq G *_Q \mathbb{P}^1$, where $Q \subseteq G$ is a parabolic acting on $\mathbb{P}^1$ via a character. Otherwise $X \setminus O$ is a single orbit and X is a wonderful embedding of O.

Proof. The implication (1) $\Longrightarrow$ (2) and the properties of X easily stem from the Luna–Vust theory: the colored space $\mathscr{E}$ is a line, whence there exists a unique smooth complete toroidal embedding X, which is obtained by adding two homogeneous divisors (corresponding to the two rays of $\mathscr{E}$) if $\mathscr{V} = \mathscr{E}$ and is wonderful if $\mathscr{V}$ is a ray.

To prove (2) $\Longrightarrow$ (1), we consider the local structure of X in a neighborhood of a closed orbit Y. Let $P = P_u \rightthreetimes L$ be a Levi decomposition of $P = P(Y)$. There is a B-chart $\mathring{X} \simeq P_u \times Z$ such that Z is an L-stable affine curve intersecting Y transversally at the unique P^--fixed point. Note that $T \subseteq L$ cannot fix Z pointwise for otherwise O^T would be infinite, which is impossible. Hence $T : Z$ has an open orbit, whence (1). □

Remark 30.18. A similar reasoning proves an embedding characterization of arbitrary rank 1 spaces, due to Panyushev [Pan5]: $r(O) = 1$ if and only if there exists a complete embedding $X \hookleftarrow O$ such that $X \setminus O$ is a divisor consisting of closed G-orbits. Here Z is an affine L-stable subvariety with a pointwise L-fixed divisor $Z \setminus O$ (provided that Y is a general closed orbit), which readily implies that general orbits of $L : Z$ are one-dimensional, whence $r(X) = r(Z) = 1$. On the other hand, it is easy to construct a desired embedding X for a homogeneous space O parabolically induced from SL_2 modulo a finite subgroup, cf. Proposition 10.3.

Spherical homogeneous spaces G/H of rank 1 were classified by Akhiezer [Akh1] and Brion [Bri5]. It is easy to derive the classification from a regular embedding of H into a parabolic $Q \subseteq G$. In the notation of Theorem 9.4 we have an alternative: either $r(M/K) = 1$, $r_{M_*}(Q_u/H_u) = 0$, or vice versa.

In the first case $H_u = Q_u$, i.e., G/H is parabolically induced from an affine spherical homogeneous rank 1 space M/K. Except for the trivial case $M/K \simeq \Bbbk^\times$ (where H is horospherical), K is sober in M and H in G.

In the second case $M = K = M_*$, and $Q_u/H_u \simeq \mathfrak{q}_u/\mathfrak{h}_u$ is an M-module such that $(\mathfrak{q}_u/\mathfrak{h}_u) \setminus \{0\}$ is a single M-orbit. Indeed, $\Bbbk[\mathfrak{q}_u/\mathfrak{h}_u]^{U(M)}$ is generated by one $B(M)$-eigenfunction, namely a highest covector in $(\mathfrak{q}_u/\mathfrak{h}_u)^*$, whence $\mathfrak{q}_u/\mathfrak{h}_u$ is an HV-cone.

Therefore M acts on $\mathfrak{q}_u/\mathfrak{h}_u \simeq \Bbbk^n$ as $\mathrm{GL}_n(\Bbbk)$ or $\Bbbk^\times \cdot \mathrm{Sp}_n(\Bbbk)$ and the highest weight of $\mathfrak{q}_u/\mathfrak{h}_u$ is a negative simple root.

We deduce that every spherical homogeneous space of rank 1 is either horospherical or parabolically induced from a *primitive* rank 1 space $O = G/H$ with G semisimple and H sober. Primitive spaces are of the two types:

(1) H is reductive.
(2) H is regularly embedded in a maximal parabolic $Q \subseteq G$ which shares a Levi subgroup M with H and $\mathfrak{q}_u/\mathfrak{h}_u$ is a simple M-module of type $\mathrm{GL}_n(\Bbbk) : \Bbbk^n$ or $\Bbbk^\times \cdot \mathrm{Sp}_n(\Bbbk) : \Bbbk^n$ generated by a simple root vector.

Primitive spherical homogeneous spaces of rank 1 are listed in Table 30.1. Those of the first type are easy to classify, e.g., by inspection of Tables 10.1, 10.3, and 26.3. We indicate the embedding $H \hookrightarrow G$ by referring to Table 26.3 (10.1) in the (non-)symmetric case. Primitive spaces of the second type are classified by choosing a Dynkin diagram and its node corresponding to a short simple root α which is adjacent to an extreme node of the remaining diagram, the latter being of type $\mathbf{A}_l$ or $\mathbf{C}_l$. The diagrams are presented in the column "$H \hookrightarrow G$", with the white node corresponding to α.

The wonderful embeddings of spherical homogeneous space of rank 1 are parabolically induced from those of primitive spaces. The latter are easy to describe. For type 1 the construction of 30.3 works whenever $N_G(H) = H$: the wonderful embedding of G/H is realized as $X = \overline{G[v]} \subseteq \mathbb{P}(V(\lambda))$, where $v \in V(\lambda)^{(\widetilde{H})}$, $\lambda \in \Lambda_+(\widetilde{G}/\widetilde{H}^0)$, and $\widetilde{H}$ is the preimage of H in $\widetilde{G}$. If $N_G(H) \neq H$ and λ spans $\Lambda_+(G/H)$, then X is the projective closure of Gv in $\mathbb{P}(V(\lambda) \oplus \Bbbk)$. The simple minimal root of X is the generator of $\Lambda_+(G/H)$.

For type 2 the wonderful embedding is $X = G *_Q \mathbb{P}(Q_u/H_u \oplus \Bbbk)$. Indeed, Q_u acts on the M-module Q_u/H_u by affine translations, whence the projective closure of Q_u/H_u consists of two Q-orbits—the affine part and the hyperplane at infinity. Here $\Pi_X^{\min} = \{w_M \alpha\}$.

Simple minimal roots of arbitrary wonderful G-varieties of rank 1 are called *spherical roots* of G. They are non-negative linear combinations of Π_G. Let Σ_G denote the set of all spherical roots. It is a finite set, which is easy to find from the classification of wonderful varieties of rank 1.

Spherical roots of reductive groups of simply connected type are listed in Table 30.2. Namely, $\lambda \in \Sigma_G$ if and only if it is a spherical root of a simple factor, or a product of two simple factors, indicated in the first column of the table. For each spherical root λ, we indicate the Dynkin diagram of the simple roots occurring in the decomposition of λ with positive coefficients. The numbering of the simple roots α_i is according to [OV], and α_i, α_j' denote simple roots of different simple factors. For arbitrary G, Σ_G is obtained from $\Sigma_{\widetilde{G}}$ by removing the spherical roots that are not in the weight lattice of G. Note that if $\lambda, \mu \in \Sigma_G$ are proportional, then $\lambda = 2\mu$ or $\mu = 2\lambda$, and also if $\lambda \in \Sigma_G \setminus \mathbb{Z}\Delta_G$, then $2\lambda \in \Sigma_G \cap \mathbb{Z}\Delta_G$.

More generally, two-orbit complete (normal) G-varieties were classified by Cupit-Foutou [C-F1] and Smirnov [Sm-A]. All of them are spherical.

Wonderful varieties of rank 2 were classified by Wasserman [Wa].

Table 30.1 Wonderful varieties of rank 1

No.	G	H	$H \hookrightarrow G$	$\Pi^{\min}_{G/H}$	Wonderful embedding
1	$\mathrm{SL}_2 \times \mathrm{SL}_2$	SL_2	diagonal	$\omega + \omega'$	$X = \{(x:t) \mid \det x = t^2\} \subset \mathbb{P}(\mathrm{L}_2 \oplus \Bbbk)$
2	$\mathrm{PSL}_2 \times \mathrm{PSL}_2$	PSL_2		$2\omega + 2\omega'$	$\mathbb{P}(\mathrm{L}_2)$
3	SL_n	$\mathrm{S}(\mathrm{L}_1 \times \mathrm{L}_{n-1})$	symmetric No. 1 ○—●— ⋯ —●	$\omega_1 + \omega_{n-1}$	$\mathbb{P}^{n-1} \times (\mathbb{P}^{n-1})^*$
4	PSL_2	PO_2	symmetric No. 3	4ω	$\mathbb{P}(\mathfrak{sl}_2)$
5	Sp_{2n}	$\mathrm{Sp}_2 \times \mathrm{Sp}_{2n-2}$	symmetric No. 4	ω_2	$\mathrm{Gr}_2(\Bbbk^{2n})$
6	Sp_{2n}	$B(\mathrm{Sp}_2) \times \mathrm{Sp}_{2n-2}$	○—●— ⋯ —●⇐●	ω_2	$\mathrm{Fl}_{1,2}(\Bbbk^{2n})$
7	SO_n	SO_{n-1}	symmetric No. 6	ω_1	$X = \{(x:t) \mid (x,x) = t^2\} \subset \mathbb{P}^n$
8	SO_n	$\mathrm{S}(\mathrm{O}_1 \times \mathrm{O}_{n-1})$		$2\omega_1$	$\mathbb{P}^{n-1}$
9	SO_{2n+1}	$\mathrm{GL}_n \rightthreetimes \bigwedge^2 \Bbbk^n$	●— ⋯ —●⇒○	ω_1	$X = \{(V_1, V_2) \mid V_1 \subset V_1^{\perp}\} \subset \mathrm{Fl}_{n,2n}(\Bbbk^{2n+1})$
10	Spin_7	$\mathbf{G}_2$	non-symmetric No. 10	ω_3	$X = \{(x:t) \mid (x,x) = t^2\} \subset \mathbb{P}(V(\omega_3) \oplus \Bbbk)$
11	SO_7	$\mathbf{G}_2$		$2\omega_3$	$\mathbb{P}(V(\omega_3))$
12	$\mathbf{F}_4$	$\mathbf{B}_4$	symmetric No. 17	ω_1	
13	$\mathbf{G}_2$	SL_3	non-symmetric No. 12	ω_1	$X = \{(x:t) \mid (x,x) = t^2\} \subset \mathbb{P}(V(\omega_1) \oplus \Bbbk)$
14	$\mathbf{G}_2$	$N(\mathrm{SL}_3)$		$2\omega_1$	$\mathbb{P}(V(\omega_1))$
15	$\mathbf{G}_2$	$\mathrm{GL}_2 \rightthreetimes (\Bbbk \oplus \Bbbk^2) \otimes \bigwedge^2 \Bbbk^2$	○⇐●	$\omega_2 - \omega_1$	

30.9 Localization of Wonderful Varieties. For arbitrary wonderful varieties, many questions can be reduced to the case of rank ≤ 2 via the procedure of localization [Lu5], [Lu6, 3.2].

Given a wonderful variety X with the open G-orbit O, there is a bijection $D_i \leftrightarrow \lambda_i$ $(i = 1, \dots, r)$ between the component set of ∂X and $\Pi_X^{\min}$. Namely λ_i is orthogonal to the facet of $\mathscr{V}$ complementary to the ray which corresponds to D_i. Also, λ_i is the T-weight of $T_z X / T_z D_i$ at the unique B^--fixed point z.

For any subset $\Sigma \subset \Pi_X^{\min}$, put $X^{\Sigma} = \bigcap_{\lambda_i \notin \Sigma} D_i$, the *localization* of X at Σ. It is a wonderful variety with $\Pi_{X^{\Sigma}}^{\min} = \Sigma$, and all colors in $\mathscr{D}^B(X^{\Sigma})$ are obtained as irreducible components of $\overline{D} \cap X^{\Sigma}$, $D \in \mathscr{D}^B$. (To see the latter, observe that every color on X^{Σ} is contained in the zeroes of a B-eigenform in projective coordinates, which extends to X by complete reducibility of G-modules.) In particular, the wonderful subvarieties X_i, X_{ij} of ranks 1,2 considered in the proof of Theorem 30.15 are the localizations of X at $\{\lambda_i\}$, $\{\lambda_i, \lambda_j\}$, respectively.

Another kind of localization is defined by choosing a subset $I \subset \Pi$. Let P_I be the respective standard parabolic in G, with the standard Levi subgroup L_I, and $T_I = Z(L_I)^0$. Denote by $Z^I, \mathring{X}^I, X^I$ the sets of T_I-fixed points in $Z, \mathring{X}$, and $X_I := P_I \mathring{X} = L_I \mathring{X}$, respectively.

Lemma 30.19. (1) *The contraction by a general dominant one-parameter subgroup $\gamma \in \mathfrak{X}^*(T_I)$ gives a P_I-equivariant retraction $\pi_I : X_I \to X^I$, $\pi_I(x) = \lim_{t \to 0} \gamma(t) x$ (where P_I is assumed to act on X^I via its quotient L_I modulo $(P_I)_{\mathrm{u}}$).*

Table 30.2 Spherical roots

G	Σ_G
$\mathbf{A}_l$	$\alpha_i+\cdots+\alpha_j$ $(\mathbf{A}_{j-i+1}, i\le j)$, $2\alpha_i$ $(\mathbf{A}_1)$, $\alpha_i+\alpha_j$ $(\mathbf{A}_1\times\mathbf{A}_1, i\le j-2)$, $(\alpha_1+\alpha_3)/2$ $(\mathbf{A}_1\times\mathbf{A}_1, l=3)$, $\alpha_{i-1}+2\alpha_i+\alpha_{i+1}$ $(\mathbf{D}_3, 1<i<l)$, $(\alpha_1+2\alpha_2+\alpha_3)/2$ $(\mathbf{D}_3, l=3)$
$\mathbf{B}_l$	$\alpha_i+\cdots+\alpha_j$ $(\mathbf{A}_{j-i+1}, i\le j<l)$, α_l $(\mathbf{A}_1)$, $2\alpha_i$ $(\mathbf{A}_1)$, $\alpha_i+\alpha_j$ $(\mathbf{A}_1\times\mathbf{A}_1, i\le j-2)$, $(\alpha_1+\alpha_3)/2$ $(\mathbf{A}_1\times\mathbf{A}_1, l=3,4)$, $\alpha_{i-1}+2\alpha_i+\alpha_{i+1}$ $(\mathbf{D}_3, 1<i<l-1)$, $(\alpha_1+2\alpha_2+\alpha_3)/2$ $(\mathbf{D}_3, l=4)$, $\alpha_i+\cdots+\alpha_l$ $(\mathbf{B}_{l-i+1}, i<l)$, $2\alpha_i+\cdots+2\alpha_l$ $(\mathbf{B}_{l-i+1}, i<l)$, $\alpha_{l-2}+2\alpha_{l-1}+3\alpha_l$ $(\mathbf{B}_3)$, $(\alpha_1+2\alpha_2+3\alpha_3)/2$ $(\mathbf{B}_3, l=3)$
$\mathbf{C}_l$	$\alpha_i+\cdots+\alpha_j$ $(\mathbf{A}_{j-i+1}, i\le j<l)$, α_l $(\mathbf{A}_1)$, $2\alpha_i$ $(\mathbf{A}_1)$, $\alpha_i+\alpha_j$ $(\mathbf{A}_1\times\mathbf{A}_1, i\le j-2)$, $\alpha_{i-1}+2\alpha_i+\alpha_{i+1}$ $(\mathbf{D}_3, 1<i<l-1)$, $\alpha_i+2\alpha_{i+1}+\cdots+2\alpha_{l-1}+\alpha_l$ $(\mathbf{C}_{l-i+1}, i<l)$, $2\alpha_{l-1}+2\alpha_l$ $(\mathbf{C}_2)$
$\mathbf{D}_l$	$\alpha_{i_1}+\cdots+\alpha_{i_k}$ $(\mathbf{A}_k, k\ge 1)$, $2\alpha_i$ $(\mathbf{A}_1)$, $\alpha_i+\alpha_j$ $(\mathbf{A}_1\times\mathbf{A}_1)$, $2\alpha_{i_1}+\alpha_{i_2}+\alpha_{i_3}$ $(\mathbf{D}_3)$, $(2\alpha_{i_1}+\alpha_{i_2}+\alpha_{i_3})/2$ $(\mathbf{D}_3, l=4)$, $2\alpha_i+\cdots+2\alpha_{l-2}+\alpha_{l-1}+\alpha_l$ $(\mathbf{D}_{l-i+1}, i<l-1)$, $\alpha_i+\cdots+\alpha_{l-2}+(\alpha_{l-1}+\alpha_l)/2$ $(\mathbf{D}_{l-i+1}, i<l-1)$, $(\alpha_{l-1}+\alpha_l)/2$ $(\mathbf{A}_1\times\mathbf{A}_1)$, $(\alpha_1+\alpha_3)/2$ $(\mathbf{A}_1\times\mathbf{A}_1, l=4)$, $(\alpha_1+\alpha_4)/2$ $(\mathbf{A}_1\times\mathbf{A}_1, l=4)$
$\mathbf{E}_l$	$\alpha_{i_1}+\cdots+\alpha_{i_k}$ $(\mathbf{A}_k, k\ge 1)$, $2\alpha_i$ $(\mathbf{A}_1)$, $\alpha_i+\alpha_j$ $(\mathbf{A}_1\times\mathbf{A}_1)$, $2\alpha_{i_1}+\cdots+2\alpha_{i_{k-2}}+\alpha_{i_{k-1}}+\alpha_{i_k}$ $(\mathbf{D}_k, k\ge 3)$
$\mathbf{F}_4$	α_i $(\mathbf{A}_1)$, $2\alpha_i$ $(\mathbf{A}_1)$, $\alpha_i+\alpha_j$ $(\mathbf{A}_1\times\mathbf{A}_1, i\le j-2)$, $\alpha_i+\alpha_{i+1}$ $(\mathbf{A}_2, i\ne 2)$, $\alpha_2+\alpha_3$ $(\mathbf{C}_2)$, $2\alpha_2+2\alpha_3$ $(\mathbf{C}_2)$, $\alpha_1+2\alpha_2+\alpha_3$ $(\mathbf{C}_3)$, $\alpha_2+\alpha_3+\alpha_4$ $(\mathbf{B}_3)$, $2\alpha_2+2\alpha_3+2\alpha_4$ $(\mathbf{B}_3)$, $3\alpha_2+2\alpha_3+\alpha_4$ $(\mathbf{B}_3)$, $2\alpha_1+3\alpha_2+2\alpha_3+\alpha_4$ $(\mathbf{F}_4)$
$\mathbf{G}_2$	α_i $(\mathbf{A}_1)$, $2\alpha_i$ $(\mathbf{A}_1)$, $\alpha_1+\alpha_2$ $(\mathbf{G}_2)$, $2\alpha_1+\alpha_2$ $(\mathbf{G}_2)$, $4\alpha_1+2\alpha_2$ $(\mathbf{G}_2)$
$\mathbf{X}_l\times\mathbf{Y}_m$	$\alpha_i+\alpha'_j$ $(\mathbf{A}_1\times\mathbf{A}_1)$, $(\alpha_l+\alpha'_m)/2$ $(\mathbf{A}_1\times\mathbf{A}_1, \mathbf{X}=\mathbf{Y}=\mathbf{C}, l,m\ge 1)$

(2) *X^I is a wonderful L_I-variety with $P(X^I)=P\cap L_I$, $\Pi^{\min}_{X^I}=\Pi^{\min}_X\cap\langle I\rangle$, and the colors of X^I are in bijection, given by the pullback along π_I, with the P_I-unstable colors of X.*
(3) *$\mathring{X}^I\simeq(P_{\mathrm{u}}\cap L_I)\times Z^I$ is the $(B\cap L_I)$-chart of X^I intersecting all orbits.*
(4) *$(P_I^-)_{\mathrm{u}}$ fixes X^I pointwise.*

Proof. It is obvious that $\gamma(t)$ contracts $\mathring{X}\simeq P_{\mathrm{u}}\times Z$ onto $\mathring{X}^I\simeq(P_{\mathrm{u}}\cap L_I)\times Z^I$, while the conjugation by $\gamma(t)$ contracts P_I to L_I, as $t\to\infty$. Hence π_I extends to a retraction of X_I onto $X^I=L_I\mathring{X}^I$, and $\pi_I^{-1}(\mathring{X}^I)=\mathring{X}$ since $X_I\setminus\mathring{X}$ is closed and γ-stable. Thus the P_I-unstable colors on X intersect X_I and are the pullbacks of the colors on X^I.

Since $(P_I^-)_{\mathrm{u}}$-orbits are connected, it suffices to prove in (4) that $(P_I^-)_{\mathrm{u}}x\cap\mathring{X}=\{x\}$, $\forall x\in\mathring{X}^I$. If $gx\in\mathring{X}$ for some $g\in(P_I^-)_{\mathrm{u}}$, then $\gamma(t)gx=\gamma(t)g\gamma(t)^{-1}x\to x$ as $t\to\infty$, whence $gx=x$, because $\gamma(t)$ contracts $\mathring{X}$ to $\mathring{X}^I$ as $t\to 0$.

Now (4) implies that X^I is closed in X, whence complete: otherwise $\overline{X^I}\setminus X^I$ would contain a B^--fixed point distinct from z. The structure of $\mathring{X}^I$ readily implies the remaining assertions on X^I in (2): both G- and B-orbits intersect $\mathring{X}^I$ in the orbits of $(P_u \cap L_I)T$, and $\Pi_X^{\min} \cap \langle I \rangle$ is the set of T-weights of Z^I. □

The wonderful variety X^I is called the *localization* of X at I. It is easy to see that $X^I \subseteq X^\Sigma = GX^I$, where $\Sigma = \Pi_X^{\min} \cap \langle I \rangle$. If $I \supseteq \Pi_L$, then $X^\Sigma = G *_{P_I^-} X^I$.

It is helpful to extend localization to an arbitrary spherical homogeneous space $O = G/H$ using an arbitrary complete toroidal embedding $X \hookleftarrow O$ instead of the wonderful one. For any components $D_1, \dots, D_s$ of ∂X, the intersection $X' = D_1 \cap \dots \cap D_s$ is either empty or a complete toroidal embedding of the G-orbit $O' \subseteq X$ corresponding to the minimal cone $\mathscr{C}'$ in the fan of X containing $v_{D_1}, \dots, v_{D_s}$. If X is smooth, then X' is smooth, too. If X' contains a unique closed G-orbit Y or, equivalently, $\mathscr{C}'$ is a face of a unique maximal cone $\mathscr{C}_Y$ in the fan of X, then X' is the standard embedding of O'. We have $\Pi_{X'}^{\min} = \Sigma := \Pi_X^{\min} \cap (\mathscr{V}')^\perp$, where $\mathscr{V}'$ is the minimal face of $\mathscr{V} = \mathscr{V}(O)$ containing $\mathscr{C}'$. For wonderful X we have $X' = X^\Sigma$. In particular, if $\mathscr{C}'$ is a solid subcone in the facet of $\mathscr{V}$ orthogonal to $\lambda \in \Pi_O^{\min}$, then X' is a wonderful variety of rank 1 with $\Pi_{X'}^{\min} = \{\lambda\}$, whence $\Pi_O^{\min} \subseteq \Sigma_G$.

Also for any $I \subset \Pi$ such that $\Lambda(O) \not\subset \langle I \rangle$ one can find $X \hookleftarrow O$ and a general dominant one-parameter subgroup $\gamma \in \mathfrak{X}^*(T_I)$ such that the image of $-\gamma$ is contained in a unique solid cone $\mathscr{C}_Y$ in the fan of X. (It suffices to take care lest the image of $\langle I \rangle^\perp$ in $\mathscr{E}(O)$ should lie in a hyperplane which separates two neighboring solid cones in the fan.) Starting with $\mathring{X} = \mathring{X}_Y$, one defines X^I as above and generalizes Lemma 30.19, except that in (2) one may only assert that X^I is standard, but it may be no longer wonderful (i.e., be singular if X is so). If $\Lambda(O) \subset \langle I \rangle$, then X^I can be defined for any toroidal embedding $X \hookleftarrow O$ using $\mathring{X} = X \setminus \bigcup_{D \in \mathscr{D}^B} D$. Lemma 30.19 extends to this setup except that X^I is wonderful (resp. standard, complete, smooth) if and only if X is so.

30.10 Types of Simple Roots and Colors. In particular, the localization of a complete toroidal variety X at a single root $\alpha \in \Pi$ yields a smooth complete subvariety X^α of rank ≤ 1 acted on by $S_\alpha = L'_\alpha \simeq \mathrm{SL}_2(\Bbbk)$ or $\mathrm{PSL}_2(\Bbbk)$. The classification of complete varieties of rank ≤ 1, together with Lemma 30.19, allows all simple roots to be subdivided into four types:

(p) $\alpha \in \Pi_L$. Here X^α is a point and P_α leaves all colors stable.

(b) $\alpha \notin \mathbb{Q}_+\Pi_O^{\min} \cup \Pi_L$. If X^α is wonderful, then $r(X^\alpha) = 0$ whence $X^\alpha = S_\alpha / B \cap S_\alpha \simeq \mathbb{P}^1$; otherwise $r(X^\alpha) = 1$ and $X^\alpha \simeq S_\alpha *_{B \cap S_\alpha} \mathbb{P}^1$, where $B \cap S_\alpha$ acts on $\mathbb{P}^1$ via a character. There is a unique P_α-unstable color $D_\alpha = \overline{\pi_\alpha^{-1}(o)}$ or $\overline{\pi_\alpha^{-1}(e * \mathbb{P}^1)}$.

(a) $\alpha \in \Pi_O^{\min}$. Here $r(X^\alpha) = 1$ and $X^\alpha \simeq \mathbb{P}^1 \times \mathbb{P}^1$. There are two P_α-unstable colors $D_\alpha^+ = \overline{\pi_\alpha^{-1}(\mathbb{P}^1 \times \{o\})}$ and $D_\alpha^- = \overline{\pi_\alpha^{-1}(\{o\} \times \mathbb{P}^1)}$.

(a') $2\alpha \in \Pi_O^{\min}$. Here $r(X^\alpha) = 1$ and $X^\alpha \simeq \mathbb{P}^2 = \mathbb{P}(\mathfrak{s}_\alpha)$. There is a unique P_α-unstable color $D_\alpha = \overline{\pi_\alpha^{-1}(\mathbb{P}(\mathfrak{b} \cap \mathfrak{s}_\alpha))}$.

The type of a color $D \in \mathscr{D}^B$ is defined as the type of $\alpha \in \Pi$ such that P_α moves D. Using Lemma 30.20 below, the localization at $\{\alpha, \beta\} \subseteq \Pi$, and the classification of wonderful varieties of rank ≤ 1, one verifies that, as a rule, each $D \in \mathscr{D}^B$ is moved by a unique P_α, with the following exceptions: $D_\alpha = D_\beta$ if and only if α, β are pairwise orthogonal simple roots of type b such that $\alpha+\beta \in \Pi_O^{\min} \sqcup 2\Pi_O^{\min}$; two sets $\{D_\alpha^\pm\}$ and $\{D_\beta^\pm\}$ may intersect in one color for distinct α,β of type a. (In fact, $D=D'$ for $D \in \{D_\alpha^\pm\}$, $D' \in \{D_\beta^\pm\}$ if and only if $\varkappa(D)=\varkappa(D')$, by axiom (A1) in Definition 30.21.) In particular, each color belongs to exactly one type. We obtain disjoint partitions $\Pi = \Pi^a \sqcup \Pi^{a'} \sqcup \Pi^b \sqcup \Pi^p$, $\mathscr{D}^B = \mathscr{D}^a \sqcup \mathscr{D}^{a'} \sqcup \mathscr{D}^b$ according to the types of simple roots and colors.

Lemma 30.20. *For any $\lambda \in \Lambda(O)$ we have*

$$\begin{aligned} \langle D_\alpha^+,\lambda\rangle + \langle D_\alpha^-,\lambda\rangle &= \langle \alpha^\vee,\lambda\rangle, && \forall \alpha \in \Pi^a, \\ \langle D_\alpha,\lambda\rangle &= \langle \tfrac{\alpha^\vee}{2},\lambda\rangle, && \forall \alpha \in \Pi^{a'}, \\ \langle D_\alpha,\lambda\rangle &= \langle \alpha^\vee,\lambda\rangle, && \forall \alpha \in \Pi^b. \end{aligned}$$

Proof. We use the localization at α of a smooth complete toroidal embedding $X \hookleftarrow O$. Let $Y^\alpha \simeq S_\alpha/(B^- \cap S_\alpha) \simeq \mathbb{P}^1$ be a closed S_α-orbit in X^α. Namely Y^α is the diagonal of $X^\alpha \simeq \mathbb{P}^1 \times \mathbb{P}^1$ in type a, a conic in $X^\alpha \simeq \mathbb{P}^2$ in type a', and a section of $X^\alpha \to \mathbb{P}^1$ in type b. Put $\delta_\lambda = \sum_{D\in\mathscr{D}^B} \langle D,\lambda\rangle D$. From the description of P_α-stable and unstable colors, we readily derive $\langle Y^\alpha,\delta_\lambda\rangle = \langle D_\alpha^+,\lambda\rangle + \langle D_\alpha^-,\lambda\rangle$, $2\langle D_\alpha,\lambda\rangle$, or $\langle D_\alpha,\lambda\rangle$, depending on the type of α.

On the other hand, $\delta_\lambda \sim -\sum \langle v_i,\lambda\rangle D_i$, where D_i runs over all G-stable prime divisors in X and $v_i \in \mathscr{V}$ is the corresponding G-valuation. Since $\mathscr{O}(D_i)|_{D_i}$ is the normal bundle to D_i, the fiber of $\mathscr{O}(D_i)$ at the B^--fixed point $z \in Y^\alpha$ is T_zX/T_zD_i for each $D_i \supseteq Y^\alpha$. Note that the T-weights λ_i of these fibers form the basis of $-\mathscr{C}_Y^\vee$ dual to the basis of $-\mathscr{C}_Y$ formed by the $-v_i$, where $Y=Gz$ is the closed G-orbit in X containing Y^α. Hence $\mathscr{O}(\delta_\lambda)|_{Y^\alpha} = \mathscr{L}(-\sum_{D_i\supseteq Y}\langle v_i,\lambda\rangle\lambda_i) = \mathscr{L}(\lambda)$ and $\langle Y^\alpha,\delta_\lambda\rangle = \deg\mathscr{L}(\lambda) = \langle\alpha^\vee,\lambda\rangle$. The lemma follows. □

30.11 Combinatorial Classification of Spherical Subgroups and Wonderful Varieties. These results show that $\mathscr{D}^{a'}, \mathscr{D}^b$ as abstract sets and their representation in $\mathscr{E}(O)$ are determined by Π^p and $\Pi_O^{\min}$. The colors of type a, together with the weight lattice, the parabolic P, and the simple minimal roots, form a collection of combinatorial invariants supposed to identify O up to isomorphism. Namely $(\Lambda(O),\Pi^p,\Pi_O^{\min},\mathscr{D}^a)$ is a homogeneous spherical datum in the sense of the following

Definition 30.21 ([Lu6, §2]). A *homogeneous spherical datum* is a collection $(\Lambda,\Pi^p,\Sigma,\mathscr{D}^a)$, where Λ is a sublattice in $\mathfrak{X}(T)$, $\Pi^p \subseteq \Pi_G$, $\Sigma \subseteq \Sigma_G \cap \Lambda$ is a linearly independent set consisting of indivisible vectors in Λ, and $\mathscr{D}^a$ is a finite set equipped with a map $\varkappa : \mathscr{D}^a \to \Lambda^*$, which satisfies the following axioms:

(A1) $\langle\varkappa(D),\lambda\rangle \leq 1$, $\forall D \in \mathscr{D}^a$, $\lambda \in \Sigma$, and the equality is reached if and only if $\lambda = \alpha \in \Sigma\cap\Pi$ and $D = D_\alpha^\pm$, where $D_\alpha^+, D_\alpha^- \in \mathscr{D}^a$ are two distinct elements depending on α.
(A2) $\varkappa(D_\alpha^+) + \varkappa(D_\alpha^-) = \alpha^\vee$ on Λ for any $\alpha \in \Sigma\cap\Pi$.
(A3) $\mathscr{D}^a = \{D_\alpha^\pm \mid \alpha \in \Sigma\cap\Pi\}$
(Σ1) If $\alpha \in \Pi\cap\frac{1}{2}\Sigma$, then $\langle\alpha^\vee,\Lambda\rangle \subseteq 2\mathbb{Z}$ and $\langle\alpha^\vee,\Sigma\setminus\{2\alpha\}\rangle \leq 0$.

(Σ2) If $\alpha, \beta \in \Pi$, $\alpha \perp \beta$, and $\alpha + \beta \in \Sigma \sqcup 2\Sigma$, then $\alpha^\vee = \beta^\vee$ on Λ.

(S) $\langle \alpha^\vee, \Lambda \rangle = 0$, $\forall \alpha \in \Pi^p$, and the pair (λ, Π^p) comes from a wonderful variety of rank 1 for any $\lambda \in \Sigma$.

A *spherical system* is a triple $(\Pi^p, \Sigma, \mathscr{D}^a)$ satisfying the above axioms with $\Lambda = \mathbb{Z}\Sigma$.

The homogeneous spherical datum of the open orbit in a wonderful variety amounts to its spherical system. It is easy to see that there are finitely many spherical systems for given G. There is a transparent graphical representation of spherical systems by *spherical diagrams* [Lu6, 4.1], [BraL, 1.2.4], which are obtained from the Dynkin diagram of G by adding some supplementary data describing types of simple roots, colors, and spherical roots.

For the homogeneous spherical datum of O, most of the axioms (A1)–(A3), (Σ1)–(Σ2), (S) are verified using the above results together with some additional general arguments. For instance, the inequality in (Σ1) stems from the fact that $\Sigma = \Pi_O^{\min}$ is a base of a root system $\Delta_O^{\min}$. On the other hand, each axiom involves at most two simple or spherical roots, like the axioms of classical root systems. Thus the localizations at one or two simple or spherical roots reduce the verification to wonderful varieties of rank ≤ 2.

Actually the list of axioms was obtained by inspecting the classification of wonderful varieties of rank ≤ 2, which leads to the following conclusion: spherical systems (homogeneous data) with $|\Sigma| \leq 2$ bijectively correspond to wonderful varieties of rank ≤ 2 (resp. to spherical homogeneous spaces $O = G/H$ with $r(G/N_G(H)) \leq 2$). It is tempting to extend this combinatorial classification to arbitrary wonderful varieties and spherical spaces. Now this program is fulfilled after a decade of joint efforts of several researchers.

Theorem 30.22. *For any connected reductive group G, there are natural bijections:*

$$\left\{ \begin{matrix} \text{spherical homogeneous} \\ G\text{-spaces} \end{matrix} \right\} \longleftrightarrow \left\{ \begin{matrix} \text{homogeneous spherical} \\ \text{data for } G \end{matrix} \right\} \tag{30.2}$$

$$\{\text{wonderful } G\text{-varieties}\} \longleftrightarrow \{\text{spherical systems for } G\} \tag{30.3}$$

Example 30.23. Solvable spherical subgroups or, more precisely, spherical subgroups contained in a Borel subgroup of G were classified by Luna [Lu3]. Spherical data arising here satisfy $\Sigma = \Pi^a$, $\Pi^{a'} = \Pi^p = \emptyset$, $D_\alpha^- \neq D_\beta^\pm$, and $\langle \varkappa(D_\alpha^+), \beta \rangle < 0 \implies \langle \varkappa(D_\beta^+), \alpha \rangle = 0$ $(\forall \alpha, \beta \in \Pi^a$, $\alpha \neq \beta)$.

Indeed, by Example 15.13 a spherical subgroup $H \subset G$ is contained in a Borel subgroup if and only if there exists a subset $\mathscr{R}_0 \subset \mathscr{D}^B$ such that $\mathscr{V} \cup \varkappa(\mathscr{R}_0)$ generates $\mathscr{E}$ as a cone and $P[\mathscr{D}^B \setminus \mathscr{R}_0] = B$. This means that $\mathscr{R}_0 \subset \mathscr{D}^a$ contains a unique element, say D_α^+, from each pair $D_\alpha^\pm$, $\Sigma = \Pi^a$ (because $\langle \mathscr{V} \cup \varkappa(\mathscr{R}_0), \Sigma \setminus \Pi^a \rangle \leq 0$ by (A1)), and $\Pi^{a'} = \Pi^p = \emptyset$. Since $\mathscr{D}^B \setminus \mathscr{R}_0$ consists of the preimages of the Schubert divisors, each of these divisors is moved by a unique minimal parabolic, whence the condition on D_α^-. The condition on pairings holds, because otherwise $\langle \mathscr{V} \cup \varkappa(\mathscr{R}_0), \alpha + \beta \rangle \leq 0$ by (A1), a contradiction.

An explicit description of connected solvable spherical subgroups was recently obtained by Avdeev [Avd].

30.12 Proof of the Classification Theorem.

Stage 1. We may assume that G is of simply connected type. The bijection (30.2) was proved by Luna provided that $G/Z(G)$ satisfies (30.3) [Lu6, §7]. The basic idea is to replace $O = G/H$ by $\overline{O} = G/\overline{H}$. This passage preserves the types of simple roots and colors, and $\Pi_{\overline{O}}^{\min}$ is obtained from $\Pi_{O}^{\min}$ by a dilation: some $\lambda \in \Pi_{O}^{\min} \setminus (\Pi \sqcup 2\Pi)$ are replaced by 2λ. It is not hard to prove that spherical subgroups H with fixed very sober hull $\overline{H}$ bijectively correspond to homogeneous spherical data $(\Lambda, \Pi^p, \Sigma, \mathscr{D}^a)$ such that $(\Pi^p, \Pi_{\overline{O}}^{\min}, \mathscr{D}^a)$ is the spherical system of $\overline{O}$, $\Lambda \supset \Pi_{\overline{O}}^{\min}$, and Σ is obtained from $\Pi_{\overline{O}}^{\min}$ by replacing $\lambda \in \Pi_{\overline{O}}^{\min} \setminus (\Pi \sqcup 2\Pi)$ with $\lambda/2$ whenever $\lambda/2 \in \Lambda$ [Lu6, §6].

Indeed, let H_0 denote the common kernel of all characters in $\mathfrak{X}(\overline{H})$. In the notation of 30.5, $T_{\overline{O}} = \overline{H}/H_0$ is a diagonalizable group with $\mathfrak{X}(T_{\overline{O}}) = \mathfrak{X}(\overline{H})$ and $H_0 \subseteq H \subseteq \overline{H}$. (Note that H_0 itself may be not spherical and even if it is spherical, it may happen that $\overline{H_0} \neq \overline{H}$. Examples are: $G = \mathrm{SL}_2(\Bbbk)$, $\overline{H} = T$, $H_0 = \{e\}$, and $\overline{H} = N(T)$, $H_0 = T$, respectively.) So the problem is to classify intermediate spherical subgroups between H_0 and $\overline{H}$ with very sober hull $\overline{H}$. Intermediate subgroups H bijectively correspond to sublattices $\mathfrak{X}(\overline{H}/H) \subseteq \mathfrak{X}(T_{\overline{O}})$. It remains to determine which of them are spherical and have very sober hull $\overline{H}$.

We shall use the notation of 30.5. The equations $\eta_D \in \Bbbk[G]^{(B\times\overline{H})}$ of colors $D \subset \overline{O}$ can be chosen $Z(G)^0$-invariant, so that λ_D, χ_D vanish on $Z(G)^0$. There is an equivariant version of the commutative diagram (30.1):

$$\begin{array}{ccccc}
\mathrm{Pic}_G \overline{O} & \longleftarrow & \mathrm{Pic}_G \overline{X} & \longrightarrow & \mathrm{Pic}_G \overline{Y} \\
\| & & \| & & \| \\
\mathfrak{X}(T_{\overline{O}}) & \longleftarrow & \mathbb{Z}\overline{\mathscr{D}}^B \oplus \mathfrak{X}(G) & \longrightarrow & \mathfrak{X}(P) \\
\chi_D + \mu & \longleftarrow & (D,\mu) & \longrightarrow & \lambda_D - \mu,
\end{array}$$

where $\overline{\mathscr{D}}^B$ is the set of colors of $\overline{O}$ and $\overline{X} \hookleftarrow \overline{O}$ is the wonderful embedding with the closed G-orbit $\overline{Y}$. The weights λ_D are easy to determine.

Lemma 30.24 ([Fo]). *For any color D of a spherical homogeneous space $O = G/H$ one has*

$$\lambda_D = \begin{cases} \sum_{D = D_{\alpha_i}^{\pm}} \omega_i, & D \in \mathscr{D}^a, \\ 2\omega_i, & D = D_{\alpha_i} \in \mathscr{D}^{a'}, \\ \sum_{D = D_{\alpha_i}} \omega_i \quad (\leq 2 \text{ summands}), & D \in \mathscr{D}^b, \end{cases}$$

where $\omega_i \in \mathfrak{X}_+$ denote the fundamental weights corresponding to the simple roots $\alpha_i \in \Pi$.

Proof. Clearly, λ_D is a positive linear combination of the ω_i such that P_{α_i} moves D. In order to determine the coefficient at ω_i, it suffices to localize at α_i and consider the respective spherical SL_2-variety X^{α_i}. □

The subgroup H is not spherical if and only if there exists an eigenspace $\Bbbk[G]^{(B\times H)}_{(\lambda,\chi)}$ of dimension >1. Since $\overline{H}$ acts on $\mathfrak{X}(H)$ via a finite quotient group, we may multiply this eigenspace by $(B\times H)$-eigenfunctions whose H-weights run over the $\overline{H}$-orbit of χ (except χ itself) and thus assume that χ is $\overline{H}$-invariant, i.e., $\overline{H}$ acts on $\Bbbk[G]^{(B\times H)}_{(\lambda,\chi)}$ by right translations of an argument. Taking an $\overline{H}$-eigenbasis of $\Bbbk[G]^{(B\times H)}_{(\lambda,\chi)}$, we obtain at least two $(B\times\overline{H})$-eigenfunctions of distinct eigenweights (λ,χ_i) $(i=1,2)$ such that $\chi_i|_H=\chi$. This means that the preimage $\widehat{\Lambda}$ of $\mathfrak{X}(\overline{H}/H)$ in $\operatorname{Pic}_G(\overline{X})$ maps to $\mathfrak{X}(B)$ non-injectively. Note that the image of $\widehat{\Lambda}$ is nothing else but the weight lattice $\Lambda=\Lambda(O)$.

Thus H is spherical if and only if $\widehat{\Lambda}$ injects into $\mathfrak{X}(B)$. In other words, for each $\lambda\in\Lambda$ there must be a unique presentation $\lambda=\sum m_D\lambda_D-\mu$ such that $\sum m_D\chi_D+\mu\in\mathfrak{X}(\overline{H}/H))$. This allows a map $\varkappa:\overline{\mathscr{D}}^B\to\Lambda^*$ to be defined such that $\langle\varkappa(D),\lambda\rangle=m_D$. This map is well defined on colors of types a',b for any H, because $m_{D_\alpha}=\langle\alpha^\vee/2,\lambda\rangle$ or $\langle\alpha^\vee,\lambda\rangle$ for $\alpha\in\Pi^{a'}$ or Π^b, respectively, by Lemma 30.24, and so the existence of $\varkappa$ is essential only for colors of type a. Clearly, $\varkappa$ is compatible with the respective map for the spherical system of $\overline{O}$. Also by Lemma 30.24 the axioms (A2), (Σ1)–(Σ2), (S) hold. The group $\mathfrak{X}(\overline{H}/H)$ is recovered from $\Lambda,\varkappa$ as the set of all $\chi=\sum\langle\varkappa(D),\lambda\rangle\chi_D-\lambda|_{Z(G)^0}$ with $\lambda\in\Lambda$.

Now $\Sigma=\Pi_O^{\min}$ is obtained from $\Pi_{\overline{O}}^{\min}$ by replacing spherical roots with proportional indivisible vectors in Λ. The map $O\to\overline{O}$ may patch some colors together, namely $D_\alpha^\pm\subset O$ map onto $D_\alpha\subset\overline{O}$ whenever $\alpha\in\Pi\cap\Pi_O^{\min}$, $2\alpha\in\Pi_{\overline{O}}^{\min}$. So $\overline{H}$ is the very sober hull of H if and only if $\alpha\notin\Lambda$ whenever $2\alpha\in\Pi_{\overline{O}}^{\min}$. We conclude that spherical subgroups with very sober hull $\overline{H}$ bijectively correspond to homogeneous spherical data such that Σ is proportional to $\Pi_{\overline{O}}^{\min}$ and $\alpha\notin\Sigma$ whenever $2\alpha\in 2\Pi\cap\Pi_{\overline{O}}^{\min}$, which is our claim.

If (30.3) holds for the adjoint group of G, then $\Pi_{\overline{O}}^{\min}$ coincides with the set $\overline{\Sigma}$ obtained from $\Sigma=\Pi_O^{\min}$ by the "maximal possible" dilation: every $\lambda\in\Sigma\setminus(\Pi\sqcup 2\Pi)$ such that $2\lambda\in\Sigma_G$ and $(2\lambda,\Pi^p)$ corresponds to a wonderful variety of rank 1 is replaced by 2λ. The spherical system $(\Pi^p,\overline{\Sigma},\mathscr{D}^a)$ is said to be the *spherical closure* of $(\Pi^p,\Sigma,\mathscr{D}^a)$. Indeed, by (30.3) this spherical closure corresponds to a certain very sober subgroup $\overline{\overline{H}}\subseteq G$. By the above, the spherical system of $\overline{O}$ corresponds to a certain subgroup with very sober hull $\overline{\overline{H}}$. Again by (30.3), this subgroup is conjugate to $\overline{H}$, and hence coincides with $\overline{\overline{H}}$. It follows that the spherical homogeneous datum of O determines the spherical system of $\overline{O}$ in a pure combinatorial way. Conversely, this spherical system together with $\Lambda,\varkappa$ determines $\overline{O}$ and O by the above, which proves (30.2).

Stage 2. The proof of (30.3) for adjoint G is much more difficult. Luna proposed the following strategy. The first stage is to prove that certain geometric operations on wonderful varieties (localization, parabolic induction, direct product, etc) are expressed in a pure combinatorial language of spherical systems. Every spherical system is obtained by these combinatorial operations from a list of *primitive* systems.

The next stage is to classify primitive spherical systems. And finally, for primitive systems, the existence and uniqueness of a geometric realization is proved case by case. This strategy was implemented by Luna in the case where all simple factors of G are of type **A** [Lu6]. Later on, this approach was extended by Bravi and Pezzini to the groups with the simple factors of types **A** and **D** [Bra1], [BraP1] or **A** and **C** (with some technical restrictions) [Pez]. In [Bra2] Bravi settled the case of arbitrary G with simply laced Dynkin diagram, and the case $\mathbf{F}_4$ was considered in [BraL]. Recently the reduction to primitive spherical systems was justified in [BraP2] for any G and the complete list of primitive spherical systems was given in [Bra3]. Using this list, the case of G with classical factors was settled in [BraP2].

On the other hand, the uniqueness of a geometric realization was proved by Losev by a general argument [Los2]. It remained to prove the existence. A new conceptual approach was suggested by Cupit-Foutou, who completed the proof of the theorem in [C-F2]. We give an outline of her proof.

Instead of assuming that G is adjoint, it is more convenient to suppose that G is semisimple simply connected. An idea of how to reconstruct a wonderful G-variety X from its spherical system is inspired by Brion's description of the Cox ring $\mathscr{R}(X)$, see 30.5. By Theorem 30.12(3), $\mathscr{X} = \operatorname{Spec}\mathscr{R}(X)$ is the total space of a flat family π_G of $\widehat{G}$-varieties with categorical quotient by U isomorphic to $\mathbb{A}^d$, $d = |\mathscr{D}^B|$, where $\widehat{T}$ acts by weights $-\widehat{\lambda}_D$, $D \in \mathscr{D}^B$. Hence $\pi_G : \mathscr{X} \to \mathbb{A}^r$ is the pullback of the universal family $\mathscr{X}^{\mathrm{univ}} \to \mathrm{Hilb}^{\widehat{G}}_{\mathbb{A}^d}$ along a T-equivariant map $\mathbb{A}^r \to \mathrm{Hilb}^{\widehat{G}}_{\mathbb{A}^d}$ (see Appendix E.3). It turns out that $\mathrm{Hilb}^{\widehat{G}}_{\mathbb{A}^d} \simeq \mathbb{A}^r$ as T-varieties if the spherical system of X is spherically closed (i.e., coincides with its spherical closure), see below. Moreover, the T-orbit in $\mathrm{Hilb}^{\widehat{G}}_{\mathbb{A}^d}$ of a typical fiber $G/\!/H_0$ of π_G is open by Theorem E.14(3), since $-\Pi_X^{\min}$, $-\Pi_{G/H_0}^{\min}$, and the tails of $G/\!/H_0$ generate one and the same cone. Therefore $\mathbb{A}^r$ is mapped to $\mathrm{Hilb}^{\widehat{G}}_{\mathbb{A}^d}$ dominantly, whence isomorphically. Thus $\mathscr{X} \simeq \mathscr{X}^{\mathrm{univ}}$ depends (as a spherical $(G \times T_X)$-variety) only on the spherical system of X (see below). Now $\mathring{\mathscr{X}}$ is obtained from $\mathscr{X}$ by removing all $(G \times T_X)$-orbits contained in colors. Indeed, removing these orbits yields the regularity locus of the rational map $\mathscr{X} \dashrightarrow X$. This open set $\mathscr{X}'$ cannot be larger than $\mathring{\mathscr{X}}$, because $\mathring{\mathscr{X}} \to X$ is an affine morphism, whence $\operatorname{codim}(\mathscr{X}' \setminus \mathring{\mathscr{X}}) = 1$, while $\operatorname{codim}(\mathscr{X} \setminus \mathring{\mathscr{X}}) > 1$. Finally, $X = \mathring{\mathscr{X}}/T_X$.

This argument also suggests a way to construct a wonderful variety from any given spherically closed spherical system $(\Pi^p, \Sigma, \mathscr{D}^a)$. Let $\mathscr{D} = \mathscr{D}^a \sqcup \mathscr{D}^{a'} \sqcup \mathscr{D}^b$ denote the set of colors of the spherical system obtained by adding to $\mathscr{D}^a$ the elements D_α, $\alpha \in \Pi^{a'} \sqcup \Pi^b$, with identifications as in 30.10. Consider a torus $T_X = (\mathbb{k}^\times)^{\mathscr{D}}$ and a subgroup $T_O \subseteq T_X$ defined by equations $\prod_{D \in \mathscr{D}} t_D^{\langle \varkappa(D), \lambda \rangle} = 1$, $\forall \lambda \in \Sigma$. (This notation is used for consistency, though there are no X and O at the moment.) Let $\varepsilon_D(t) = t_D$ denote the basic characters of T_X and $\chi_D = \varepsilon_D|_{T_O}$. Define the weights λ_D by the formulæ of Lemma 30.24. Note that $\sum \langle \varkappa(D), \lambda \rangle \lambda_D = \lambda$, $\forall \lambda \in \Sigma$, and the biweights $\widehat{\lambda}_D = (\lambda_D, \chi_D)$ are linearly independent. We shall freely use other notation from 30.5.

Consider the invariant Hilbert scheme $\mathrm{Hilb}^{\widehat{G}}_{\widehat{\lambda}}$ parameterizing affine $\widehat{G}$-varieties Z with $Z/\!/U \simeq \mathbb{A}^d$, $d = |\mathscr{D}|$, where $\widehat{T}$ acts by weights $-\widehat{\lambda}_D$, $D \in \mathscr{D}$. The isomorphism $Z/\!/U \xrightarrow{\sim} \mathbb{A}^d$ gives rise to a unique $\widehat{G}$-equivariant closed embedding $Z \hookrightarrow V = \bigoplus V(\widehat{\lambda}_D^*)$. It follows that $\mathrm{Hilb}^{\widehat{G}}_{\widehat{\lambda}}$ is an open subset of the invariant Hilbert scheme $\mathrm{Hilb}^{\widehat{G}}_m(V)$ parameterizing affine spherical $\widehat{G}$-subvarieties of V with rank semigroup $\widehat{\Lambda}_+ = \sum \mathbb{Z}_+ \widehat{\lambda}_D$ which are not contained in any proper $\widehat{G}$-submodule of V. (Here m is the indicator function of $\widehat{\Lambda}_+$.)

The $\widehat{T}$-action on $\mathrm{Hilb}^{\widehat{G}}_{\widehat{\lambda}}$, which is induced by $\widehat{T} : V$, where $\widehat{T}$ acts on $V(\widehat{\lambda}_D^*)$ by weight $-\widehat{\lambda}_D$, contracts $\mathrm{Hilb}^{\widehat{G}}_{\widehat{\lambda}}$ to the unique $\widehat{T}$-fixed point corresponding to an S-variety $Z_0 = \overline{\widehat{G}v}$, where $v = \sum v_{-\widehat{\lambda}_D}$ is the sum of lowest vectors, see Theorems E.14 and 28.3.

The tangent space and the obstruction space of $\mathrm{Hilb}^{\widehat{G}}_{\widehat{\lambda}}$ at $[Z_0]$ are given by Proposition E.11. It turns out that $T_{[Z_0]}\mathrm{Hilb}^{\widehat{G}}_{\widehat{\lambda}} \simeq \mathbb{A}^r$, $r = |\Sigma|$, where T acts with eigenweight set $-\Sigma$ and $T^2(Z_0)^{\widehat{G}} = 0$. The proof involves computation of cohomologies of non-reductive groups and Lie algebras; this is the most technical part of [C-F2]. It then follows that $\mathrm{Hilb}^{\widehat{G}}_{\widehat{\lambda}} \simeq \mathbb{A}^r$.

Let $\pi_G : \mathscr{X} \subseteq V \times \mathbb{A}^r \to \mathbb{A}^r$ be the universal family. The T_O-action on V extends to T_X by letting it act on each $V(\widehat{\lambda}_D^*)$ by weight $-\varepsilon_D$, and π_G is clearly T_X-equivariant.

Lemma 30.25. (1) *$\mathscr{X}$ is a factorial affine spherical $(G \times T_X)$-variety defined by a supported colored cone $(\mathscr{C}, \widehat{\mathscr{D}})$, where $\mathscr{C} = (\mathbb{Q}_+\Lambda_+(\mathscr{X}))^\vee$ and $\widehat{\mathscr{D}}$ is the set of colors of $\mathscr{X}$, which is identified with $\mathscr{D}$. The spherical roots and the types of simple roots and colors for $\mathscr{X}$ are the same as for the given spherical system.*
(2) *The algebra $\Bbbk[\mathscr{X}]^U$ (resp. the semigroup $\Lambda_+(\mathscr{X})$) is freely generated by the restrictions of linear $(B \times T_X)$-eigenfunctions on V (resp. by their biweights $(\lambda_D, \varepsilon_D)$) and by the coordinate functions on $\mathbb{A}^r$ (resp. by their biweights $(0, \sum \langle \varkappa(D), \lambda \rangle \varepsilon_D)$, $\lambda \in \Sigma$).*
(3) *These functions are the equations of the colors and of the $(G \times T_X)$-stable divisors on $\mathscr{X}$, respectively, or, equivalently, these divisors are mapped by $\varkappa$ to the basis dual to the basis of $\Lambda_+(X)$.*

Proof. $\mathscr{X}$ is spherical, because the fibers of π_G are spherical $\widehat{G}$-varieties and general fibers are transitively permuted by T_X. The description of the colored data of $\mathscr{X}$ stems from the results of 15.1 observing that $\mathscr{X}$ contains a fixed point 0, whose colored cone is solid by Proposition 15.14. The assertion (2) immediately follows from the definition of the invariant Hilbert scheme and the universal family.

By Theorem E.14(3), the tails of $\mathscr{X}$ or, equivalently, of a general fiber Z of π_G span a free semigroup with basis $-\Sigma$, whence, by ($\mathrm{T_O}$), $\Pi^{\min}_{\mathscr{X}}$ is proportional to Σ. But it is easy to check that $\Lambda(\mathscr{X}) \cap \mathfrak{X}(T) = \mathbb{Z}\Sigma$, whence Σ consists of indivisible vectors in $\Lambda(\mathscr{X})$, i.e., $\Pi^{\min}_{\mathscr{X}} = \Sigma$.

It follows from (2) that the simple roots α of type p for $\mathscr{X}$ are those satisfying $\langle\alpha^\vee,\lambda_D\rangle=0$, $\forall D\in\mathscr{D}$, i.e., $\alpha\in\Pi^p$. Hence the types of simple roots with respect to $\mathscr{X}$ are the same as for our spherical system. In particular, $\widehat{\mathscr{D}}^{a'}$ and $\widehat{\mathscr{D}}^b$ are in bijection with $\mathscr{D}^{a'}$ and $\mathscr{D}^b$, respectively. Moreover, it stems from the definition of λ_D that for any $D\in\mathscr{D}^{a'}\sqcup\mathscr{D}^b$ the respective color $\widehat{D}\in\widehat{\mathscr{D}}^{a'}\sqcup\widehat{\mathscr{D}}^b$ is mapped by $\varkappa$ to the vector of the dual basis corresponding to the vector $(\lambda_D,\varepsilon_D)$ in the basis of $\Lambda_+(\mathscr{X})$.

As for colors of type a, we use an observation of R. Camus: for any $\alpha\in\Pi^a$, at least one of the two P_α-unstable colors $\widehat{D}_\alpha^\pm\in\widehat{\mathscr{D}}^a$ is mapped to an edge of $\mathscr{C}$. (Otherwise $-\alpha$ is non-negative on $\mathscr{C}$, because it is non-negative on $\mathscr{V}(\mathscr{X})\cup\varkappa(\widehat{\mathscr{D}}\setminus\{\widehat{D}_\alpha^\pm\})$ by Lemma 30.20 and axiom (A1), whence $-\alpha\in\mathbb{Z}_+\Lambda_+(\mathscr{X})$ is dominant, a contradiction.) Since $\varkappa(\widehat{D}_\alpha^+)+\varkappa(\widehat{D}_\alpha^-)=\overline{\alpha^\vee}$ takes the value 1 on $(\lambda_{D_\alpha^\pm},\varepsilon_{D_\alpha^\pm})$ and vanishes on the other vectors in the basis of $\Lambda_+(X)$, we deduce that $\varkappa(\widehat{D}_\alpha^\pm)$ are the vectors of the dual basis corresponding to $(\lambda_{D_\alpha^\pm},\varepsilon_{D_\alpha^\pm})$.

Thus $\widehat{\mathscr{D}}$ is in bijection with $\mathscr{D}$ preserving types of colors, and the colors of $\mathscr{X}$ are represented by the vectors of the dual basis of $\Lambda(\mathscr{X})^*$ corresponding to the vectors $(\lambda_D,\varepsilon_D)$ $(D\in\mathscr{D})$ in the basis of $\Lambda_+(\mathscr{X})$. Clearly, the remaining vectors in the dual basis (which spans $\mathscr{C}$) correspond to the $(G\times T_X)$-stable divisors, whence $\mathscr{X}$ is factorial. This completes the proof of (1) and (3). □

Let $\mathring{\mathscr{X}}\subseteq\mathscr{X}$ be an open subset obtained by removing all $(G\times T_X)$-orbits contained in a color. Alternatively, $\mathring{\mathscr{X}}$ can be described as the set of points having non-zero projection to each $V(\widehat{\lambda}_D^*)$ or the union of open $\widehat{G}$-orbits in the fibers of π_G. It is a smooth simple toroidal $(G\times T_X)$-variety defined by the face $\mathscr{C}'\subseteq\mathscr{C}$ spanned by the edges orthogonal to $(\lambda_D,\varepsilon_D)$ $(D\in\mathscr{D})$. Now $\mathring{\mathscr{X}}\subseteq\prod(V(\widehat{\lambda}_D^*)\setminus\{0\})\times\mathbb{A}^r$ admits a geometric quotient $X=\mathring{\mathscr{X}}/T_X$ by a free T_X-action, which is a smooth simple toroidal G-variety. It is easy to see that X is complete (because $\mathscr{C}'$ maps onto $\mathscr{V}(X)$), whence wonderful, and the spherical system of X coincides with $(\Pi^p,\Sigma,\mathscr{D}^a)$.

If $(\Pi^p,\Sigma,\mathscr{D}^a)$ is not spherically closed, then one may consider its spherical closure $(\Pi^p,\overline{\Sigma},\mathscr{D}^a)$ and construct the respective wonderful embedding $\overline{X}\hookleftarrow\overline{O}=G/\overline{H}$. A subgroup $H\subset\overline{H}$ such that $O=G/H$ has a wonderful embedding X with spherical system $(\Pi^p,\Sigma,\mathscr{D}^a)$ was constructed at Stage 1.

In fact, it is proved in [C-F2] that $\mathrm{Hilb}^{\widehat{G}}_{\widehat{\lambda}}$ is an r-dimensional vector space on which T acts with eigenweight set $-\overline{\Sigma}$, for any spherical system $(\Pi^p,\Sigma,\mathscr{D}^a)$. Arguing as above, we see that the Cox family $\pi_G:\mathscr{X}=\operatorname{Spec}\mathscr{R}(X)\to\mathbb{A}^r$ is pulled back from $\mathscr{X}^{\mathrm{univ}}$ along a unique T-equivariant finite map $\mathbb{A}^r\to\mathrm{Hilb}^{\widehat{G}}_{\widehat{\lambda}}$, and therefore X is uniquely determined by $(\Pi^p,\Sigma,\mathscr{D}^a)$. This concludes the proof of Theorem 30.22.

31 Frobenius Splitting

Frobenius splitting is a powerful tool of modern algebraic geometry which allows various geometric and cohomological results to be proved by reduction to positive characteristic. This notion was introduced by Mehta and Ramanathan [MRa] in their study of Schubert varieties.

31.1 Basic Properties. Let X be an algebraic variety over an algebraically closed field $\Bbbk$ of characteristic $p > 0$. The Frobenius endomorphism $f \mapsto f^p$ of $\mathscr{O}_X$ gives rise to the Frobenius morphism $F : X^{1/p} \to X$, where $X^{1/p} = X$ as ringed spaces but the $\Bbbk$-algebra structure on $\mathscr{O}_{X^{1/p}}$ is defined as $c * f = c^p f$, $\forall c \in \Bbbk$. (We emphasize here that F acts identically on points, but non-trivially on functions.)

If X is a subvariety in $\mathbb{A}^n$ or $\mathbb{P}^n$, then $X^{1/p}$ is, too. The defining equations of $X^{1/p}$ are obtained from those of X by replacing all coefficients with their p-th roots. The Frobenius morphism F is given by raising all coordinates to the power p.

The Frobenius endomorphism may be regarded as an injection of $\mathscr{O}_X$-modules $\mathscr{O}_X \hookrightarrow F_*\mathscr{O}_{X^{1/p}}$, where $F_*\mathscr{O}_{X^{1/p}} = \mathscr{O}_X$ is endowed with another $\mathscr{O}_X$-module structure: $f * h = f^p h$ for any local sections f of $\mathscr{O}_X$ and h of $F_*\mathscr{O}_{X^{1/p}}$.

Definition 31.1. The variety X is said to be *Frobenius split* if the Frobenius homomorphism has an $\mathscr{O}_X$-linear left inverse $\sigma : F_*\mathscr{O}_{X^{1/p}} \to \mathscr{O}_X$, called a *Frobenius splitting*. In other words, σ is a $\mathbf{Z}_p$-linear endomorphism of $\mathscr{O}_X$ such that $\sigma(1) = 1$ and $\sigma(f^p h) = f\sigma(h)$.

For any closed subvariety $Y \subset X$ one has $\sigma(\mathscr{I}_Y) \supseteq \mathscr{I}_Y$, because $\mathscr{I}_Y \supseteq \mathscr{I}_Y^p$. The splitting σ is *compatible* with Y if $\sigma(\mathscr{I}_Y) = \mathscr{I}_Y$. Clearly, a compatible splitting induces a splitting of Y.

More generally, let δ be an effective Cartier divisor on X, with the canonical section $\eta_\delta \in \mathrm{H}^0(X, \mathscr{O}(\delta))$, $\operatorname{div} \eta_\delta = \delta$. We say that X is *Frobenius split relative to* δ if there exists an $\mathscr{O}_X$-module homomorphism, called a *δ-splitting*, $\sigma_\delta : F_*\mathscr{O}_{X^{1/p}}(\delta) \to \mathscr{O}_X$ such that $\sigma(h) = \sigma_\delta(h\eta_\delta)$ is a Frobenius splitting or, equivalently, $\sigma_\delta(\eta_\delta) = 1$ and $\sigma_\delta(f^p\eta) = f\sigma_\delta(\eta)$ for any local section η of $\mathscr{O}(\delta)$. The δ-splitting σ_δ is *compatible* with Y if the support of δ contains no component of Y (i.e., δ restricts to a divisor on Y) and σ is compatible with Y. Then σ_δ induces a $(\delta \cap Y)$-splitting of Y.

For a systematic treatment of Frobenius splitting and its applications, we refer to a monograph of Brion and Kumar [BKu]. Here we recall some of its most important properties.

Clearly, a Frobenius splitting of X (compatible with Y, relative to δ) restricts to a splitting of every open subvariety $U \subset X$ (compatible with $Y \cap U$, relative to $\delta \cap U$). Conversely, if X is normal and $\operatorname{codim}(X \setminus U) > 1$, then any splitting of U extends to X. In applications it is often helpful to consider $U = X^{\mathrm{reg}}$.

If $\varphi : X \to Z$ is a morphism such that $\varphi_*\mathscr{O}_X = \mathscr{O}_Z$, then a Frobenius splitting of X descends to a splitting of Z. If the splitting of X is compatible with $Y \subset X$, then the splitting of Z is compatible with $\overline{\varphi(Y)}$. For instance, one obtains a splitting of a normal variety X from that of a desingularization of X (if any exists).

It is not hard to prove that Frobenius split varieties are *weakly normal*, i.e., every bijective finite birational map onto a Frobenius split variety has to be an isomorphism [BKu, 1.2.5].

Proposition 31.2. (1) *Suppose that X is a Frobenius split projective variety; then $\mathrm{H}^i(X,\mathscr{L})=0$ for any ample line bundle $\mathscr{L}$ on X and all $i>0$.*
(2) *If $Y\subset X$ is a compatibly split subvariety, then the restriction map $\mathrm{H}^0(X,\mathscr{L})\to\mathrm{H}^0(Y,\mathscr{L})$ is surjective.*
(3) *If the splittings above are relative to an ample divisor δ, then the assertions of* (1)–(2) *hold for any numerically effective (e.g., globally generated) line bundle, i.e., $\mathscr{L}$ such that $\langle\mathscr{L},C\rangle\geq 0$ for any closed curve $C\subseteq X$.*
(4) *There are relative versions of assertions* (1)–(3) *for a proper morphism $\varphi:X\to Z$ stating that $\mathrm{R}^i\varphi_*\mathscr{L}=0$ and $\varphi_*\mathscr{L}\to\varphi_*(\iota_*\iota^*\mathscr{L})$ is surjective under the same assumptions, with $\iota:Y\hookrightarrow X$.*

Proof. The idea is to embed the cohomology of $\mathscr{L}$ as a direct summand in the cohomology of a sufficiently big power of $\mathscr{L}$. Namely the canonical homomorphism $\mathscr{L}\to F_*F^*\mathscr{L}=\mathscr{L}\otimes_{\mathscr{O}_X}F_*\mathscr{O}_{X^{1/p}}$ has a left inverse $\mathbf{1}\otimes\sigma$, whence $\mathscr{L}$ is a direct summand in $F_*F^*\mathscr{L}$. Taking the cohomology yields a split injection

$$\mathrm{H}^i(X,\mathscr{L})\hookrightarrow\mathrm{H}^i(X,F_*F^*\mathscr{L})\simeq\mathrm{H}^i(X^{1/p},F^*\mathscr{L})\simeq\mathrm{H}^i(X,\mathscr{L}^{\otimes p}),\qquad\forall i\geq 0.$$

(The right isomorphism is only $\mathbf{Z}_p$-linear.) Iterating this procedure yields a split $\mathbf{Z}_p$-linear injection $\mathrm{H}^i(X,\mathscr{L})\hookrightarrow\mathrm{H}^i(X,\mathscr{L}^{\otimes p^k})$ compatible with the restriction to Y. Thus the assertions (1) and (2) are reduced to the case of the line bundle $\mathscr{L}^{\otimes p^k}$, $k\gg 0$, where the Serre theorem applies [Har2, III.5.3].

Similar reasoning applies to (3) making use of a split injection $\mathrm{H}^i(X,\mathscr{L})\hookrightarrow\mathrm{H}^i(X,\mathscr{L}^{\otimes p}\otimes\mathscr{O}(\delta))$ together with ampleness of $\mathscr{L}^{\otimes p}\otimes\mathscr{O}(\delta)$. The relative assertions are proved by the same arguments. □

Among other cohomology vanishing results for Frobenius split varieties we mention the extension of the Kodaira vanishing theorem [BKu, 1.2.10(i)]: if X is smooth projective and Frobenius split, then $\mathrm{H}^i(X,\mathscr{L}\otimes\omega_X)=0$ for ample $\mathscr{L}$ and $i>0$.

31.2 Splitting via Differential Forms. Now we reformulate the notion of Frobenius splitting for smooth varieties in terms of differential forms.

The de Rham derivation of $\Omega_X^\bullet$ may be considered as an $\mathscr{O}_X$-linear derivation of $F_*\Omega_{X^{1/p}}^\bullet$. Let $\mathscr{H}_X^k$ denote the respective cohomology sheaves. It is easy to check that $f\mapsto[f^{p-1}\mathrm{d}f]$ is a $\Bbbk$-derivation of $\mathscr{O}_X$ taking values in $\mathscr{H}_X^1$ (where $[\cdot]$ denotes the de Rham cohomology class). By the universal property of Kähler differentials, it induces a homomorphism of graded $\mathscr{O}_X$-algebras

$$c:\Omega_X^\bullet\to\mathscr{H}_X^\bullet,\qquad c(f_0\,\mathrm{d}f_1\wedge\cdots\wedge\mathrm{d}f_k)=[f_0^p(f_1\cdots f_k)^{p-1}\mathrm{d}f_1\wedge\cdots\wedge\mathrm{d}f_k],$$

called the *Cartier operator*. Cartier proved that c is an isomorphism for smooth X. (Using local coordinates, the proof is reduced to the case $X=\mathbb{A}^n$, where the verification is straightforward [BKu, 1.3.4].)

Now suppose that X is smooth. Then we have the *trace map*

$$\tau : F_*\omega_{X^{1/p}} \to \omega_X, \qquad \tau(\omega) = c^{-1}[\omega].$$

In local coordinates $x_1,\dots,x_n$, the trace map can be characterized as the unique $\mathscr{O}_X$-linear map taking $(x_1\cdots x_n)^{p-1}\mathrm{d}x_1\wedge\cdots\wedge\mathrm{d}x_n \mapsto \mathrm{d}x_1\wedge\cdots\wedge\mathrm{d}x_n$ and $x_1^{k_1}\cdots x_n^{k_n}\,\mathrm{d}x_1\wedge\cdots\wedge\mathrm{d}x_n \mapsto 0$ unless $k_1\equiv\cdots\equiv k_n\equiv p-1 \pmod p$.

Using the trace map, it is easy to establish an isomorphism

$$\mathscr{H}om(F_*\mathscr{O}_{X^{1/p}},\mathscr{O}_X) \simeq F_*\omega_{X^{1/p}}^{1-p}, \qquad \sigma\leftrightarrow\widehat{\sigma},$$

such that $\sigma(h)\omega = \tau(h\omega^{\otimes p}\otimes\widehat{\sigma})$ for any local sections h of $F_*\mathscr{O}_{X^{1/p}}$ and ω of ω_X. Similarly, for any divisor δ on X we have

$$\mathscr{H}om(F_*\mathscr{O}_{X^{1/p}}(\delta),\mathscr{O}_X) \simeq F_*\omega_{X^{1/p}}^{1-p}(-\delta).$$

This leads to the following conclusion.

Proposition 31.3 ([BKu, 1.3.8, 1.4.10]). *Suppose that X is smooth and irreducible. Then $\sigma\in\mathrm{Hom}(F_*\mathscr{O}_{X^{1/p}},\mathscr{O}_X)$ is a splitting of X if and only if the Taylor expansion of $\widehat{\sigma}$ at some (hence any) $x\in X$ has the form*

$$\Bigl((x_1\cdots x_n)^{p-1}+\sum c_{k_1,\dots,k_n}x_1^{k_1}\cdots x_n^{k_n}\Bigr)(\partial_1\wedge\cdots\wedge\partial_n)^{\otimes(p-1)},$$

where the sum is taken over all multiindices $(k_1,\dots,k_n)$ such that $\exists k_i\not\equiv p-1 \pmod p$. (Here x_i denote local coordinates and ∂_i the vector fields dual to $\mathrm{d}x_i$.) If X is complete, then it suffices to have

$$\widehat{\sigma} = ((x_1\cdots x_n)^{p-1}+\cdots)(\partial_1\wedge\cdots\wedge\partial_n)^{\otimes(p-1)}.$$

The splitting σ is relative to any effective divisor $\delta\le\operatorname{div}\widehat{\sigma}$.

By abuse of language, we shall say that $\widehat{\sigma}$ splits X if σ does. Also, X is said to be *split by a $(p-1)$-th power* if $\alpha^{\otimes(p-1)}$ splits X for some $\alpha\in\mathrm{H}^0(X,\omega_X^{-1})$. This splitting is compatible with the zero set of α. For instance, a smooth complete variety X is split by the $(p-1)$-th power of α if the divisor of α in a neighborhood of some $x\in X$ is a union of $n=\dim X$ smooth prime divisors intersecting transversally at x.

Example 31.4. Every smooth toric variety X is Frobenius split by a $(p-1)$-th power compatibly with ∂X. For complete X, this stems from the structure of its canonical divisor, given by Proposition 30.8 (which extends to positive characteristic in the toric case). The general case follows by passing to a smooth toric completion. Now toric resolution of singularities readily implies that all normal toric varieties are Frobenius split compatibly with their invariant subvarieties.

Example 31.5 ([Ram], [BKu, Ch. 2–3]). Generalized flag varieties are Frobenius split by a $(p-1)$-th power. For $X=G/B$, $\omega_X^{-1}=\mathscr{L}(-2\rho)$ and the splitting is provided by $\alpha = f_\rho \cdot f_{-\rho} \in V^*(2\rho)$, where $f_{\pm\rho} \in V^*(\rho)$ are T-weight vectors of weights $\pm\rho$.

Moreover, this splitting is compatible with all Schubert subvarieties $S_w = \overline{B(wo)} \subset X$, $w \in W$. Using the weak normality of S_w and the Bott–Samelson resolution of singularities

$$\varphi : \check{S} = \check{S}_{\alpha_1,\dots,\alpha_l} := P_{\alpha_1} *_B \cdots *_B P_{\alpha_l}/B \to S_w,$$
$$w = r_{\alpha_1} \cdots r_{\alpha_l}, \quad \alpha_i \in \Pi, \quad l = \dim S_w,$$

with connected fibers and $\mathrm{R}^i\varphi_*\mathscr{O}_{\check{S}} = 0$, $\forall i > 0$, one deduces that S_w are normal (Demazure, Seshadri) and have rational resolution of singularities (Andersen, Ramanathan). These properties descend to Schubert subvarieties in G/P, $\forall P \supset B$.

Splitting by a $(p-1)$-th power has further important consequences. For instance, the Grauert–Riemenschneider theorem extends to this situation, due to Mehta–van der Kallen [MK]:

> If $\varphi : X \to Y$ is a proper birational morphism, X is smooth and split by $\alpha^{\otimes(p-1)}$ such that φ is an isomorphism on X_α, then $\mathrm{R}^i\varphi_*\omega_X = 0$, $\forall i > 0$.

31.3 Extension to Characteristic Zero. Although the concept of Frobenius splitting is defined in characteristic $p > 0$, it successfully applies to algebraic varieties in characteristic zero via reduction mod p.

Namely let X be an algebraic variety over an algebraically closed field $\Bbbk$ of characteristic 0. One can find a finitely generated subring $R \subset \Bbbk$ such that X is defined over R, i.e., is obtained from an R-scheme $\mathscr{X}$ by extension of scalars. One may assume that $\mathscr{X}$ is flat over R, after replacing R by a localization. For any maximal ideal $\mathfrak{p} \lhd R$ we have $R/\mathfrak{p} \simeq \mathbb{F}_{p^k}$. The variety $X_\mathfrak{p}$ obtained from the fiber $\mathscr{X}_\mathfrak{p}$ of $\mathscr{X} \to \operatorname{Spec} R$ over $\mathfrak{p}$ by an extension of scalars $\mathbb{F}_{p^k} \to \mathbb{F}_{p^\infty}$ is called a *reduction* mod p of X and sometimes denoted simply by X_p (by abuse of notation).

Reductions mod p exist and share geometric properties of X (affinity, projectivity, completeness, smoothness, normality, rational resolution of singularities, etc) for all sufficiently large p. Conversely, a local geometric property of open type (e.g., smoothness, normality, rational resolution of singularities) holds for X if it holds for X_p whenever $p \gg 0$. Replacing R by an appropriate localization, one may always assume that a given finite collection of algebraic and geometric objects on X (subvarieties, line bundles, coherent sheaves, morphisms, etc) is defined over R, and hence specializes to X_p for $p \gg 0$; coherent sheaves may be supposed to be flat over R.

Cohomological applications of reduction mod p are based on the semicontinuity theorem [Har2, III.12.8], which may be reformulated in our setup as follows:

> If X is complete and $\mathscr{F}$ is a coherent sheaf on X, then $\dim \mathrm{H}^i(X,\mathscr{F}) = \dim \mathrm{H}^i(X_p,\mathscr{F}_p)$ for all $p \gg 0$.

This implies, for instance, that the assertions of Proposition 31.2 hold in characteristic zero provided that X_p are Frobenius split for $p \gg 0$. This is the case, e.g., for Fano varieties. Another case, which is important in the scope of this chapter, are spherical varieties.

31.4 Spherical Case.

Theorem 31.6 ([BI]). *If X is a spherical G-variety in characteristic 0, then X_p is Frobenius split by a $(p-1)$-th power compatibly with all G-subvarieties and relative to any given B-stable effective divisor, for $p \gg 0$.*

Proof. Using an equivariant completion of X and a toroidal desingularization of this completion, we may assume that X is smooth, complete, and toroidal. Consider the natural morphism $\varphi : X \to X(\mathfrak{h})$, where $\mathfrak{h}$ is a generic isotropy subalgebra for $G:X$. By Proposition 30.8, $\omega_X^{-1} = \mathscr{O}(\partial X + \varphi^*\mathscr{H})$, where $\mathscr{H}$ is a hyperplane section of $X(\mathfrak{h})$.

The restriction of $\mathscr{O}(\partial X)$ to a closed G-orbit $Y \subset X$ is the top exterior power of the normal bundle to Y, whence $\omega_Y^{-1} = \omega_X^{-1}|_Y \otimes \mathscr{O}(-\partial X)|_Y = \mathscr{O}(\varphi^*\mathscr{H})|_Y$. Since Y is a generalized flag variety, Y_p is split by the $(p-1)$-th power of (the reduction mod p of) some $\alpha_Y \in \mathrm{H}^0(Y, \omega_Y^{-1})$. The G-module $\mathrm{H}^0(Y, \omega_Y^{-1})$ being irreducible and $\mathscr{O}(\varphi^*\mathscr{H})$ globally generated, the restriction map $\mathrm{H}^0(X, \mathscr{O}(\varphi^*\mathscr{H})) \to \mathrm{H}^0(Y, \omega_Y^{-1})$ is surjective and α_Y extends to $\alpha_0 \in \mathrm{H}^0(X, \mathscr{O}(\varphi^*\mathscr{H}))$.

We have $\partial X = D_1 \cup \cdots \cup D_k$, where D_i runs over all G-stable prime divisors of Y. It is easy to see from Proposition 31.3 that $\alpha = \alpha_0 \otimes \alpha_1 \otimes \cdots \otimes \alpha_k$ provides a splitting for X_p, where $\alpha_i \in \mathrm{H}^0(X, \mathscr{O}(D_i))$, $\operatorname{div} \alpha_i = D_i$. Moreover, this splitting is compatible with all $(D_i)_p$ and therefore with all G-subvarieties in X_p, because the latter are unions of transversal intersections of some $(D_i)_p$.

Finally, for any B-stable effective divisor δ we have $\delta \leq (1-p)K_X$ for $p \gg 0$, by Proposition 30.8. Hence the splitting is relative to δ_p by Proposition 31.3. □

It is worth noting that not all spherical varieties in positive characteristic are Frobenius split. Counterexamples are provided by some complete homogeneous spaces with non-reduced isotropy group subschemes [La].

Frobenius splitting of spherical varieties provides short and conceptual proofs for a number of important geometric and cohomological properties. In particular, Theorem 15.20 can be deduced in the following way.

Consider a resolution of singularities $\psi : X' \to X$, where X' is toroidal and quasiprojective. Choose an ample B-stable effective divisor δ on X'; then X'_p is split relative to δ_p for $p \gg 0$. By semicontinuity and Proposition 31.2(4) applied to the trivial line bundle over X'_p, $\mathrm{R}^i\psi_*\mathscr{O}_{X'} = 0$ for $i > 0$, whence X has rational singularities. By the same reason, $\mathscr{O}_X = \psi_*\mathscr{O}_{X'}$ surjects onto $\psi_*\mathscr{O}_{Y'}$ for any irreducible closed G-subvariety $Y' \subset X'$, whence $\psi_*\mathscr{O}_{Y'} = \mathscr{O}_Y$ for $Y = \psi(Y')$. Since Y' is smooth, Y is normal and has rational singularities by the above.

For any line bundle $\mathscr{L}$ on X denote $\mathscr{L}' = \psi^*\mathscr{L}$. The Leray spectral sequence

$$\mathrm{H}^{i+j}(X', \mathscr{L}') \Longleftarrow \mathrm{H}^i(X, \mathrm{R}^j\psi_*\mathscr{L}') = \mathrm{H}^i(X, \mathscr{L} \otimes \mathrm{R}^j\psi_*\mathscr{O}_{X'})$$

degenerates to $\mathrm{H}^i(X', \mathscr{L}') = \mathrm{H}^i(X, \mathscr{L})$, $\forall i \geq 0$. The same holds for direct images instead of cohomology. Together with Proposition 31.2, applied to X'_p and $\mathscr{L}'_p$, this proves the following

Corollary 31.7. *Suppose that* $\operatorname{char}\Bbbk = 0$. *If X is a complete spherical G-variety, $Y \subset X$ a G-subvariety, and $\mathscr{L}$ a numerically effective line bundle on X, then* $\mathrm{H}^i(X,\mathscr{L}) = 0$, $\forall i > 0$, *and the restriction map* $\mathrm{H}^0(X,\mathscr{L}) \to \mathrm{H}^0(Y,\mathscr{L})$ *is surjective. More generally, if X is spherical and $\varphi : X \to Z$ is a proper morphism, then* $\mathrm{R}^i\varphi_*\mathscr{L} = 0$, $\forall i > 0$, *and $\varphi_*\mathscr{L} \to \varphi_*(\iota_*\iota^*\mathscr{L})$ is surjective, where $\iota : Y \hookrightarrow X$ is a closed G-embedding.*

See [Bri7], [Bri12] for other proofs.

More precise results on Frobenius splitting of spherical varieties and their subvarieties (usually G- or B-orbit closures) are obtained in special cases.

As noted above, generalized flag varieties are Frobenius split compatibly with their Schubert subvarieties, and the latter have rational resolution of singularities in positive, hence any (by semicontinuity), characteristic.

Equivariant normal embeddings of G (see §27) are Frobenius split compatibly with their $(G \times G)$-subvarieties, in all positive characteristics. For wonderful completions of adjoint semisimple groups, this was established by Strickland [Str1]. The general case is due to Rittatore [Rit2], see also [BKu, Ch. 6]. This implies that normal reductive group embeddings have rational resolution of singularities (in particular, they are Cohen–Macaulay) and that the coordinate algebras of normal reductive monoids have "good" filtration [Rit2, §4], [BKu, 6.2.13].

Brion and Polo proved that the closures of the Bruhat double cosets in wonderful completions of adjoint semisimple groups (called *large Schubert varieties*) are compatibly split and deduced that they are normal and Cohen–Macaulay [BPo].

De Concini and Springer proved that wonderful embeddings of symmetric spaces for adjoint G are Frobenius split compatibly with their G-subvarieties in odd characteristics [CS, 5.9]. However this splitting is not always compatible with B-orbit closures; in fact, the latter may be neither normal nor Cohen–Macaulay [Bri16].

See [Bri16] for a detailed study of B-orbits in spherical varieties and their closures. This is an area of active current research, with many open questions.

Appendices

A Algebraic Geometry

Here we collect several issues which are not covered by our standard sources in algebraic geometry.

A.1 Rational Singularities.

Definition A.1. A *resolution of singularities* or *desingularization* of a variety X is a proper birational morphism $\varphi : X' \to X$, where X' is smooth.

The existence of a desingularization is very important. In characteristic zero it was proved by Hironaka in 1960's. One can even construct a resolution of singularities in a canonical way, which commutes with open embeddings [BM]. In particular, there exists an equivariant resolution of singularities. In positive characteristic, the existence of a desingularization is known in dimension ≤ 3 (Abhyankar, 1950–60's).

The following definition is due to Kempf [KKMS, Ch. I, §3].

Definition A.2. A resolution of singularities $\varphi : X' \to X$ is said to be *rational* if the following two conditions are satisfied:

(1) $\varphi_* \mathcal{O}_{X'} = \mathcal{O}_X$, $\mathrm{R}^i \varphi_* \mathcal{O}_{X'} = 0$, $\forall i > 0$;
(2) $\mathrm{R}^i \varphi_* \omega_{X'} = 0$, $\forall i > 0$.

Varieties having a rational desingularization are normal and Cohen–Macaulay (with dualizing sheaf $\varphi_* \omega_{X'}$) [KKMS, Ch. I, §3], [BKu, 3.4.2].

In characteristic zero, (2) always holds by the Grauert–Riemenschneider theorem [GR]. Moreover, the rationality property does not depend on a chosen resolution of singularities. Indeed, given two desingularizations $\varphi' : X' \to X$ and $\varphi'' : X'' \to X$, one can construct a desingularization $\varphi : \check{X} \to X$ dominating them both, i.e., such that there is a commutative diagram of proper birational maps

D.A. Timashev, *Homogeneous Spaces and Equivariant Embeddings*,
Encyclopaedia of Mathematical Sciences 138, DOI 10.1007/978-3-642-18399-7,

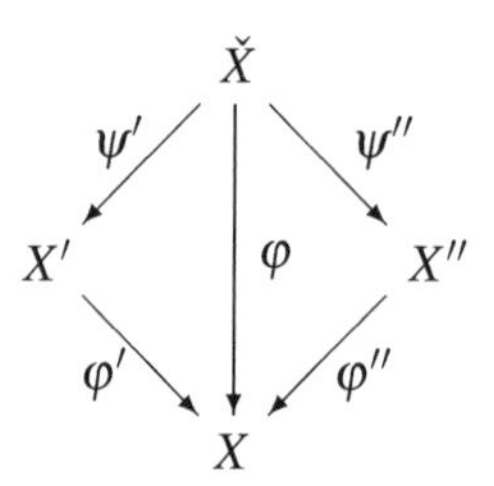

(One may take for $\check{X}$ a desingularization of the closure in $X' \times X''$ of the graph of the birational map $X' \dashrightarrow X''$.) Now the local duality theorem [Har1, VII.3.3] yields a spectral sequence $\mathrm{R}^{i+j}\psi'_*\mathscr{H}om(\mathscr{F},\omega_{\check{X}}) \Longleftarrow \mathscr{E}xt^i(\mathrm{R}^j\psi'_*\mathscr{F},\omega_{X'})$ for any locally free sheaf $\mathscr{F}$ on $\check{X}$. For $\mathscr{F} = \omega_{\check{X}}$, this spectral sequence degenerates by (2) to $\mathrm{R}^i\psi'_*\mathscr{O}_{\check{X}} = \mathscr{E}xt^i(\psi'_*\omega_{\check{X}},\omega_{X'})$, and $\psi'_*\omega_{\check{X}} = \omega_{X'}$ implies that ψ' is a rational resolution (although there is nothing to resolve). Hence the Leray spectral sequence $\mathrm{R}^{i+j}\varphi_*\mathscr{O}_{\check{X}} \Longleftarrow \mathrm{R}^i\varphi'_*\mathrm{R}^j\psi'_*\mathscr{O}_{\check{X}}$ degenerates to $\mathrm{R}^i\varphi_*\mathscr{O}_{\check{X}} = \mathrm{R}^i\varphi'_*\mathscr{O}_{X'}$; similarly for φ'',ψ'' instead of φ',ψ'. Thus the resolutions $\varphi,\varphi',\varphi''$ are rational or not simultaneously. This leads to the following

Definition A.3. A variety X in characteristic zero has *rational singularities* if some, and hence any, desingularization of X is rational.

The notion of rational singularities is local, so that one may speak about rational singularity at a point.

Theorem A.4 ([Elk, Th. 4]). *Rationality of singularities is an open property, i.e., for any flat family $\varphi : X \to Y$ (see Appendix E) the set of $x \in X$ such that the fiber $X_y = \varphi^{-1}(y)$ over $y = \varphi(x)$ has rational singularity at x is open in X.*

Theorem A.5 ([Elk, Th. 5]). *If $\varphi : X \to Y$ is a flat family whose base Y and all fibers X_y have rational singularities, then X has rational singularities, too. In particular, rationality of singularities is preserved by products.*

Theorem A.6 ([Bout]). *The categorical quotient $X/\!\!/G$ of any affine variety X modulo a reductive group G (or a good quotient of any G-variety, see Appendix D) has rational singularities whenever X has rational singularities.*

A.2 Mori Theory. One of the ultimate goals of algebraic geometry is to classify algebraic varieties. Applying the Chow lemma, completion, and desingularization, we reduce the problem in a sense to irreducible smooth projective varieties. The first stage is to classify them up to birational equivalence. The second stage is to classify varieties in a given birational class up to isomorphism. Here the strategy is to find a good model (one or more) in each birational class and explain how to obtain all other models from good ones.

For curves, the second stage is trivial: there is a unique smooth projective curve in each birational class. The first stage is performed by fixing a discrete invariant—the genus of a smooth projective curve—and constructing the moduli space of curves with given genus [MFK, Ch. 5–7], [PV, 4.6]. For surfaces, the classification in a given birational class was performed by classics of the Italian school of algebraic geometry. The classification of birational classes proceeds again by constructing moduli spaces with fixed discrete invariants, see [IS]. A program of classification of higher dimensional varieties extending the cases of dimension ≤ 2 was proposed by S. Mori. We overview it here. Our basic sources in the Mori theory are [KMM] and [M].

We assume that $\operatorname{char}\mathbb{k} = 0$. Given a smooth projective model X of a fixed function field $K/\mathbb{k}$, one can obtain plenty of other models by blowing up various closed subvarieties or subschemes in X. Thus it is natural to hope that X can be obtained by this process from some "most economical" model, which in its turn is obtained from X by an inverse process of "blowing down", i.e., by a proper birational map.

For surfaces, the blowing down contracts (-1)-curves (i.e., smooth rational curves with self-intersection -1) to points. This contraction reduces $\operatorname{rk}\operatorname{Pic}X$, and finally one obtains a surface having no (-1)-curves. This is the case, e.g., if the canonical class K_X is numerically effective. In the latter case, the resulting "minimal" surface is unique in the given birational class; otherwise it is $\mathbb{P}^2$ or a ruled surface, and there are infinitely many non-isomorphic ruled surfaces in the given birational class, which are obtained from each other by so-called elementary transformations.

In arbitrary dimension, we say that X is a *minimal model* if a canonical divisor K_X is numerically effective. Otherwise it appears that there exists a very special surjective morphism $\varphi : X \to Y$ with connected fibers (contraction of an extremal ray, see below). If φ is not birational, then X is called a *Mori fiber space*. The structure of Mori fiber spaces is well understood, and they can serve as good models of K. If φ is birational, then we replace X by Y and continue the process.

However, unlike the 2-dimensional case, this program, called the *minimal model program*, encounters several obstacles. First, even if X were smooth, Y may have singularities. Thus, one has to allow "good" singularities.

Definition A.7. X has *terminal* (*canonical*) *singularities* if it is normal, $\mathbb{Q}$-Gorenstein (i.e., a multiple of K_X is Cartier), and admits a resolution of singularities $\nu : \check{X} \to X$ such that $K_{\check{X}} - \nu^* K_X$ is equivalent in $\operatorname{Pic}(X) \otimes \mathbb{Q}$ to a sum of all exceptional divisors (contracted by ν) with coefficients in $\mathbb{Q}^+$ ($\mathbb{Q}_+$, respectively).

Terminal singularities are rational.

There is a natural pairing $\operatorname{Pic}X \times A_1(X) \to \mathbb{Z}$ given by the degree of a line bundle restricted to a curve in X and pulled back to its normalization. It gives rise to a duality between the quotient spaces $N^1(X)$ and $N_1(X)$ of $\operatorname{Pic}(X) \otimes \mathbb{R}$ and $A_1(X) \otimes \mathbb{R}$, respectively, modulo the kernels of the pairing. Let $NE(X) \subset N^1(X)$ be the closed cone generated by effective 1-cycles, called the *Mori cone*. It is a strictly convex, because each ample divisor is positive on $\mathrm{NE}(X) \setminus \{0\}$ by Kleiman's ampleness criterion [KMM, 0-1-2].

Cone Theorem. *Suppose that X has terminal singularities. Then the set of extremal rays of* $\mathrm{NE}(X)$ *contained in the open half-space* $\{\langle K_X, \cdot \rangle < 0\}$ *is discrete. Furthermore, these rays are generated by irreducible curves C_j and the limit rays may lie only on the boundary hyperplane $K_X^{\perp}$.*

If $\varphi : X \to Y$ is a morphism onto another normal projective variety, then the set of curves contracted by φ generates an extremal set ("face") $F \subseteq \mathrm{NE}(X)$ (as a closed cone). Indeed, if δ is an ample divisor on Y, then a curve $C \subset X$ is contracted if and only if $\langle \varphi^*\delta, C\rangle = 0$, otherwise $\langle \varphi^*\delta, C\rangle > 0$. Also, for any $\delta' \in \mathrm{CaDiv}\,X$ positive on $F \setminus \{0\}$, $\delta' + n\varphi^*\delta$ is ample for $n \gg 0$ by Kleiman's criterion, whence $\langle \delta', \gamma\rangle \geq 0$ for any $\gamma \in \mathrm{NE}(X) \cap (\varphi^*\delta)^{\perp}$, which implies $F = \mathrm{NE}(X) \cap (\varphi^*\delta)^{\perp}$. By Zariski's Main Theorem, φ is characterized by F provided that it has connected fibers; in this case φ is called the *contraction* of F. In general, not every face of $\mathrm{NE}(X)$ is contractible.

Contraction Theorem. *If X has terminal singularities and δ is a numerically effective Cartier divisor such that K_X is negative on $F = \mathrm{NE}(X) \cap \delta^{\perp}$ (excluding 0), then F is contractible.*

Note that, by the Cone Theorem, F is a true face generated by finitely many extremal rays. By Kleiman's criterion, $-K_X$ is ample relative to the contraction morphism $\varphi : X \to Y$. Also, $\varphi^* \operatorname{Pic} Y \xrightarrow{\sim} F^{\perp} \subseteq \operatorname{Pic} X$ and $N_1(Y) \simeq N_1(X)/\langle F\rangle$.

In particular, extremal rays spanned by curves having negative pairing with K_X are contractible. Suppose that X is also $\mathbb{Q}$-factorial. Then contractions of extremal rays are divided into three types:

(1) φ is not birational. Then X is called a *Mori fiber space*. General fibers of φ are $\mathbb{Q}$-Fano varieties with terminal singularities, X is uniruled, and Y is $\mathbb{Q}$-factorial.
(2) φ is birational and its exceptional locus D has codimension 1 in X. Then D is a uniruled prime divisor and Y is $\mathbb{Q}$-factorial with terminal singularities.
(3) φ is an isomorphism in codimension 1. Here Y may have "bad" singularities.

The existence of contractions of the third type is the second obstacle for the minimal model program. In this case, one proceeds by replacing X not with Y but with another model X^+ called a flip of X.

Definition A.8. Suppose that $\varphi : X \to Y$ is a contraction of an extremal ray which is an isomorphism in codimension 1. A *flip* of (X, φ) is a pair (X^+, φ^+), where X^+ is a normal projective variety and $\varphi^+ : X^+ \to Y$ is a birational morphism which is an isomorphism in codimension 1 and K_{X^+} is ample relative to φ^+.

A flip exists if and only if $\mathscr{R} = \bigoplus_{n\geq 0} \varphi_*\mathscr{O}(nK_X)$ is a sheaf of finitely generated $\mathscr{O}_Y$-algebras, and in this case $X^+ = \operatorname{Proj}_{\mathscr{O}_Y} \mathscr{R}$. The variety X^+ is also $\mathbb{Q}$-factorial with terminal singularities, and $\operatorname{Pic}(X^+) \otimes \mathbb{Q} = (\varphi^+)^* \operatorname{Pic}(Y) \otimes \mathbb{Q} \oplus \langle K_{X^+}\rangle$.

The main problems are the existence of flips and termination of sequences of directed flips

$$(X, \varphi) \rightsquigarrow (X^+, \varphi^+) \rightsquigarrow (X^+, \psi) \rightsquigarrow (X^{++}, \psi^+) \rightsquigarrow \dots.$$

The existence of flips was proved by Hacon and McKernan [HM] by induction on dimension in the minimal model program; see also a recent preprint [CL], where the existence of flips (together with other main theorems of the Mori theory) is derived from finite generation of canonical rings. The termination of flips was proved in dimension ≤ 4 and in some other cases, but is still conjectural in general.

If the flip conjecture is true, then the minimal model program works well, and every birational class contains either a Mori fiber space or a minimal model. There may be several (presumably finitely many) minimal models in a given birational class; they are pairwise isomorphic in codimension 1 and connected with each other by sequences of birational transformations called *flops*, which are closely related to flips. As for Mori fiber spaces, birational maps between them are decomposed into a sequence of elementary transformations called *links* (the Sarkisov program). There are also logarithmic and relative versions of the Mori program.

A.3 Schematic Points. Given a $\Bbbk$-scheme X, it is often instructive to consider the respective representable functor associating with any $\Bbbk$-scheme S the set $X(S)$ of $\Bbbk$-morphisms $S \to X$, called *S-points* of X. If $S = \operatorname{Spec} A$ is affine, then S-points are called *A-points* and the notation $X(A) := X(S)$ is used.

Example A.9. If $X \subseteq \mathbb{A}^n$ is an (embedded) affine scheme of finite type, then an A-point of X is given by an algebra homomorphism

$$\Bbbk[X] = \Bbbk[t_1,\dots,t_n]/\mathscr{I}(X) \to A,$$

i.e., by an n-tuple $x = (x_1,\dots,x_n) \in A^n$ satisfying the defining equations of X.

We require a closer look at this notion in case where X is an algebraic variety over $\Bbbk$ and A is a local $\Bbbk$-algebra with the maximal ideal $\mathfrak{m}$. Given $\chi \in X(A)$, the closed point of $\operatorname{Spec} A$ is mapped by χ to the generic point of an irreducible subvariety $Y \subseteq X$ called the *center* of χ. If $\mathring{X} \subseteq X$ is an affine chart meeting Y, then $\chi \in \mathring{X}(A)$. Thus $X(A) = \bigcup \mathring{X}(A)$ over all affine open subsets $\mathring{X} \subseteq X$. From the algebraic viewpoint, an A-point of X is given by an irreducible subvariety $Y \subseteq X$ and a local algebra homomorphism $\mathscr{O}_{X,Y} \to A$, $\mathfrak{m}_{X,Y} \to \mathfrak{m}$, or by a homomorphism $\Bbbk[\mathring{X}] \to A$, where $\mathring{X} \subseteq X$ is an affine chart (intersecting Y).

Example A.10. The *generic point* of an irreducible variety X over $\Bbbk(X)$ has the center X, and $\mathscr{O}_{X,X} \to \Bbbk(X)$ is the identity map. Informally, the coordinates of the generic point are indeterminates bound only by relations that hold identically on X.

Example A.11. If v is a valuation of $\Bbbk(X)$ with center $Y \subseteq X$, then the inclusion $\mathscr{O}_{X,Y} \subseteq \mathscr{O}_v$ yields an $\mathscr{O}_v$-point of X with center Y.

Example A.12. Any A-point of a quasiprojective scheme $X \subseteq \mathbb{P}^n$ is at the same time an A-point of $X \cap \mathbb{A}^n$ for a certain affine chart $\mathbb{A}^n \subseteq \mathbb{P}^n$. In view of Example A.9, A-points of X are identified with tuples $x = (x_0 : \dots : x_n)$, $x_i \in A$, considered up to proportionality, satisfying the defining equations of X, and such that at least one x_i is invertible.

It is quite common in algebraic geometry to consider the case where A is a field. For applications in §24, we consider points over the function field of an algebraic curve or its formal analogue.

Definition A.13. A *germ of a curve* in X is a pair (χ, θ_0), where $\chi \in X(\Bbbk(\Theta))$, Θ is a smooth projective curve, and $\theta_0 \in \Theta$. In other words, a germ of a curve is given by a rational map from a curve to X and a fixed base point on the curve.

The germ is said to be *convergent* if $\chi \in X(\mathscr{O}_{\Theta,\theta_0})$, i.e., the rational map $\chi : \Theta \dashrightarrow X$ is regular at θ_0. The point $x_0 = \chi(\theta_0)$ is the *limit* of the germ.

There is a formal analytic analogue of this notion.

Definition A.14. A *germ of a formal curve* (or simply a *formal germ*) in X is a $\Bbbk((t))$-point of X. A $\Bbbk[[t]]$-point is called a *convergent formal germ*, and its center $x_0 \in X$ is the *limit* of the formal germ.

It is natural to think of a formal germ as a "parameterized formal analytic curve" $x(t)$ in X. In local coordinates, $x(t)$ is a tuple of Laurent series satisfying the defining equations. If $x(t)$ converges, then its coordinates are power series, and their constant terms are the coordinates of the limit $x_0 =: x(0) = \lim_{t\to 0} x(t)$.

With any germ of a curve $(\theta_0 \in \Theta \dashrightarrow X)$ one can associate a formal germ via the inclusions $\mathscr{O}_{\Theta,\theta_0} \subset \widehat{\mathscr{O}}_{\Theta,\theta_0} \simeq \Bbbk[[t]]$, $\Bbbk(\Theta) \subset \Bbbk((t))$, depending on the choice of a formal uniformizing parameter $t \in \widehat{\mathscr{O}}_{\Theta,\theta_0}$. (Here $\widehat{\mathscr{O}}$ denotes the completion of a local ring $\mathscr{O}$.)

Proposition A.15. *In characteristic zero, a formal germ is induced by a germ of a curve if and only if its center has dimension* ≤ 1.

Proof. The "only if" direction and the case where the center is a point, are clear. Suppose that the center of a formal germ is a curve $C \subseteq X$. Then $\Bbbk(C) \hookrightarrow \Bbbk((t))$. Choose any $f \in \Bbbk(C)$, $\operatorname{ord}_t f = k > 0$, and consider $s \in \Bbbk[[t]]$, $s^k = f$. Then $\Bbbk(C)(s) = \Bbbk(\Theta)$ is a function field of a smooth projective curve Θ, and $\Bbbk(\Theta) \cap \Bbbk[[t]] = \mathscr{O}_{\Theta,\theta_0}$ for a certain $\theta_0 \in \Theta$, so that $\widehat{\mathscr{O}}_{\Theta,\theta_0} = \Bbbk[[s]] = \Bbbk[[t]]$. □

There is a *t-adic topology* on $X(\Bbbk((t)))$ thinner than the Zariski topology. For $X = \mathbb{A}^n$, a basic t-adic neighborhood of $x(t) = (x_1(t), \dots, x_n(t))$ consists of all $y(t) = (y_1(t), \dots, y_n(t))$ such that $\operatorname{ord}_t(y_i(t) - x_i(t)) \geq N$, $\forall i = 1, \dots, n$, where $N \in \mathbb{N}$. The t-adic topology on arbitrary varieties is induced from that on affine spaces using affine charts.

An important approximation result is due to Artin:

Theorem A.16 ([Art, Th. 1.10]). *The set of formal germs induced by germs of curves is dense in* $X(\Bbbk((t)))$ *with respect to the t-adic topology.*

B Geometric Valuations

Let K be a function field, i.e., a finitely generated field extension of $\Bbbk$. By a valuation v of K we always mean a discrete $\mathbb{Q}$-valued valuation of $K/\Bbbk$, i.e., assume the following properties:

(1) $v : K^\times \to \mathbb{Q}$, $v(0) = \infty$;
(2) $v(K^\times) \simeq \mathbb{Z}$ or $\{0\}$;
(3) $v(\Bbbk^\times) = 0$;
(4) $v(fg) = v(f) + v(g)$;

(5) $v(f+g) \geq \min(v(f), v(g))$.

Remark B.1. If v is defined only on a $\Bbbk$-algebra A with $\operatorname{Quot} A = K$, then it is extended to K in a unique way by putting $v(f/g) = v(f) - v(g)$, $f, g \in A$.

Our main source in the valuation theory is [ZS, Ch. 6, App. 2].

Definition B.2. A valuation v is called *geometric* if there exists a normal (pre)variety X with $\Bbbk(X) = K$ (a *model* of K) and a prime divisor $D \subset X$ such that $v(f) = c \cdot v_D(f)$, $\forall f \in K$, for some $c \in \mathbb{Q}_+$. Here $v_D(f)$ is the order of f along D.

To any valuation corresponds a *(discrete) valuation ring (DVR)* $\mathscr{O}_v = \{f \in K \mid v(f) \geq 0\}$, which is a local ring with the maximal ideal $\mathfrak{m}_v = \{f \in K \mid v(f) > 0\}$ and quotient field K. The *residue field* of v is $\Bbbk(v) = \mathscr{O}_v/\mathfrak{m}_v$.

Example B.3. If $v \neq 0$ is geometric, then $\mathscr{O}_v = \mathscr{O}_{X,D}$, $\Bbbk(v) = \Bbbk(D)$.

Properties. (1) *$\mathscr{O}_v$ is a maximal subring of K.*
(2) *$\mathscr{O}_v$ determines v up to proportionality.*

Definition B.4. Let X be a model of K. A closed irreducible subvariety $Y \subseteq X$ is a *center* of v on X if $\mathscr{O}_v$ dominates $\mathscr{O}_{X,Y}$ (i.e., $\mathscr{O}_v \supseteq \mathscr{O}_{X,Y}$, $\mathfrak{m}_v \supseteq \mathfrak{m}_{X,Y}$, which implies that $\Bbbk(v) \supseteq \Bbbk(Y)$).

Example B.5. A prime divisor $D \subset X$ is a center of the respective geometric valuation.

If $\varphi : X \to X'$ is a dominant morphism and v has a center $Y \subseteq X$, then the restriction v' of v to $K' = \Bbbk(X')$ has a center $Y' = \overline{\varphi(Y)} \subseteq X'$.

Valuative criterion of separation. *X is separated if and only if any (geometric) valuation has at most one center on X.*

Valuative criterion of properness. *The map $\varphi : X \to X'$ is proper if and only if any (geometric) valuation of K has the center on X provided that its restriction to K' has a center on X'.*

Valuative criterion of completeness. *X is complete if and only if any (geometric) valuation has a center on X.*

Proposition B.6. *If X is affine, then v has the center $Y \subseteq X$ if and only if $v|_{\Bbbk[X]} \geq 0$, and then $\mathscr{I}(Y) = \Bbbk[X] \cap \mathfrak{m}_v$.*

Proposition B.7. *A valuation $v \neq 0$ is geometric if and only if* $\operatorname{tr.deg} \Bbbk(v) = \operatorname{tr.deg} K - 1$.

Proof. Assume that $\operatorname{tr.deg} K = n$ and that the residues of $f_1, \dots, f_{n-1} \in \mathscr{O}_v$ form a transcendence base of $\Bbbk(v)/\Bbbk$. Take a nonzero $f_n \in \mathfrak{m}_v$; then $f_1, \dots, f_n$ are easily seen to be a transcendence base of $K/\Bbbk$. Consider an affine variety X such that $\Bbbk[X]$ is the integral closure of $\Bbbk[f_1, \dots, f_n]$ in K. It is easy to show that $v|_{\Bbbk[X]} \geq 0$, whence v has the center $D \subset X$ and $f_1, \dots, f_{n-1} \in \Bbbk[D]$ are algebraically independent. Hence D is a prime divisor, and $\mathscr{O}_v = \mathscr{O}_{X,D}$ implies $v = v_D$ up to a multiple. The converse implication is obvious. $\square$

Proposition B.8. *Let $\Bbbk \subseteq K' \subseteq K$ be a subfield.*

(1) *If v is a geometric valuation of K, then $v' = v|_{K'}$ is geometric.*
(2) *Any geometric valuation v' of K' extends to a geometric valuation v of K.*

Proof. (1) Take $f_1, \dots, f_k \in \mathscr{O}_v$ whose residues form a transcendence base of $\Bbbk(v)/\Bbbk(v')$. They are algebraically independent over K' (otherwise one can take an algebraic dependence of $f_1, \dots, f_k$ over $\mathscr{O}_{v'}$ with at least one coefficient not in $\mathfrak{m}_{v'}$, and pass to residues obtaining a contradiction). Hence $\operatorname{tr.deg} \Bbbk(v') = \operatorname{tr.deg} \Bbbk(v) - \operatorname{tr.deg}_{\Bbbk(v')} \Bbbk(v) \geq \operatorname{tr.deg} K - 1 - \operatorname{tr.deg}_{K'} K = \operatorname{tr.deg} K' - 1$, and we conclude by Proposition B.7.
(2) Take a complete normal variety X' with a prime divisor $D' \subset X'$ such that v' is proportional to $v_{D'}$. We may construct a complete normal variety X with $\Bbbk(X) = K$ mapping onto X': take any complete model X of K and replace it by the normalization of the closure in $X \times X'$ of the graph of the rational map $X \dashrightarrow X'$. Let $D \subset X$ be a component of the preimage of D' mapping onto D'. Then we may take $v = v_D$ up to a multiple. □

C Rational Modules and Linearization

Rational modules are representations of algebraic groups compatible with the structure of an algebraic variety.

Definition C.1. Let G be a linear algebraic group. A finite-dimensional G-module M is called *rational* if the representation map $R : G \to \mathrm{GL}(M)$ is a homomorphism of algebraic groups. The terminology is explained by observing that for $G \subseteq \mathrm{GL}_n(\Bbbk)$ the matrix entries of $R(g)$ are rational functions in the matrix entries of $g \in G$ (the denominator being a power of $\det g$). Generally, a *rational G-module* is a union of finite-dimensional rational submodules.

A G-algebra A is said to be *rational* if it is a rational G-module and G acts on A by algebra automorphisms.

If a rational G-module M is at the same time an A-module and $g(am) = (ga)(gm)$, $\forall g \in G, a \in A, m \in M$, then M is called a *rational G-A-module*.

Let $\mathrm{Mor}(X,M)$ denote the set of all morphisms of an algebraic variety X to a vector space M. (If $\dim M = \infty$, then a morphism $X \to M$ is by definition a morphism to a finite-dimensional subspace of M.) It is a free $\Bbbk[X]$-module: $\mathrm{Mor}(X,M) \simeq \Bbbk[X] \otimes M$. If X is a G-variety and M is a rational G-module, then $\mathrm{Mor}(X,M)$ is a rational G-$\Bbbk[X]$-module.

The $\Bbbk[X]^G$-submodule $\mathrm{Mor}_G(X,M) \simeq (\Bbbk[X] \otimes M)^G$ of equivariant morphisms is called the *module of covariants* on X with values in M. If G is reductive, X is affine, and $\dim M < \infty$, then $\mathrm{Mor}_G(X,M)$ is finite over $\Bbbk[X]^G$ [PV, 3.12].

More generally, rational G-modules are formed by global sections of sheaves on G-varieties.

Let X be a G-variety, let α and $\pi_X : G \times X \to X$ be the action morphism and the projection, and let $\mathscr{F}$ be a quasicoherent sheaf on X.

Definition C.2. A *G-linearization* of $\mathscr{F}$ is an isomorphism $\widehat{\alpha} : \pi_X^* \mathscr{F} \xrightarrow{\sim} \alpha^* \mathscr{F}$ inducing a G-action on the set of local sections of $\mathscr{F}$ via isomorphisms $\widehat{\alpha}|_{g \times X} : \mathscr{F}(U) \xrightarrow{\sim} \mathscr{F}(gU)$ over all $g \in G$, U open in X.

A *G-sheaf* is a quasicoherent sheaf equipped with a G-linearization.

Theorem C.3 ([Kem2]). *Given a G-variety X and a G-sheaf $\mathscr{F}$ on X, $\Bbbk[X]$ is a rational G-algebra and $\mathrm{H}^i(X, \mathscr{F})$ are rational G-$\Bbbk[X]$-modules.*

If $\mathscr{F}$ is the sheaf of sections of a vector bundle $F \to X$, then a G-linearization of $\mathscr{F}$ is given by a fiberwise linear action $G : F$ compatible with the projection onto X.

By abuse of language we often make no terminological difference between vector bundles and the respective locally free sheaves of sections since they determine each other.

An important problem is to construct G-linearizations for line bundles on G-varieties. A treatment of this problem goes back to Mumford. Here we follow [KKLV].

Assume that G is connected.

Theorem C.4 ([KKLV, 2.4]). *If G is factorial, i.e., $\operatorname{Pic} G = 0$, then any line bundle $\mathscr{L}$ on a normal G-variety X is G-linearizable.*

We say that an algebraic group $\widetilde{G}$ is a *universal cover* of G if $\widetilde{G}/\mathrm{R_u}(\widetilde{G})$ is a product of a torus and a simply connected semisimple group, and there is an epimorphism $\widetilde{G} \to G$ with finite kernel. Every connected group has a universal cover: it is well known for reductive groups [Hum, §§32,33], [Sp3, Ch. 12], and generally we may put $\widetilde{G} = G \times_{G_{\mathrm{red}}} \widetilde{G_{\mathrm{red}}}$, where $G_{\mathrm{red}} = G/\mathrm{R_u}(G)$. By [Po2] or [KKLV, §4], $\operatorname{Pic} \widetilde{G} = 0$ and $\operatorname{Pic} G$ is finite.

Corollary C.5. *Any line bundle $\mathscr{L}$ on X is $\widetilde{G}$-linearizable.*

Corollary C.6. *A certain power $\mathscr{L}^{\otimes d}$ of $\mathscr{L}$ is G-linearizable.*

For d one may take the degree of a universal covering or the order of $\operatorname{Pic} G$ [KKLV, 2.4].

Let $\operatorname{Pic}_G(X)$ denote the group of G-linearized invertible sheaves, up to a G-equivariant isomorphism. The kernel of the natural homomorphism $\operatorname{Pic}_G(X) \to \operatorname{Pic}(X)$ (forgetting G-linearization) consists of all G-linearizations of the trivial line bundle $X \times \Bbbk \to X$. A G-linearization of the trivial bundle, i.e., a fiberwise linear G-action on $X \times \Bbbk$ is given by multiplication by an algebraic cocycle $c : G \times X \to \Bbbk^\times$, $c(g_1 g_2, x) = c(g_1, g_2 x) c(g_2, x)$, $\forall g_1, g_2 \in G$, $x \in X$. For connected G and irreducible X we have $c(g,x) = \chi(g) h(x)$, because an invertible function on a product of two irreducible varieties is a product of invertible functions on factors [KKV, 1.1]. Now it is easy to deduce from the cocycle property that $h(x) \equiv 1$ and $\chi \in \mathfrak{X}(G)$. Thus any two G-linearizations of a given line bundle differ by a character of G, and for factorial G and normal irreducible X we have an exact sequence

$$\mathfrak{X}(G) \longrightarrow \mathrm{Pic}_G(X) \longrightarrow \mathrm{Pic}(X) \longrightarrow 0. \tag{C.1}$$

The existence of a G-linearization has fundamental consequences in the local description of G-varieties, due to Sumihiro:

Theorem C.7 ([Sum], [KKLV, §1]). *Let G be a connected group acting on a normal variety X. Then any point $x \in X$ has an open G-stable neighborhood U which admits a locally closed G-equivariant embedding $U \hookrightarrow \mathbb{P}(V)$ for some G-module V.*

Proof. Take an affine neighborhood $U_0 \ni x$. The complement $D = X \setminus U_0$ may support no effective Cartier divisor. If however we remove $\bigcap_{g \in G} gD$ from X, then any effective Weil divisor with support D becomes base point free, hence Cartier (cf. Lemma 17.3).

Let $\mathscr{L}$ denote the respective line bundle. Take $\sigma_0 \in \mathrm{H}^0(X, \mathscr{L})$ such that $U_0 = X_{\sigma_0}$. Then

$$\Bbbk[U_0] = \bigcup_{d \geq 0} \mathrm{H}^0(X, \mathscr{L}^{\otimes d})/\sigma_0^d = \Bbbk\left[\frac{\sigma_1}{\sigma_0^{d_1}}, \dots, \frac{\sigma_m}{\sigma_0^{d_m}}\right]$$

for some $\sigma_i \in \mathrm{H}^0(X, \mathscr{L}^{\otimes d_i})$, $d_i \in \mathbb{N}$. Replacing $\mathscr{L}$ by a power, we may assume it to be a G-bundle and all $d_i = 1$. Include $\sigma_0, \dots, \sigma_m$ in a finite-dimensional G-submodule $M \subseteq \mathrm{H}^0(X, \mathscr{L})$. The induced rational map $X \dashrightarrow \mathbb{P}(V)$, $V = M^*$, is a locally closed embedding on $U = GU_0$. □

Remark C.8. If X is itself quasiprojective, then one may take $U = X$. Indeed, a certain power of an ample line bundle on X is G-linearizable, and we can find a finite-dimensional G-stable space of sections inducing a projective embedding of X.

D Invariant Theory

Let G be a linear algebraic group and A be a rational G-algebra. The subject of algebraic invariant theory is the structure of the subalgebra A^G of G-invariant elements.

A geometric view on the subject is to consider an affine G-variety $X = \operatorname{Spec} A$, provided that A is finitely generated. (Note that each rational G-algebra is a union of finitely generated G-stable subalgebras.) If A^G is finitely generated too, then one may consider $X/\!\!/G := \operatorname{Spec} A^G$ and the natural dominant morphism $\pi = \pi_G : X \to X/\!\!/G$. The variety $X/\!\!/G$, considered together with π, is called the *categorical quotient* of $G : X$, because it is the universal object in the category of G-invariant morphisms from X to affine varieties. This means that every morphism $\varphi : X \to Y$ (with affine Y) which is constant on G-orbits fits into a unique commutative triangle:

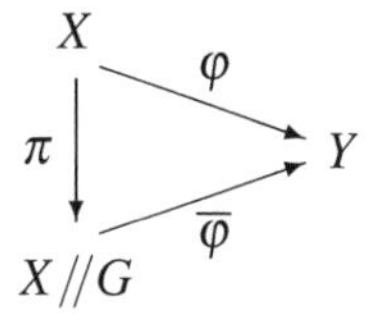

In particular, if Z is an affine G-variety and $\psi : X \to Z$ is a G-equivariant morphism, then we have a commutative square

$$\begin{array}{ccc} X & \xrightarrow{\psi} & Z \\ \downarrow & & \downarrow \\ X/\!\!/G & \xrightarrow{\psi/\!\!/G} & Z/\!\!/G. \end{array}$$

(Here $Y = Z/\!\!/G$, $\varphi = \pi_G \psi$.)

Geometric properties of $X/\!\!/G$ and π_G translate into algebraic properties of A^G and of its embedding into A, and vice versa. The case of reductive G is considered by Geometric Invariant Theory (GIT). We collect basic results on invariants and quotients of affine varieties by reductive groups in the following theorem.

Main Theorem of GIT. *Let G be a reductive group and let A be a rational G-algebra.*

(1) *If A is finitely generated, then so is A^G.*

Under this assumption, put $X = \operatorname{Spec} A$. Then $\pi : X \to X/\!\!/G$ is well defined and has the following properties:

(2) *π is surjective and maps closed G-stable subsets of X to closed subsets of $X/\!\!/G$.*
(3) *$X/\!\!/G$ carries the quotient topology with respect to π, and $\mathscr{O}_{X/\!\!/G} = \pi_* \mathscr{O}_X^G$.*
(4) *If $Z_1, Z_2 \subset X$ are disjoint closed G-stable subsets, then $\pi(Z_1) \cap \pi(Z_2) = \emptyset$. In particular, each fiber of π contains a unique closed orbit.*

Thus $X/\!\!/G$ may be regarded as the "variety of closed orbits" for $G : X$. It is not hard to show that $\pi_G : X \to X/\!\!/G$ is the categorical quotient in the category of *all* algebraic varieties.

Finite generation of G-invariants goes back to Hilbert and Weyl (in characteristic zero), the general case is due to Nagata and Haboush. Other assertions are due to Mumford.

If G is *linearly reductive*, i.e., all rational G-modules are completely reducible (e.g., $\operatorname{char} \Bbbk = 0$ or G is a torus), then the proof is considerably simplified [PV, 3.4, 4.4] by using the G-A^G-module decomposition $A = A^G \oplus A_G$, where A_G is the sum of all nontrivial irreducible G-submodules. The respective projection

$$A \mapsto A^G, \qquad f \mapsto f^\natural,$$

is known as the *Reynolds operator*. For finite G and $\operatorname{char} \Bbbk = 0$, it is just the group averaging:

$$f^\natural = \frac{1}{|G|} \sum_{g \in G} gf, \qquad \forall f \in A.$$

For a complex reductive group G, the Reynolds operator may be defined by averaging over a compact real form of G.

The proof in positive characteristic may be found in [MFK, App. 1A, 1C].

The Main Theorem of GIT is a base for other constructions of quotients for reductive group actions. A localized version of categorical quotient was introduced by Seshadri: a *good quotient* of a G-variety X (where G is reductive) is a variety Y together with a dominant affine morphism $\pi : X \to Y$ such that $\pi_* \mathcal{O}_X^G = \mathcal{O}_Y$. If a good quotient exists, then it is unique and is the categorical quotient in the category of algebraic varieties. Mumford proved that any projective G-variety X contains an open subset X^{ss} (semistable locus) admitting a good quotient $X^{ss}/\!/G \simeq \operatorname{Proj} \Bbbk[\widehat{X}]^G$, where $\widehat{X}$ is the affine cone over X [MFK, §1.4], [PV, Th. 4.16].

For non-reductive groups the situation is not so nice—even finite generation of invariants fails due to famous Nagata's counterexample [Nag] and results of Popov [Po4]. However for subgroups of reductive groups acting on algebras or affine varieties, there are positive results on finite generation and the structure of invariant algebras and categorical quotients.

Lemma D.1. *Let G be a reductive group, $H \subseteq G$ an algebraic subgroup, A a rational G-algebra, and $I \lhd A$ a G-stable ideal. Then $(A/I)^H$ is a purely inseparable integral extension of A^H/I^H.*

The lemma is obvious for $\operatorname{char} \Bbbk = 0$, since A/I lifts to a G-submodule of A, whence $(A/I)^H = A^H/I^H$. The proof for $H = G$ may be found in [MFK, Lemma A.1.2] and the general case follows by the transfer principle (Remark 2.12).

Corollary D.2. *Let M be a G-module and let $N \subset M$ be a G-submodule. For any $\overline{m} \in (M/N)^{(H)}$ there exist $q = p^n$ and $m \in (\mathrm{S}^q M)^{(H)}$ such that $m \mapsto \overline{m}^q \in \mathrm{S}^q(M/N)$.*

Proof. Just apply Lemma D.1 to $A = \mathrm{S}^\bullet M$, $I = AN$, replacing H by the common kernel H_0 of all $\chi \in \mathfrak{X}(H)$, and use the fact that H/H_0 is diagonalizable. □

Corollary D.3. *If A carries a G-stable filtration, then $(\operatorname{gr} A)^H$ is a purely inseparable extension of $\operatorname{gr}(A^H)$.*

Proof. Each homogeneous component of $\operatorname{gr} A$ has the form M/N, where M, N are two successive members of the filtration. It remains to apply Corollary D.2 to M and N. □

The most important case is $H = U$, a maximal unipotent subgroup in G. U-invariants of rational G-algebras were studied by Hadzhiev, Vust, Popov (in characteristic zero), Donkin, Grosshans (in arbitrary characteristic), et al. We refer to [Gr2] for systematic exposition of the theory.

Lemma D.4 ([Gr2, 14.3]). *A rational G-algebra A is integral over its subalgebra $\langle G \cdot A^U \rangle$.*

The proof relies on Lemma D.1. In characteristic zero, Lemma D.4 is trivial, since $A = \langle G \cdot A^U \rangle$ by complete reducibility of G-modules and highest weight theory.

The fundamental importance of U-invariants is explained by the fact that A and A^U share a number of properties.

Theorem D.5. *Let G be a connected reductive group.*

(1) *A rational G-algebra A is finitely generated (resp. has no nilpotents, is an integral domain) if and only if* A^U *is so.*
(2) *In particular, for any affine G-variety X, the categorical quotient* $\pi_U : X \to X//U$ *is well defined.*
(3) *In characteristic zero, X is normal (has rational singularities) if and only if* $X//U$ *is so.*

In characteristic zero, finite generation of U-invariants is due to Hadzhiev; other assertions were partially proved by Brion (nilpotents and zero divisors), Vust (normality), Kraft (rationality of singularities), and by Popov in full generality [Po5]. The theorem was extended to arbitrary characteristic by Grosshans. We shall give an outline of the proof scattered in [Gr2].

Finite generation of $A^U \simeq (\Bbbk[G/U] \otimes A)^G$ (see Remark 2.12) stems from that of A and of $\Bbbk[G/U]$. The latter is proved by a representation-theoretic argument (Lemma 2.23) or by providing an explicit embedding of G/U into a G-module, with the boundary of codimension ≥ 2 [Gr2, 5.6] (cf. Theorem 28.3).

The other assertions are proved using horospherical contraction (cf. 7.3).

The algebra A is endowed with a G-stable increasing filtration $A^{(n)}$ such that $\operatorname{gr} A$ has an integral extension $S = (\Bbbk[G/U^-] \otimes A^U)^T$ and $A^U \simeq \operatorname{gr}(A^U) = (\operatorname{gr} A)^U \simeq S^U$, where T acts on G/U^- by right translations. In characteristic zero this filtration is described in 7.3 (for $A = \Bbbk[X]$) and the general case is considered in [Gr2, §15].

Now finite generation of A^U implies that of S, and hence of $\operatorname{gr} A$ (both are finite modules over $\langle G \cdot S^U \rangle$), and finally of A by a standard argument. Moreover, the algebra $R = \bigoplus_{n=0}^{\infty} A^{(n)} t^n \subseteq A[t]$ is finitely generated, too (because $R^U \simeq A^U[t]$) [Gr2, 16.5].

The remaining assertions may be proved for finitely generated A and $X = \operatorname{Spec} A$. As in 7.3, put $E = \operatorname{Spec} R$ and consider the natural $(G \times \Bbbk^\times)$-morphism $\delta : E \to \mathbb{A}^1$ with the zero fiber $X_0 = \operatorname{Spec}(\operatorname{gr} A)$ and other fibers isomorphic to X. Note that $\Bbbk^\times$ contracts E to X_0 (i.e., $\forall x \in E\ \exists \lim_{t \to 0} t \cdot x \in X_0$), because the grading on R is non-negative.

Since δ is flat, the set of $x \in E$ such that the schematic fiber $\delta^{-1}(\delta(x))$ has a given local property of open type (e.g., is reduced, irreducible, normal, or has rational singularity) at x is open in E. The complementary closed subset of E is $\Bbbk^\times$-stable, and hence it intersects X_0 whenever it is non-empty. It follows that X has the property of open type whenever X_0 has this property.

If an affine $\Bbbk$-scheme Z of finite type is reduced (resp. irreducible or normal), then so is $Z//H$ for any algebraic group H acting on Z. (Only normality requires some explanation.) In particular, these properties are inherited by $X//U$ from X.

Conversely, if $X//U$ has one of these properties, then $\operatorname{Spec} S = (G//U^- \times X//U)//T$ and X_0 have it, too. (For normality, we use the isomorphism $S \simeq \operatorname{gr} A$ in characteristic zero.) By the above, X has this property. More elementary (and lengthy) arguments are given in [Gr2, §18].

The same reasoning works for rational singularities in characteristic zero, using the facts that $G//U$ has rational singularities [Kem1], [BKu, 3.4.7] and that rational singularities are preserved by products (Theorem A.5) and categorical quotients modulo reductive groups (Theorem A.6), which guarantees that $X/\!/U = (G/\!/U \times X)/\!/G$ has rational singularities provided that so has X.

Lemma D.6 (M. Brion). *If X is an affine irreducible G-variety and $\widetilde{X}$ is its normalization, then $\widetilde{X}/\!/U$ is the normalization of $X/\!/U$.*

Proof. Put $R = \Bbbk[X]$, $S = \Bbbk[\widetilde{X}]$. The conductor $(R:S) := \{f \in R \mid fS \subseteq R\}$ is a G-stable ideal both in R and S. Since $\operatorname{Quot} R = \operatorname{Quot} S$ and S is finite over R, we have $(R:S) \neq 0$. Then $0 \neq (R:S)^U \subseteq (R^U : S^U)$. For any nonzero $f \in (R^U : S^U)$, $S^U \simeq fS^U \subseteq R^U$ is a finite R^U-module and $S^U \subseteq \operatorname{Quot} R^U$, whence the claim. □

We conclude this section by a simple lemma on semiinvariants.

Lemma D.7 (cf. [PV, Th. 3.3]). *Let G be a connected solvable algebraic group, let A be a rational G-algebra without zero divisors, and let $K = \operatorname{Quot} A$. Then for every $f \in K^{(G)}$ there exist $f_1, f_2 \in A^{(G)}$ such that $f = f_1/f_2$.*

Proof. We have $f = p_1/p_2$, $p_i \in A$. By the Lie–Kolchin theorem, there exists a nonzero linear combination $f_2 = \sum c_j(g_j p_2) \in A^{(G)}$, $c_j \in \Bbbk$, $g_j \in G$. Then $f = f_1/f_2$, where $f_i = \sum c_j(g_j p_i) \in A^{(G)}$. □

E Hilbert Schemes

One of important problems in algebraic geometry and its applications is to classify algebraic or geometric objects of a certain type with fixed discrete invariants. It is natural to expect that the remaining "continuous" parameters (*moduli*) are given by coordinates on a certain algebraic variety, called a *moduli space*, whose points parameterize the objects of our class. A stronger version of a moduli space (*fine moduli space*) is obtained by considering a certain class of "continuous" families of our objects parameterized by points of varieties or schemes, which is closed under base change, and defining the moduli space as the scheme which represents the contravariant functor associating with a scheme S the set of all families over the base S. In other words, every family is obtained from the universal family over the moduli space by a unique base change.

In particular, it is quite common to consider *flat families* of schemes, i.e., flat morphisms of finite type $\pi : \mathscr{X} \to S$, where $\mathscr{X}, S$ are Noetherian $\Bbbk$-schemes. It is well known that many discrete invariants of schemes do not vary in flat families. In this context, various versions of Hilbert schemes are defined.

E.1 Classical Case. The classical Grothendieck's Hilbert scheme parameterizes projective subschemes of a projective space $\mathbb{P}(V)$ with a fixed Hilbert polynomial.

Definition E.1. The *Hilbert scheme* $\operatorname{Hilb}_\Phi(\mathbb{P}(V))$ is the scheme representing a functor which associates with each scheme S the set of flat families $\pi : \mathscr{X} \to S$ such that $\mathscr{X} \subseteq \mathbb{P}(V) \times S$ is a closed subscheme, π is the natural projection, and all fibers $X_s = \pi^{-1}(s)$ have Hilbert polynomial Φ. More directly, there exists a *universal* flat family $\mathscr{X}^{\mathrm{univ}} \to \mathrm{Hilb}_\Phi(\mathbb{P}(V))$ of the above type such that every flat family is obtained from the universal one by the pullback along a unique morphism $S \to \mathrm{Hilb}_\Phi(\mathbb{P}(V))$, so that $\mathscr{X} = \mathscr{X}^{\mathrm{univ}} \times_{\mathrm{Hilb}_\Phi(\mathbb{P}(V))} S$.

Theorem E.2 ([Gro], [Ser, 4.3.4]). $\mathrm{Hilb}_\Phi(\mathbb{P}(V))$ *exists and is a projective scheme.*

Clearly, the Hilbert scheme is uniquely defined being an object representing a functor. Geometric properties of the Hilbert scheme and the universal family are derived from the properties of this representable "functor of points", cf. Appendix A.3. For instance, all fibers of the universal family over closed points are pairwise distinct subschemes of $\mathbb{P}(V)$ and all closed subschemes with Hilbert polynomial Φ occur as fibers (because every such subscheme X may be considered as a family over a single point, and hence is induced by a unique map of the point to $\mathrm{Hilb}_\Phi(\mathbb{P}(V))$). Another manifestation of this principle is a computation of the tangent space to the Hilbert scheme.

Proposition E.3 ([Gro]). *The tangent space to* $\mathrm{Hilb}_\Phi(\mathbb{P}(V))$ *at the closed point* $[X]$ *corresponding to a subscheme* $X \subseteq \mathbb{P}(V)$ *is*

$$T_{[X]}\mathrm{Hilb}_\Phi(\mathbb{P}(V)) = \mathrm{H}^0(X, \mathscr{N}_{X/\mathbb{P}(V)}),$$

where $\mathscr{N}_{X/Y} = \mathscr{H}om_{\mathscr{O}_X}(\mathscr{I}_X/\mathscr{I}_X^2, \mathscr{O}_X)$ *is the normal sheaf of a closed subscheme* X *in a scheme* Y.

This proposition easily stems from the interpretation of tangent vectors as morphisms $S = \mathrm{Spec}\,\Bbbk[t]/(t^2) \to \mathrm{Hilb}_\Phi(\mathbb{P}(V))$ mapping the closed point to $[X]$, or, equivalently, as flat families $\mathscr{X} \to S$ with special fiber X, see [Har3, §2] for details.

A more subtle question is the smoothness of the Hilbert scheme at a closed point. The tangent space at $[X]$ classifies infinitesimal first order deformations of X, so the problem may be reformulated as whether every infinitesimal deformation is extendible to a local deformation. Obstructions to such extensions lie in the cohomology of the so-called higher tangent sheaves of Lichtenbaum and Schlessinger.

To define these sheaves, consider first the affine case. Let X be an affine $\Bbbk$-scheme of finite type. Consider a closed embedding in a vector space $X \subseteq V$ and put $I = \mathscr{I}(X) \lhd \Bbbk[V] = R$. Choose a short exact sequence of R-modules

$$0 \longrightarrow Q \longrightarrow F \xrightarrow{\;q\;} I \longrightarrow 0,$$

where F is finite and free, and consider the submodule $K \subseteq Q$ spanned by the Koszul syzygies $q(f)g - q(g)f$, $f, g \in F$. Tensoring by $A = \Bbbk[X] \simeq R/I$ yields an exact sequence

$$Q/K \longrightarrow F/IF \longrightarrow I/I^2 \longrightarrow 0.$$

Recall the second exact sequence for Kähler differentials:

$$I/I^2 \longrightarrow \Omega_{R/\Bbbk} \otimes_R A \longrightarrow \Omega_{A/\Bbbk} \longrightarrow 0.$$

Now define the *cotangent complex*

$$\cdots \longrightarrow 0 \longrightarrow L_2 \longrightarrow L_1 \longrightarrow L_0 \longrightarrow 0 \longrightarrow \cdots$$

of X as follows: $L_2 = Q/K$, $L_1 = F/IF$, $L_0 = \Omega_{R/\Bbbk} \otimes_R A$, with obvious differentials. The *i-th tangent module* $T^i(X)$ is the i-th cohomology of the dual complex $\mathrm{Hom}_A(L_\bullet, A)$. Clearly, $T^0(X) = \mathrm{Hom}_A(\Omega_{A/\Bbbk}, A) = \mathrm{H}^0(X, \mathscr{T}_X)$ and $T^{1,2}(X)$ are the cokernels of the maps $\mathrm{Hom}_R(\Omega_{R/\Bbbk}, A) = \mathrm{H}^0(X, \mathscr{T}_V) \to \mathrm{H}^0(X, \mathscr{N}_{X/V}) = \mathrm{Hom}_A(I/I^2, A)$ and $\mathrm{Hom}_A(F/IF, A) \to \mathrm{Hom}_A(Q/K, A)$, respectively. The functors T^i do not depend on the choice of V and F and commute with localization. Thus one can globalize them and define the *i-th tangent sheaf* $\mathscr{T}_X^i$ on an arbitrary scheme X, see [Har3, §3] for details. Note that $\mathscr{T}_X^0 = \mathscr{T}_X$.

The following result explains when an infinitesimal deformation of X can be extended to a thicker infinitesimal deformation.

Proposition E.4 ([Har3, 10.2]). *Let $\mathscr{O}$ be a finite-dimensional local $\Bbbk$-algebra and let $\mathscr{O}'$ be a quotient of $\mathscr{O}$ modulo a one-dimensional ideal. Denote by S, S' the spectra of $\mathscr{O}, \mathscr{O}'$ and suppose that there is a flat family $\mathscr{X}' \to S'$ with special fiber X. There are three successive obstructions for the existence of an extension of $\mathscr{X}'$ to $\mathscr{X} \to S \supset S'$, lying in $\mathrm{H}^0(X, \mathscr{T}_X^2)$, $\mathrm{H}^1(X, \mathscr{T}_X^1)$, and $\mathrm{H}^2(X, \mathscr{T}_X^0)$. This means that one can construct an element in the first space (depending on S, S') which has to be zero in order that an extension may exist and, if this first obstruction vanishes, then one can construct an element in the second space which has to be zero, and so on.*

Corollary E.5 ([Har3, 6.3, 10.4]). *If $\mathrm{H}^0(X, \mathscr{T}_X^2) = \mathrm{H}^1(X, \mathscr{N}_{X/\mathbb{P}(V)}) = 0$, then $\mathrm{Hilb}_\Phi(\mathbb{P}(V))$ is smooth at $[X]$.*

Proof. A morphism of a punctual scheme $S' \to \mathrm{Hilb}_\Phi(\mathbb{P}(V))$ mapping the closed point to $[X]$ corresponds to a flat closed subfamily $\mathbb{P}(V) \times S' \supseteq \mathscr{X}' \to S'$ with special fiber X. By Proposition E.4, the unique obstruction for the existence of local extensions (on affine charts) of $\mathscr{X}'$ to any larger punctual scheme S (such that $\mathscr{I}(S') \lhd \Bbbk[S]$ is one–dimensional) lies in $\mathrm{H}^0(X, \mathscr{T}_X^2)$. Hence $\mathscr{X}'$ can be extended locally and, moreover, these extensions can be performed inside $\mathbb{P}(V) \times S$ by the infinitesimal lifting property [Har3, Exs. 4.7, 10.1], since $\mathbb{P}(V)$ is smooth. The obstruction for gluing these local deformations together lies in $\mathrm{H}^1(X, \mathscr{N}_{X/\mathbb{P}(V)})$ [Har3, 6.2(b)]. Hence $\mathscr{X}' \to S'$ can be extended globally to a flat closed subfamily $\mathbb{P}(V) \times S \supseteq \mathscr{X} \to S$, i.e., the morphism $S' \to \mathrm{Hilb}_\Phi(\mathbb{P}(V))$ can be extended to S. Now the converse of the infinitesimal lifting property [Har3, 4.6] implies that $\mathrm{Hilb}_\Phi(\mathbb{P}(V))$ is smooth at $[X]$. □

Now we discuss some other versions of Hilbert schemes which are important for this survey.

E.2 Nested Hilbert Scheme. The first one is nested (or flag) Hilbert scheme parameterizing tuples of closed subschemes in $\mathbb{P}(V)$ with fixed Hilbert polynomials and inclusion relations [Ser, 4.5].

Definition E.6. The *nested Hilbert scheme* $\mathrm{Hilb}_{\Phi_1,\dots,\Phi_k}(\mathbb{P}(V))$ is the scheme representing a functor which associates with each scheme S the set of k-tuples of flat families $\mathscr{X}_i \to S$ $(i=1,\dots,k)$ such that $\mathscr{X}_i \subseteq \mathbb{P}(V)\times S$ are closed subschemes with prescribed inclusion relations between them and all fibers of $\mathscr{X}_i \to S$ have Hilbert polynomial Φ_i.

It is easy to see that $\mathrm{Hilb}_{\Phi_1,\dots,\Phi_k}(\mathbb{P}(V))$ is a closed subscheme in the product $\mathrm{Hilb}_{\Phi_1}(\mathbb{P}(V))\times\cdots\times\mathrm{Hilb}_{\Phi_k}(\mathbb{P}(V))$. Similarly to Proposition E.3, one proves

Proposition E.7 ([Ser, Pr. 4.5.3]).

$$T_{(X_1,\dots,X_k)}\mathrm{Hilb}_{\Phi_1,\dots,\Phi_k}(\mathbb{P}(V)) = \mathrm{H}^0(\mathbb{P}(V),\mathscr{N}),$$

where $\mathscr{N}\subseteq\mathscr{N}_{X_1/\mathbb{P}(V)}\oplus\cdots\oplus\mathscr{N}_{X_k/\mathbb{P}(V)}$ *is formed by tuples* $(\xi_1,\dots,\xi_k)$ *of normal vector fields such that* $\xi_i|_{X_j}=\xi_j \bmod \mathscr{N}_{X_j/X_i}$ *whenever* $X_i\supseteq X_j$.

E.3 Invariant Hilbert Schemes. There are several equivariant versions of Hilbert schemes. From now on, we assume that $\operatorname{char}\Bbbk=0$. Let G be a connected reductive group or, more generally, a product of such a group and a finite Abelian group. For such groups, the highest weight theory works well. Fix a multiplicity function $m:\mathfrak{X}_+\to\mathbb{Z}_+$, $\lambda\mapsto m_\lambda$.

Definition E.8. The *invariant Hilbert scheme* $\mathrm{Hilb}^G_m(V)$ is the scheme representing a functor which associates with each scheme S the set of flat families $\mathscr{X}\to S$ of affine G-subschemes of a G-module V such that $\mathscr{X}\subseteq V\times S$ is a closed G-subscheme and all fibers X_s satisfy $m_\lambda(X_s)=m_\lambda$, $\forall\lambda\in\mathfrak{X}_+$.

Theorem E.9. $\mathrm{Hilb}^G_m(V)$ *exists and is a quasiprojective scheme.*

This theorem was proved by Haiman–Sturmfels [HS] for diagonalizable G (in which case the assumption $\operatorname{char}\Bbbk=0$ can be waived) and by Alexeev–Brion [AB3] in general. In fact, the general case can be deduced from the diagonalizable one as follows. First, observe that for any flat family $\pi:\mathscr{X}\to S$ of affine G-schemes, i.e., a flat affine G-invariant morphism of finite type, the G-isotypic components of $\pi_*\mathscr{O}_{\mathscr{X}}$ have the following structure: $(\pi_*\mathscr{O}_{\mathscr{X}})_{(\lambda)}\simeq(\pi_*\mathscr{O}^U_{\mathscr{X}})_\lambda\otimes V(\lambda)$, where the T-isotypic components $(\pi_*\mathscr{O}^U_{\mathscr{X}})_\lambda\subseteq\pi_*\mathscr{O}^U_{\mathscr{X}}$ are locally free sheaves of rank m_λ. Hence the good quotient $\mathscr{X}/\!\!/U=\mathrm{Spec}_{\mathscr{O}_S}\pi_*\mathscr{O}^U_{\mathscr{X}}$ is a flat family of affine T-schemes over S. Now it is not hard to prove that $\mathrm{Hilb}^G_m(V)$ is realized as a closed subscheme of $\mathrm{Hilb}^T_m(V/\!\!/U)$, the scheme parameterizing affine T-subschemes of $V/\!\!/U$ (embedded in some T-module), see [AB3, 1.2] for details.

Remark E.10. The classical Hilbert scheme is a particular case of invariant Hilbert scheme for $G=\Bbbk^\times$ acting on V by homotheties [HS, 4.1].

Here is an analogue of Propositions E.3 and E.4.

Proposition E.11 ([AB3, Pr. 1.13], [C-F2, 3.5]). *The tangent space to the invariant Hilbert scheme is*

$$T_{[X]}\mathrm{Hilb}^G_m(V)=\mathrm{H}^0(X,\mathscr{N}_{X/V})^G=\mathrm{Hom}_{G,A}(I/I^2,A),$$

where $I = \mathscr{I}(X) \lhd \Bbbk[V] = R$ *and* $A = \Bbbk[X] = R/I$. *The obstruction space at* $[X]$ *is* $T^2(X)^G$.

There is yet another version of invariant Hilbert scheme parameterizing affine G-schemes with fixed algebra of U-invariants.

Definition E.12. Let Y be an affine T-scheme with finite multiplicity function m. The *invariant Hilbert scheme* Hilb_Y^G is the scheme representing a functor which associates with each scheme S the set of flat families $\mathscr{X} \to S$ of affine G-schemes together with T-equivariant isomorphisms $\mathscr{X}/\!/U \xrightarrow{\sim} Y \times S$.

Theorem E.13 ([AB3, Th. 1.12]). Hilb_Y^G *exists and is an affine scheme of finite type.*

To prove the theorem, consider an open subscheme $\mathrm{Hilb}_m^G(V)^0 \subseteq \mathrm{Hilb}_m^G(V)$ representing those families $\mathscr{X} \subseteq V \times S$ for which $\pi_*\mathscr{O}_{\mathscr{X}}^U$ is generated by $\pi_*(V^*)^U|_{\mathscr{X}}$ as an $\mathscr{O}_S$-algebra. Put $V_U = V/\langle Uv - v \mid v \in V\rangle = \operatorname{Spec} \mathrm{S}^\bullet(V^*)^U$. There is a natural morphism $\mathrm{Hilb}_m^G(V)^0 \to \mathrm{Hilb}_m^T(V_U)$ associating with each $X \subseteq V$ its categorical quotient $X/\!/U$ canonically embedded in V_U. We may choose V in such a way that Y admits a T-equivariant closed embedding in V_U. Then it is easy to see that the schematic fiber of the above morphism over $[Y]$ is the desired Hilb_Y^G.

In an important particular case, where $Y = V_U$ is a multiplicity-free T-module with linearly independent weights, $\mathrm{Hilb}_Y^G = \mathrm{Hilb}_m^G(V)^0$, where $m_\lambda = 1$ or 0, depending on whether $\lambda \in \Lambda_+(Y)$ or not.

The coordinate algebra $\Bbbk[\mathrm{Hilb}_Y^G]$ can be explicitly described. Consider a horospherical variety $X_0 = (G/\!/U^- \times Y)/\!/T$, the unique one with $X_0/\!/U \simeq Y$. For any closed point $[X] \in \mathrm{Hilb}_Y^G$, there is a canonical G-module isomorphism $\Bbbk[X] \simeq \Bbbk[X_0]$ extending $\Bbbk[X]^U \simeq \Bbbk[Y] \simeq \Bbbk[X_0]^U$. Thus $[X]$ is determined by a G-equivariant multiplication law on $\Bbbk[X_0] \simeq \bigoplus_{\lambda \in \Lambda_+(Y)} V(\lambda) \otimes \Bbbk[Y]_\lambda$ extending the multiplication in $\Bbbk[Y]$, i.e., by a collection of linear maps

$$m_{\lambda\mu}^\nu \in M_{\lambda\mu}^\nu := \mathrm{Hom}_G(V(\lambda) \otimes V(\mu), V(\nu)) \otimes \mathrm{Hom}(\Bbbk[Y]_\lambda \otimes \Bbbk[Y]_\mu, \Bbbk[Y]_\nu)$$

satisfying commutativity and associativity conditions. These $m_{\lambda\mu}^\nu$ may be viewed as morphisms $\mathrm{Hilb}_Y^G \to M_{\lambda\mu}^\nu$, and their matrix entries are regular functions on Hilb_Y^G. Note that $m_{\lambda\mu}^\nu([X]) \neq 0$ if and only if $\nu - \lambda - \mu$ is a tail of $\Bbbk[X]$.

There is a T-action on Hilb_Y^G by shifting the isomorphisms $X/\!/U \xrightarrow{\sim} Y$ via the T-action on Y. This action can be described in terms of the isotypic decomposition $V = V_{(\lambda_1)} \oplus \cdots \oplus V_{(\lambda_d)}$. Namely, define a new T-action on V so that $V_{(\lambda_i)}$ become T-eigenspaces of weights $w_G\lambda_i$. This action induces the action on $\mathrm{Hilb}_m^G(V) \supseteq \mathrm{Hilb}_Y^G$ coinciding with the one defined above. Under this action, the matrix entries of $m_{\lambda\mu}^\nu$ have T-eigenweight $\lambda + \mu - \nu$.

Theorem E.14 ([AB3, 2.4]).

(1) *The T-action contracts* Hilb_Y^G *to* $[X_0]$.
(2) $\Bbbk[\mathrm{Hilb}_Y^G]$ *is generated by the matrix entries of all* $m_{\lambda\mu}^\nu$.

(3) *For any closed point* $[X] \in \mathrm{Hilb}^G_Y$, *the coordinate algebra of its* T*-orbit closure is* $\Bbbk[\overline{T[X]}] = \Bbbk[\Xi(X)]$, *where the semigroup* $\Xi(X)$ *is generated by the* T*-weights opposite to the tails of* $\Bbbk[X]$.

Proof. Since the T-action on $Y \simeq X_0/\!/U$ lifts G-equivariantly to X_0, $[X_0]$ is a T-fixed point. We claim that the differentials of the matrix entries of $m^{\nu}_{\lambda\mu}$ generate $T^*_{[X]}\mathrm{Hilb}^G_Y$ for any X. Indeed, otherwise they would vanish on a certain nonzero tangent vector, which would imply that the family $\mathscr{X} \to S$ over $S = \mathrm{Spec}\,\Bbbk[t]/(t^2)$ pulled back along the respective map $S \to \mathrm{Hilb}^G_Y$ would have the same multiplication law in $\Bbbk[\mathscr{X}]$ as the trivial family $X \times S$, i.e., $\mathscr{X} \simeq X \times S$ would correspond to a map $S \to \{[X]\}$, a contradiction. Hence the T-eigenweights of $T_{[X_0]}\mathrm{Hilb}^G_Y$ are tails, and therefore negative integer combinations of positive roots. By the graded Nakayama lemma, the matrix entries of $m^{\nu}_{\lambda\mu}$ generate the maximal ideal $\mathscr{I}([X_0]) \lhd \Bbbk[\mathrm{Hilb}^G_Y]$ and the algebra $\Bbbk[\mathrm{Hilb}^G_Y]$. This implies all other claims. □

Corollary E.15. *The tails of any affine* G*-variety span a finitely generated semigroup.*

Proof. Since tails are preserved under restriction to general G-orbits, the question is reduced to G-orbit closures, which have finite multiplicity function by (2.2). Now the claim stems from Theorem E.14(3). □

References

[Akh1] D. N. Akhiezer, *Equivariant completion of homogeneous algebraic varieties by homogeneous divisors*, Ann. Global Anal. Geom. **1** (1983), no. 1, 49–78.

[Akh2] D. N. Akhiezer, *Actions with a finite number of orbits*, Funct. Anal. Appl. **19** (1985), no. 1, 1–4. **Russian original:** Д. Н. Ахиезер, *О действиях с конечным числом орбит*, Функц. анализ и его прил. **19** (1985), № 1, 1–5.

[Akh3] D. N. Akhiezer, *On modality and complexity of actions of reductive groups*, Russian Math. Surveys **43** (1988), no. 2, 157–158. **Russian original:** Д. Н. Ахиезер, *О модальности и сложности действий редуктивных групп*, Успехи мат. наук **43** (1988), № 2, 129–130.

[Akh4] D. N. Akhiezer, *Spherical varieties*, Schriftenreihe, Heft Nr. 199, Bochum, 1993.

[AV] D. N. Akhiezer, E. B. Vinberg, *Weakly symmetric spaces and spherical varieties*, Transform. Groups **4** (1999), no. 1, 3–24.

[AVS] D. V. Alekseevskij, E. B. Vinberg, A. S. Solodovnikov, *Geometry of spaces of constant curvature*, Geometry II, Encyclopædia Math. Sci., vol. 29, pp. 1–138, Springer, Berlin, 1993. **Russian original:** Д. В. Алексеевский, Э. Б. Винберг, А. С. Солодовников, *Геометрия пространств постоянной кривизны*, Соврем. пробл. мат. Фундам. направл., т. 29, с. 5–146, ВИНИТИ, Москва, 1988.

[AB1] V. Alexeev, M. Brion, *Stable reductive varieties, I: Affine varieties*, Invent. Math. **157** (2004), no. 2, 227–274.

[AB2] V. Alexeev, M. Brion, *Stable reductive varieties, II: Projective case*, Adv. Math. **184** (2004), no. 2, 380–408.

[AB3] V. Alexeev, M. Brion, *Moduli of affine schemes with reductive group action*, J. Algebraic Geom. **14** (2005), no. 1, 83–117.

[AG] V. I. Arnold, A. B. Givental, *Symplectic geometry*, Dynamical systems IV, Encyclopædia Math. Sci., vol. 4, pp. 1–138, Springer, Berlin, 2001. **Russian original:** В. И. Арнольд, А. Б. Гивенталь, *Симплектическая геометрия*, Соврем. пробл. мат. Фундам. направл., т. 4, с. 5–139, ВИНИТИ, Москва, 1985.

[Art] M. Artin, *Algebraic approximation of structures over complete local rings*, Inst. Hautes Études Sci. Publ. Math. (1969), no. 36, 23–58.

[Arzh] I. V. Arzhantsev, *On SL_2-actions of complexity one*, Izv. Math. **61** (1997), no. 4, 685–698. **Russian original:** И. В. Аржанцев, *О действиях сложности один группы* SL_2, Изв. РАН, сер. мат. **61** (1997), № 4, 3–18.

[AC] I. V. Arzhantsev, O. V. Chuvashova, *Classification of affine homogeneous spaces of complexity one*, Sb. Math. **195** (2004), no. 6, 765–782. **Russian original:** И. В. Аржанцев, О. В. Чувашова, *Классификация аффинных однородных пространств сложности один*, Мат. сб. **195** (2004), № 6, 3–20.

[Avd] R. Avdeev, *On solvable spherical subgroups of semisimple algebraic groups*, Oberwolfach Reports **7** (2010), no. 2, 1105–1108.

D.A. Timashev, *Homogeneous Spaces and Equivariant Embeddings*, Encyclopaedia of Mathematical Sciences 138, DOI 10.1007/978-3-642-18399-7,

[Bar] I. Barsotti, *Structure theorems for group-varieties*, Ann. Mat. Pura Appl. (4) **38** (1955), 77–119.

[BH] F. Berchtold, J. Hausen, *Homogeneous coordinates for algebraic varieties*, J. Algebra **266** (2003), no. 2, 636–670.

[BGGN] F. A. Berezin, I. M. Gelfand, M. I. Graev, M. A. Naimark, *Group representations*, Amer. Math. Soc. Transl. **16** (1960), 325–353. **Russian original:** Ф. А. Березин, И. М. Гельфанд, М. И. Граев, М. А. Наймарк, *Представления групп*, Успехи мат. наук **11** (1956), № 6, 13–40.

[Ber] D. I. Bernstein, *The number of roots of a system of equations*, Funct. Anal. Appl. **9** (1975), no. 3, 183–185. **Russian original:** Д. И. Бернштейн, *О числе решений системы уравнений*, Функц. анализ и его прил. **9** (1975), № 3, 1–4.

[BGG] I. N. Bernstein, I. M. Gelfand, S. I. Gelfand, *Schubert cells and cohomology of the spaces G/P*, Russian Math. Surveys **28** (1973), no. 3, 1–26. **Russian original:** И. Н. Бернштейн, И. М. Гельфанд, С. И. Гельфанд, *Клетки Шуберта и когомологии пространств G/P*, Успехи мат. наук **28** (1973), № 3, 3–26.

[B-B1] A. Białynicki-Birula, *Some properties of the decompositions of algebraic varieties determined by actions of a torus*, Bull. Acad. Polon. Sci. Sér. Sci. Math. Astronom. Phys. **24** (1976), no. 9, 667–674.

[B-B2] A. Białynicki-Birula, *On induced actions of algebraic groups*, Ann. Inst. Fourier (Grenoble) **43** (1993), no. 2, 365–368.

[BHM] A. Białynicki-Birula, G. Hochschild, G. D. Mostow, *Extensions of representations of algebraic linear groups*, Amer. J. Math. **85** (1963), 131–144.

[BiB] F. Bien, M. Brion, *Automorphisms and local rigidity of regular varieties*, Compositio Math. **104** (1996), no. 1, 1–26.

[BM] E. Bierstone, P. D. Milman, *Canonical desingularization in characteristic zero by blowing up the maximum strata of a local invariant*, Invent. Math. **128** (1997), no. 2, 207–302.

[BCP] E. Bifet, C. de Concini, C. Procesi, *Cohomology of regular embeddings*, Adv. Math. **82** (1990), no. 1, 1–34.

[Bol] A. V. Bolsinov, *Completeness of families of functions in involution related to compatible Poisson brackets*, Tensor and vector analysis (O. V. Manturov, A. T. Fomenko, V. V. Trofimov, eds.), pp. 3–24, Gordon and Breach, Amsterdam, 1998. **Russian original:** А. В. Болсинов, *О полноте семейств функций в инволюции, связанных с согласованными скобками Пуассона*, Тр. семинара по вект. и тенз. анализу **23** (1988), 18–38.

[Bor1] A. Borel, *Sur la cohomologie des espaces fibrés principaux et des espaces homogènes des groupes de Lie compacts*, Ann. of Math. (2) **57** (1953), no. 1, 115–207.

[Bor2] A. Borel, *Linear algebraic groups*, Graduate Texts in Math., vol. 126, Springer-Verlag, New York, 2nd edn., 1991.

[BoB] W. Borho, J.-L. Brylinski, *Differential operators on homogeneous spaces, II: Relative enveloping algebras*, Bull. Soc. Math. France **117** (1989), no. 2, 167–210.

[Bou1] N. Bourbaki, *Groupes et algèbres de Lie*, chap. IV–VI, Éléments de Math., fasc. XXXIV, Hermann, Paris, 1968.

[Bou2] N. Bourbaki, *Groupes et algèbres de Lie*, chap. VII–VIII, Éléments de Math., fasc. XXXVIII, Hermann, Paris, 1975.

[Bout] J.-F. Boutot, *Singularités rationnelles et quotients par les groupes réductifs*, Invent. Math. **88** (1987), no. 1, 65–68.

[Bra1] P. Bravi, *Wonderful varieties of type* **D**, Ph.D. thesis, Università di Roma "La Sapienza", Dec. 2003.

[Bra2] P. Bravi, *Wonderful varieties of type* **E**, Represent. Theory **11** (2007), 174–191.

[Bra3] P. Bravi, *Primitive spherical systems*, Preprint, 2009, `arXiv:math.RT/0909.3765`.

[BraL] P. Bravi, D. Luna, *An introduction to wonderful varieties with many examples of type* $\mathbf{F}_4$, J. Algebra **329** (2011), no. 1, 4–51.

[BraP1] P. Bravi, G. Pezzini, *Wonderful varieties of type* **D**, Represent. Theory **9** (2005), 578–637.

[BraP2] P. Bravi, G. Pezzini, *Wonderful varieties of type* **B** *and* **C**, Preprint, 2009, `arXiv:math.AG/0909.3771`.

[Bri1] M. Brion, *Quelques propriétés des espaces homogènes sphériques*, Manuscripta Math. **55** (1986), no. 2, 191–198.

[Bri2] M. Brion, *Classification des espaces homogènes sphériques*, Compositio Math. **63** (1987), no. 2, 189–208.

[Bri3] M. Brion, *Sur l'image de l'application moment*, Séminaire d'algèbre Paul Dubreil et Marie-Paule Malliavin, Lect. Notes in Math., vol. 1296, pp. 177–192, 1987.

[Bri4] M. Brion, *Groupe de Picard et nombres charactéristiques des variétés sphériques*, Duke Math. J. **58** (1989), no. 2, 397–424.

[Bri5] M. Brion, *On spherical varieties of rank one*, Group actions and invariant theory, CMS Conf. Proc., vol. 10, pp. 31–41, AMS, Providence, 1989.

[Bri6] M. Brion, *Spherical varieties: an introduction*, Topological methods in algebraic transformation groups (H. Kraft, T. Petrie, G. Schwarz, eds.), Progress in Math., vol. 80, pp. 11–26, Birkhäuser, Basel–Boston–Berlin, 1989.

[Bri7] M. Brion, *Une extension du théorème de Borel–Weil*, Math. Ann. **286** (1990), no. 4, 655–660.

[Bri8] M. Brion, *Vers une généralisation des espaces symétriques*, J. Algebra **134** (1990), no. 1, 115–143.

[Bri9] M. Brion, *Sur la géométrie des variétés sphériques*, Comment. Math. Helv. **66** (1991), no. 2, 237–262.

[Bri10] M. Brion, *Variétés sphériques et théorie de Mori*, Duke Math. J. **72** (1993), no. 2, 369–404.

[Bri11] M. Brion, *Piecewise polynomial functions, convex polytopes and enumerative geometry*, Parameter spaces, Banach Center Publ., vol. 36, pp. 25–44, Inst. of Math., Polish Acad. Sci., Warszawa, 1996.

[Bri12] M. Brion, *Curves and divisors in spherical varieties*, Algebraic groups and Lie groups; a volume of papers in honor of the late R. W. Richardson (G. Lehrer et al., ed.), Austral. Math. Soc. Lect. Ser., vol. 9, pp. 21–34, Cambridge University Press, Cambridge, 1997.

[Bri13] M. Brion, *Variétés sphériques*, Notes de la session de la S. M. F. "Opérations hamiltoniennes et opérations de groupes algébriques", Grenoble, 1997, `http://www-fourier.ujf-grenoble.fr/~mbrion/spheriques.pdf`.

[Bri14] M. Brion, *The behavior at infinity of the Bruhat decomposition*, Comment. Math. Helv. **73** (1998), no. 1, 137–174.

[Bri15] M. Brion, *Equivariant cohomology and equivariant intersection theory*, Representation theories and algebraic geometry (A. Broer, ed.), Nato ASI Series C, vol. 514, pp. 1–37, Kluwer, Dordrecht, 1998.

[Bri16] M. Brion, *On orbit closures of spherical subgroups in flag varieties*, Comment. Math. Helv. **76** (2001), no. 2, 263–299.

[Bri17] M. Brion, *Group completions via Hilbert schemes*, J. Algebraic Geom. **12** (2003), no. 4, 605–626.

[Bri18] M. Brion, *The total coordinate ring of a wonderful variety*, Duke Math. J. **313** (2007), no. 1, 61–99.

[BI] M. Brion, S. P. Inamdar, *Frobenius splitting of spherical varieties*, Algebraic groups and their generalizations: classical methods (W. J. Haboush, B. J. Parshall, eds.), Proc. Symp. Pure Math., vol. 56, part 1, pp. 207–218, AMS, Providence, 1994.

[BKn] M. Brion, F. Knop, *Contractions and flips for varieties with group actions of small complexity*, J. Math. Sci. Univ. Tokyo **1** (1994), no. 3, 641–655.

[BKu] M. Brion, S. Kumar, *Frobenius splitting methods in geometry and representation theory*, Progress in Math., vol. 231, Birkhäuser, Boston, 2005.

[BL] M. Brion, D. Luna, *Sur la structure locale des variétés sphériques*, Bull. Soc. Math. France **115** (1987), no. 2, 211–226.

[BLV] M. Brion, D. Luna, Th. Vust, *Espaces homogènes sphériques*, Invent. Math. **84** (1986), no. 3, 617–632.

[BPa] M. Brion, F. Pauer, *Valuations des espaces homogènes sphériques*, Comment. Math. Helv. **62** (1987), no. 2, 265–285.

[BPo] M. Brion, P. Polo, *Large Schubert varieties*, Represent. Theory **4** (2000), 97–126.

[BR] M. Brion, A. Rittatore, *The structure of normal algebraic monoids*, Semigroup Forum **74** (2007), no. 3, 410–422.

[Car1] É. Cartan, *Sur une classe remarquable d'espaces de Riemann*, Bull. Soc. Math. France **54** (1926), 214–264.

[Car2] É. Cartan, *Sur une classe remarquable d'espaces de Riemann, II*, Bull. Soc. Math. France **55** (1927), 114–134.

[Car3] É. Cartan, *Sur la détermination d'un système orthogonal complet dans un espace de Riemann symétrique clos*, Rend. Circ. Mat. Palermo **53** (1929), 217–252.

[CX] E. Casas-Alvero, S. Xambó-Descamps, *The enumerative theory of conics after Halphen*, Lect. Notes in Math., vol. 1196, 1986.

[CM] J.-Y. Charbonnel, A. Moreau, *The index of centralizers of elements of reductive Lie algebras*, Doc. Math. **15** (2010), 387–421.

[Cha] M. Chasles, *Determination du nombre des sections coniques qui doivent toucher cinq courbes d'ordre quelconque, ou satisfaire à diverses autres conditions*, C. R. Acad. Sci. Paris Sér. I Math. **58** (1864), 222–226.

[CF] A. Collino, W. Fulton, *Intersection rings of spaces of triangles*, Mém. Soc. Math. France (N.S.) **38** (1989), 75–117.

[Con] C. de Concini, *Normality and non-normality of certain semigroups and orbit closures*, Invariant theory and algebraic transformation groups III (R. V. Gamkrelidze, V. L. Popov, eds.), Encyclopædia Math. Sci., vol. 132, pp. 15–35, Springer-Verlag, Berlin, 2004.

[CGMP] C. de Concini, M. Goresky, R. MacPherson, C. Procesi, *On the geometry of quadrics and their degenerations*, Comment. Math. Helv. **63** (1988), no. 3, 337–413.

[CP1] C. de Concini, C. Procesi, *Complete symmetric varieties*, Invariant theory (F. Gherardelli, ed.), Lect. Notes in Math., vol. 996, pp. 1–44, Springer, Berlin, 1983.

[CP2] C. de Concini, C. Procesi, *Complete symmetric varieties, II: Intersection theory*, Algebraic groups and related topics (R. Hotta, ed.), Adv. Studies in Pure Math., vol. 6, pp. 481–513, North-Holland, Amsterdam, 1985.

[CP3] C. de Concini, C. Procesi, *Cohomology of compactifications of algebraic groups*, Duke Math. J. **53** (1986), no. 3, 585–594.

[CS] C. de Concini, T. A. Springer, *Compactification of symmetric varieties*, Transform. Groups **4** (1999), no. 2–3, 273–300.

[CL] A. Corti, V. Lazić, *Finite generation implies the Minimal Model Program*, Preprint, 2010, `arXiv:math.AG/1005.0614`.

[Cox] D. Cox, *The homogeneous coordinate ring of a toric variety*, J. Algebraic Geom. **4** (1995), no. 1, 17–50.

[C-F1] S. Cupit-Foutou, *Classification of two-orbit varieties*, Comment. Math. Helv. **78** (2003), no. 2, 245–265.

[C-F2] S. Cupit-Foutou, *Wonderful varieties: a geometrical realization*, Preprint, 2009, `arXiv:math.AG/0907.2852`.

[Dan] V. I. Danilov, *The geometry of toric varieties*, Russian Math. Surveys **33** (1978), no. 2, 97–154. **Russian original:** В. И. Данилов, *Геометрия торических многообразий*, Успехи мат. наук **33** (1978), № 2, 85–134.

[Dem1] M. Demazure, *Sur la formule des caractères de H. Weyl*, Invent. Math. **9** (1969/1970), no. 3, 249–252.

[Dem2] M. Demazure, *Désingularisation des variétés de Schubert généralisées*, Ann. Sci. École Norm. Sup. (4) **7** (1974), 53–88.

[Dem3] M. Demazure, *A very simple proof of Bott's theorem*, Invent. Math. **33** (1976), no. 3, 271–272.

[Dem4] M. Demazure, *Limites de groupes orthogonaux ou symplectiques*, Preprint, Paris, 1980.

[DG] M. Demazure, P. Gabriel, *Groupes algébriques I*, Masson/North-Holland, Paris/Amsterdam, 1970.

[Don] S. Donkin, *Good filtrations of rational modules for reductive groups*, The Arcata Conf. on Representations of Finite Groups, Proc. Symp. Pure Math., vol. 47, part 1, pp. 69–80, AMS, Providence, 1987.

[Do] S. Doty, *Representation theory of reductive normal algebraic monoids*, Trans. Amer. Math. Soc. **351** (1999), no. 6, 2539–2551.

[Ela1] A. G. Elashvili, *Canonical form and stationary subalgebras of points of general position for simple linear Lie groups*, Funct. Anal. Appl. **6** (1972), no. 1, 44–53. **Russian original:** А. Г. Элашвили, *Канонический вид и стационарные подалгебры точек общего положения для простых линейных групп Ли*, Функц. анализ и его прил. **6** (1972), № 1, 51–62.

[Ela2] A. G. Elashvili, *Stationary subalgebras of points in general position for irreducible linear Lie groups*, Funct. Anal. Appl. **6** (1972), no. 2, 139–148. **Russian original:** А. Г. Элашвили, *Стационарные подалгебры точек общего положения для неприводимых линейных групп Ли*, Функц. анализ и его прил. **6** (1972), № 2, 65–78.

[EKW] E. J. Elizondo, K. Kurano, K. Watanabe, *The total coordinate ring of a normal projective variety*, J. Algebra **276** (2004), no. 2, 625–637.

[Elk] R. Elkik, *Singularités rationnelles et déformations*, Invent. Math. **47** (1978), no. 2, 139–147.

[Fo] A. Foschi, *Variétés magnifiques et polytopes moment*, Thèse de doctorat, Institut Fourier, Université J. Fourier, Grenoble, Oct. 1998.

[Ful1] W. Fulton, *Intersection theory*, Ergeb. Math. Grenzgeb. (3), vol. 2, Springer-Verlag, Berlin–Heidelberg–New York–Tokyo, 1984.

[Ful2] W. Fulton, *Introduction to toric varieties*, Princeton University Press, Princeton, 1993.

[FMSS] W. Fulton, R. MacPherson, F. Sottile, B. Sturmfels, *Intersection theory on spherical varieties*, J. Algebraic Geom. **4** (1995), no. 1, 181–193.

[FS] W. Fulton, B. Sturmfels, *Intersection theory on toric varieties*, Topology **36** (1997), no. 2, 335–353.

[Gel] I. M. Gelfand, *Spherical functions on symmetric Riemannian spaces*, Amer. Math. Soc. Transl. **37** (1964), 39–43. **Russian original:** И. М. Гельфанд, *Сферические функции на симметрических римановых пространствах*, Докл. АН СССР **70** (1950), 5–8.

[Gin] V. Ginzburg, *Admissible modules on a symmetric space*, Astérisque **173–174** (1989), 199–255.

[GOV] V. V. Gorbatsevich, A. L. Onishchik, E. B. Vinberg, *Structure of Lie groups and Lie algebras*, Lie groups and Lie algebras III, Encyclopædia Math. Sci., vol. 41, Springer, Berlin, 1994. **Russian original:** Э. Б. Винберг, В. В. Горбацевич, А. Л. Онищик, *Строение групп и алгебр Ли*, Соврем. пробл. мат. Фундам. направл., т. 41, ВИНИТИ, Москва, 1990.

[Gra] W. A. de Graaf, *Computing with nilpotent orbits in simple Lie algebras of exceptional type*, London Math. Soc. J. Comput. Math. **11** (2008), 280–297.

[GR] H. Grauert, O. Riemenschneider, *Verschwindungssätze für analytische Kohomologiegruppen auf komplexen Räumen*, Invent. Math. **11** (1970), no. 4, 263–292.

[Gr1] F. D. Grosshans, *Constructing invariant polynomials via Tschirnhaus transformations*, Invariant theory (S. S. Koh, ed.), Lect. Notes in Math., vol. 1278, pp. 95–102, Springer, Berlin–Heidelberg–New York, 1987.

[Gr2] F. D. Grosshans, *Algebraic homogeneous spaces and Invariant theory*, Lect. Notes in Math., vol. 1673, 1997.

[Gro] A. Grothendieck, *Techniques de construction et théorèmes d'existence en géométrie algébrique, IV: Les schémas de Hilbert*, Séminaire Bourbaki **13** (1960/61), no. 221.

[GS] V. Guillemin, S. Sternberg, *Multiplicity-free spaces*, J. Differential Geom. **19** (1984), no. 1, 31–56.

[HM] C. D. Hacon, J. McKernan, *Extension theorems and the existence of flips*, Flips for 3-folds and 4-folds, Oxford Lecture Ser. Math. Appl., vol. 35, pp. 76–110, Oxford Univ. Press, Oxford, 2007.

[HS] M. Haiman, B. Sturmfels, *Multigraded Hilbert schemes*, J. Algebraic Geom. **13** (2004), no. 4, 725–769.

[Har1] R. Hartshorne, *Residues and duality*, Lect. Notes in Math., vol. 20, 1966.

[Har2] R. Hartshorne, *Algebraic geometry*, Springer, New York, 1977.

[Har3] R. Hartshorne, *Deformation theory*, Graduate Texts in Math., vol. 257, Springer, New York, 2010.

[Hau] J. Hausen, *Cox rings and combinatorics, II*, Moscow Math. J. **8** (2008), no. 4, 711–757.

[Hel1] S. Helgason, *Differential geometry, Lie groups, and symmetric spaces*, Academic Press, New York, 1978.

[Hel2] S. Helgason, *Groups and geometric analysis. Integral geometry, invariant differential operators, and spherical functions*, Academic Press, Orlando, 1984.

[Hir] A. Hirschowitz, *Le groupe de Chow équivariant*, C. R. Acad. Sci. Paris Sér. I Math. **298** (1984), no. 5, 87–89.

[Ho] R. E. Howe, *The First Fundamental Theorem of Invariant Theory and spherical subgroups*, Algebraic groups and their generalizations: classical methods (W. J. Haboush, B. J. Parshall, eds.), Proc. Symp. Pure Math., vol. 56, part 1, pp. 333–346, AMS, Providence, 1994.

[HW] A. T. Huckleberry, T. Wurzbacher, *Multiplicity-free complex manifolds*, Math. Ann. **286** (1990), 261–280.

[Hum] J. E. Humphreys, *Linear algebraic groups*, Springer-Verlag, New York, 1975.

[IS] V. A. Iskovskikh, I. R. Shafarevich, *Algebraic surfaces*, Algebraic geometry II, Encyclopædia Math. Sci., vol. 35, pp. 127–262, Springer, Berlin, 1996. **Russian original:** В. А. Исковских, И. Р. Шафаревич, *Алгебраические поверхности*, Соврем. пробл. мат. Фундам. направл., т. 35, с. 131–263, ВИНИТИ, Москва, 1989.

[IM] N. Iwahori, H. Matsumoto, *On some Bruhat decomposition and the structure of the Hecke rings of p-adic Chevalley groups*, Inst. Hautes Études Sci. Publ. Math. (1965), no. 25, 5–48.

[Jan] J. C. Jantzen, *Representations of algebraic groups*, Academic Press, Boston, 1987.

[Ka] S. Kato, *A Borel–Weil–Bott type theorem for group completions*, J. Algebra **259** (2003), no. 2, 572–580.

[KMM] Y. Kawamata, K. Matsuda, K. Matsuki, *Introduction to the minimal model program*, Algebraic geometry (T. Oda, ed.), Adv. Studies in Pure Math., vol. 10, pp. 283–360, North-Holland, Amsterdam, 1987.

[Kaz] B. Ya. Kazarnovskii, *Newton polyhedra and the Bézout formula for matrix-valued functions of finite-dimensional representations*, Funct. Anal. Appl. **21** (1987), no. 4, 319–321. **Russian original:** Б. Я. Казарновский, *Многогранники Ньютона и формула Безу для матричных функций конечномерных представлений*, Функц. анализ и его прил. **21** (1987), № 4, 73–74.

[Kem1] G. Kempf, *On the collapsing of homogeneous bundles*, Invent. Math. **37** (1976), no. 3, 229–239.

[Kem2] G. Kempf, *The Grothendieck–Cousin complex of an induced representation*, Adv. Math. **29** (1978), no. 3, 310–396.

[KKMS] G. Kempf, F. Knudsen, D. Mumford, B. Saint-Donat, *Toroidal embeddings, I*, Lect. Notes in Math., vol. 339, 1973.

[KeR] G. R. Kempf, A. Ramanathan, *Multi-cones over Schubert varieties*, Invent. Math. **87** (1987), no. 2, 353–363.

[Kle] S. L. Kleiman, *The transversality of a general translate*, Compositio Math. **28** (1974), 287–297.

[Kl] F. Klein, *Vergleichende Betrachtungen über neuere geometrische Forschungen*, Math. Ann. **43** (1893), no. 1, 63–100.

[Kn1] F. Knop, *Weylgruppe und Momentabbildung*, Invent. Math. **99** (1990), no. 1, 1–23.

[Kn2] F. Knop, *The Luna–Vust theory of spherical embeddings*, Proc. Hyderabad Conf. on Algebraic Groups (S. Ramanan, ed.), pp. 225–249, Manoj Prakashan, Madras, 1991.

[Kn3] F. Knop, *Über Bewertungen, welche unter einer reductiven Gruppe invariant sind*, Math. Ann. **295** (1993), no. 2, 333–363.

[Kn4] F. Knop, *Über Hilberts vierzehntes Problem für Varietäten mit Kompliziertheit eins*, Math. Z. **213** (1993), no. 1, 33–36.

[Kn5] F. Knop, *The asymptotic behavior of invariant collective motion*, Invent. Math. **116** (1994), 309–328.

[Kn6] F. Knop, *A Harish-Chandra homomorphism for reductive group actions*, Ann. of Math. (2) **140** (1994), no. 2, 253–288.

[Kn7] F. Knop, *On the set of orbits for a Borel subgroup*, Comment. Math. Helv. **70** (1995), no. 2, 285–309.

[Kn8] F. Knop, *Automorphisms, root systems, and compactifications of homogeneous varieties*, J. Amer. Math. Soc. **9** (1996), no. 1, 153–174.

[KKLV] F. Knop, H. Kraft, D. Luna, Th. Vust, *Local properties of algebraic group actions*, Algebraische Transformationsgruppen und Invariantentheorie (H. Kraft, P. Slodowy, T. A. Springer, eds.), DMV Seminar, vol. 13, pp. 63–75, Birkhäuser, Basel–Boston–Berlin, 1989.

[KKV] F. Knop, H. Kraft, Th. Vust, *The Picard group of a G-variety*, Algebraische Transformationsgruppen und Invariantentheorie (H. Kraft, P. Slodowy, T. A. Springer, eds.), DMV Seminar, vol. 13, pp. 77–87, Birkhäuser, Basel–Boston–Berlin, 1989.

[Kob] T. Kobayashi, *Proper action on a homogeneous space of reductive type*, Math. Ann. **285** (1989), no. 2, 249–263.

[KoR] B. Kostant, S. Rallis, *Orbits and representations associated with symmetric spaces*, Amer. J. Math. **93** (1971), 753–809.

[Kou] A. G. Kouchnirenko, *Polyèdres de Newton et nombres de Milnor*, Invent. Math. **32** (1976), no. 1, 1–31.

[Krä] M. Krämer, *Sphärische Untergruppen in kompakten zusammenhängenden Liegruppen*, Compositio Math. **38** (1979), no. 2, 129–153.

[LT] G. Lancaster, J. Towber, *Representation-functors and flag-algebras for the classical groups, I*, J. Algebra **59** (1979), no. 1, 16–38.

[Lau] J. Lauret, *Gelfand pairs attached to representations of compact Lie groups*, Transform. Groups **5** (2000), no. 4, 307–324.

[La] N. Lauritzen, *Splitting properties of complete homogeneous spaces*, J. Algebra **162** (1993), no. 1, 178–193.

[Lit] P. Littelmann, *On spherical double cones*, J. Algebra **166** (1994), no. 1, 142–157.

[LP] P. Littelmann, C. Procesi, *Equivariant cohomology of wonderful compactifications*, Operator algebras, unitary representations, enveloping algebras, and invariant theory, Progress in Math., vol. 92, pp. 219–262, Birkhäuser, Boston, 1990.

[Los1] I. V. Losev, *Demazure embeddings are smooth*, Internat. Math. Res. Notices (2009), no. 14, 2588–2596.

[Los2] I. V. Losev, *Uniqueness property for spherical homogeneous spaces*, Duke Math. J. **147** (2009), no. 2, 315–343.

[Lu1] D. Luna, *Sur les orbites fermées des groupes algébriques réductifs*, Invent. Math. **16** (1972), no. 1–2, 1–5.

[Lu2] D. Luna, *Slices étales*, Bull. Soc. Math. France Suppl. Mém. **33** (1973), 81–105.

[Lu3] D. Luna, *Sous-groupes sphériques résolubles*, Prépublication de l'Institut Fourier no. 241, Grenoble, 1993.

[Lu4] D. Luna, *Toute variété magnifique est sphérique*, Transform. Groups **1** (1996), no. 3, 249–258.

[Lu5] D. Luna, *Grosses cellules pour les variétés sphériques*, Algebraic groups and Lie groups, Austral. Math. Soc. Lect. Ser., vol. 9, pp. 267–280, Cambridge University Press, Cambridge, 1997.

[Lu6] D. Luna, *Variétés sphériques de type A*, Inst. Hautes Études Sci. Publ. Math. (2001), no. 94, 161–226.

[LV] D. Luna, Th. Vust, *Plongements d'espaces homogènes*, Comment. Math. Helv. **58** (1983), no. 2, 186–245.

[M] K. Matsuki, *Introduction to the Mori program*, Universitext, Springer-Verlag, New York, 2002.

[Ma] H. Matsumura, *Commutative ring theory*, Cambridge University Press, Cambridge, 1986.

[Mat] Y. Matsushima, *Espaces homogènes de Stein des groupes de Lie complexes*, Nagoya Math. J. **16** (1960), 205–218.

[MRo] J. C. McConnell, J. C. Robson, *Noncommutative Noetherian rings*, Graduate Studies in Math., vol. 30, AMS, Providence, 2001.

[McG] W. M. McGovern, *The adjoint representation and the adjoint action*, Invariant theory and algebraic transformation groups II (R. V. Gamkrelidze, V. L. Popov, eds.), Encyclopædia Math. Sci., vol. 131, pp. 159–238, Springer, Berlin, 2002.

[MK] V. B. Mehta, W. van der Kallen, *On a Grauert–Riemenschneider vanishing theorem for Frobenius split varieties in characteristic p*, Invent. Math. **108** (1992), no. 1, 11–13.

[MRa] V. B. Mehta, A. Ramanathan, *Frobenius splitting and cohomology vanishing for Schubert varieties*, Ann. of Math. (2) **122** (1985), no. 1, 27–40.

[Mik] I. V. Mikityuk, *On the integrability of invariant Hamiltonian systems with homogeneous configuration spaces*, Math. USSR-Sb. **57** (1987), no. 2, 527–546. **Russian original:** И. В. Микитюк, *Об интегрируемости инвариантных гамильтоновых систем с однородными конфигурационными пространствами*, Мат. сб. **129** (1986), № 4, 514–534.

[MF1] A. S. Mishchenko, A. T. Fomenko, *Euler equations on finite-dimensional Lie groups*, Math. USSR-Izv. **12** (1978), no. 2, 371–389. **Russian original:** А. С. Мищенко, А. Т. Фоменко, *Уравнения Эйлера на конечномерных группах Ли*, Изв. АН СССР, сер. мат. **42** (1978), № 2, 396–415.

[MF2] A. S. Mishchenko, A. T. Fomenko, *Generalized Liouville method of integration of Hamiltonian systems*, Funct. Anal. Appl. **12** (1978), no. 2, 113–121. **Russian original:** А. С. Мищенко, А. Т. Фоменко, *Обобщённый метод Лиувилля интегрирования гамильтоновых систем*, Функц. анализ и его прил. **12** (1978), № 2, 46–56.

[Mon] P.-L. Montagard, *Une nouvelle propriété de stabilité du pléthysme*, Comment. Math. Helv. **71** (1996), no. 3, 475–505.

[M-J1] L. Moser-Jauslin, *Normal embeddings of* $\mathrm{SL}(2)/\Gamma$, Thesis, University of Geneva, 1987.

[M-J2] L. Moser-Jauslin, *The Chow rings of smooth complete* $SL(2)$*-embeddings*, Compositio Math. **82** (1992), no. 1, 67–106.

[Mum] D. Mumford, *Abelian varieties*, Tata Inst. Fund. Research Studies in Math., no. 5, Oxford University Press, London, 1974.

[MFK] D. Mumford, J. Fogarty, F. Kirwan, *Geometric Invariant Theory*, Springer-Verlag, Berlin–Heidelberg–New York, 3rd edn., 1994.

[Nag] M. Nagata, *On the 14-th problem of Hilbert*, Amer. J. Math. **81** (1959), 766–772.

[Oda] T. Oda, *Convex bodies and algebraic geometry: an introduction to the theory of toric varieties*, Springer-Verlag, Berlin–Heidelberg–New York, 1988.

[Oni1] A. L. Onishchik, *Complex hulls of compact homogeneous spaces*, Soviet Math. Dokl. **1** (1960), 88–91. **Russian original:** А. Л. Онищик, *Комплексные оболочки компактных однородных пространств*, Докл. АН СССР **130** (1960), № 2, 726–729.

[Oni2] A. L. Onishchik, *Inclusion relations among transitive compact transformation groups*, Amer. Math. Soc. Transl. **50** (1966), 5–58. **Russian original:** А. Л. Онищик, *Отношения включения между транзитивными компактными группами преобразований*, Тр. ММО **11** (1962), 199–242.

[OV] A. L. Onishchik, E. B. Vinberg, *Lie groups and algebraic groups*, Springer, Berlin–Heidelberg–New York, 1990. **Russian original:** Э. Б. Винберг, А. Л. Онищик, *Семинар по группам Ли и алгебраическим группам*, Наука, Москва, 1988.

[Pan1] D. I. Panyushev, *Complexity and rank of homogeneous spaces*, Geom. Dedicata **34** (1990), no. 3, 249–269.

[Pan2] D. I. Panyushev, *Complexity of quasiaffine homogeneous varieties, t-decompositions, and affine homogeneous spaces of complexity* 1, Lie groups, their discrete subgroups and invariant theory (E. B. Vinberg, ed.), Adv. Soviet Math., vol. 8, pp. 151–166, AMS, Providence, 1992.

[Pan3] D. I. Panyushev, *Complexity and rank of double cones and tensor product decompositions*, Comment. Math. Helv. **68** (1993), no. 3, 455–468.

[Pan4] D. I. Panyushev, *Complexity and nilpotent orbits*, Manuscripta Math. **83** (1994), 223–237.

[Pan5] D. I. Panyushev, *On homogeneous spaces of rank one*, Indag. Math. **6** (1995), no. 3, 315–323.

[Pan6] D. I. Panyushev, *A restriction theorem and the Poincaré series for U-invariants*, Math. Ann. **301** (1995), no. 4, 655–675.

[Pan7] D. I. Panyushev, *Complexity and rank of actions in Invariant theory*, J. Math. Sci. (New York) **95** (1999), no. 1, 1925–1985.

[Pau1] F. Pauer, *Normale Einbettungen von G/U*, Math. Ann. **257** (1981), no. 3, 371–396.

[Pau2] F. Pauer, *Glatte Einbettungen von G/U*, Math. Ann. **262** (1983), no. 3, 421–429.

[Pau3] F. Pauer, *Plongements normaux de l'espace homogène $SL(3)/SL(2)$*, C. R. du 108ième Congrès Nat. Soc. Sav., vol. 3, pp. 87–104, Grenoble, 1983.

[Pau4] F. Pauer, *"Caracterisation valuative" d'une classe de sous-groupes d'un groupe algébrique*, C. R. du 109ième Congrès Nat. Soc. Sav., vol. 3, pp. 159–166, Grenoble, 1984.

[Pez] G. Pezzini, *Wonderful varieties of type* **C**, Ph.D. thesis, Università di Roma "La Sapienza", Dec. 2003.

[Po1] V. L. Popov, *Quasihomogeneous affine algebraic varieties of the group* SL(2), Math. USSR-Izv. **7** (1973), no. 4, 793–831. **Russian original:** В. Л. Попов, *Квазиоднородные аффинные алгебраические многообразия группы* SL(2), Изв. АН СССР, сер. мат. **37** (1973), № 4, 792–832.

[Po2] V. L. Popov, *Picard groups of homogeneous spaces of linear algebraic groups and one-dimensional homogeneous vector bundles*, Math. USSR-Izv. **8** (1974), no. 2, 301–327. **Russian original:** В. Л. Попов, *Группы Пикара однородных пространств линейных алгебраических групп и одномерные однородные векторные расслоения*, Изв. АН СССР, сер. мат. **38** (1974), № 2, 294–322.

[Po3] V. L. Popov, *Algebraic curves with an infinite automorphism group*, Math. Notes **23** (1978), no. 2, 102–108. **Russian original:** В. Л. Попов, *Алгебраические кривые с бесконечной группой автоморфизмов*, Мат. Заметки **23** (1978), № 2, 183–195.

[Po4] V. L. Popov, *Hilbert's theorem on invariants*, Soviet Math. Dokl. **20** (1979), 1318–1322. **Russian original:** В. Л. Попов, *О теореме Гильберта об инвариантах*, Докл. АН СССР **249** (1979), № 3, 551–555.

[Po5] V. L. Popov, *Contraction of the actions of reductive algebraic groups*, Math. USSR-Sb. **58** (1987), no. 2, 311–335. **Russian original:** В. Л. Попов, *Стягивание действий редуктивных алгебраических групп*, Мат. сб. **130** (1986), № 3, 310–334.

[PV] V. L. Popov, E. B. Vinberg, *Invariant theory*, Algebraic geometry IV, Encyclopædia Math. Sci., vol. 55, pp. 123–278, Springer, Berlin, 1994. **Russian original:** Э. Б. Винберг, В. Л. Попов, *Теория инвариантов*, Соврем. пробл. мат. Фундам. направл., т. 55, с. 137–309, ВИНИТИ, Москва, 1989.

[PK] A. V. Pukhlikov, A. G. Khovanskii, *Finitely additive measures of virtual polytopes*, St. Petersburg Math. J. **4** (1993), no. 2, 337–356. **Russian original:** А. В. Пухликов, А. Г. Хованский, *Конечно-аддитивные меры виртуальных многогранников*, Алгебра и Анализ **4** (1992), № 2, 161–185.

[Put] M. S. Putcha, *Linear algebraic monoids*, London Math. Soc. Lect. Note Ser., no. 133, Cambridge University Press, Cambridge, 1988.

[Ram] A. Ramanathan, *Equations defining Schubert varieties and Frobenius splitting of diagonals*, Inst. Hautes Études Sci. Publ. Math. (1987), no. 65, 61–90.

[Ren1] L. E. Renner, *Quasi-affine algebraic monoids*, Semigroup Forum **30** (1984), 167–176.

[Ren2] L. E. Renner, *Classification of semisimple algebraic monoids*, Trans. Amer. Math. Soc. **292** (1985), no. 1, 193–223.

[Ren3] L. E. Renner, *Linear algebraic monoids*, Invariant theory and algebraic transformation groups V (R. V. Gamkrelidze, V. L. Popov, eds.), Encyclopædia Math. Sci., vol. 134, Springer-Verlag, Berlin, 2005.

[Ri1] R. W. Richardson, *Affine coset spaces of reductive algebraic groups*, Bull. London Math. Soc. **9** (1977), no. 1, 38–41.

[Ri2] R. W. Richardson, *Orbits, invariants and representations associated to involutions of reductive groups*, Invent. Math. **66** (1982), no. 2, 287–312.

[RS1] R. W. Richardson, T. A. Springer, *The Bruhat order on symmetric varieties*, Geom. Dedicata **35** (1990), 389–436.

[RS2] R. W. Richardson, T. A. Springer, *Combinatorics and geometry of K-orbits on the flag manifold*, Linear algebraic groups and their representations, Contemp. Math., vol. 153, pp. 109–142, AMS, Providence, 1993.

[Rit1] A. Rittatore, *Algebraic monoids and group embeddings*, Transform. Groups **3** (1998), no. 4, 375–396.

[Rit2] A. Rittatore, *Reductive embeddings are Cohen–Macaulay*, Proc. Amer. Math. Soc. **131** (2003), no. 3, 675–684.

[Rit3] A. Rittatore, *Algebraic monoids with affine unit group are affine*, Transform. Groups **12** (2007), no. 3, 601–605.

[Ros] M. Rosenlicht, *Some basic theorems on algebraic groups*, Amer. J. Math. **78** (1956), no. 2, 401–443.

[Ryb] L. G. Rybnikov, *On the commutativity of weakly commutative Riemannian homogeneous spaces*, Funct. Anal. Appl. **37** (2003), no. 2, 114–122. **Russian original:** Л. Г. Рыбников, *О коммутативности слабо коммутативных римановых однородных пространств*, Функц. анализ и его прил. **37** (2003), № 2, 41–51.

[Sch1] H. C. H. Schubert, *Kalkül der abzählenden Geometrie*, 1879, reprinted by Springer-Verlag, Berlin–New York, 1979.

[Sch2] H. C. H. Schubert, *Anzahl-Bestimmungen für lineare Räume beliebiger Dimension*, Acta Math. **8** (1886), no. 1, 97–118.

[Sch3] H. C. H. Schubert, *Allgemeine Anzahlfunctionen für Kegelschnitte, Flächen und Räume zweiten Grades in n Dimensionen*, Math. Ann. **45** (1894), no. 2, 153–206.

[Sel] A. Selberg, *Harmonic analysis and discontinuous groups in weakly symmetric Riemannian spaces with application to Dirichlet series*, J. Indian Math. Soc. (N.S.) **20** (1956), 47–87.

[Sem1] J. G. Semple, *On complete quadrics*, J. London Math. Soc. **23** (1948), no. 4, 258–267.

[Sem2] J. G. Semple, *On complete quadrics, II*, J. London Math. Soc. **27** (1952), no. 3, 280–287.

[Ser] E. Sernesi, *Deformations of algebraic schemes*, Grundlehren der Mathematischen Wissenschaften, vol. 334, Springer-Verlag, Berlin, 2006.

[Se1] J.-P. Serre, *Espaces fibrés algébriques*, Anneaux de Chow et applications, Séminaire C. Chevalley, vol. 3, exp. n°1, E. N. S., Paris, 1958.

[Se2] J.-P. Serre, *On the fundamental group of a unirational variety*, J. London Math. Soc. **34** (1959), no. 4, 481–484.

[Sha] I. R. Shafarevich, *Basic algebraic geometry*, vol. 1, 2, Springer-Verlag, Berlin, 2nd edn., 1994. **Russian original:** И. Р. Шафаревич, *Основы алгебраической геометрии*, т. 1, 2, Наука, Москва, 2-е изд., 1988.

[Sm-A] A. V. Smirnov, *Quasi-closed orbits in projective representations of semisimple complex Lie groups*, Trans. Moscow Math. Soc. (2003), 193–247. **Russian original:** А. В. Смирнов, *Квазизамкнутые орбиты в проективных представлениях полупростых комплексных групп Ли*, Тр. ММО **64** (2003), 213–270.

[Sm-E] E. Yu. Smirnov, *On systems of generators for ideals of S-varieties*, Moscow Univ. Math. Bull. **60** (2005), no. 3, 1–4. **Russian original:** Е. Ю. Смирнов, *О системах порождающих для идеалов S-многообразий*, Вестник МГУ, сер. 1, мат. мех. (2005), № 3, 3–6.

[Sp1] T. A. Springer, *Some results on algebraic groups with involutions*, Algebraic groups and related topics (R. Hotta, ed.), Adv. Studies in Pure Math., vol. 6, pp. 525–543, North-Holland, Amsterdam, 1985.

[Sp2] T. A. Springer, *The classification of involutions of simple algebraic groups*, J. Fac. Sci. Univ. Tokyo Sect. IA Math. **34** (1987), no. 3, 655–670.

[Sp3] T. A. Springer, *Linear algebraic groups*, Progress in Math., vol. 9, Birkhäuser, Boston, 2nd edn., 1998.

[Sp4] T. A. Springer, *Intersection cohomology of $B \times B$-orbit closures in group compactifications*, J. Algebra **258** (2002), no. 1, 71–111.

[St] R. Steinberg, *Endomorphisms of linear algebraic groups*, Mem. Amer. Math. Soc. **80** (1968).

[Ste] J. R. Stembridge, *Multiplicity-free products and restrictions of Weyl characters*, Represent. Theory **7** (2003), 404–439.

[Str1] E. S. Strickland, *A vanishing theorem for group compactifications*, Math. Ann. **277** (1987), no. 1, 165–171.

[Str2] E. S. Strickland, *Computing the equivariant cohomology of group compactifications*, Math. Ann. **291** (1991), no. 2, 275–280.

[Sukh] A. A. Sukhanov, *Description of the observable subgroups of linear algebraic groups*, Math. USSR-Sb. **65** (1990), no. 1, 97–108. **Russian original:** А. А. Суханов, *Описание наблюдаемых подгрупп линейных алгебраических групп*, Мат. сб. **137** (1988), № 1, 90–102.

[Sum] H. Sumihiro, *Equivariant completion*, J. Math. Kyoto Univ. **14** (1974), no. 1, 1–28.

[Tho] E. G. F. Thomas, *An infinitesimal characterization of Gelfand pairs*, Conference in modern analysis and probability, Contemp. Math., vol. 26, pp. 379–385, AMS, Providence, 1984.

[Tim1] D. A. Timashev, *Generalization of the Bruhat decomposition*, Russian Acad. Sci. Izv. Math. **45** (1995), no. 2, 339–352 (English). **Russian original:** Д. А. Тимашёв, *Обобщение разложения Брюа*, Изв. РАН, сер. мат. **58** (1994), № 5, 110–123.

[Tim2] D. A. Timashev, *Classification of G-varieties of complexity* 1, Izv. Math. **61** (1997), no. 2, 363–397. **Russian original:** Д. А. Тимашёв, *Классификация G-многообразий сложности* 1, Изв. РАН, сер. мат. **61** (1997), № 2, 127–162.

[Tim3] D. A. Timashev, *Cartier divisors and geometry of normal G-varieties*, Transform. Groups **5** (2000), no. 2, 181–204.

[Tim4] D. A. Timashev, *Equivariant compactifications of reductive groups*, Sb. Math. **194** (2003), no. 4, 589–616. **Russian original:** Д. А. Тимашёв, *Эквивариантные компактификации редуктивных групп*, Мат. сб. **194** (2003), № 4, 119–146.

[Tim5] D. A. Timashev, *Complexity of homogeneous spaces and growth of multiplicities*, Transform. Groups **9** (2004), no. 1, 65–72.

[Tim6] D. A. Timashev, *Equivariant embeddings of homogeneous spaces*, Surveys in geometry and number theory: reports on contemporary Russian mathematics, London Math. Soc. Lect. Note Ser., no. 338, pp. 226–278, Cambridge University Press, Cambridge, 2007.

[Tim7] D. A. Timashev, *Torus actions of complexity one*, Toric topology (M. Harada, Y. Karshon, M. Masuda, T. Panov, eds.), Contemp. Math., vol. 460, pp. 349–364, AMS, Providence, 2008.

[Tyr] J. A. Tyrrell, *Complete quadrics and collineations in S_n*, Mathematika **3** (1956), no. 1, 69–79.

[Vin1] E. B. Vinberg, *Complexity of actions of reductive groups*, Funct. Anal. Appl. **20** (1986), no. 1, 1–11. **Russian original:** Э. Б. Винберг, *Сложность действий редуктивных групп*, Функц. анализ и его прил. **20** (1986), № 1, 1–13.

[Vin2] E. B. Vinberg, *On reductive algebraic semigroups*, Lie groups and Lie algebras: E. B. Dynkin seminar (S. G. Gindikin, E. B. Vinberg, eds.), Amer. Math. Soc. Transl., vol. 169, pp. 145–182, AMS, Providence, 1995.

[Vin3] E. B. Vinberg, *Commutative homogeneous spaces and co-isotropic symplectic actions*, Russian Math. Surveys **56** (2001), no. 1, 1–60. **Russian original:** Э. Б. Винберг, *Коммутативные однородные пространства и коизотропные симплектические действия*, Успехи мат. наук **56** (2001), № 1, 3–62.

[Vin4] E. B. Vinberg, *Commutative homogeneous spaces of Heisenberg type*, Trans. Moscow Math. Soc. (2003), 45–78. **Russian original:** Э. Б. Винберг, *Коммутативные однородные пространства гейзенбергова типа*, Тр. ММО **64** (2003), 54–89.

[VK] E. B. Vinberg, B. N. Kimelfeld, *Homogeneous domains on flag manifolds and spherical subgroups of semisimple Lie groups*, Funct. Anal. Appl. **12** (1978), no. 3, 168–174. **Russian original:** Э. Б. Винберг, Б. Н. Кимельфельд, *Однородные области на флаговых многообразиях и сферические подгруппы полупростых групп Ли*, Функц. анализ и его прил. **12** (1978), № 3, 12–19.

[VP] E. B. Vinberg, V. L. Popov, *On a class of quasihomogeneous affine varieties*, Math. USSR-Izv. **6** (1972), no. 4, 743–758. **Russian original:** Э. Б. Винберг, В. Л. Попов, *Об одном классе квазиоднородных аффинных многообразий*, Изв. АН СССР, сер. мат. **36** (1972), № 4, 749–764.

[Vog] D. A. Vogan, *Irreducible characters of semisimple Lie groups, III: Proof of Kazhdan–Lusztig conjecture in the integral case*, Invent. Math. **71** (1983), no. 2, 381–417.

[Vu1] Th. Vust, *Opération de groupes réductifs dans un type de cônes presque homogènes*, Bull. Soc. Math. France **102** (1974), 317–333.

[Vu2] Th. Vust, *Plongements d'espaces symétriques algébriques: une classification*, Ann. Scuola Norm. Sup. Pisa Cl. Sci. (4) **17** (1990), no. 2, 165–195.

[Wa] B. Wasserman, *Wonderful varieties of rank two*, Transform. Groups **1** (1996), no. 4, 375–403.

[Wei] B. Yu. Weisfeiler, *On a class of unipotent subgroups in semisimple algebraic groups*, Uspekhi Mat. Nauk **21** (1966), no. 2, 222–223. **Russian original:** Б. Ю. Вейсфейлер, *Об одном классе унипотентных подгрупп полупростых алгебраических групп*, Успехи мат. наук **21** (1966), № 2, 222–223.

[Yak1] O. S. Yakimova, *Weakly symmetric spaces of semisimple Lie groups*, Moscow Univ. Math. Bull. **57** (2002), no. 2, 37–40. **Russian original:** О. С. Якимова, *Слабо симметрические пространства полупростых групп Ли*, Вестник МГУ, сер. 1, мат. мех. (2002), № 2, 57–60.

[Yak2] O. S. Yakimova, *Gelfand pairs*, Bonner Mathematische Schriften **374** (2005), `http://hss.ulb.uni-bonn.de/2004/0513/0513.htm`.

[Yak3] O. S. Yakimova, *The index of centralizers of elements in classical Lie algebras*, Funct. Anal. Appl. **40** (2006), no. 1, 42–51. **Russian original:** О. С. Якимова, *Индекс централизаторов элементов в классических алгебрах Ли*, Функц. анализ и его прил. **40** (2006), № 1, 52–64.

[ZS] O. Zariski, P. Samuel, *Commutative algebra*, vol. 1, 2, Van Nostrand, Princeton, 1958, 1960.

[Zor] A. A. Zorin, *On a relation between sphericity and weak symmetry of homogeneous spaces of reductive groups*, Moscow Univ. Math. Bull. **60** (2005), no. 4, 15–18. **Russian original:** А. А. Зорин, *О связи сферичности и слабой симметричности однородных пространств редуктивных групп*, Вестник МГУ, сер. 1, мат. мех. (2005), № 4, 14–18.

Name Index

D.A. Timashev, *Homogeneous Spaces and Equivariant Embeddings*, Encyclopaedia of Mathematical Sciences 138, DOI 10.1007/978-3-642-18399-7,

E

F

G

H

I

J

K

L

W

Y

Z

Subject Index

D.A. Timashev, *Homogeneous Spaces and Equivariant Embeddings*,
Encyclopaedia of Mathematical Sciences 138, DOI 10.1007/978-3-642-18399-7,

H

I

K

L

Notation Index

Symbols

D.A. Timashev, *Homogeneous Spaces and Equivariant Embeddings*, Encyclopaedia of Mathematical Sciences 138, DOI 10.1007/978-3-642-18399-7,